Second Edition

CYTOKINES
and the
CNS

T0174096

Second Edition

CYTOKINES
and the
CNS

Edited by
Richard M. Ransohoff
Etty N. Benveniste

CRC Press
Taylor & Francis Group
Boca Raton London New York

CRC Press is an imprint of the
Taylor & Francis Group, an **informa** business

A TAYLOR & FRANCIS BOOK

CRC Press
Taylor & Francis Group
6000 Broken Sound Parkway NW, Suite 300
Boca Raton, FL 33487-2742

First issued in paperback 2019

© 2006 by Taylor & Francis Group, LLC
CRC Press is an imprint of Taylor & Francis Group, an Informa business

No claim to original U.S. Government works

ISBN-13: 978-0-8493-1622-7 (hbk)
ISBN-13: 978-0-367-39193-5 (pbk)

Library of Congress Cataloging-in-Publication Data

Cytokines and the CNS / edited by Richard M. Ransohoff, Etty Benveniste.--2nd ed.
    p. cm.
Includes bibliographical references and index.
ISBN 0-8493-1622-7
1. Neuroimmunology. 2. Cytokines. I. Ransohoff, Richard M. II. Benveniste, Etty N.

QP356.47.C96 2005
616.8'0479--dc22   2005046283

Visit the Taylor & Francis Web site at
http://www.taylorandfrancis.com

and the CRC Press Web site at
http://www.crcpress.com

Dedication

I dedicate this book to my wife, Margaret S. Ransohoff, to my children, Amy and Lena Ransohoff, and to my parents, Sue and Jerry Ransohoff.

Richard Ransohoff

This book is dedicated to my mother, Oro Benveniste.

Etty Benveniste

Preface

It has been almost exactly 10 years since we began work on the first edition of this book. We were strongly motivated by the explosive growth in knowledge about cytokines/chemokines and cytokine/chemokine signaling, and the potential for application of this information to understanding the development, physiology, and pathology of the central nervous system (CNS). Our efforts were gratified by considerable interest in the book, which remains in print today (selling three units in the first half of 2004). Using the prosaic descriptor afforded by counting PubMed citations ("Cytokines" AND "Central Nervous System"), there were 2225 papers in this field during more than 28 years captured in the current PubMed index (from 1966 to September 1994), with less than 1000 in the period 1966–1989 and 1244 (56%) appearing in the five years before our first edition (1989–1994). Indeed, the MESH entries for "cytokine" (1991) and "cytokine receptor" (1994) were then still newly-minted. Since then, the pace has only accelerated, with another 6189 papers (74% of all articles on the subject) in the past 10 years, meaning that, during the past decade, every 18 months brought on average as many reports in this field as in the first 23 years.

The first edition was organized to present information that would serve overlapping purposes for neurobiologists, immunobiologists, and other interested parties. Basic cytokine biology (receptors, signaling, cellular responses) took up the first three chapters, with the fourth devoted to introducing the unique cellular elements of the CNS. Three chapters addressed how these CNS cells respond to cytokine stimulation, with the last six addressing how cytokine interactions play out in the setting of CNS pathology.

When preparing the first edition, our feeling was that cytokine biology needed an introduction to neurobiologists for two reasons: (1) receptor structure and post-receptor signaling had achieved a level of coherence and mechanistic detail that could be readily communicated; and (2) there was enough similarity (at times identity) between factors involved in inflammation or immunity and those implicated in neural development, physiology, and repair that familiarity with cytokines was an essential element in the neurobiologist's cognitive tool kit. Conversely, for immunobiologists, the concepts elaborated on by neurobiologists to understand developmental patterning and networked organ function were so elegant that comprehension of cytokine action in the CNS could only enhance appreciation of ways the immune system might work. These principles remain unaltered. The design of this second edition embraces three chapters on basic cytokine biology, following one chapter on the CNS (cells, tissues, basic pathological reactions to insult). Next are four chapters that integrate cytokine biology into basic CNS processes (development, inflammation, immunity, degeneration/repair). There are five chapters that apply this body of knowledge to disease or pathology (neurodegeneration, neoplasia, multiple sclerosis, trauma, infection). Finally, we look forward in a last chapter to the translation of cytokine biology to treatment, an entirely new and welcome topic.

Fortunately, the organizing concepts for the first volume remain largely valid. We put forth that the newly-proposed notion of grouping cytokines according to receptor structure would be enduringly useful, and the groups I and II cytokine receptors were presented as illustrative examples. The ongoing utility of this approach can be deduced from following the expanding (but not exploding) field of interleukins. Our first volume indexed 10 interleukins; there are now at least 29. Reassuringly, the newer interleukins fall into families defined by older members (for example, IL-22, IL-24, IL-26, IL-28 and IL-29 are members of the IL-10 family; IL-23 is a member of the IL-12 family; IL-25 is a member of the IL-17 family; and IL-27 is an IL-6 family member), forestalling a bit of 'shock of the new.' Remarkably, the families remain functionally defined, as proposed more than a decade ago, by shared receptor subunit utilization. Challenges to usable core memory remain, however: Online Mendelian Inheritance in Man (OMIM) lists no less than 18 TNF superfamily ligands (TNFSFs), signaling to 25 receptors (TNFSFRs). This nomenclature has been adopted haltingly and spottily, making the relevant literature dense for the casual user. Nevertheless, the critical importance of this group of molecules for neuroimmunology (comprising components such as the p75/LNGFR in addition to TNF, its receptors; Fas and FasL; CD154 and CD40) dictates that the effort must be made. Ready availability of Web-based resources is a welcome aide memoire, growing at about the same pace as new nomenclatures and molecules.

The current edition of *Cytokines and the CNS* is therefore continuous with its predecessor. However, in another sense, it is entirely original, since each chapter was prepared by a new contributor. This evolutionary change was not dictated by any one of the previous authors leaving the field, but rather by our desire to have a fresh look at each topic. The chapters of this second edition cover much of the same ground as previously, but focus more on processes (immunity, inflammation, development) than individual cell types (oligodendroglia, astrocytes; neurons). The major pathological processes that concern us today (infection, trauma, neurodegeneration; multiple sclerosis) are those discussed previously. We are endlessly grateful to our contributors, whose scholarship and hard work were continuously amazing, and hope that their efforts translate seamlessly to benefit for readers. The advance of knowledge in these fields will be apparent to cursory comparison of the first and second editions. Tasks that remain undone are formidable indeed: foremost among them is to understand how cells and tissues integrate simultaneous or sequential multiple cytokine signals to produce biological outcomes. These types of questions demand new methods of analysis, for which the very first drafts have only recently been described. We hope that the biology described in the current book helps to persuade the contemporary and future community of CNS/cytokine researchers that the goal is worth the effort.

Editors

Etty (Tika) Benveniste is professor and chairman of the Department of Cell Biology at the University of Alabama at Birmingham. She received her Ph.D. in immunology from the University of California, Los Angeles, in 1983, and was a postdoctoral fellow in the laboratory of Dr. Jean Merrill, Department of Neurology, UCLA, from 1983 to 1986.

Dr. Benveniste has received numerous honors and awards, including NIH Training Grant Fellowships (1982–1983, 1984–1985); a postdoctoral fellowship award from the National Multiple Sclerosis Society (1986–1987); Plenary Lecturer, Fourth International Congress of Neuroimmunology (1994); Plenary Lecturer, UCLA Neurobiology of Disease Conference (1995); Member and Plenary Lecturer, Sixth International Congress on TNF and Related Cytokines (1996); Distinguished Scientist Lecturer, University of Arkansas (1998); Keynote Speaker, Great Lakes Glia Meeting (1999); Chair, FASEB Summer Conference, Neural-Immune Interactions (2000, 2002); Symposium Speaker, Oklahoma Center for Neuroscience (2003); and Executive Chair, NIH Workshop on Glial Inflammation (2003).

Dr. Benveniste has served on numerous review and advisory boards. These include Member, NIH Special Section for AIDS and Related Research Review Group (1998–1991); Member, American Cancer Society: Advisory Committee for Cell Biology (1992–1995); Member, NIH Neurosciences Program Project Review Committee B (1993–1995); Chair, NIH Neurosciences Program Project Review Committee B (1995–1997); External Advisory Board, Center for Neurovirology, University of Nebraska Medical Center (1997–present); Member, National Multiple Sclerosis Society Grant Review Committee (1998–2003); Member, NIH Training Grant and Career Development Review Committee (1999–2002); Member; NIH Clinical Neuroimmunology and Brain Tumors (2000–2004); Member, Sontag Foundation Scientific Advisory Board (2003–present); Member, National Multiple Sclerosis Society, Research Programs Advisory Committee (2004–present); Chair, NIH Clinical Neuroimmunology and Brain Tumors (2004–present); and Member, Oklahoma Medical Research Foundation Board of Scientific Visitors (2004–present). She is a member of the Editorial Boards of *the Journal of Biological Chemistry, Journal of Neuroscience, GLIA, Journal of Neuroimmunology,* and *Journal of Neurovirology.*

For the past 18 years, Dr. Benveniste's research has focused on the function of cytokines and signal transduction pathways operative in glial cells, and their contributions to CNS disease. Dr. Benveniste has received continuous research support from the NIH, NMSS, and AmFAR since 1988. She has published over 150 scientific reports and reviews, numerous book chapters, and edited two books.

Dr. Benveniste is a member of the International Society of Neuroimmunology, the American Society for Neurochemistry, the American Association of Immunologists, the Society for Neuroscience, and the American Society of Cell Biology.

Dr. Benveniste is married to Dr. Casey Morrow, and they have one son, Jackson Morrow (12).

Richard M. Ransohoff is professor of molecular medicine at the Cleveland Clinic Lerner College of Medicine, a staff scientist in the Department of Neurosciences of the Lerner Research Institute, and a staff neurologist in the Mellen Center for MS Treatment and Research, both at the Cleveland Clinic Foundation (CCF), Cleveland, Ohio. He is also a professor of pathology (Adjunct) at Case Medical School. Dr. Ransohoff graduated with honors from Bard College, Annandale, New York, with a B.A. in literature and received the M.D. degree with honors from Case School of Medicine, Cleveland, Ohio. He completed residencies in internal medicine (Mt. Sinai Medical Center, Cleveland, Ohio; Board Certified 1981) and neurology (CCF; Board Certified 1985). From 1984 to 1989, Dr. Ransohoff was a postdoctoral fellow in the laboratory of Dr. Timothy Nilsen, Department of Molecular Biology and Microbiology, Case School of Medicine.

Among other honors and awards, he received a Physician's Research Training Award from the American Cancer Society (1984–1986); a Harry Weaver Neuroscience Scholarship from the National Multiple Sclerosis Society (NMSS; 1987–1992); a Clinical Investigator Development Award from the National Institutes of Health (NIH; 1988–1993); a Heritage Scholar at the University of Alberta (1998); Distinguished Lecturer at the University of Arkansas School of Medicine (2002); a recipient of the John and Samuel Bard Award in Science and Medicine (2002); and a speaker at the American Academy of Neurology's Plenary Symposium "Frontiers in Clinical Neuroscience" in 2004. He has been cited from 1996 through the present (2004) in the *Best Doctors in America* for his expertise in the clinical care of patients with multiple sclerosis (MS).

Dr. Ransohoff served as a regular member on NIH and National Multiple Sclerosis Society (NMSS) Study Sections; on numerous Special Emphasis Panels; and as Chair of the NMSS Peer Review Committee B from October 2004. He is a member of the editorial boards of *The Journal of Immunology* (where he is presently Section Editor); *Trends in Immunology; Current Immunology Reviews,* and the *Journal of Neuroimmunology.* From 1998 to 2000, Dr. Ransohoff was a member of the NINDS Director's Planning Panel on "The Neural Environment." He is a member of the Steering Committee for the NIH/NINDS Spinal Muscular Atrophy Project; the International Advisory Boards for the 7th (2004) and 8th (2006) Congresses on Neuroimmunology; and the Scientific Advisory Board for Chemocentryx, San Carlos, California. He serves on External Advisory Boards for CHARTER (CNS HIV Anti-Retroviral Therapy Effects Research; MH22005); a Program Project on Alexander's Disease (NS 42803); the MS Lesion Project (NMSS RG 3185); the University of Nebraska's Center for Neurovirology & Neurodegenerative Disorders (NS43985) and is the External Advisor for the European Union's Project on 'Mechanisms of Brain Inflammation" (QLG3-00612). He is a member of the National MS Society's Medical Advisory Board. He is a Co-director of the Marine Biological Laboratory's special topics course on "Pathogenesis of neuroimmunological disease" held biennially at Wood's Hole, Massachusetts.

For the past decade, Dr. Ransohoff's research has focused on the functions of chemokines and chemokine receptors in development and pathology of the nervous system. He also has a longstanding and continuing interest in the mechanisms of action of interferon-beta. Dr. Ransohoff has received continuous research support from the NIH and the NMSS since 1988. He has published more than 130 scientific reports, more than 35 reviews and book chapters, and three edited books.

Dr. Ransohoff is a member of the American Academy of Neurology, the American Neurological Association, the American Association for the Advancement of Science, and the American Association of Immunologists.

Dr. Ransohoff is married to Margaret Ransohoff. They have two daughters, Amy (14) and Lena (10).

Contributors

Wendy Smith Begolka
Northwestern University
Feinberg School of Medicine
Chicago, Illinois, USA

John R. Bethea
University of Miami
School of Medicine
Miami, Florida, USA

Valerie Bracchi-Ricard
University of Miami
School of Medicine
Miami, Florida, USA

Roberta Brambilla
University of Miami
School of Medicine
Miami, Florida, USA

Celia F. Brosnan
Albert Einstein College of Medicine
Bronx, New York, USA

Valéry Combes
Université de la Méditerranée
Aix-Marseille II, France

Melissa A. Cosenza
Albert Einstein College of Medicine
Bronx, New York, USA

Michael David
University of California San Diego
La Jolla, California, USA

Nandini Dey
Emory University School of Medicine
Atlanta, Georgia, USA

Donald L. Durden
Emory University School of Medicine
Atlanta, Georgia, USA

Limin Gao
Case Western Reserve University
Cleveland, Ohio, USA

Georges E. Grau
Université de la Méditerranée
Aix-Marseille II, France

Reinhard Hohlfeld
Ludwig Maximilians University
Munich, Germany

Sunhee C. Lee
Albert Einstein College of Medicine
Bronx, New York, USA

Margot Mayer-Pröschel
University of Rochester Medical Center
Rochester, New York, USA

Eileen J. McMahon
Northwestern University Feinberg
School of Medicine
Chicago, Illinois, USA

Erwin G. Van Meir
Emory University School of Medicine
Atlanta, Georgia, USA

Robert H. Miller
Case Western Reserve University
Cleveland, Ohio, USA

Stephen D. Miller
Northwestern University
Feinberg School of Medicine
Chicago, Illinois, USA

David Mock
University of Rochester Medical Center
Rochester, New York, USA

Greer M. Murphy, Jr.
Stanford University School of Medicine
Stanford, California, USA

Mark Noble
University of Rochester Medical Center
Rochester, New York, USA

Trevor Owens
Montreal Neurological Institute
Montreal, Quebec, Canada

Chris Pröschel
University of Rochester Medical Center
Rochester, New York, USA

P. Rieckmann
Clinical Research Group for Multiple
  Sclerosis and Neuroimmunology
Würzburg, Germany

Mark Rivieccio
Albert Einstein College of Medicine
Bronx, New York, USA

Stefan Rose-John
Biochemisches Institut der Christian
  Albrechts Universität zu Kiel
Kiel, Germany

Parvathy Saravanapavan
Stanford University School of Medicine
Stanford, California, USA

Heidi Schooltink
Biochemisches Institut der Christian
  Albrechts Universität zu Kiel
Kiel, Germany

Qiusheng Si
Albert Einstein College of Medicine
Bronx, New York, USA

Jan Vilćek
New York University School of
  Medicine
New York, New York, USA

Rachel D. Wheeler
Montreal Neurological Institute
Montreal, Quebec, Canada

Heinz Wiendl
University of Tübingen
Tübingen, Germany

Simone P. Zehntner
Montreal Neurological Institute
Montreal, Quebec, Canada

Contents

1 The CNS: Cells, Tissues, and Reactions to Insult

Sunhee C. Lee, Melissa A. Cosenza, Qiusheng Si, Mark Rivieccio, and Celia F. Brosnan

CONTENTS

I. INTRODUCTION

There are four major cell groups within the central nervous system (CNS): neurons, oligodendrocytes, astrocytes, and microglia. In the context of cytokines and the CNS, neurons have been predominantly considered as either targets of the cytokine response or as initiators of a cytokine response in other cell types following injury. The literature that documents neurons, particularly primary cell neurons, as a source of cytokines is scant and remains controversial. A similar situation exists for oligodendrocytes, the myelin-forming cell of the CNS. Again, these cells have been studied predominantly as targets of cytokine activity rather than sources of these factors in the CNS. This is clearly not the case for microglia and astrocytes. These glial cell populations are widely recognized as both important sources of cytokines in the CNS, as well as major responders to cytokines.

Insults to the CNS are characterized by reactive changes in microglia and astrocytes, a response that is collectively referred to as a reactive gliosis. Depending on the pathological context, this reactive gliosis can function either as a reparative event leading to the production of neurotrophins and growth factors that permit some degree of recovery of CNS activity or as a degenerative response that leads to loss of function and even to the death of neurons, resulting in permanent injury to the CNS.

Both astrocytes and microglia reside ubiquitously in the brain parenchyma as highly ramified cells and react to insults in several ways: by changes in cell morphology that alter their interactions with adjacent cell populations; by upregulating surface antigens and receptors that modulate responses to the surrounding environment, and by producing soluble mediators such as cytokines and chemokines that activate a range of downstream signaling cascades. Microglia, the resident macrophages of the brain, play a central role in this response through the expression of a wide range of cell surface receptors, such as the Toll-like receptors, scavenger receptors, and Fc receptors, that enable a rapid response to changes in the CNS microenvironment. The products of activated microglia can initiate an inflammatory cascade by activating other glial cells and in some disorders by recruiting systemic inflammatory cells into the CNS. Activated immunocompetent cells, particularly Th1-type T cells that secrete IFN, also target microglial cells, leading to their activation and involvement in immune-mediated responses in the CNS. In most cases, microglial activation is considered a secondary process following insult to neurons and other parenchymal elements. Alternatively, microglia may be the primary target of the disease process, as is the case in HIV-1 encephalitis (HIVE). In HIVE, microglial cells are productively infected by the virus, expressing both the viral and host cell factors that can secondarily affect the function and survival of neurons and other glia.

Glial-derived cytokines, chemokines, enzymes, and other inflammatory mediators have been shown to modulate neuronal toxicity in several CNS disorders, suggesting that disorders with widely differing etiologies (e.g., viral infections, autoimmune disorders, stroke and neurodegenerative diseases such as Alzheimer's disease [AD] and Parkinson's disease) may have shared pathogenetic mechanisms. In this chapter, we will review evidence that supports common immune mechanisms of neurodegeneration, with special emphasis on the role of microglial IL-1. We will also discuss the mechanism of microglial activation involved in innate immunity and present evidence that IL-1 activates parallel mechanisms in astrocytes.

II. BIOLOGY OF MICROGLIA

A. MICROGLIA IN NORMAL AND INJURED CNSs

Microglia constitute a distinct glial population in the CNS [1–4]. Unlike neurons and macroglia that are of neuroepithelial origin, microglia are mesodermal (i.e., bone marrow) in origin and seed the brain early in embryogenesis: during development, monocytes migrate to the brain through the vessels located in specific regions of the brain (called "glial fountains" in humans). These areas are concentrated around the subventricular zones where active neurogenesis occurs. These ameboid tissue macrophages then migrate throughout the entire brain parenchyma and differentiate into resident microglial cells. In the mature CNS, microglia are ubiquitously present as highly ramified cells ("resting" microglia) [5,6]. They respond to changes in the CNS microenvironment in a variety of disorders with or without the participation of the systemic monocytes. Although in degenerative disorders such as AD and Parkinson's disease there is little evidence to support recruitment of monocytes from the periphery, in infectious and autoimmune diseases such as HIVE and multiple sclerosis (MS) and in stroke, there is frank infiltration of monocyte-derived macrophages as well as other inflammatory cells. Even in these diseases in which monocytes are known to contribute significantly to the disease process, studies using sensitive markers of microglia invariably demonstrate that parenchymal microglia are one of the earliest reacting cell types in the brain [7,8]. Unlike perivascular macrophages, the highly ramified cell processes of parenchymal and juxtavascular microglia (see below) make them ideal candidates for intercellular interactions with other glia and with neurons. Consequently, they are more likely to be implicated in neurodegenerative disease processes in which there is no overt disruption of the blood–brain barrier (BBB). Their close contact with neurons and other glia makes them a leading suspect as the source of inflammatory soluble mediators that contribute to neurotoxicity in neurodegenerative conditions.

Determining the relative contribution of monocyte-derived macrophages and intrinsic microglia to the disease process, although conceptually important, is hampered by the fact that no single surface marker reliably differentiates intrinsic microglia from blood-borne monocytes [2] (see also below). There are several animal models that have contributed significantly to an understanding of microglial cell biology. The bone marrow chimera studies by Hickey and colleagues provided the first evidence that perivascular microglia (macrophages) in normal brain are a distinct population of CNS macrophages, different from parenchymal ramified microglia. Perivascular macrophages turn over more rapidly, and they are capable of presenting antigen in immune response reactions although intrinsic microglia represent a stable population with limited antigen presenting potential [9,10].

Contrary to these views, somewhat different results are presented in more recent studies using stem cell transplantation. When bone marrow stem cells (expressing the green fluorescent protein, GFP) are transplanted into the systemic circulation of irradiated mice, the GFP+ cells are shown to migrate into the adult CNS across the BBB and differentiate into the ramified parenchymal microglial cells [11]. These results are similar to those obtained earlier, which showed that up to a quarter of the regional microglial population was donor-derived by 4 months after transplantation [12]. However, a very similar study reported previously by Vallieres and Sawchenko demonstrated that the vast majority of the transplanted cells become perivascular macrophages in the CNS [13]. The latter study agrees with an earlier one on human subjects who received sex-mismatched bone marrow transplantation, which showed that Y chromosome marker-bearing CD45+ cells (donor-derived mononuclear leukocytes) entered the normal-appearing brains of female recipients and transformed into "perivascular cells" [10]. Therefore, the potential for blood-borne macrophages to migrate into the brain parenchyma and differentiate into microglia in mature brain is still an issue open for debate.

The third type of resident brain macrophages that has received relatively little attention is the juxtavascular microglia. Juxtavascular microglia are characterized by the parenchymal location of the cell body, ramified cell processes, and direct contact with the basal lamina of blood vessels by the cell processes. According to Dailey and colleagues, who studied them in rat hippocampal slice cultures, approximately 10–30 % of total brain microglia belong to this population of juxtavascular microglia [14]. The authors identified them as a mobile subpopulation of parenchymal microglia that activate rapidly and that are preferentially recruited to the surfaces of blood vessels following brain tissue injury. As such, this particular subpopulation of microglia may represent cells specialized to facilitate signaling between the injured brain parenchyma and components of the bloodñbrain barrier *in vivo* [14].

Regardless of their ontogeny and relationship to blood–borne monocytes, it is now well established that microglial cells are a distinct brain macrophage population capable of mounting various reactive and reparative responses. Using the rat facial nerve axotomy model, Kreutzberg and colleagues have elegantly illustrated the various cellular and molecular changes that occur in microglial cells within the degenerating facial nucleus [15,16]. They demonstrated that microglial activation is a key factor in the defense of the neural parenchyma against various insults and have shown that microglia function as scavenger cells as well as in tissue repair and neural regeneration, ultimately facilitating the return to tissue homeostasis.

B. MICROGLIA AND INNATE IMMUNITY

The early phase of an effective immune response to invading pathogens is essential for the survival of organisms and is known as the innate immune response. It is a type of immunity that is not dependent upon memory T- or B-cells but is dependent on secretory factors and macrophages [17]. The inflammation that is characteristic of neurodegenerative disorders (referred to as *neuroinflammation*) has been compared to innate immunity. The innate immune response can be driven through specific recognition systems, the best examples being the interactions between microbial components and the Toll-like receptors (TLR). In the CNS, the cells that bear the appropriate receptors to interact with these microbial components are monocyte-derived macrophages and resident brain microglia. Innate immunity is characterized by the *de novo* production of mediators that directly contribute to antimicrobial activity or set off secondary inflammatory cascades that could ultimately result in inflammation and host injury. Some of the best characterized gene products that are induced as a result of innate immune response are cytokines.

C. ACTIVATION OF MICROGLIA THROUGH SURFACE IMMUNE RECEPTORS

By far the most potent activator of microglia and macrophages is lipopolysaccharide (LPS) from the Gram (−) bacterial cell wall. LPS complexed with LPS-binding protein (LBP) has been shown to bind cells through the specific receptor CD14 [18], but because CD14 lacks a functional intracellular domain, it has been unclear how the receptor signal is transduced within cells. Recently, members of the microbial pattern recognition receptor (PRR) family called Toll-like receptors have been found to interact with specific microbial components [19–21]; TLR4 is now shown to be the signaling partner for CD14, whereas similar cell signaling pathways have been found to be activated following binding of the double stranded (ds) viral RNA to TLR3. Peptidoglycan, a Gram (+) bacterial cell wall component, triggers signaling through TLR2, whereas the bacterial nucleotide CpG sequence specifically binds to and triggers signaling through TLR9. The primary function of microbial PRRs is in innate immunity and studies implicate these receptors in defense against bacteria, yeast, and viruses and possibly factors released by dead and dying cells.

1. TLR4 Signal Transduction

LPS induces responses in leukocytes by interacting with a soluble binding protein present in serum, LBP (Figure 1.1). The LPS/LBP complex then initiates its biological activities through a heteromeric receptor complex containing CD14, TLR4, and at least one other protein, myeloid differentiation protein-2 (MD-2) [22]. The intracellular signaling domain (termed TIR [Toll-IL-1 receptor homologous region]) of TLR4 is homologous to that of the IL-1 receptor, and TIR involves downstream activators such as the receptor-associated adapter protein, myeloid differentiation factor 88 (MyD88), IL-1 receptor-associated protein kinase (IRAK), and TNF receptor activated factor 6 (TRAF6), leading to activation of nuclear factor κ-B (NFκ-B) and mitogen-activated protein (MAP) kinases [19]. This signaling cascade leads to the robust production of cytokines such as IL-1, TNFα, or IL-6 or chemokines such as IL-8/CXCL8. Although the activation of NFκ-B is a conserved response following activation of most TLRs, activation of interferon regulatory factor-3 (IRF-3) is a response unique to TLR3 and TLR4 pathways. Specifically, activation of IRF3 through TLR3/TLR4, together with NFκ-B, results in the induction of IFNβ (and several other primary response genes); IFNβ then activates a group of secondary response genes (Figure 1.1 and Figure 1.2). Activation of the IRF3 pathway through TLR3/TLR4 is shown to be MyD88-independent but dependent on the MyD88-like adaptor known as TIR-domain containing adaptor inducing IFNβ (TRIF) [23]. IRF3 had been known as a constitutively expressed transcription factor that can be activated by phosphorylation by certain RNA viruses [24]. In Sendai virus-infected cells, both IFNβ and the chemokine RANTES have been found to be induced in an IRF3-dependent manner. The identities of the kinases that are responsible for IRF3 phosphorylation have only recently been identified. They are now known to be the Iκ-B kinase (IKK)-related kinases, IKK and TANK-binding kinase-1 (TBK1), kinases previously implicated in NFκ-B activation [25,26]. Thus, IKK and TBK1 have a pivotal role in coordinating the activation of IRF3 and NFκ-B in the innate immune response.

2. Activation of Microglia through CD14/TLR4 *In Vivo*

Although microglial activation by bacterial products like LPS is well documented *in vitro*, whether intrinsic microglia in the brain can respond to similar signals has

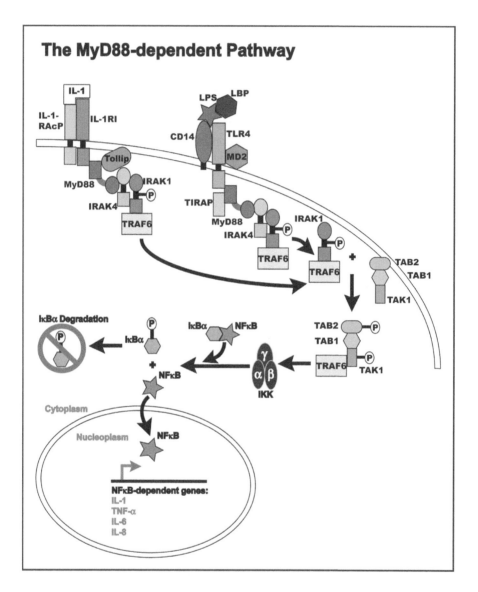

FIGURE 1.1 TLR and IL-1 receptor signaling: the MyD88-dependent pathway. The MyD88-dependent pathway of NFκ-B activation occurs through the IL-1R or TLRs [99,100]. The IL-1 receptor complex is composed of the type 1 receptor (IL-1RI) and the receptor accessory protein (IL-1RAcp). LPS/LBP binds to a heteromeric receptor complex containing CD14, TLR4, and MD2 [22]. IL-1/TLR signaling pathways arise from intracytoplasmic TIR domains and TIR domain-containing adaptors such as MyD88, Toll-interacting protein (Tollip), or TIR domain-containing adaptor protein (TIRAP). Tollip is specifically involved in IL-1R signaling [101], whereas TIRAP is specifically involved in MyD88-dependent TLR4 (and TLR2) signaling [102,103]. Upon ligand binding, the cytosolic adaptor proteins are recruited to the receptor complex. IRAK is recruited and phosphorylated by IRAK4, and then TRAF6 is subsequently recruited to the complex. IRAK brings TRAF6 to TGFβ-activated kinase

been questioned due to the lack of appropriate receptor expression in the normal brain. In the periphery, CD14 is expressed by monocytes and neutrophils. Monocytes isolated from the blood show time-dependent loss of CD14 in culture, whereas normal microglia isolated from the brain acquire CD14 expression in culture [27,28]. These results suggest that downregulation of surface CD14 is a feature that accompanies monocyte differentiation into tissue macrophages. Microglia are derived from CD14+ monocytes during embryogenesis with subsequent loss of CD14 and other myeloid-lineage markers during brain maturation. Upon injury, however, many of the macrophage surface proteins reappear in reactive microglial cells; reexpression of CD14 in human CNS has been documented following stroke and HIVE [8,29]. The ability of parenchymal ramified microglia to express CD14 *de novo* has been elegantly illustrated by Rivest and colleagues [30]: adult rats that received LPS peripherally (thus leaving the BBB intact) show diffuse parenchymal induction of CD14 mRNA following the waves of TNFα induction in the circumventricular organs, areas of the brain that lack a BBB.

Although the principal ligand for CD14 is the LPS/LBP complex, other less well-known ligands might also trigger CD14/TLR4 cell activation pathways. For instance, Fassbender et al. [31] found that CD14 interacts with fibrils of Alzheimer amyloid peptide (Aβ): anti-CD14 antibodies as well as mice with a genetic deficiency for CD14 showed reduced Aβ peptide-induced microglial activation and neuronal death in a mouse model of AD. CD14 mRNA was strongly induced in activated microglia in this model. These results suggest that CD14 may contribute to the neuroinflammatory response to amyloid peptide and highlighted the possibility that the progress being made in the field of innate immunity could be extended to the research of neurodegenerative disorders.

Gene profiling studies are useful in determining the cell response programs activated by specific ligand/receptors. To gain insight into the transcriptional machinery activated by TLR4 in CNS glial cells, we performed microarray analyses of LPS-activated human microglial cells and compared their response with that of IL-1-activated human astrocytes in culture. These studies revealed that the mRNAs induced in the two cell populations are very similar: the genes that are activated include those involved in TLR4 signal transduction such as IFN-β, IRF-3, and NFκ-B-related genes, as well as those that are known end products of TLR4 signal transduction, that is, cytokines, chemokines, and other inflammatory products (Table 1.1) [32]. These results suggest a synergism between microglia and astrocytes during

---

(TAK1), a member of the MAPK kinase kinase (MAPKKK) family, and TAB1 and TAB2, two proteins that bind to TAK1, which are preassociated on the membrane. This leads to phosphorylation of TAK1 and TAB2 followed by dissociation of TRAF6-TAK1-TAB1-TAB2 from IRAK and subsequent translocation to the cytosol. IRAK stays in the membrane and is degraded. Through ubiquitin E2 ligases and other unknown mechanisms, TRAF6 activates TAK1 [100]. Activated TAK1 then activates the IKK complex consisting of IKKα, IKKβ, and IKKγ, which induces phosphorylation and degradation of IκBα releasing NFκ-B. NFκ-B translocates to the nucleus and activates transcription of NFκ-B-dependent genes.

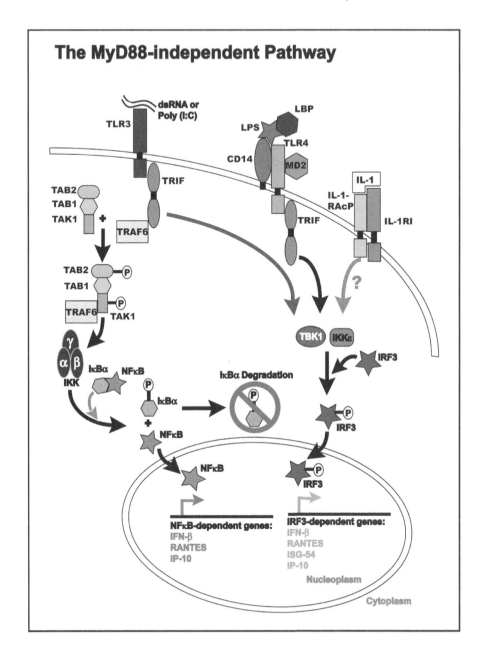

FIGURE 1.2 TLR and IL-1 receptor signaling: the MyD88-independent pathway. The MyD88-independent pathway is activated by ligand binding to TLR3 or TLR4 involving TRIF as the TIR domain-containing adaptor protein [23,104]. LPS stimulation of MyD88-null macrophages leads to delayed activation of NF-κB, as well as activation of the transcription factor IRF-3, thereby inducing IFNβ (primary response gene). IFNβ, in turn, activates the induction of several IFN-inducible genes such as RANTES/CCL5, IP-10/CXCL10, and ISG-54 (secondary response genes). Virus and viral-derived dsRNA are potent activators of IRF-3,

innate immune response. Furthermore, IL-1 produced by microglia in virtually all human CNS pathologies (see below) could activate TLR-like signaling cascades in astrocytes contributing to neuroinflammation that is indistinguishable from an innate as immune response (see Section IVB).

3. Activation of Microglia through Scavenger Receptors

Scavenger receptors (SR) are a class of heterogeneous macrophage receptors that bind acetylated or other modified lipoproteins with high affinity and are involved in lipid metabolism, phagocytosis, and cell signaling (for review, see [33]). The SR family includes class A SR (SR-A), class B type 1 SR (SR-B1), CD36, receptor for advanced glycation product (RAGE), macrophage receptor containing collagenous domain (MARCO), and CD68. Most SRs are highly expressed in macrophages and microglia during development and injury, but their expression is downregulated in normal microglia *in vivo*. Amyloid plaque-associated microglia demonstrate enhanced expression of surface antigens, including class B SRs [34,35]. Additionally, astrocytes and endothelial cells have been shown to express SR-B1 [33]. In human macrophages, CD36 is involved in reactive oxygen radical formation and cytokine production following binding of acetylated low-density lipoprotein (LDL). Aβ has been shown to activate microglia through the receptor complexes that include CD36 and to trigger intracellular signaling cascades involving tyrosine kinase activation and expression of IL-1β [36]. Another SR that is relevant to microglial function is MARCO. MARCO is a type of SR-A that is involved in cytoskeletal rearrangement and the downregulation of antigen uptake function during dendritic cell and microglial cell maturation [37]. MARCO appears to be important in innate immunity, and its expression has been shown to be upregulated by LPS and GM-CSF [38]. CD68 (the human homologue of mouse macrosialin) is closely related to the family of lysosomal-associated mucin-like membrane proteins (lamps). The biological function of CD68 is not fully understood, but it is thought to play a role in cholesterol metabolism and foam cell (lipid-laden macrophage) formation by binding to oxidized LDL [39]. In human CNS, CD68 is widely used as a marker of resting and activated microglia (Figure 1.3).

4. Activation of Microglia through the Fc Receptors

The receptors for the immunoglobulin (Ig) constant region (Fc) are expressed primarily by myeloid-lineage cells. In human brain, resting microglia express FcR with further upregulation during diseases such as AD, multiple sclerosis (MS) [40,41],

---

which leads to the initial phase of IFNβ induction. The noncanonical IKKs, TBK-1, and IKK ε (also called inducible IKK or IKKi) are shown to be IRF-3 kinases [25,26]. It has been shown that LPS can activate the similar mechanism of IRF-3 activation in macrophages [105,106], but whether this pathway is activated following IL-1 stimulation is yet to be determined. Studies with TBK1 knockout cells demonstrate that the expression of TBK1 and IKK ε is cell type-dependent and that TBK1 plays a more general role [105,106].

TABLE 1.1
Gene Profiling in LPS-Stimulated Microglia Microarray*

Cytokines	Fold Induction	GeneBank#
IL-1α	96.0	AA936768
IL-1β	152.6	W47101
IL-1 receptor antagonist	41.7	BG288796
IL-1 receptor accessory protein	3.6	AA256132
Caspase-1	2.2	T95052
IL-6	42.9	N98591
TNFα	114.2	AA699697
Chemokines		
CCL2 (MCP-1)	2.9	AA425102
CCL4 (MIP-1β)	35.3	H62985
CCL5 (RANTES)	47.2	AA486072
CCL7 (MCP-3)	3.9	XM_012649
CCL8 (MCP-2)	27.9	AV734258
CXCL8 (IL-8)	12.7	AV717082
CXCL9 (MIG)	38.0	NM_002416
CXCL10 (IP-10)	4.5	AA878880
CXCL11 (I-TAC)	18.9	BC005292
Interferon-related genes		
IFN-stimulated protein 15 kD (ISG 15)	36.2	AA406020
Interleukin-15	3.8	N59270
Guanylate binding protein 1 (GBP-1), 67kD	13.5	AA486849
2'-5'oligoadenylate synthetase (2-5OAS) 1	7.1	AL582281
IFN-induced transmembrane protein 1	33.4	AA419251
Myxovirus resistance 1	10.2	AA456886
IFN-induced, hepatitis C-associated microtubular aggregate protein (44kD)	9.4	XM_001628
IFN consensus sequence binding protein 1 (ICSBP 1)	4.1	AW515838
IRF-7	13.3	AA877255
IRF-9	2.8	AA291577
Protein kinase, IFN-inducible dsRNA-dependent (PKR)	6.5	W42587
Growth and hypoxia related factors		
VEGF	2.3	AI827301
Gravin	0.8	XM_004539
Hypoxia-inducible factor 1	1.4	AU134078
Angiopoietin-1	0.7	BG496539

*Primary human fetal microglial cultures were prepared as previously described [53,98] and stimulated with 100 ng/ml LPS or medium alone (control) for 6 hours. Total RNA was extracted, amplified and hybridized to cDNA microarrays containing 21K human transcripts. Data processing and analyses were performed as described [84].

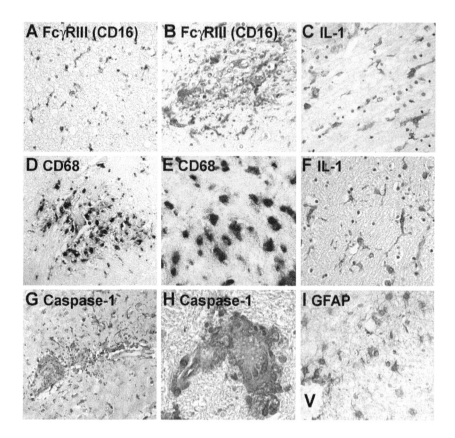

FIGURE 1.3 (Color insert follows page 208). Microglia and astrocytes in human CNS. Immunohistochemistry on human postmortem brain tissues with diverse pathologies demonstrate microglia and astrocytes with distinct morphology and antigen expression. Expression of FcRIII α-chain (CD16) protein in ramified microglia in normal brain (A) and in activated microglia and macrophages in HIVE in a microglial nodule (B). IL-1 expression in ramified microglial cells in an acute infarct (C): the tissue is otherwise normal appearing on hematoxylin and eosin stain (not shown). CD68 is a macrophage-lineage antigen expressed in lysosomal membrane. A perivascular focus of CD68+ macrophages and microglia in HIVE is shown in (D). Lipid-laden macrophages in an active MS lesion are positive for CD68 (E). IL-1α expression in activated microglia in HIVE: cerebral white matter shows IL-1+ cells with a small bipolar cell body and delicate processes characteristic of microglia (F). Caspase-1, an enzyme that cleaves IL-1 to a biologically active form, is expressed in activated microglia and macrophages in human CNS: caspase-1 in HIVE in a microglial nodule (G) and in a multinucleated giant cell, a hallmark of HIV-1-infected cells (H). Reactive astrocytes in an active MS lesion display increased immunoreactivity for GFAP (I): a vessel (V) in the lower left corner is surrounded by astrocyte processes that define perivascular space (glia limitans). Immunocytochemistry on paraffin-embedded postmortem sections employing methods illustrated in references [8,57,107]. Chromogens were either diaminobenzidine (all except D and E) or nitroblue tetrazolium (D and E): all diaminobenzidine slides were also counterstained with hematoxylin. Scale bar represents 200 μm for B, D, and G; 50 μm for A, C, E, F and I; and 20 μm for H.

and HIVE (see Figure 1.3). Antibody production against pathogens or intrinsic brain components may occur in infectious or autoimmune disorders of the CNS. In addition, vaccines or passive antibody therapies are being developed against several human CNS diseases including AD [42] and *Cryptococcus neoformans* meningoencephalitis [43,44]. Furthermore, a recent report indicated that FcR is one of the most highly induced genes in MS lesions, and blocking FcR signaling ameliorated clinical disease in an animal model of MS [45]. Thus, understanding the mechanism by which immune complexes activate microglial FcR may have practical implications.

Immune (antigen–antibody) complexes trigger activation of FcR signaling by cross-linking the surface FcR. The binding of Ig to human FcR varies depending on the Ig isotype (α, δ, ϵ, γ, and μ), and the species from which Ig is derived. FcR (type I and III) consist of the ligand binding α-chain and the common signaling γ-chain, whereas FcRII consists of a single chain that contains both domains. We studied primary human microglia and murine microglia deficient in the expression of the receptors for IgG (FcR) to establish the signaling pathways triggered by *C. neoformans*–murine IgG immune complexes [46–48]. The data showed that blocking of either the high affinity FcRI (α-chain: CD64) or FcRIII (α-chain: CD16) mildly reduced FcR signal transduction, whereas blocking the common signaling γ-chain completely abolished phagocytosis and chemokine production induced by immune complex. Blocking FcRII had little effect. Importantly, it was determined that different signal transduction pathways were involved in phagocytosis and microglial chemokine gene induction: whereas the Ras-MEK-ERK kinase pathway led to chemokine production, the PI3K/Akt pathway was involved in phagocytosis of the immune complexes. The robust induction of chemokine genes by FcR activation was intriguing and suggested that IgG can trigger macrophage FcRs to produce local inflammatory cell infiltration. These results also suggested that antibody therapy could result in brain inflammation as a side effect. Understanding the FcR signaling pathways could allow for refinement of therapy, i.e., blocking some and preserving other functions of FcRs.

III. INTERLEUKIN-1 AND NEURODEGENERATION

IL-1 is a monokine (macrophage cytokine) whose expression is regulated at multiple steps [49,50]. Following transcriptional activation and translation, generation of mature (active) IL-1 requires processing of proIL-1 by IL-1 converting enzyme (ICE: caspase-1). Microglial expression of IL-1 is readily induced *in vitro* by a number of different stimuli, including HIV-1, hypoxia, β-amyloid, LPS, and cytokines (including IL-1 itself) [51–53] (and unpublished observations), as well as activation through the TLRs and scavenger receptors discussed above. IL-1 is unique in the sense that its own antagonist, IL-1 receptor antagonist (IL-1ra), is produced endogenously by macrophages. In the CNS, microglial cells express IL-1ra and recombinant IL-1ra is a potent and specific inhibitor of IL-1 action [54]. Microglial IL-1ra expression is subject to regulation by microbial components and cytokines. In general, proinflamma-tory cytokines that induce IL-1 also induce IL-1ra; however, anti-inflammatory cytokines such as IL-4 and IFNβ have opposing effects on the expression of IL-1 and IL-1ra: they inhibit IL-1 while inducing IL-1ra, shifting the balance to an IL-1ra-predominant state [54].

Microglial IL-1 expression is found associated with pathologic lesions in stroke, AD, HIVE, and other inflammatory conditions. In human stroke, microglial IL-1 expression is one of the earliest measurable changes in the brain (Figure 1.3) [55]. IL-1 expression is specifically induced in amyloid plaque-associated microglial cells in AD [56]. IL-1 expression can be induced as a result of primary insult but also by secondary activation of microglia due to a number of factors (hypoxia, cytokines, and growth factors, etc.) that often coexist in these brains. For example, in HIVE, IL-1 expression is found not only in HIV-1-infected microglia and macrophages, but also in uninfected, activated microglia found throughout the brain (Figure 1.3) [57].

The role of IL-1 in neurodegeneration has been extensively studied in acute and chronic neurodegenerative diseases. In animal models of stroke or trauma, proinflammatory cytokines are induced rapidly after acute CNS insult and are expressed in a temporal and spatial pattern consistent with their involvement in subsequent neuronal death. Modulation of exogenous and endogenous cytokines *in vivo* and *in vitro* has yielded conflicting results: overall, IL-1 appears to contribute directly to neurodegeneration, whereas TNFα and IL-6 can both enhance and inhibit neuronal injury (for review, see [58]). Administration of IL-1ra has been shown to reduce the infarct size in animal models of stroke [59,60]. Caspase-1 knock out mice demonstrate reduced amounts of neuronal damage from ischemic and excitotoxic origin compared to wild-type mice [61]. In mice deficient in the expression of the IL-1 receptor (IL-1RI), the emergence of ameboid microglia and the expression of inflammatory mediators (IL-1, IL-6, and cyclooxygenase-2) are suppressed and delayed following penetrating brain injury [62]. Collectively, these *in vivo* results support the role of IL-1 in brain inflammation, reactive gliosis, and neuronal damage.

Alzheimer's disease is a chronic neurodegenerative disease in which the role of IL-1 in neuronal damage has been extensively investigated. The intimate association of IL-1-expressing microglia with β-amyloid in senile plaques suggested that IL-1 might trigger an inflammatory cascade that ultimately contributes to neuronal demise [56,63,64]. Exposure of cultured cells to insoluble aggregates of β-amyloid leads to production of a number of inflammatory molecules, including proinflammatory cytokines and nitric oxide. IL-1 mediates the pathological effects of microglia on neurons such as abnormal phosphorylation of tau and reduced synthesis of synaptophysin [65]. Polymorphisms in the IL-1α and IL-1β genes have also been linked to the increased risk for AD [64].

IL-1's role in neuronal damage is firmly established in a number of *in vitro* systems. Several mechanisms (direct and indirect) may underlie IL-1-mediated neurotoxicity, with some notable differences between human and rodent systems. For example, IL-1 may contribute to glutamate-mediated neurodegeneration by modulating NMDA receptor function. In rat hippocampal neurons, pretreatment with IL-1 increases NMDA receptor function through activation of tyrosine kinases and subsequent NR2A/B subunit phosphorylation [66]. In a series of *in vivo* and *in vitro* experiments using the HIV-1 glycoprotein gp120 as the insult, Corasaniti and colleagues identified IL-1 as the neurotoxic cytokine [67]. They showed that injection of gp120 to rat brain or stimulation of the chemokine receptors CXCR4 or CCR5 *in vitro* by gp120 led to the induction of IL-1β and the activation of caspase-1 and COX-2 expression, culminating in neuronal death. Using recombinant IL-1β in primary cultures of human neurons and glia, we and others have found that IL-1 is toxic to neurons [68,69]. The IL-1-induced neurotoxicity was enhanced by the Th1 cytokine IFN but reduced by inhibitors of TNFα or nitric oxide (NO) synthase. These studies support the idea that NO alone is a weak neurotoxin and that the conditions in which both NO and a proinflammatory cytokine (TNFα) are coexpressed lead to significant damage to the CNS [70]. Importantly, these studies have delineated striking species differences in the cytokine-induced neurotoxicity cascade. Whereas LPS, IL-1, or TNFα can all trigger neurotoxicity in rodent cultures, the induction of neurotoxicity in human brain cell cultures appears to be dependent on the presence of IL-1 [68,69,71,72]. Furthermore, IL-1-activated astrocytes are an important mediator of neurotoxicity in the human system by producing NO and TNFα [73]. The redundancy in the IL-1-inducing signals for human microglia, the lack of expression of IL-1 in human astrocytes, and the number of genes regulated by IL-1 in human astrocytes [32] all point to the unique role of microglial IL-1 in human pathologies.

IV. BIOLOGY OF ASTROCYTES

A. REACTIVE ASTROCYTES AND CNS INJURY

The term *astrocytes* encompasses a diverse population of cells. Although different subpopulations of astrocytes were originally identified based on morphological criteria alone, it is now well recognized that astrocytes, even within well-defined areas of the brain such as the hippocampus, possess qualitatively different ion current phenotypes, display differences in neurotransmitter responsiveness, and may be coupled by different gap junction proteins [74,75]. These emerging data, much of which have been collected using glial fibrillary acidic protein (GFAP)/enhanced green fluorescence protein (EGFP) transgenic mice, argue strongly that astrocytes are comprised of groups of cells with a much larger functional heterogeneity that has heretofore been recognized [76].

What is perhaps of greater import in the context of the focus of this volume on cytokines is that astrocytes are involved in every facet of brain function [77]. As such, the response of these cells to injury is likely to play a pivotal role in defining the subsequent outcome of the pathologic process. Consistent with this concept is the observation that changes in the properties of astrocytes is one of the best recognized markers of damage to the CNS, characterized by astrocytic hypertrophy, proliferation, and altered gene expression [78]. The identification of reactive astrocytes is relatively straightforward because these cells are usually strongly immunoreactive for astrocyte-specific markers such as GFAP [79], but determining the initiating factors and functional consequences of a reactive astrogliosis is a much more complex process. The diffuse nature of a reactive astrogliosis in the CNS suggests a role for either soluble mediators, such as cytokines, or the presence of an integrated astrocyte-to-astrocyte syncytium that permits transfer of information across extended distances. There is ample evidence from studies both *in vivo* and *in vitro* that both pathways may be operative in the injured CNS and further that signaling via these pathways may intersect. Studies suggest that activated astrocytes produce proinflammatory factors that facilitate leukocyte infiltration and damage to the neural elements, but they also produce growth factors that promote neuronal survival. Recently, the dual nature of the reactive astrogliosis has been illustrated using mice in which reactive astrocytes were selectively targeted for ablation in the injured CNS [80,81].

B. ASTROCYTE GENE EXPRESSION INDUCED BY IL-1

We have used a transcriptional gene-profiling technique to define the role of cytokines in regulating astrocytic cell function and interactions in human fetal astrocytes. Because IL-1 has been shown to be a critical initiating cytokine in the early stages of astrocyte activation both *in vivo* and *in vitro* [82,83], we have focused on the IL-1-induced gene activation program in astrocytes. The results are similar to those obtained in adult astrocytes [84,85] and suggest that IL-1 is a potent modulator of human fetal astrocyte gene expression resulting in activation/repression of a large number of factors involved in inflammation, development and growth, cell–matrix interactions, and the maintenance of the BBB. For example, application of exogenous IL-1 to such cultures induces the expression of multiple acute phase genes involved in the early stages of an immune response, including chemokines, cytokines, and adhesion molecules, the majority of which are known to be strongly proinflammatory [86,87]. Astrocytes also provide trophic functions during development and injury. Accordingly, in cultured astrocytes, IL-1 also induces a subset of genes primarily associated with neuronal and glial growth and survival, including IL-6, ciliary neurotrophic factor, and GM-CSF [88,89].

Another set of genes that are strongly induced in IL-1-activated human astrocytes are the IFN-inducible genes. This result was unexpected because astrocytes lack CD14, and not being immune cells themselves, they were not expected to express such a large number of IFN-inducible genes involved in antiviral or innate immune responses. In fact, the data show that a striking similarity exists in the gene expression profile of LPS-activated microglia (Table 1.1) and IL-1-activated astrocytes as reported earlier [32]. Underlying this phenomenon might be the common intracellular signaling pathway shared by TLR4 and IL-1 receptor leading to the activation of transcription factors such as NF-B and AP-1 (see Figure 1.1). Interestingly, IL-1 activated astrocytes also demonstrate evidence for the activation of the IRF-3 pathway [108] that is implicated in antiviral immunity [25]. From these data, we hypothesize that astrocytes are endowed with a specific function in the brain resembling that of pathogen-activated brain macrophages. IL-1-activated astrocytes may function as the second line of defense in the brain that amplifies antimicrobial immune responses as well as neurodegenerative mechanisms. Alternatively, the microglial IL-1-initiated activation cascades in both microglia and astrocytes may represent a universal glial activation mechanism and may in part offer an explanation for the nonspecific pathologies and overlapping and redundant mediator expression found in CNS diseases of diverse etiologies.

C. ASTROCYTE RESPONSE TO LPS

The failure of astrocytes to respond robustly to LPS is due to their lack of CD14, which is known to increase sensitivity to LPS by several orders of magnitude [90]. Thus, whereas CD14+ cells respond strongly to nanogram concentrations of LPS, CD14-negative cells respond weakly to high concentrations (μg/ml) of LPS. On the other hand, even the weak response would require TLR4, the signaling receptor for LPS. Astrocytes *in vivo* and *in vitro* express TLR2 and TLR3, but the expression of TLR4 (in both human and rodent astrocytes) is controversial [91–93]. Furthermore, because TLR2 could also transduce an LPS signal, some of the astrocyte LPS responses could be attributed to TLR2 rather than TLR4. A large body of literature documents that rodent astrocytes respond to LPS and upregulate a number of inflammatory genes such as cytokines, chemokines, and iNOS. On the other hand, we have not been able to activate gene expression by LPS in human fetal astrocytes except in rare instances (low level M-CSF induction, for example). Whether these discrepancies are due to species differences in CD14/TLR expression and signal transduction or due to differences in tissue culture techniques is currently unresolved. Regardless, because of the exquisitely sensitive response of microglia to LPS (and astrocytes to IL-1), the biological significance of an astrocyte LPS response is probably limited.

D. ASTROCYTES AND THE BLOOD–BRAIN BARRIER

Astrocytes play an important role in establishing and maintaining the BBB. Anatomically, astrocytic foot processes wrap around the blood vessels (glia limitans) and are separated from the latter only by a layer of basal lamina. This unique relationship between astrocytes and CNS blood vessels can be best visualized by GFAP immunocytochemistry (Figure 1.3). IL-1-activated astrocytes recapitulate the hypoxia-induced astrocyte phenotype with respect to the expression of genes implicated in BBB function. One of the most robustly induced genes in IL-1-activated human astrocyte microarrays is vascular endothelial cell growth factor (VEGF) [32,85]. VEGF is a mitogen for endothelial cells and is induced by hypoxia through activation of the transcription factor, hypoxia-inducible factor [94]. Recently, Src-suppressed C kinase substrate (SSeCKS)/gravin [95], a scaffolding protein that integrates cytoskeletal dynamics with cellular signaling cascades, has been shown to coordinate the astrocyte response to oxygen tension [96]. SSeCKS expression is reduced during hypoxia; SSeCKS suppresses angiogenesis but increases endothelial tight junction formation. Similar to hypoxia, IL-1 decreases astrocyte transcript levels for gravin and angiopoietin-1 but increases those for VEGF [32]. The expected *in vivo* outcomes are increased endothelial proliferation, increased vascular permeability and disruption of the endothelial tight junctions [96], ultimately compromising the BBB [97]. Together, these findings indicate that astrocytes are a significant driving force for the maturation and maintenance of the BBB and that IL-1 might disrupt this important astrocytic function during brain inflammation. Curiously, the microarray data demonstrate a similar gene expression profile in LPS-activated microglia (Table 1.1) verifying the notion that the TLR-activated macrophage transcription program is preserved in IL-1-activated astrocytes.

V. CONCLUDING REMARKS

Substantial progress has been made in recent years in the identification of several key mechanisms that are involved in neuroinflammation. Advancing knowledge in the neuronal and glial signaling pathways has also refined approaches toward understanding and treating CNS disorders. More and more evidence supports the conclusion that macrophage/microglial receptors that are involved in innate immune response are also activated in neurodegenerative diseases and that products of these cells leads to an associated activation of astrocytes. However, in spite of the rapid progress in the field, fundamental questions remain unanswered. For example, what are the determinants that convert the beneficial "tolerance" inducing signal in the CNS to a detrimental "priming" signal? If common pathways of neuronal damage exist in inflammatory neurodegenerative diseases, how can disease specificity be determined? The extent and tempo of neuronal damage also varies greatly among disorders associated with neuroinflammation, yet our knowledge of how these processes are controlled remains rudimentary. The diverse disease manifestations must result in part from the complexity of the action of inflammatory molecules such as cytokines. The net cellular consequences of a particular molecule depend upon the environmental context in which it is expressed, as well as the expression of its receptors on target cells. It is also important to remember that not all "neurotoxins" are toxic in every situation; under certain conditions the same molecules can be neuroprotective. Indeed, several recent studies from mice with targeted deletions of specific cytokine genes or their receptors have shown that the beneficial effects of cytokines for CNS repair are, in some instances, more notable than their deleterious effects. The development of model systems that permit the inducible knock-in and knock-out of specific cytokines and their receptors at different phases of the disease process are, therefore, likely to provide even better insights into their specific roles in the CNS. In humans, cytokines or their inhibitors are now part of the therapeutic armamentarium for diseases such as multiple sclerosis, rheumatoid arthritis, and ulcerative colitis. The success of these treatments provides a strong impetus for continued research in this direction and demonstrates that the importance of continuing work on the role of cytokines in human diseases is now beyond question.

ACKNOWLEDGMENTS

The authors thank Gareth R. John, Xianyuan Song, Meng-Liang Zhao, and Dennis W. Dickson for their helpful input and the work described in this chapter. We apologize to those whose work could not be adequately referenced because of the space limitations.

This work was supported by NIH RO1s: MH55477, AI44641, and NS40137; NIH Training Grant NS07098; and the Einstein CFAR Grant P30 AI51519.

REFERENCES

1. Perry V.H., Gordon S. Macrophages and microglia in the nervous system, *Trends Neurosci.* 11, 273, 1988.
2. Dickson D.W., Lee S.C. Microglia. In: *Textbook of neuropathology* 3rd edition (eds. Davis R.L., Robertson D.M.). Williams & Wilkins: Baltimore, 165. 1997.

3. Lassman H., Hickey W.F. Dynamics of microglia in brain pathology: radiation bone marrow chimeras as a tool to study microglia turnover in normal brain and inflammation, *Clin. Neuropathol.* 12, 284, 1993.

4. Gehrmann J., Matsumoto Y., Kreutzberg G.W. Microglia: intrinsic immune effector cell of the brain, *Brain Res. - Brain Res. Rev.* 20, 269, 1995.

5. Lawson L.J. et al. Heterogeneity in the distribution and morphology of microglia in the normal adult mouse brain, *Neuroscience* 39, 151, 1990.

6. Dickson D.W. et al. Biology of disease: microglia in human disease, with an emphasis on the acquired immunodeficiency syndrome, *Lab. Invest.* 64, 135, 1991.

7. Barnett M.H., Prineas J.W. Relapsing and remitting multiple sclerosis: pathology of the newly forming lesion, *Ann. Neurol.* 55, 458, 2004.

8. Cosenza M. et al. Human brain parenchymal microglia express CD14 and CD45 and are productively infected by HIV-1 in HIV-1 encephalitis, *Brain Pathol* 12, 442, 2002.

9. Hickey W.F., Kimura H. Perivascular microglial cells of the CNS are bone marrow-derived and present antigen *in vivo, Science* 239, 290, 1988.

10. Unger E.R. et al. Male donor-derived cells in the brains of female sex-mismatched bone marrow transplant recipients: a Y-chromosome specific *in situ* hybridization study, *J. Neuropathol. Exp. Neurol.* 52, 460, 1993.

11. Simard A.R., Rivest S. Bone marrow stem cells have the ability to populate the entire central nervous system into fully differentiated parenchymal microglia, *FASEB J.* 18, 998–1000, 2004.

12. Priller J. et al. Targeting gene-modified hematopoietic cells to the central nervous system: use of green fluorescent protein uncovers microglial engraftment, *Nat. Med.* 7, 1356, 2001.

13. Vallieres L., Sawchenko P.E. Bone marrow-derived cells that populate the adult mouse brain preserve their hematopoietic identity, *J. Neurosci.* 23, 5197, 2003.

14. Grossmann R. et al. Juxtavascular microglia migrate along brain microvessels following activation during early postnatal development, *Glia* 37, 229, 2002.

15. Kreutzberg G.W. Microglia: a sensor for pathological events in the CNS, *Trends Neurosci.* 19, 312, 1996.

16. Kaul M., Garden G.A., Lipton S.A. Pathways to neuronal injury and apoptosis in HIV-associated dementia, *Nature* 410, 988, 2001.

17. Nguyen M.D., Julien J.P., Rivest S. Innate immunity: the missing link in neuroprotection and neurodegeneration? *Nat. Rev. Neurosci.* 3, 216, 2002.

18. Wright S.D. et al. CD14, a receptor for complexes of lipopolysaccharide (LPS) and LPS binding protein, *Science* 249, 1431, 1990.

19. Akira S., Takeda K., Kaisho T. Toll-like receptors: critical proteins linking innate and acquired immunity, *Nat. Immunol.* 2, 675, 2001.

20. Underhill D.M., Ozinsky A. Toll-like receptors: key mediators of microbe detection, *Curr. Opin. Immunol.* 14, 103, 2002.

21. Haynes L.M. et al. Involvement of Toll-like receptor 4 in innate immunity to respiratory syncytial virus, *J Virol.* 75, 10730, 2001.

22. da Silva C.J. et al. Lipopolysaccharide is in close proximity to each of the proteins in its membrane receptor complex. transfer from CD14 to TLR4 and MD-2, *J Biol. Chem.* 276, 21129, 2001.

23. Yamamoto M. et al. Role of adaptor TRIF in the MyD88-independent Toll-like receptor signaling pathway, *Science* 301, 640, 2003.

24. Servant M.J. et al. Identification of distinct signaling pathways leading to the phosphorylation of interferon regulatory factor 3, *J Biol. Chem.* 276, 355, 2001.

25. Sharma S. et al. Triggering the interferon antiviral response through an IKK-related pathway, *Science* 300, 1148, 2003.
26. Fitzgerald K.A. et al. IKKepsilon and TBK1 are essential components of the IRF3 signaling pathway, *Nat. Immunol.* 4, 491, 2003.
27. Williams K. et al. Biology of adult human microglia in culture: comparisons with peripheral blood monocytes and astrocytes, *J. Neuropathol. Exp. Neurol.* 51, 538, 1992.
28. Becher B., Fedorowicz V., Antel J.P. Regulation of CD14 expression on human adult central nervous system-derived microglia, *J. Neurosci. Res.* 45, 375, 1996.
29. Beschorner R. et al. Infiltrating CD14+ monocytes and expression of CD14 by activated parenchymal microglia/macrophages contribute to the pool of CD14+ cells in ischemic brain lesions, *J. Neuroimmunol.* 126, 107, 2002.
30. Nadeau S., Rivest S. Role of microglial-derived tumor necrosis factor in mediating CD14 transcription and nuclear factor kappa B activity in the brain during endotoxemia, *J Neurosci.* 20, 3456, 2000.
31. Fassbender K. et al. The LPS receptor (CD14) links innate immunity with Alzheimer's disease, *FASEB J.* 18, 203, 2004.
32. John, G. R. et al. IL-1 regulated responses in astrocytes: relevance to injury and recovery. *Glia.* 49, 161–176, 2005.
33. Husemann J. et al. Scavenger receptors in neurobiology and neuropathology: their role on microglia and other cells of the nervous system, *Glia* 40, 195, 2002.
34. Christie R.H., Freeman M., Hyman B.T. Expression of the macrophage scavenger receptor, a multifunctional lipoprotein receptor, in microglia associated with senile plaques in Alzheimer's disease, *Am. J. Pathol.* 148, 399, 1996.
35. Coraci I.S. et al. CD36, a class B scavenger receptor, is expressed on microglia in Alzheimer's disease brains and can mediate production of reactive oxygen species in response to beta-amyloid fibrils, *Am. J. Pathol.* 160, 101, 2002.
36. Amat J.A. et al. Phenotypic diversity and kinetics of proliferating microglia and astrocytes following cortical stab wounds, *Glia* 16, 368, 1996.
37. Granucci F. et al. The scavenger receptor MARCO mediates cytoskeleton rearrangements in dendritic cells and microglia, *Blood* 102, 2940, 2003.
38. Re F. et al. Granulocyte-macrophage colony-stimulating factor induces an expression program in neonatal microglia that primes them for antigen presentation, *J. Immunol.* 169, 2264, 2002.
39. de Villiers W.J., Smart E.J. Macrophage scavenger receptors and foam cell formation, *J. Leukoc. Biol.* 66, 740, 1999.
40. Peress N.S. et al. Identification of Fc gamma RI, II and III on normal human brain ramified microglia and on microglia in senile plaques in Alzheimer's disease, *J. Neuroimmunol.* 48, 71, 1993.
41. Ulvestad E. et al. Reactive microglia in multiple sclerosis lesions have an increased expression of receptors for the Fc part of IgG, *J. Neurol. Sci.* 121, 125, 1994.
42. Greenberg S.M., Bacskai B.J., Hyman B.T. Alzheimer disease's double-edged vaccine, *Nat. Med.* 9, 389, 2003.
43. Casadevall A. et al. Characterization of a murine monoclonal antibody to Cryptococcus neoformans polysaccharide that is a candidate for human therapeutic studies, *Antimicrobial Agents and Chemotherapy* 42, 1437, 1998.
44. Casadevall A., Goldman D., Lee S.C. Cryptococcal meningoencephalitis. In: *New concepts on the immunopathogenesis of CNS infections* (eds. Peterson P.K., Remington J.S.). Blackwell Science, Inc., Malden, MA. 2000.
45. Lock C. et al. Gene-microarray analysis of multiple sclerosis lesions yields new targets validated in autoimmune encephalomyelitis, *Nat. Med.* 8, 500, 2002.

46. Goldman D. et al. *Cryptococcus neoformans* indices macrophage inflammatory protein 1α (MIP-1α) and MIP-1β in human microglia: role of specific antibody and soluble capsular polysaccharide, *Infect. Immun.* 69, 1808, 2001.

47. Song X. et al. FcγRI- and FcγRIII-mediated MIP-1α induction in primary human and murine microglia, *Infect. Immun.* 70, 5177, 2002.

48. Song X. et al. Fcγ receptor signaling in primary human microglia: differential roles of PI-3K and Ras/ERK MAPK pathways in phagocytosis and chemokine induction, *J. Leukoc. Biol.*, 2004.

49. Dinarello C.A. The interleukin-1 family: 10 years of discovery, *FASEB J.* 8, 1314, 1994.

50. Dinarello C.A. Biologic basis for interleukin-1 in disease, *Blood* 87, 2095, 1996.

51. Lee Y.B., Nagai A., Kim S.U. Cytokines, chemokines, and cytokine receptors in human microglia, *J. Neurosci. Res.* 69, 94, 2002.

52. Meda L. et al. Proinflammatory profile of cytokine production by human monocytes and murine microglia stimulated with beta-amyloid [25–35], *J. Neuroimmunol.* 93, 45, 1999.

53. Lee S.C. et al. Cytokine production by human fetal microglia and astrocytes: differential induction by LPS and IL-1β, *J. Immunol.* 150, 2659, 1993.

54. Liu J.S.H. et al. Interferons are critical regulators of IL-1b and IL-1ra expression in human fetal microglia., *J. Immunol.* 161, 1989, 1998.

55. Downen, M., and Lee S. C. Glial cytokine expression in human CNS culture following hypoxia, *Soc. Neurosci.* Abstract 229, 3, 1997.

56. Griffin W.S. et al. Brain interleukin-1 and S-100 immunoreactivity are elevated in Down syndrome and Alzheimer's disease, *Proc. Natl. Acad. Sci. U.S.A.* 86, 7611, 1989.

57. Zhao M.L. et al. Expression of iNOS, IL-1 and caspase-1 in HIV-1 encephalitis, *J. Neuroimmunol.* 115, 182, 2001.

58. Allan S.M., Rothwell N.J. Cytokines and acute neurodegeneration, *Nat. Rev. Neurosci.* 2, 734, 2001.

59. Rothwell N.J., Allan S., Toulmond S. The role of interleukin 1 in acute neurodegeneration and stroke: pathophysiological and therapeutic implications, *J. Clin. Invest.* 100, 2648, 1997.

60. Rothwell N.J., Luheshi G.N. Interleukin 1 in the brain: biology, pathology and therapeutic target, *TINS* 23, 618, 2000.

61. Hara H. et al. Inhibition of interleukin-1 beta converting enzyme family proteases reduces ischemic and excitotoxic neuronal damage, *Proc. Natl. Acad. Sci. U.S.A.* 94, 2007, 1997.

62. Basu A. et al. The type 1 interleukin-1 receptor is essential for the efficient activation of microglia and the induction of multiple proinflammatory mediators in response to brain injury, *J. Neurosci.* 22, 6071, 2002.

63. Griffin W.S. et al. Glial-neuronal interactions in Alzheimer's disease: the potential role of a 'cytokine cycle' in disease progression, *Brain Pathol.* 8, 65, 1998.

64. Griffin W.S., Mrak R.E. Interleukin-1 in the genesis and progression of and risk for development of neuronal degeneration in Alzheimer's disease, *J Leukoc. Biol.* 72, 233, 2002.

65. Li Y. et al. Interleukin-1 mediates pathological effects of microglia on tau phosphorylation and on synaptophysin synthesis in cortical neurons through a p38-MAPK pathway, *J. Neurosci.* 23, 1605, 2003.

66. Viviani B. et al. Interleukin-1beta enhances NMDA receptor-mediated intracellular calcium increase through activation of the Src family of kinases, *J. Neurosci.* 23, 8692, 2003.

67. Corasaniti M.T. et al. Neurobiological mediators of neuronal apoptosis in experimental neuroAIDS, *Toxicol. Lett.* 139, 199, 2003.
68. Downen M. et al. Neuronal death in cytokine-activated primary human brain cell culture: role of tumor necrosis factor-alpha, *Glia* 28, 114, 1999.
69. Chao C.C. et al. Cytokine-stimulated astrocytes damage human neurons via a nitric oxide mechanism, *Glia* 16, 276, 1996.
70. Blais V., Rivest S. Effects of TNF-alpha and IFN-gamma on nitric oxide-induced neurotoxicity in the mouse brain, *J. Immunol.* 172, 7043, 2004.
71. Chao C.C. et al. Activated microglia mediate neuronal cell injury via a nitric oxide mechanism, *J. Immunol.* 149, 2736, 1992.
72. Lee S.C., Dickson D.W., Brosnan C.F. Interleukin-1, nitric oxide and reactive astrocytes, *Brain, Behavior, and Immunity* 9, 345, 1995.
73. Liu J. et al. Expression of type II nitric oxide synthase in primary human astrocytes and microglia: role of IL-1β and IL-1 receptor antagonist, *J. Immunol.* 157, 3569, 1996.
74. Nolte C. et al. GFAP promoter-controlled EGFP-expressing transgenic mice: a tool to visualize astrocytes and astrogliosis in living brain tissue, *Glia* 33, 72, 2001.
75. Blomstrand F. et al. Extent of intercellular calcium wave propagation is related to gap junction permeability and level of connexin-43 expression in astrocytes in primary cultures from four brain regions, *Neuroscience* 92, 255, 1999.
76. Matthias K. et al. Segregated expression of AMPA-type glutamate receptors and glutamate transporters defines distinct astrocyte populations in the mouse hippocampus, *J. Neurosci.* 23, 1750, 2003.
77. Ransom B., Behar T., Nedergaard M. New roles for astrocytes (stars at last), *Trends Neurosci.* 26, 520, 2003.
78. Ridet J.L. et al. Reactive astrocytes: cellular and molecular cues to biological function, *TINS* 20, 570, 1997.
79. Lee S.C., Brosnan C.F. Molecular biology of glia: astrocytes. In: *Molecular biology of multiple sclerosis.* (ed. Russell W.C.). John Wiley & Sons, Ltd, Chichester, England, 71. 1997.
80. Bush T.G. et al. Leukocyte infiltration, neuronal degeneration, and neurite outgrowth after ablation of scar-forming, reactive astrocytes in adult transgenic mice, *Neuron* 23, 297, 1999.
81. Faulkner J.R. et al. Reactive astrocytes protect tissue and preserve function after spinal cord injury, *J. Neurosci.* 24, 2143, 2004.
82. Giulian D. et al. Interleukin-1 injected into mammalian brain stimulates astrogliosis and neovascularization, *J. Neurosci.* 8, 2485, 1988.
83. Herx L.M., Rivest S., Yong V.W. Central nervous system-initiated inflammation and neurotrophism in trauma: interleukin-1β is required for the production of ciliary neurotrophic factor, *J. Immunol.* 165, 2232, 2000.
84. John G.R. et al. Multiple sclerosis: re-expression of a developmental pathway that restricts oligodendrocyte maturation, *Nat. Med.* 8, 1115, 2002.
85. Meeuwsen S. et al. Cytokine, chemokine and growth factor gene profiling of cultured human astrocytes after exposure to proinflammatory stimuli, *Glia* 43, 243, 2003.
86. Weiss J.M. et al. Astrocyte-derived monocyte-chemoattractant protein-1 directs the transmigration of leukocytes across a model of the human blood-brain barrier, *J. Immunol.* 161, 6896, 1998.
87. Hua L.L., Lee S.C. Distinct patterns of stimulus-inducible chemokine mRNA accumulation in human fetal astrocytes and microglia, *Glia* 30, 74, 2000.
88. Kamiguchi H. et al. Release of ciliary neurotrophic factor from cultured astrocytes and its modulation by cytokines, *Neurochem. Res.* 20, 1187, 1995.

89. Friedman W.J. et al. Regulation of nerve growth factor mRNA by interleukin-1 in rat hippocampal astrocytes is mediated by NFkB, *J. Biol. Chem.* 271, 31115, 1996.
90. Ulevitch R.J., Tobias P.S. Receptor-dependent mechanisms of cell stimulation by bacterial endotoxin, *Annu. Rev. Immunol.* 13, 437, 1995.
91. Bsibsi M. et al. Broad expression of Toll-like receptors in the human central nervous system, *J. Neuropathol. Exp. Neurol.* 61, 1013, 2002.
92. Lehnardt S. et al. The Toll-like receptor TLR4 is necessary for lipopolysaccharide-induced oligodendrocyte injury in the CNS, *J Neurosci.* 22, 2478, 2002.
93. Bowman C.C. et al. Cultured astrocytes express Toll-like receptors for bacterial products, *Glia* 43, 281, 2003.
94. Maxwell P.H., Ratcliffe P.J. Oxygen sensors and angiogenesis, *Semin. Cell Dev. Biol.* 13, 29, 2002.
95. Gelman I.H. et al. Control of cytoskeletal architecture by the src-suppressed C kinase substrate, SSeCKS, *Cell Motil. Cytoskeleton* 41, 1, 1998.
96. Lee S.W. et al. SSeCKS regulates angiogenesis and tight junction formation in blood-brain barrier, *Nat. Med.* 9, 900, 2003.
97. Rieckmann P., Engelhardt B. Building up the blood-brain barrier, *Nat. Med.* 9, 828, 2003.
98. Lee S.C. et al. Characterization of human fetal dissociated CNS cultures with an emphasis on microglia, *Lab. Invest.* 67, 465, 1992.
99. Takeda K., Akira S. TLR signaling pathways, *Semin. Immunol.* 16, 3, 2004.
100. Jiang Z. et al. Interleukin-1 (IL-1) receptor-associated kinase-dependent IL-1-induced signaling complexes phosphorylate TAK1 and TAB2 at the plasma membrane and activate TAK1 in the cytosol, *Mol. Cell Biol.* 22, 7158, 2002.
101. Burns K. et al. Tollip, a new component of the IL-1RI pathway, links IRAK to the IL-1 receptor, *Nat. Cell Biol.* 2, 346, 2000.
102. Horng T. et al. The adaptor molecule TIRAP provides signalling specificity for Toll-like receptors, *Nature* 420, 329, 2002.
103. Yamamoto M. et al. Essential role for TIRAP in activation of the signalling cascade shared by TLR2 and TLR4, *Nature* 420, 324, 2002.
104. Jiang Z. et al. Toll-like receptor 3-mediated activation of NF-kappaB and IRF3 diverges at Toll-IL-1 receptor domain-containing adapter inducing IFN-beta, *Proc. Natl. Acad. Sci. U. S. A.* 101, 3533, 2004.
105. Hemmi H. et al. The roles of two Ikappab kinase-related kinases in lipopolysaccharide and double stranded RNA signaling and viral infection, *J. Exp. Med.* 199, 1641, 2004.
106. Perry A.K. et al. Differential requirement for TANK-binding Kinase-1 in Type I interferon responses to Toll-like receptor activation and viral infection, *J. Exp. Med.* 199, 1651, 2004.
107. Liu J.S.H. et al. Expression of inducible nitric oxide synthase and nitrotyrosine in multiple sclerosis lesions, *Am. J. Pathol.* 158, 2057, 2001.
108. Rivieccio et al. The cytokine IL-1 activates IFN response factor 3 in human fetal astrocytes in culture, *J. Immunol.* 174, 3719–3726, 2005.

2 Cytokines: Wherefrom and Whereto

Jan Vilček

CONTENTS

I. WHAT ARE CYTOKINES?

Over 200 distinct cytokine genes and proteins have been identified to date [1,2]. The names of many cytokines are based on their first-identified biological action, e.g., interferons were first found to interfere with virus replication, and tumor necrosis factor caused necrosis of tumor cells. The use of descriptive names can be misleading because the first-identified activity of a cytokine may not be the most significant one. For example, the pleiotropic cytokine, transforming growth factor-β (TGF-β) plays important roles in the regulation of the immune system and in development, but it does not produce malignant transformation of cells. Most cytokines described after 1980 have been named "interleukin" and numbered sequentially; the series has reached IL-31 at the time of this writing [3]. Although the interleukin nomenclature is more "neutral," the traditional descriptive names make it somewhat easier to recall specific features of individual cytokines. To facilitate the study of cytokines, it is helpful to divide them into distinct groups. Various systems of classification have been designed, but none are perfect. The most widely used classification is based primarily on the structural features of the cytokine proteins and their interacting receptors. The major cytokine families, grouped on the basis of their molecular structure, will be briefly surveyed later in this chapter. (The nature of cytokine receptors, which is another important basis for cytokine classification, will be reviewed in one of the subsequent chapters in this volume.)

What are cytokines, anyway? Cytokines are important orchestrators of host defenses (innate and adaptive immune responses, inflammatory responses), they regulate diverse cellular functions (proliferation, survival, maturation), and they are important for the development and functions of the hematopoietic and nervous systems. Cytokines tend to be produced preferentially by white blood cells, especially lymphocytes and monocytes or macrophages, but many cytokines are also readily produced by other somatic cells. Since no short definition can encompass all of their features, cytokines are best defined by a set of characteristic properties (Table 2.1). The broader characteristic features of cytokines can be best understood in their historical context, which will be tackled in the next section of this chapter.

Most cytokines are secreted proteins, usually glycosylated. However, a large number of cytokines, specifically most members of the tumor necrosis factor (TNF) family, are synthesized as type II transmembrane proteins, which may or may not be subsequently cleaved and released from the cell surface [4]. When cleavage does take place, the extracellular portions of the transmembrane proteins are enzymatically processed by metalloproteases, resulting in the release of soluble active cytokine molecules. Alternatively, biological activity is mediated by the integral transmembrane form of the cytokine proteins expressed on the cell surface, which can bind to receptors on neighboring cells. It is interesting that the biological action produced by the transmembrane form can be qualitatively different from the action produced by the released, soluble form of the cytokine [4]. Other cytokines that are not classical secretory proteins (i.e., lack an N-terminal cleavable signal peptide sequence) include members of the IL-1 family; they are released from cells by an incompletely understood mechanism, and they require activation by proteolytic cleavage.

Because cytokines are powerful agents and their prolonged actions in the intact organism can be harmful, it is not surprising that they are usually produced only in response to appropriate stimuli and then only transiently (Table 2.1). As usual, there are exceptions. For example, low levels of some interferon proteins are produced constitutively in some cells and these small quantities of constitutively produced interferon appear to have a physiological role, e.g., in the maintenance of expression of major histocompatibility class I genes in lymphoid cells [5]. The synthesis of most cytokines is tightly regulated at the transcriptional level. Additionally, cytokine production can also be regulated by RNA processing at the level of translation or posttranslationally. Members of the TGF-β family (isoforms TGF-β1, TGF-β2, and TGF-β3) are examples of cytokines whose availability is regulated in a complex manner. TGF-β genes (especially TGF-β1) can be upregulated in response to stress and injury or by some oncogenes and viral transactivators. Stabilization of TGF-β mRNA transcripts also contributes to the upregulated expression under some circumstances. However, the major form of regulation is by proteolytic cleavages, some of which occur intracellularly and result in the release of an inactive TGF-β propeptide, and some that occur outside the cell, leading to the release of the biologically active form.

In general, sustained and elevated cytokine production in the body is often associated with pathologies. Thus, increased levels of TNF, IL-1, and IL-6 are hallmarks of chronic inflammatory disorders, such as rheumatoid arthritis, Crohn's disease or psoriasis, and the administration of agents that neutralize excess cytokine activity (e.g., monoclonal antibodies or soluble receptor constructs that neutralize TNF activity) has been used with great success in the treatment of chronic inflammatory disorders [6,7].

TABLE 2.1
Characteristic Features of Cytokines

Most cytokines are simple polypeptides or glycoproteins ≤30 kDa in size (many cytokines form
larger homodimers or homotrimers. and some cytokines are heterodimers).

Most cytokines are produced as secretory proteins, but a significant number are made as integral
transmembrane proteins, which may or may not be subsequently cleaved and released.

Constitutive production of cytokines is usually low or absent; production is regulated by various
inducing stimuli at the level of transcription, translation, or posttranslationally.

Cytokine production is usually transient.

The action radius of cytokines is usually short (typical action is autocrine or paracrine, not
endocrine).

Cytokines act by binding to specific high affinity cell surface receptors (Kd in the range of
$10^{-9} - 10^{-12}$ M). Cytokine receptors can be homodimers, heterodimers, homotrimers, or hetero-
trimers.

Most cytokine actions can be attributed to an altered pattern of gene expression in the target cells.
Phenotypically, cytokine actions lead to an increase or decrease in the rate of cell proliferation,
change in cell differentiation state, or a change in the expression of some differentiated functions.

Although cytokines have typically a broad and diverse range of actions, at least some actions of
each cytokine are targeted at hematopoietic cells.

Another characteristic feature of cytokines is that their actions tend to be autocrine or, more typically, paracrine, i.e., their action radius is short (Table 2.1). This mode of action distinguishes cytokines from classical hormones, most of which act by an endocrine mode. Like hormones, cytokines act by binding to specific cell surface receptors, with receptor ligation generally followed by the triggering of characteristic intracellular signaling cascades. Activation of signaling cascades is most typically followed by the activation or silencing of specific target genes.

II. EVOLUTION OF THE CYTOKINE FIELD

Although he term *cytokine* was first used only in the 1970s [8], the origins of cytokine research can be traced to the middle of the 20th century (reviewed by Oppenheim and Feldmann [9]; Vilcek [10]). Among the earliest described factors are endogenous pyrogen [11] (which is probably identical to the cytokine later termed IL-1), nerve growth factor (NGF) [12], and interferon [13]. The present field of cytokine research is the result of a fusion of several separate areas of investigation. The arguably most important foundation for the cytokine field was research in cellular immunology. In the 1960s, several groups investigating lymphocyte interactions in cultures reported that soluble factors produced by activated lymphocytes affected various functions of white blood cells [14–18]. Several other fields of investigation contributed to the development of cytokine research. Interferons were originally described in the 1950s as selective antiviral agents [13], but it gradually became apparent that they also exert a broad range of actions on the growth and differentiation of cells of the immune system, in addition to other somatic cells. Initially few ties had existed between investigators studying soluble factors affecting immune responses and inter-feron researchers, but the dividing line has gradually dwindled. Two other originally distinct research areas that have become closely linked to the cytokine field are the investigation of hematopoietic growth factors (also called colony stimulating factors) and the study of growth factors acting on nonhematopoietic cells. As early as 1953, Levi-Montalcini and Hamburger reported that mouse sarcoma cells implanted in chick embryos produced a factor, now known as NGF, that induced growth and differentiation of sympathetic nerve ganglia [12]. (Parenthetically, the study of Levi-Montalcini, for which she was awarded the Nobel Prize, marks the beginning of the investigation of cytokines affecting the CNS.) As for other growth factors acting on nonhematopoietic cells, such as platelet-derived growth factor (PDGF), epidermal growth factor (EGF), and fibroblast growth factor (FGF), it is clear that many of these agents display cytokine-like properties. At least one agent originally considered a growth factor, that is, transforming growth factor-β (TGF-β), is now viewed primarily as a cytokine. I will elaborate on some of the milestones in the development of cytokine research.

Among the first observations in the field of cellular immunology relevant to the development of the cytokine field was the demonstration that migration of normal macrophages is inhibited by factors released from sensitized lymphocytes upon exposure to antigen [14,15]. One putative factor responsible for this action was named *macrophage migration inhibitory factor* (MIF). At about the same time, the presence of a factor mitogenic for lymphocytes was described in supernatants of mixed leukocyte cultures [16]. These observations were followed by the discovery of lymphotoxin in supernatants of activated lymphocyte cultures, which was defined as a factor causing cytotoxicity for some target cells [17,18]. Eventually, Dumonde et al. [19] coined the term *lymphokine* to designate soluble factors responsible for cell-mediated immunological reactions that are generated during interaction of sen-sitized lymphocytes with specific antigen. However, it soon became apparent that lymphocytes are not the only cell type capable of producing soluble proteins with immunoregulatory activity. Specifically, many soluble mediators were shown to be produced by monocytes and macrophages, and such agents were sometimes termed *monokines*. Later it became apparent that many other types of cells in the body, including nonhematopoietic cells, had the capacity to produce similar soluble medi-ators, which led to the general acceptance of the term *cytokine* for mediators pro-duced by lymphocytes, monocytes as well as other cells in the body [8].

Among the earliest known monocyte-derived cytokines was lymphocyte activa-tion factor (LAF), now known as IL-1. (There are two related but distinct active IL-1 proteins, encoded by separate genes — IL-1α and IL-1β [20].) LAF activity, defined as a mitogenic signal for thymocytes, was originally detected in supernatants of adherent cells isolated from human peripheral blood [21]. Other investigators described activities that are now known to be mediated by IL-1 under a variety of other names, for example, mitogenic protein, leukocytic pyrogen, endogenous pyro-gen, B cell-activating factor, leukocyte endogenous mediator, etc. (reviewed by Aarden et al. [22]). Another important monocyte/macrophage-derived cytokine is tumor necrosis factor (TNF), originally identified by Carswell et al. [23] in 1975 as a cytotoxic protein present in the serum of animals sensitized with Bacillus Calmette-Guérin and challenged with bacterial lipopolysaccharide (LPS). TNF was postulated to be the mediator of LPS-induced hemorrhagic necrosis of Meth A sarcoma tumors in mice; it also showed direct cytotoxicity for some tumor cells *in vitro*. Today we appreciate the fact that TNF is much more important as a regulator of inflammatory responses than as a "tumor necrosis factor." (An irreverent student in my laboratory had once suggested that TNF should stand for "too numerous functions.") I should add that both IL-1 and TNF can be produced by many diverse cell types, in addition to monocytes and macrophages.

It has been known since the early 1970s that lymphocytes produce one or more factors mitogenic for other lymphocytes. Morgan et al. [24] found that supernatants of mitogen-activated human mononuclear cells contain a growth factor that could support the continuous growth of human bone marrow-derived T cells. The responsible mitogenic factor is IL-2, then known as T cell growth factor (TCGF) and by a variety of other names that by now are largely forgotten [22]. In 1979 a group of investigators proposed to adopt the name *interleukin* for soluble lymphokines and monokines that act as communication signals between different populations of leukocytes. As a first step, the group introduced the names IL-1 and IL-2 for two important cytokines that up until then had been described under a variety of different descriptive names. Although the name *interleukin* was originally used for factors that act as communication signals between different populations of leukocytes, the term is now broadly applied also to cytokines that are produced by nonleukocytic cells or that act on nonleukocytic cells.

Whereas the early studies of lymphokines and monokines were largely the domain of immunologists who were seeking a better understanding of delayed-type hypersensitivity and other cell-mediated immune responses, interferons were discovered by virologists. Interferon was first described in 1957 by Isaacs and Lindenmann as a factor produced by virus-infected cells capable of inducing cellular resistance to infection [13]. That interferons would affect immune responses and other cellular functions was initially not known. However, several years later, Wheelock [25] described a virus-inhibitory protein (today known as IFN-γ or type II IFN) produced by mitogen-activated T lymphocytes. It is now known that IFN-γ, produced mainly by T cells, and NK cells are structurally completely distinct from the large family of IFN-α/β (or type I IFN) proteins, which are produced by a variety of cell types, including dendritic cells, monocytes, NK cells, and B cells, in addition to many nonhematopoietic cells [26–28].

A large and important group of cytokines are the chemokines—chemotactic cytokines whose primary function is to regulate the migration of leukocytes. All chemokines are structurally related 8–12 kDa polypeptides. The first identified member of this family is IL-8, now also known under the name CXCL8. [29]. About 50 distinct members of the chemokine family have been described to date.

The cytokine family now also includes hematopoietic growth factors or colony stimulating factors (CSF), proteins whose major function is to support the growth and differentiation of hematopoietic cells. Their name reflects the early observation that CSFs promote the formation of granulocyte or monocyte colonies in semisolid medium [30,31]. In addition to promoting the growth and differentiation of hematopoietic stem cells, CSFs have been shown to regulate some functions of fully differentiated hematopoietic cells, thus blurring the dividing line between these agents and cytokines.

Many proteins that promote the growth and development of nonhematopoietic cells have been identified; best known among these are NGF (already mentioned above), epidermal growth factor (EGF), platelet-derived growth factor (PDGF), fibroblast growth factors (FGF), and vascular endothelial growth factor (VEGF). Whether these and other growth factors should be considered cytokines is a matter of personal opinion. What is clear, however, is that these growth factors share many important properties with *bona fide* cytokines. A case in point is transforming growth factor-β (TGF-β). In 1978, de Larco and Todaro described a factor, originally termed *sarcoma growth factor,* that promoted the growth of normal rat fibroblasts in soft agar [32]. Subsequently, two distinct families of "transforming growth factors" with different spectrums of biological activities have been identified and molecularly characterized, termed *TGF-α* and *TGF-β* [33,34]. TGF-α is closely related to EGF, whereas the family of TGF-β proteins plays important roles not only in cell growth, differentiation, and morphogenesis but also in inflammation and immune regulation. Among the important actions of TGF-β proteins are the recruitment and activation of mononuclear cells, promotion of wound healing, fibrosis and angiogenesis, and a potent immunosuppressive action on T lymphocyte functions [35,36]. The wide spectrum of TGF-β actions, along with the many activities important in inflammation and immune responses, underscores the close relationship of TGF-β proteins to the cytokine family.

III. BRIEF SURVEY OF MAJOR CYTOKINE FAMILIES

One way to classify cytokines is to divide them on the basis of their principal functions into mediators of innate immunity, mediators of adaptive immunity, proinflammatory cytokines, regulators of hematopoiesis, etc. The difficulty with this approach is that most cytokines are pleiotropic, i.e., they generally mediate more than one important function. A more rational and practical basis for the grouping of cytokines is based on their molecular structure and the nature of interacting receptors (Table 2.2).

A number of important cytokines form the type I cytokine or hematopoietin family [37]. Although these cytokines show little or no significant homology in their primary structure, they share the presence of four α-helical bundles in a spatially similar arrangement [38–40]. (These features are also shared by erythropoietin and human growth hormone.) In addition, many of these cytokines show a similar genomic organization and many of their structural genes are located in close proximity to one another on human chromosome 5 and mouse chromosome 11 [41]. Furthermore, these and some other cytokines bind to the so-called class I cytokine or hematopoietic receptors [42,43]. The characteristic feature of these receptors is that their extracellular regions contain a domain with two conserved pairs of cysteine residues and a membrane proximal tryptophan–serine doublet separated by an intervening amino acid (WSXWS motif). This group can be subdivided into type I "long" (further subdivided in groups 1, 2, and 3) and type I "short" receptors and corresponding cytokine ligands [37].

TABLE 2.2
Major Cytokine Groups and Families Based on Structural
Features of Cytokines and Their Interacting Receptors

Group	Family	Representative Members
Type I cytokines	IL-2	IL-2
		IL-7
		IL-9
		IL-15
		IL-21
	IL-4	IL-3
		IL-4
		IL-5
		IL-13
		GM-CSF
	IL-6	IL-6
		IL-11
		Oncostatin M (OSM)
		Leukemia inhibitory factor (LIF)
		Ciliary neurotrophic factor (CNTF)
		Cardiotrophin 1 (CT-1)
	IL-12	IL-12
		IL-23
		IL-27
Type II cytokines	IFN-α/β	IFN-α (many subtypes)
		IFN-β
		IFN-ω
		IFN-τ
		IFN-κ
		Limitin
	IFN-γ	IFN-γ
	IL-10	IL-10
		IL-19
		IL-20
		IL-22
		IL-24 (MDA-7)
		IL-26
		IL-28A&28B (IFN-λ1&λ2)
		IL-29 (IFN-λ3)
Ungrouped	IL-17	IL-17
		IL-25
	Tumor necrosis factor	TNF (also termed TNF-α)
		Lymphotoxin α (LT-α)
		LT-β
		Fas ligand

(continued)

TABLE 2.2 (Continued)
Major Cytokine Groups and Families Based on Structural
Features of Cytokines and Their Interacting Receptors

Group	Family	Representative Members
		CD40 ligand
		OX40 ligand
		TRAIL
		BAFF
		APRIL
		RANK ligand
		LIGHT
		TWEAK
	Interleukin-1	IL-1α
		IL-1β
		IL-1 receptor antagonist
		IL-18
		IL-1F5-10
	TGF-β	TGF-β1-3
		Bone morphogenetic proteins
		Inhibins
		Activins
	Chemokines	CXC subfamily (CXCL1-16)
		CC subfamily (CCL1-28)
		C subfamily (CL1/Lymphotactin)
		CX3C subfamily
		(CX3CL1/Fractalkin)

Another group is formed by the type II cytokine receptors and corresponding cytokine ligands, which includes the interferons and the IL-10 family [44] (Table 2.2). All members of the IFN-α/β (type I IFN) family show at least 30% homology to one another in their amino acid sequences and they bind to the same heterodimeric receptor [45] In contrast, IFN-γ (type II IFN), though sharing some biological activities with members of the IFN-α/β family, is structurally distinct and binds to a different heterodimeric type II cytokine receptor. More recently characterized is the IL-10 family, which, in addition to IL-10, includes numerous other cytokines [44,46,47]. The recently identified members of this family, IL-28A, IL-28B, and IL-29 (also known as IFN-λ1, IFN-λ2, and IFN-λ3), inhibit virus replication and have other interferon-like activities, indicating that the IFN-α/β and IL-10 families are related both structurally and functionally. Members of the type I and type II cytokine families bind to receptors that commonly signal through the JAK (Janus kinase)-STAT (signal transducer and activator of transcription) signaling pathway.

Another important cytokine family comprises the IL-1 and IL-18 proteins (Table 2.2). IL-1 is one of the longest-known cytokines [20], whereas IL-18 was discovered much more recently as "IFN-γ-inducing factor" [48]. All members of the IL-1 family show some homology at the nucleotide and amino acid levels, and they also share a related intron–exon structure. A common characteristic of IL-1α, IL-1β, and IL-18 is that they are produced as precursor proteins lacking leader sequences and they are processed into their mature, active forms by the IL-1β converting enzyme (caspase 1) or some other proteases. The tertiary structure of IL-1 family members consists of 12 β-strands packed in a domain termed β-trefoil. These structural features are shared by members of the fibroblast growth factor (FGF) family, even though FGF ligands bind to different receptors and have functions distinct from the IL-1 family members. A new nomenclature was recently proposed for the growing IL-1 family genes and proteins [49]. IL-1α is termed IL-1F1, IL-1β is IL-1F2, IL-1 receptor antagonist is IL-1F3, and IL-18 is IL-1F4 (the letter "F" stands for family). Newer members of the IL-1 family, whose functions are still incompletely understood, were designated IL-1F5 through IL-1F10. IL-1 family receptors are heterodimers; the presence of the Toll-homology domain and the activation of characteristic intracellular signaling pathways indicates that these receptors are members of the growing family of Toll-like receptors [50].

Members of the chemokine superfamily show limited sequence homology to one another, but they contain two or, more commonly, four conserved cysteines [51]. They are subdivided into four groups, termed CXC, CC, C, and CX_3C, based on the number of cysteine (C) residues and the presence and number of intervening amino acids (X; Table 2.2). Chemokine receptors are unique among cytokine receptors in that they are seven-transmembrane-domain, G-protein linked receptors, related to rhodopsin-like receptors that mediate neurotransmission and responses to sensory stimuli [52]. New numerical designations have been adopted for the four subfamilies of chemokines (and chemokine receptors), consisting of the letters CXC, C, CX_3C, and CC plus a consecutive number.

A large and important cytokine family is the TNF family, currently comprising approximately 20 ligands and 28 receptors [53]. The ligands (with the exception of LT-α) are synthesized as type II transmembrane proteins (intracellular N-terminus). The extracellular C-terminus of these ligands contains a TNF homology domain (THD), which facilitates their assembly into homotrimers—the conformation required for receptor cross-linking and activation of signal transduction. A characteristic feature of TNF family receptors is the presence of repeating cysteine-rich domains in the extracellular regions. The intracellular domains of some of these receptors contain a "death domain," important for intracellular signaling.

The TGF-β family represents another important and extensive cytokine family comprising five TGF-β ligands (TGF-β1-5), the subfamilies of activins and inhibins, and bone morphogenetic proteins (Table 2.2). Members of this family are synthesized as larger glycosylated preproteins initially comprising a hydrophobic signal peptide sequence. All members of the TGF-β family are thought to bind to receptors as homodimers. Their receptors are transmembrane serine–threonine kinases comprising a ligand-binding extracellular domain, transmembrane domain, and large cytoplasmic domain that includes kinase domains [54,55]. Based on their structure, TGF-β receptors are divided into type I (signaling component) and type II (binding component) receptors. Active signaling involves cellular signaling proteins termed Smads, and it requires the presence of both type I and type II receptors.

IV. UNIQUE FEATURES OF CYTOKINE ACTIONS

Cytokine actions have certain characteristic features (Table 2.3). Foremost among these is *pleiotropy,* the well-known propensity of a single cytokine to affect multiple types of target cells and to elicit diverse biological actions in them. Pleiotropy is a characteristic of all major cytokines; as already mentioned TNF, IL-1, the interferons, and TGF-β all display multiple actions on different cells and tissues. Pleiotropy severely limits the therapeutic application of many cytokines because their administration often results in undesirable side effects along with the desired actions. Even more startling is the redundancy of cytokine actions, referring to the well-known ability of multiple cytokines to elicit the same or similar biological action. An example of redundancy is the long list of actions shared by TNF and IL-1, one of the main reasons being that both cytokines, though acting through distinct receptors and signaling pathways, activate the transcription factor NF-kB [56]. Because of redundancy, elimination or blocking of a single cytokine may not have a major impact as other cytokines may compensate. In view of these considerations, it is actually surprising that so many cytokines (e.g., many colony stimulating factors, IFN-α, and IFN-β) and cytokine antagonists (e.g., TNF and IL-1 inhibitors) have proven therapeutically highly effective [7].

Another characteristic of the actions of many cytokines is their propensity to act synergistically or antagonistically with other cytokines (Table 2.3). For example, there are many instances of synergy between TNF and IFN-γ — including the induction of differentiation in human myeloid cell lines [57], antiviral action [58], or induction of gene expression [59]. An example of antagonism are the actions of IFN-γ and IL-10 on macrophage activation, with the former acting as activator and the latter as inhibitor [60]. The importance of synergy and antagonism in cytokine actions were initially not fully appreciated because in most laboratory experiments investigators utilize homogeneous preparations of pure recombinant cytokines. In contrast, in real life, cells or tissue rarely, if ever, come in contact with a single

---

TABLE 2.3
Characteristics of Cytokine Actions

Pleiotropy	Cytokines often have multiple target cells and multiple actions
Redundancy	Different cytokines may have similar actions
Synergism/Antagonism	Exposure of cells to two or more cytokines at a time may lead to qualitatively different responses
Cytokine Cascade	A cytokine may increase (or decrease) the production of another cytokine
Receptor Transmodulation	A cytokine may increase (or decrease) the expression of receptors for another cytokine or growth factor
Receptor Trans-Signaling	A cytokine may increase (or decrease) signaling by receptors for other cytokines or growth factors

Source: From Vilcek, J., In *The Cytokine Handbook* 1, 4th ed., Thomson, A. W. and Lotze, M. T., Eds., Academic Press, 2003, pp. 3–18.

---

cytokine at a time. Rather, cells are likely to be exposed to a cocktail of several cytokines, with the resulting biological action reflecting various synergistic and antagonistic interactions of the cytokines present. Cytokine actions are contextual [61]. Thus, cytokine actions are profoundly affected not only by the presence or absence of other cytokines, but also by other biologically active agents (e.g., hormones, growth factors, prostaglandins, microbial components, etc.).

Another characteristic feature of cytokines is that they may act by stimulating or inhibiting the production of other cytokines (Table 2.3). As a result, many cytokine actions are indirect, e.g., due to an increase or decrease in the production of other cytokines, which then results in an altered biological response. Among the earliest discovered examples of such an indirect action was the demonstration that the mitogenic action of IL-1 in murine thymocytes involves the stimulation of IL-2 production and that IL-2 is the actual effector molecule responsible for this action [62].

Yet another mechanism important in cytokine actions involves the modulation of the level of cytokine receptor expression (Table 2.3). One of the earliest described examples of such action is the induction of high-affinity IL-2 receptors on T cells by IL-1 [63]. In other instances, receptor transmodulation by cytokines results in a reduced level of receptor expression, e.g., IL-1 was shown to downregulate TNF receptor expression [64]. There are also examples of downregulation of receptor function that do not actually result from a decreased expression of receptors but are caused by actions on receptor function, which can be termed *receptor trans-signaling* [65]. For example, TNF was shown to inhibit insulin signaling by decreasing tyrosine phosphorylation of the insulin receptor and its substrate, IRS-1 [66].

V. CONCLUSION

It is remarkable and surprising that nature has developed such an intricate, baroque system of multiple interacting cytokine and cytokine receptor networks in order to regulate essential immune and inflammatory responses and many other functions. If a scientist were given the freedom to design a new cytokine system from scratch, he/she might be able to come up with a blueprint for the regulation of all essential immune and inflammatory responses that would rely on the existence of only 30 or so cytokines and a similar number of interacting receptors, instead of hundreds. However, despite the apparently chaotic complexity of cytokine networks that frustrates many a student of cytokine biology, the outcome of most cytokine actions is orderly and compatible with homeostasis. Perhaps the much simpler system designed by our imaginary scientist would not work as reliably and break down more often than the complex system designed by nature. Chances are we will never find out.

REFERENCES

1. Oppenheim, J. J. and Feldmann, M., *Cytokine reference*, Academic Press, 2001.
2. Thomson, A. W. and Lotze, M. T., *The cytokine handbook*, 4th ed. Academic Press, 2003.

3. Dillon, S. R., Sprecher, C., Hammond, A., Bilsborough, J., Rosenfeld-Franklin, M., Presnell, S. R., Haugen, H. S., Maurer, M., Harder, B., Johnston, J., Bort, S., Mudri, S., Kuijper, J. L., Bukowski, T., Shea, P., Dong, D. L., Dasovich, M., Grant, F. J., Lockwood, L., Levin, S. D., LeCiel, C., Waggie, K., Day, H., Topouzis, S., Kramer, J., Kuestner, R., Chen, Z., Foster, D., Parrish-Novak, J., and Gross, J. A., Interleukin 31, a cytokine produced by activated T cells, induces dermatitis in mice, *Nat Immunol,* 5 (7), 752–760, 2004.

4. Wajant, H., Pfizenmaier, K., and Scheurich, P., Tumor necrosis factor signaling, *Cell Death Differ* 10 (1), 45–65, 2003.

5. Lallemand, C., Lebon, P., Rizza, P., Blanchard, B., and Tovey, M. G., Constitutive expression of specific interferon isotypes in peripheral blood leukocytes from normal individuals and in promonocytic U937 cells, *J Leukoc Biol* 60 (1), 137–46, 1996.

6. Feldmann, M., Development of anti-TNF therapy for rheumatoid arthritis, *Nat Rev Immunol* 2 (5), 364–71, 2002.

7. Vilcek, J. and Feldmann, M., Historical review: cytokines as therapeutics and targets of therapeutics, *Trends Pharmacol Sci* 25 (4), 201–9, 2004.

8. Cohen, S., Bigazzi, P. E., and Yoshida, T., Commentary. Similarities of T cell function in cell-mediated immunity and antibody production, *Cellular Immunology* 12 (1), 150–9, 1974.

9. Oppenheim, J. J. and Feldmann, M., Introduction to the role of cytokines in innate host defense and adaptive immunity, In *Cytokine Reference* 1, Oppenheim, J. J. and Feldmann, M., Eds., Academic Press, 2001, pp. 3–20.

10. Vilcek, J., The cytokines: an overview, In *The Cytokine Handbook* 1, 4th ed., Thomson, A. W. and Lotze, M. T., Eds., Academic Press, 2003, pp. 3–18.

11. Bennett, I. L., Jr. and Beeson, P. B., Studies on the pathogenesis of fever, II: characterization of fever-producing substances from polymorphonuclear leukocytes and from the fluid of sterile exudates, *J. Exp. Med.* 98, 493–508, 1953.

12. Levi-Montalcini, R. and Hamburger, V., A diffusable agent of mouse sarcoma producing hyperplasia of sympathetic ganglia and hyperneurotization of viscera in the chick embryo, *J. Exp. Zool.* 123, 233–288, 1953.

13. Isaacs, A. and Lindenmann, J., Virus interference. 1. The interferon, *Proc. R. Soc. Lond. B.* 147, 258–267, 1957.

14. Bloom, B. R. and Bennett, B., Mechanism of a reaction in vitro associated with delayed-type hypersensitivity, *Science* 153 (731), 80–2, 1966.

15. David, J. R., Delayed hypersensitivity in vitro: its mediation by cell-free substances formed by lymphoid cell-antigen interaction, *Proc Natl Acad Sci U.S.A.* 56 (1), 72–7, 1966.

16. Kasakura, S. and Lowenstein, L., A factor stimulating DNA synthesis derived from the medium of leukocyte cultures, *Nature* 208 (12), 794–5, 1965.

17. Ruddle, N. H. and Waksman, B. H., Cytotoxicity mediated by soluble antigen and lymphocytes in delayed hypersensitivity. 3. Analysis of mechanism, *J. Exp. Med.* 128 (6), 1267–79, 1968.

18. Granger, G. A. and Williams, T. W., Lymphocyte cytotoxicity in vitro: activation and release of a cytotoxic factor, *Nature* 218 (148), 1253–4, 1968.

19. Dumonde, D. C., Wolstencroft, R. A., Panayi, G. S., Matthew, M., Morley, J., and Howson, W. T., "Lymphokines": non-antibody mediators of cellular immunity generated by lymphocyte activation, *Nature* 224 (214), 38–42, 1969.

20. Dinarello, C. A., Biologic basis for interleukin-1 in disease, *Blood* 87 (6), 2095–147, 1996.

21. Gery, I., Gershon, R. K., and Waksman, B. H., Potentiation of cultured mouse thymocyte responses by factors released by peripheral leucocytes, *J. Immunol.* 107 (6), 1778–80, 1971.

22. Aarden, L. A., Brunner, T. K., Cerottini, J.-C., Dayer, J.-M., de Weck, A. L., Dinarello, C. A., Di Sabato, G., Farrar, J. J., Gery, I., Gillis, S., Handschumacher, R. E., Henney, C. S., Hoffmann, M. K., Koopman, W. J., Krane, S. M., Lachman, L. B., Lefkowits, I., Mishell, R. I., Mizel, S. B., Oppenheim, J. J., Paetkau, V., Plate, J., Röllinghoff, M., Rosenstreich, D., Rosenthal, A. S., Rosenwasser, L. J., Schimpl, A., Shin, H. S., Simon, P. L., Smith, K. A., Wagner, H., Watson, J. D., Wecker, E., and Wood, D. D., Revised nomenclature for antigen-nonspecific T cell proliferation and helper factors, *J. Immunol.* 123 (6), 2928–2929, 1979.

23. Carswell, E. A., Old, L. J., Kassel, R. L., Green, S., Fiore, N., and Williamson, B., An endotoxin-induced serum factor that causes necrosis of tumors, *Proc Natl Acad Sci U.S.A.* 72 (9), 3666–70, 1975.

24. Morgan, D. A., Ruscetti, F. W., and Gallo, R., Selective in vitro growth of T lymphocytes from normal human bone marrows, *Science* 193 (4257), 1007–8, 1976.

25. Wheelock, E. F., Interferon-like virus-inhibitor induced in human leukocytes by phytohemagglutinin, *Science* 149, 310–311, 1965.

26. Biron, C. A. and Sen, G. C., Interferons and other cytokines, In *Fields virology*, 4th ed., Knipe, D. M. and Howley, P. M., Eds., Lippincott Williams & Wilkins, 2001, pp. 321–351.

27. Vilcek, J. and Sen, G. C., Interferons and other cytokines, In *Fields virology*, 3rd ed., Fields, B. N., Knipe, D. M., and Howley, P. M., Eds., Lippincott–Raven Publishers, Philadelphia, 1996, pp. 375–99.

28. Siegal, F. P., Kadowaki, N., Shodell, M., Fitzgerald-Bocarsly, P. A., Shah, K., Ho, S., Antonenko, S., and Liu, Y. J., The nature of the principal type 1 interferon-producing cells in human blood, *Science* 284 (5421), 1835–7, 1999.

29. Yoshimura, T., Matsushima, K., Tanaka, S., Robinson, E. A., Appella, E., Oppenheim, J. J., and Leonard, E. J., Purification of a human monocyte-derived neutrophil chemotactic factor that has peptide sequence similarity to other host defense cytokines, *Proc Natl Acad Sci U.S.A.* 84 (24), 9233–7, 1987.

30. Sachs, L., The molecular control of blood cell development, *Science* 238 (4832), 1374–9, 1987.

31. Bradley, T. R. and Metcalf, D., The growth of mouse bone marrow cells in vitro, *Australian Journal of Experimental Biology & Medical Science* 44 (3), 287–99, 1966.

32. de Larco, J. E. and Todaro, G. J., Growth factors from murine sarcoma virus-transformed cells, *Proc Natl Acad Sci U.S.A.* 75 (8), 4001–5, 1978.

33. Roberts, A. B., Anzano, M. A., Lamb, L. C., Smith, J. M., and Sporn, M. B., New class of transforming growth factors potentiated by epidermal growth factor: isolation from non-neoplastic tissues, *Proc Natl Acad Sci U.S.A.* 78 (9), 5339–43, 1981.

34. Moses, H. L., Branum, E. L., Proper, J. A., and Robinson, R. A., Transforming growth factor production by chemically transformed cells, *Cancer Res* 41 (7), 2842–8, 1981.

35. Roberts, A. B. and Sporn, M. B., The transforming growth factor-βs, In *Peptide growth factors and their receptors I, Sporn,* M. B. and Roberts, A. B., Eds., Springer-Verlag, Berlin, 1990, pp. 419–472.

36. Flanders, K. C. and Roberts, A. B., TGF-beta, In *Cytokine reference* 1, Oppenheim, J. J. and Feldmann, M., Eds., Academic Press, 2001, pp. 719–746.

37. Boulay, J. L., O'Shea, J. J., and Paul, W. E., Molecular phylogeny within type I cytokines and their cognate receptors, *Immunity* 19 (2), 159–63, 2003.

38. Parry, D. A., Minasian, E., and Leach, S. J., Cytokine conformations: predictive studies, *J Mol Recognit* 4 (2–3), 63–75, 1991.

39. Boulay, J. L. and Paul, W. E., The interleukin-4 family of lymphokines, *Current Opinion in Immunology* 4 (3), 294–8, 1992.

40. Bazan, J. F., Unraveling the structure of IL-2, *Science* 257 (5068), 410–3, 1992.

41. Arai, K. I., Lee, F., Miyajima, A., Miyatake, S., Arai, N., and Yokota, T., Cytokines: coordinators of immune and inflammatory responses, *Annu Rev Biochem* 59, 783–836, 1990.

42. Kishimoto, T., Taga, T., and Akira, S., Cytokine signal transduction, *Cell* 76 (2), 253–62, 1994.

43. Bazan, J. F., A novel family of growth factor receptors: a common binding domain in the growth hormone, prolactin, erythropoietin and IL-6 receptors, and the p75 IL-2 receptor β-chain, *Biochemical & Biophysical Research Communications* 164 (2), 788–95, 1989.

44. Vilcek, J., Novel interferons, *Nat Immunol* 4 (1), 8–9, 2003.

45. De Maeyer, E. and De Maeyer-Guignard, J., *Interferons and other regulatory cytokines* John Wiley & Sons, New York, 1988.

46. Kotenko, S. V., Gallagher, G., Baurin, V. V., Lewis-Antes, A., Shen, M., Shah, N. K., Langer, J. A., Sheikh, F., Dickensheets, H., and Donnelly, R. P., IFN-lambdas mediate antiviral protection through a distinct class II cytokine receptor complex, *Nat Immunol* 4 (1), 69–77, 2003.

47. Sheppard, P., Kindsvogel, W., Xu, W., Henderson, K., Schlutsmeyer, S., Whitmore, T. E., Kuestner, R., Garrigues, U., Birks, C., Roraback, J., Ostrander, C., Dong, D., Shin, J., Presnell, S., Fox, B., Haldeman, B., Cooper, E., Taft, D., Gilbert, T., Grant, F. J., Tackett, M., Krivan, W., McKnight, G., Clegg, C., Foster, D., and Klucher, K. M., IL-28, IL-29 and their class II cytokine receptor IL-28R, *Nat Immunol* 4 (1), 63–8, 2003.

48. Okamura, H., Tsutsi, H., Komatsu, T., Yutsudo, M., Hakura, A., Tanimoto, T., Torigoe, K., Okura, T., Nukada, Y., Hattori, K., et al., Cloning of a new cytokine that induces IFN-gamma production by T cells, *Nature* 378 (6552), 88–91, 1995.

49. Sims, J. E., Nicklin, M. J., Bazan, J. F., Barton, J. L., Busfield, S. J., Ford, J. E., Kastelein, R. A., Kumar, S., Lin, H., Mulero, J. J., Pan, J., Pan, Y., Smith, D. E., and Young, P. R., A new nomenclature for IL-1-family genes, *Trends Immunol* 22 (10), 536–7, 2001.

50. Akira, S. and Takeda, K., Toll-like receptor signalling, *Nat Rev Immunol* 4 (7), 499–511, 2004.

51. Baggiolini, M., Dewald, B., and Moser, B., Human chemokines: an update, *Annu Rev Immunol* 15, 675–705, 1997.

52. Horuk, R., Chemokine receptors, *Cytokine Growth Factor Rev* 12 (4), 313–35, 2001.

53. Ware, C. F., The TNF superfamily, *Cytokine Growth Factor Rev* 14 (3–4), 181–4, 2003.

54. Massague, J., TGF-beta signal transduction, *Annu Rev Biochem* 67, 753–91, 1998.

55. Wrana, J. L. and Attisano, L., The Smad pathway, *Cytokine Growth Factor Rev* 11 (1–2), 5–13, 2000.

56. Le, J. and Vilcek, J., Tumor necrosis factor and interleukin 1: cytokines with multiple overlapping biological activities., *Laboratory Investigation* 56 (3), 234–48, 1987.

57. Trinchieri, G., Kobayashi, M., Rosen, M., Loudon, R., Murphy, M., and Perussia, B., Tumor necrosis factor and lymphotoxin induce differentiation of human myeloid cell lines in synergy with immune interferon, *J. Exp. Med.* 164 (4), 1206–25, 1986.

58. Wong, G. H. W. and Goeddel, D. V., Tumour necrosis factors α and β inhibit virus replication and synergize with interferons, *Nature* 323 (6091), 819–22, 1986.

59. Beresini, M. H., Lempert, M. J., and Epstein, L. B., Overlapping polypeptide induction in human fibroblasts in response to treatment with interferon-α, interferon-γ, interleukin 1 α, interleukin 1 β, and tumor necrosis factor, *J. Immunol.* 140 (2), 485–93, 1988.

60. Moore, K. W., O'Garra, A., de Waal Malefyt, R., Vieira, P., and Mosmann, T. R., Interleukin-10, *Annual Review of Immunology* 11, 165–90, 1993.

61. Sporn, M. B. and Roberts, A. B., Peptide growth factors are multifunctional, *Nature* 332 (6161), 217–9, 1988.

62. Smith, K. A., Lachman, L. B., Oppenheim, J. J., and Favata, M. F., The functional relationship of the interleukins, *J. Exp. Med.* 151 (6), 1551–6, 1980.

63. Kaye, J., Gillis, S., Mizel, S. B., Shevach, E. M., Malek, T. R., Dinarello, C. A., Lachman, L. B., and Janeway, C. A., Jr., Growth of a cloned helper T cell line induced by a monoclonal antibody specific for the antigen receptor: interleukin 1 is required for the expression of receptors for interleukin 2, *J. Immunol.* 133 (3), 1339–45, 1984.

64. Holtmann, H. and Wallach, D., Down regulation of the receptors for tumor necrosis factor by interleukin 1 and 4 β-phorbol-12-myristate-13-acetate, *J. Immunol.* 139 (4), 1161–7, 1987.

65. Castellino, A. M. and Chao, M. V., Trans-signaling by cytokine and growth factor receptors, *Cytokine Growth Factor Rev.* 7 (4), 297–302, 1996.

66. Hotamisligil, G. S., Murray, D. L., Choy, L. N., and Spiegelman, B. M., Tumor necrosis factor alpha inhibits signaling from the insulin receptor, *Proc Natl Acad Sci U.S.A.* 91 (11), 4854–8, 1994.

3 Cytokine Receptors

Heidi Schooltink and Stefan Rose-John

CONTENTS

I. INTRODUCTION

Besides a central control by the brain, the communication between peripheral organs and cells is essential for a coordinated interaction of a highly developed multicellular organism with its surroundings. This communication allows fast and adequate local and systemic reactions after the homeostasis has been disturbed. One group of cell–cell-communication-molecules is the cytokines. Cytokines are proteins of 100 to 200 amino acids, which are synthesized by various cells of the body including cells of the immune system. They act as extracellular mediators triggering a cell- and cytokine-specific signal on diverse target cells [1]. The tasks of cytokines are widespread including control of the immune system, regulation of survival of neuronal cells, and developmental processes like hematopoiesis.

II. CYTOKINE RECEPTOR PROTEINS

Cytokines bind to specific receptor proteins on the cellular membrane of their target cells. All cytokine receptors are type-I membrane proteins. The receptor proteins can be grouped into distinct families according to their structural features (Table 3.1).

TABLE 3.1
Cytokine-Receptors

Families	Examples
Class-I Cytokine Receptors	*Receptors of the IL-2 family*: specific ligand binding subunits for IL-2, IL-4, IL-7, IL-9, IL-15, IL-21 (in the case of IL-2, two different binding subunits: IL-2Rα, IL-2Rβ); common γ–subunit (cγ)
	Receptors of the IL-3 family: specific ligand binding subunits for: IL-3, IL-5, GM-CSF; common β-subunit
	Receptors of the IL-6 family: different combinations of receptor components for IL-6, IL-11, CNTF, LIF, OSM, CT-1, CLC, NP; common signal transducer (gp130)
	Homodimeric Receptors: EPO-R, GH-R, G-CSF-R
Class-II Cytokine Receptors	*Receptors of Type I interferons*: IFNAR-1 + IFNAR-2
	Receptors of Type II interferons: IFNGR-1 + IFNAR-2
	Receptors of the IL-10-family: IL-19, IL-20, IL-22, IL-24
TNF-Receptor Family	TNFα: p55 or p75
	Receptors for NGF, RANK, FAS-Ligand, CD 40, CD 27
IL-1 Receptor Family	IL-1 R-complex: IL-1RI or IL-1RII + IL-1RacP
	IL-18R-complex: IL-1RI + IL-18RacP
	11 different TLRs

The receptors for the hematopoietins (members include interleukins [IL] and colony stimulating factors [CSF]) are called *class I receptors* [2], those for the interferons (IFN) and IL-10 type cytokines are *class II receptors* [3,4]. class I and class II receptors have structurally related extracellular cytokine binding regions, consisting of two β-sheet-rich clusters forming two cytokine receptor homology regions (CRH). The loops between the β-sheets of the CRHs are responsible for the contact to the cytokine. The membrane distal CRH contains two pairs of cysteins, which form two disulfide bridges. Furthermore class I receptors contain the so-called WSXWS motif (an amino acid sequence of tryptophan-serine-X-tryptophan-serine, in which X can be any amino acid residue) in their membrane proximal CRH [5,6]. The TNFR belongs to a family of proteins including nerve-growth factor receptor (NGFR), FAS, CD40, CD27, and receptor activator of nuclear factor kappa B (RANK). This family is characterized by two to six cystein-rich motifs in the extracellular parts of the receptors The TNFR contains four of these cystein-rich repeats [7].

Modules found in cytokine receptors are also found in receptors belonging to other families, e.g., the tyrosine kinase receptor family. The extracellular portion of the insulin receptor contains cystein-rich repeats (like the TNFR family) and fibronectin type III domains as they can be found in some members of the class-I cytokine receptor family [8]. A modular structure can also be found in the IL-1R family. The extracellular part of the IL-1R is characterized by three immunoglobulin-like domains [9]. Interestingly, IL-1 receptors are related to the toll-like receptors (TLR).

The ligands of the receptors of the IL-1 family are cytokines, whereas TLRs bind to molecules specific for microbial invaders, like lipopolysaccharides (LPS, gram-negative bacteria), lipoproteins (gram-positive bacteria), bacterial flagellin, and bacterial DNA [10]. TLRs are found in insects (Drosophila) and in mammals. Whereas the intracellular domains of TLR and IL-1R show significant homologies, the extracellular domains are unrelated, again underlining a modular construction of these receptor molecules [10]. In mammals, TLRs are expressed in a variety of different cell types, such as monocytes, dendritic cells, and endothelial cells. Activation of these cells through the TLRs leads to the production of cytokines, which subsequently activate the innate immune system and in a second step lead, together with antigens presented by the MHC I, to the activation of T-cells [10].

Intracellular domains and signaling cascades initiated by cytokine receptors will be described in the review Cytokine Signaling Pathways (chapter 5) within this volume.

III. STRUCTURE OF CYTOKINE RECEPTOR COMPLEXES

Upon binding of one or more cytokine molecules to the extracellular portions of the membrane receptors, dimerization or oligomerization of one or more receptor components leads to the active cytokine receptor complex. Complex formation involves conformational changes in the intracellular parts of the receptor molecules, which initiate an intracellular signal cascade.

The composition of cytokine receptor complexes varies considerably [11]. The simplest case is realized for growth hormone (GH) a member of the hematopoietin family (see Figure 3.1A). One GH molecule binds to two identical receptor (GHR) molecules [12]. Interestingly, two complete different interaction sites (site I and site II) of GH interact with nearly the same residues in the two GHR molecules. Binding of the two receptor molecules happens sequentially, site I of the GH molecules first contacts one receptor molecule, followed by the contact of site II of GH and a second GHR [12].

In contrast to GH, many cytokines bind to heterodimeric receptor complexes. This is the case for IL-4 (Figure 3.1B) [13]. This cytokine first binds to a receptor protein called IL-4R then to the common gamma chain (γC), a protein that is also used as a receptor molecule by many other cytokines (see below). Also cytokines of the IL-1 family bind to heterodimeric receptor complexes [14].

An even more complicated stochiometry is realized in the active receptor complexes of IL-6 (Figure 3.1C), IL-11, and ciliary neurotrophic factor (CNTF; Figure 3.1D). These cytokines have three distinct binding epitops (site I, site II, site III). Site I contacts a specific α-receptor (IL-6R, IL-11R, CNTFR). This contact is necessary for the binding of the cytokine to the signal-transducing subunits gp130 and leukemia inhibitory factor R (LIFR). In the case of IL-6 and IL-11, the cytokine/ α-receptor complex further interacts with two gp130-proteins (via site II and site III). In the case of CNTF, site II forms the contact to one gp130 molecule, whereas site III

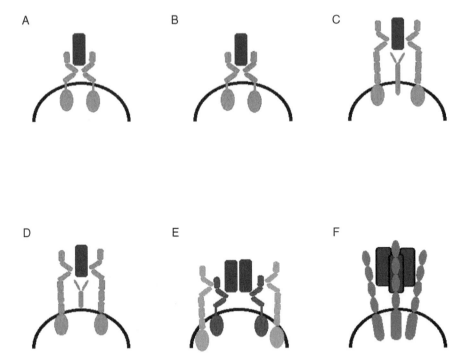

FIGURE 3.1 Possible composition of cytokine receptor complexes. A functional cytokine receptor complex consists of at least two receptor proteins. These might be identical receptor subunits, e.g., in the GH/GHR complex (A) or different receptor subunits, e.g., in the IL-4/IL-4R complex (B). Other cytokine receptor complexes consist of a specific cytokine binding subunit (pink) and a signaling dimer (green). The signalling dimer can be a homodimer, e.g., in the IL-6/IL-6R complex (C) or a heterodimer, e.g., in the CNTF/CNTFR complex (D). Even hexameric complexes are realized, containing two cytokine molecules and two pairs of different receptor molecules, e.g., in the IFNγ/IFNγ R complex (E), or three cytokine molecules and three identical receptor molecules, e.g., in the TNFα/TNFαR complex (F).

binds to the LIFR. To date, it remains unclear whether the active configuration of the receptor complex is a tetramer (as described above) or a hexameric complex of two cytokine molecules; two α-receptors and two gp130 proteins is the active IL-6 receptor complex [6,15,16].

Also for the IFN cytokine family, an active hexameric receptor complex has been proposed. In the case of IFNγ (Figure 3.1E), a complex of two cytokine molecules bound to two α-receptor proteins has been crystallized [17]. But the active receptor complex contains a second receptor subunit necessary for signal transduction. Biochemical data suggested a stochiometry of 2:2:2 [18].

The cytokine TNFα forms a trimer in solution. This trimer binds to three identical receptor molecules (p75, or p55, see below) on the surface of target cells (Figure 3.1F).

The active receptor complex is therefore also a hexamer, but with the different stochiometry of 3:3 [19].

IV. TRANSDUCING THE SIGNAL THROUGH THE MEMBRANE

The binding of the cytokines to their cognate receptors must be transmitted into intracellular signaling events. Changes in the relative positions of the transmembrane regions of the cytokine receptors bring about the transduction of the signal across the membrane. In the case of the erythropoietin receptor, it has been shown that the unliganded receptor forms a dimer via the two unoccupied ligand binding sites. Binding of the ligand erythropoietin leads to a shift of the two receptor molecules, which juxtaposes the cytoplasmic domains of the receptors, which in turn activate the JAK kinases that are constitutively bound to the cytoplasmic portion of the receptors [20].

In the case of the cytokine receptors that contain three fibronection III domains between the cytokine binding region and the transmembrane domain, the receptors are most likely not rigid enough to transmit the allosteric changes brought about by ligand binding directly through the transmembrane region. In a recently solved cryo-electron-microscopy structure of the insulin receptor [21], it is evident that the fibronection III domains of the insulin receptor are in direct contact with the intracellular tyrosine kinase domains and are therefore directly involved in the receptor activation mechanism. Since the cytokine receptors gp130 and LIFR, like the insulin receptor, contain three fibronection III domains, J. Grötzinger recently speculated that this activation mechanism might also be applicable for cytokine receptors [8].

V. RESTRICTION OF EXPRESSION AND INTERNALIZATION

A possibility to restrict cytokine effects to a defined cell population is the control of the expression of cytokine receptors. Because only a limited number of cell types express cytokine receptor on the surface, only these cell types are potential targets for this cytokine. Furthermore, many cytokine receptor proteins are not constitutively expressed, but are only synthesized after appropriate stimulation. The IL-2Rα is only expressed by activated (but not by resting) T-cells. The expression of the common signal transducer of the IL-6 family (gp130) on smooth muscle cells is increased after gp130 stimulation [22].

Furthermore, the expression of some cytokine receptors is restricted to the apical or basolateral membrane of polarized cells. In the polarized epithelial cell line Madin-Darby canine kidney (MDCK), the CNTFR is expressed exclusively on the apical membrane, whereas the common signal transducer of the IL-6 family is preferentially found on the basolateral membrane. These findings might be important for the understanding of cytokine function on polarized cells, e.g., hepatocytes and epithelial cells [23].

For a functional cytokine network, the transient nature of cytokine signals is also fundamental. Cytokine signaling is terminated by internalization of the receptor complexes. For IL-6 it has been found that the cytoplasmic domain of the common signal transducer gp130 is responsible for ligand-induced internalization [24].

VI. DIFFERENT CYTOKINES USE THE SAME RECEPTOR COMPONENTS

Different cytokine signaling pathways are connected at the level of receptors through the use of the same receptor components by structurally related cytokines. IL-3, IL-5, and GM-CSF all use a ligand-specific α-subunit. The β-subunit, however, is identical for all three cytokines [25]. The IL-2 receptor consists of three subunits. The third subunit, known as the common γ-chain (γC), is also part of the receptors for IL-7, IL-9, IL-4, IL-15, and IL-21 [26]. The family of IL-6-type cytokines (members are IL-6, LIF, IL-11, CNTF, oncostatin M [OSM], cardiotrophin-1 [CT-1], new-neurotrophin—1 also known as cardiotrophin-like cytokine [CLC]—and neuropoi-etin [NP]) is of special interest because it includes cytokines that are active on neural cells. All members of the IL-6 subfamily use gp130 as a common signal transducer [16]. The three α-receptors for IL-6, IL-11, and CNTF have been molecularly identified: The IL-11R is used only by IL-11. In contrast, it has recently been shown that the IL-6R, besides IL-6, can also bind CNTF [27] and in the case of the CNTFR, CNTF, NP, and CLC have been identified as possible ligands [28,29].

Also in other cytokine families, e.g., the interferon family (Figure 3.1E) [4], sharing of the same receptor components by different cytokines is a common scheme (Table 3.1).

VII. SOME CYTOKINES HAVE DIFFERENT, FUNCTIONAL INDEPENDENT RECEPTORS

Some cytokines possess two different, functional independent receptors. In the case of the TNFαR, these are two proteins (p55, p75). Each of these receptors binds as a homotrimeric complex to the trimeric ligand TNFα and thereby generates an intra-cellular signal [30]. TNFαRs can be found on a variety of cell types. The spectrum of activity of p55 and p75 partially overlaps, but the p55-receptor is responsible for most of the proinflammatory signals, whereas the p75 plays a role in T-cell prolifer-ation. The influence of p75 on TNF-α-p55 signals is controversially discussed. A possible role of p75 might be that it passes the ligand to p55. Alternatively an antag-onistic property of p75 in sequestering the ligand has been discussed [30].

Also the IL-1-system has two functionally independent receptors (IL-1RI, IL-1RII). But in contrast to the two TNFαRs, only the IL-1RI together with the IL-1 receptor accessory protein (IL-1RacP) transfers the IL-1-signal into the cell. The IL-1RII is unable to generate an intracellular signal because it lacks a cytoplasmic domain [31]. Therefore, the IL-1RII functions as a "decoy-receptor." In addition to sequestering the ligand IL-1, the essential coreceptor IL-1RacP is withdrawn from the IL-1 signal pathway by binding to the IL-1RII [32].

VIII. SOLUBLE RECEPTORS

Many cytokines receptors occur, besides as membrane-associated forms, as soluble proteins in the circulation. Soluble receptors (sR) are found for TNFα and many interleukins. These soluble receptors retain the ability to bind their ligands. Two different mechanisms of generation of soluble receptor proteins are described in the literature. On one hand, soluble isoforms are generated by translation from differentially spliced mRNAs. This is the case, for example, for the IL-4R. On the other hand, soluble receptor proteins are produced by limited proteolysis of the membrane-bound form of the receptor, a process that is called *shedding* [33]. Membrane-bound proteases cleave receptor proteins at defined positions (close to the transmembrane domain) within their extracellular portions. Metalloproteinases belonging to a family called ADAM (a disintegrin and metalloproteinase) are responsible for the shedding of many cytokines and cytokine receptors. Substrates for ADAM 17, also known as TACE (TNFα converting enzyme, named after the first substrate that was found), are, among others, transforming growth factor α (TGFα), IL-6R, IL-1RI, and TNF-R. The importance of the shedding process (although not understood in detail) is underlined by the fact that mice deficient in ADAM-10 and ADAM-17 genes are not viable [33].

The physiological consequences of receptor shedding might be complex, depending on the substrate. Due to the proteolytic cleavage of their membrane-bound receptors, cells lose functional cytokine receptors on their surface and, as a consequence, are no longer target cells for this cytokine. The resulting soluble receptor proteins are distributed through the circulation. By binding their ligands, they form a cytokine/receptor complex with prolonged half-life in the circulation as compared to the cytokine alone [34]. The complex of soluble receptor and cytokine might have different properties.

In the case of the IL-6 family of cytokines, the complex of cytokine and soluble receptor interacts with the gp130 homodimer or the gp130/LIFR heterodimer on the surface of target cells, thereby stimulating the cells (Figure 3.2A). This mechanism may lead to an amplification of the IL-6 signal on target cells if the amount of the IL-6R is the limiting factor. Furthermore, it might also be a mechanism for recruiting new target cells. Cells that are not responsive to an IL-6 because they do not express the IL-6R become new target cells because the missing membrane receptor subunit is substituted by the soluble receptor (Figure 3.2A). This form of signal transduction was termed *transsignaling* [35,36]. Transsignaling has been shown for the ligand-binding soluble α-subunits for IL-6, CNTF, and IL-11 in complex with their cognate ligands. The transsignaling mechanism is of physiological importance because gp130 is ubiquitously expressed, whereas the expression of the ligand binding subunit of IL-6 is restricted to a few cell types. Shedding of the α-subunits of receptors of the IL-6 family therefore leads to the expansion of the spectrum of potential target cells. In the IL-6 system it has been shown that endothelial cells, smooth muscle cells, hematopoietic progenitor cells, and neural cells need soluble receptors (sIL-6R, sIL-11R, sCNTF-R) to react to the corresponding cytokines [2,37].

For other cytokines, for example, TNFα, the complex of cytokine/soluble receptor is unable to bind to receptor components on the surface of target cells (Figure 3.2B).

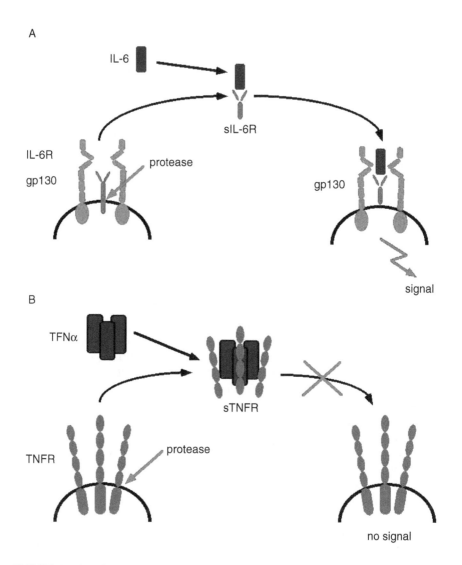

FIGURE 3.2 Possible physiological consequences of shedding of cytokine receptor molecules As a first consequence of shedding of cytokine receptor proteins from the cell surface, these cells loose their ability to react to the cognate cytokine. Further consequences depend on the characteristics of the soluble receptor/ligand complexes. (A) For the cytokine IL-6, it has been found that the soluble IL-6R acts as an agonistic soluble receptor. The soluble receptor/cytokine complex binds to gp130 receptor subunits on the surface of target cells. Because on most cells the expression of the IL-6R is the limiting step of the cytokine response, the soluble IL-6R functions as an amplifier of the cytokine response. Even the complete substitution of IL-6R by the soluble IL-6R is possible. This creates new target cells for the cytokine IL-6. This mechanism was termed *transsignaling*. (B) Antagonistic soluble receptors: In the case of TNFα, there is competition of the soluble TNFαR with the membrane-associated form of the receptor for the ligand TNFα; therefore, the soluble TNFαR acts as a competitive antagonist.

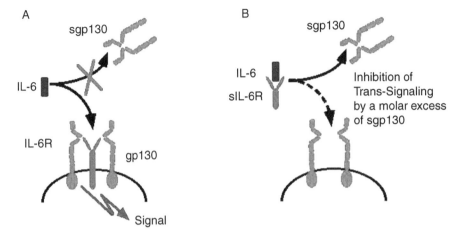

FIGURE 3.3 Specific inhibition of the IL-6/sIL-6R transsignal by soluble gp130 (A) IL-6 alone does not bind to the sgp130 dimer. Once bound to the membrane form of the IL-6R, soluble gp130 has no access to the cytokine. Therefore, soluble gp130 is not able to antagonize responses of IL-6 via the membrane form of the IL-6R. (B) Soluble gp130 competes with the membrane-bound gp130 for the binding of the complex of IL-6 and sIL-6R, thus acting as an antagonist of the IL-6/sIL-6R response.

In these cases the soluble receptor competes with the membrane-bound receptor for the cytokine and, as a consequence, acts as an antagonist (Figure 3.2B). Such antagonistic soluble receptor proteins are molecules with a high therapeutic potential (as discussed in the next section).

The soluble isoform of the common signal transducer of the IL-6 family gp130 (sgp130) can bind to the complex of IL-6 and sIL-6R (and of IL-11/IL-11R and CNTF/CNTFR), thereby preventing the contact of the soluble complex to membrane-bound gp130. Sgp130, however, does not bind to IL-6 alone. Thus sgp130 acts as competitive inhibitor protein for IL-6/sIL-6R-mediated but not for IL-6-mediated signals. Therefore sgp130 does not interfere with the signal generated by the membrane-bound IL-6R (Figure 3.3). Sgp130 has been found in the circulation at high levels (100-300 ng/ml). The physiological role might therefore be the limitation of systemic responses to IL-6/sIL-6R and related cytokines [2,37].

Furthermore, shedding may change the binding characteristics of the corresponding receptor protein. In the IL-1 system there are three ligands, two of which (IL-1α and IL-1β) elicit an intracellular signal, whereas the third ligand (IL-1Ra) is an antagonistic molecule that binds to the receptor but does not elicit a signal. The three ligands bind to two independent receptor molecules (IL-1RI and IL-1RII), of which only IL-1RI is able to transduce an IL-1 signal. The extracellular domain of IL-1RII (sIL-1RII) is shed from the surface of a variety of cell types. The affinity of sIL-1RII for the ligands IL-1α and IL-1β remains unchanged, whereas the affinity to IL-1Ra drops dramatically. Binding of the antagonistic ligand IL-1Ra to the antagonistic soluble receptor sIL-1RII would diminish the antagonistic potential of both molecules because they are no longer available to block the IL-1 signal pathway.

Therefore, the loss of the high affinity for IL-1Ra guarantees that the antagonistic effect of sIL-1RII (competition with the membrane-bound IL-1RI for the binding of the ligand) does not interfere with the antagonistic effect of IL-1Ra (occupation of IL-1R on the cell surface) [38].

IX. HETERODIMERIC CYTOKINES

Heterodimeric cytokines consist of a cytokine (member of the hematopoietins) and a soluble receptor protein (member of the class I cytokine receptors) [39]. In contrast to the complexes of soluble receptors and cytokines, which are formed in the circulation, heterodimeric cytokines are formed within the cell. In general, both components are not secreted by synthesizing cells unless they form a complex. To date, four heterodimeric cytokines have been identified: IL-12, CLC/CLF-1 (cardiotrophin-like cytokine/cytokine-like factor 1), IL-23, and IL-27. Heterodimeric cytokines bind to membrane-bound receptors, which belong to the class I receptors. CLC/CLF-1 binds to the same receptor complex (CNTFR, gp130, LIFR) as CNTF and NP. Interestingly, the heterodimeric cytokines IL-12 and IL-23 share the soluble receptor subunit p40 and bind to a receptor complex on the surface of target cells that share a common subunit (IL-12: IL-12Rbeta1/IL-12RbetaII; IL-23 [IL-12Rbeta1/IL-23R]) [40].

X. CYTOKINE RECEPTORS AS VIRAL PROTEINS

Pathogenic DNA viruses contain genes with high homology to cellular cytokines or cytokine receptors [41]. Some viruses carry genes coding for antagonistic soluble cytokine receptor molecules. By binding and antagonizing the cytokines, these viral counterparts of the IL-1R, IFNR, and TNFαR suppress the inflammatory and immune-stimulatory effects of these cytokines and thereby protect the virus harboring cell [42].

Interestingly, the genome of the human herpes virus 8 (HHV8), which is associated with Kaposi Sarcoma and Castleman Disease, encodes a cytokine with significant homology to human IL-6. This viral IL-6 (vIL-6) directly binds to gp130 with a stochiometry of two vIL-6 molecules and two gp130 molecules [43]. Therefore, vIL-6 functionally substitutes for the complex of IL-6 and IL-6R (in membrane-bound or soluble forms). As gp130 is ubiquitously expressed, vIL-6 is able to stimulate virtually all cells of the body, which might help to explain the role of vIL-6 in the pathophysiology of HHV8 [44].

XI. CYTOKINE RECEPTORS AND EVOLUTION

From an evolutionary point of view, the acquired immune system with its humoral and cellular branch is relatively new and is not found in animals more primitive than vertebrates. But in many invertebrate species such as in the slime mold (Dictostelium) and in insects (Drosophila), intracellular signal molecules found in cytokine signaling pathways like signal transducers and activators of transcription (STATs), TNFR-associated factors (TRAFs), and Janus kinases (JAKs) have been identified.

Recently, the first cytokines and cytokine receptors in Drosophila were found [45]. This gene called *domeless* encodes a membrane-bound protein that interacts with the JAK/STAT pathway and shows homology with gp130, the signal transducer of the IL-6 family [45]. Moreover, two proteins that comprise a TNF-like system in Drosophila were described [46].

The discovery of cytokines and receptors in invertebrates shows that cytokines are older than the acquired immune system. Interestingly, functional analysis of the intracellular cytokine signal molecules JAK and STAT in Drosophila has shown that these proteins are essential for the control of the self-renewal of embryonic stem cells [47]. It therefore can be speculated that this activity has predestined cytokines as signaling molecules in the immune system where clonal expansions of huge cell populations from small progenitor or stem cell populations is controlled by cytokines. On the other hand, it might be worthwhile to more rigorously analyze the mechanism by which cytokines govern the decision of adult and embryonic stem cells to undergo self-renewal or differentiation. This may bring about a more fundamental understanding of these cellular functions and might have a huge impact on future therapeutic strategies in regenerative medicine.

XII. CLINICAL USE OF CYTOKINE RECEPTORS

Cytokines as mediators of the immune system and of developmental processes are molecules with a high therapeutic potential. As shown in this review, different cytokine pathways are connected by shared receptor components. Therefore, systemic interventions in the cytokine network often have fundamental and unplanned side effects. To date, the application of CNTF as a neurotrophic effector molecule in multiple sclerosis and amyotrophic lateral sclerosis is hampered by its toxic side effect. The fact that CNTF binds to the CNTFR and the IL-6R widens the spectrum of target cells for this cytokine [27] and explains the profound side effects of high doses of CNTF. The construction of a CNTF molecule, which exclusively acts via the CNTFR, might improve the safety profile of CNTF without compromising its neurotrophic activities and therefore might allow clinical application of this cytokine.

Antagonistic soluble receptors have a high therapeutic potential, especially when they interfere only with one cytokine pathway. A protein, on the molecular basis of a soluble form of the p75-TNFα-R, called *etanercept*, is used in the clinic for a variety of different chronic inflammatory diseases [48] including rheumatoid arthritis, psoriasis, psoriatic arthritis, and ankylosing spondylitis. To date, thousands of people have been treated with etanercept and, in general, the safety profile is satisfactory. Rare side effects, however, include the development of lupus erythematosus, tuberculosis, or a demyelination syndrome [49].

Another promising candidate for future clinical applications is the soluble form of the common signal transducer of IL-6-type cytokines, gp130. As discussed in this review, sgp130 exclusively blocks transsignaling, that is, signals mediated by the IL-6/IL-6R complex (or IL-11/IL-11R complex and CNTF/CNTFR complex). Signaling of IL-6 via the membrane-bound IL-6R is not affected by sg130. Transsignaling may be a pathogenetic factor in diverse human chronic inflammatory diseases such as multiple myeloma, Castleman's disease, prostate carcinoma, Crohn's disease, systemic sclerosis, Still's disease, osteoporosis, and cardiovascular diseases [50]. The successful application of soluble gp130 in animal models of Crohn's disease, rheumatoid arthritis, and peritonitis has been described [51–53].

The molecular knowledge on cytokines and cytokine receptors has allowed us a very precise understanding of the immune system. Moreover, these molecules are a very specific target for therapeutic interventions to manipulate the immune system.

REFERENCES

1. Oppenheim, J.J., Cytokines: past, present, and future, *Int J Hematol, 74*, 3, 2002.
2. Jones, S. and Rose-John, S., The role of soluble receptors in cytokine biology: the agonistic properties of the sIL-6R/IL-6 complex, *Biochim. Biophys. Acta., 1592*, 251, 2002.
3. Fickenscher, H., et al., The interleukin-10 family of cytokines, *Trends Immunol, 23*, 89, 2002.
4. Oritani, K., et al., Type I interferons and limitin: a comparison of structures, receptors, and functions, *Cytokine and Growth Factor Reviews, 12*, 337, 2001.
5. Bazan, J.F., Neuropoietic cytokines in the hematopoietic fold, *Neuron, 7*, 197, 1991.
6. Wells, J.A., Hematopoietic receptor complexes, *Annu. Rev. Biochem., 65*, 609, 1996.
7. Idriss, H.T. and Naismith, J.H., TNF alpha and the TNF receptor superfamily: structure-function relationship(s)., *Microsc Res Tech, 50*, 184, 2000.
8. Grötzinger, J., Molecular mechanisms of cytokine receptor activation, *Biochim Biophys Acta, 1592*, 215, 2002.
9. Sims, S., IL-1 and IL-18 receptors, and their extended family, *Curr Opin Immunol, 14*, 117, 2002.
10. Akira, S., Toll-like receptor signaling, *J Biol Chem, 278*, 38105–38108, 2003.
11. Stroud, R.M. and Wells, J.A., Mechanistic diversity of cytokine receptor signaling across cell membranes, *Sci STKE, 231*, RE7, 2004.
12. De Vos, A.M., Ultsch, M., and Kossiakoff, A.A., Human growth hormone and extra-cellular domain of its receptor: crystall structure and of the complex, *Science, 255*, 306, 1992.
13. Hage, T., Sebald, W., and Reinemer, P., Crystal structure of the interleukin-4/receptor chain complex reveals a mosaic binding interface, *Cell, 97*, 271, 1999.
14. Martin, M.U. and Wesche, H., Summary and comparison of the signaling mechanisms of the Toll/interleukin-1 receptor family, *Biochim Biophys Acta, 1592*, 265, 2002.
15. Grötzinger, J., et al., IL-6 type Cytokine receptor complexes: hexamer or tetramer or both?, *Biol. Chem., 380*, 803, 1999.
16. Taga, T. and Kishimoto, T., gp130 and the interleukin-6 family of cytokines, *Annu. Rev. Immunol., 15*, 797, 1997.
17. Walter, M.R., et al., Crystal structure of a complex between interferon-gamma and its soluble high-affinity receptor, *Nature, 376*, 230, 1995.
18. Bach, E.A., Aguet, M., and Schreiber, R.D., The IFN gamma receptor: a paradigm for cytokine receptor signaling, *Annu Rev Immunol, 15*, 563, 1997.
19. Banner, D.W., et al., Crystal structure of the soluble human 55 kd TNF receptor-human TNF beta complex: implications for TNF receptor activation, *Cell, 73*, 431, 1993.
20. Livnah, O., et al., Crystallographic evidence for preformed dimers of erythropoietin receptor before ligand activation, *Science, 283*, 987, 1999.

21. Luo, R.Z., et al., Quaternary structure of the insulin-insulin receptor complex, *Science, 285*, 1077, 1999.
22. Klouche, M., et al., Novel path of activation of primary human smooth muscle cells: upregulation of gp130 creates an autocrine activation loop by IL-6 and its soluble receptor, *J. Immunol., 163*, 4583, 1999.
23. Buk, D.M., et al., Polarity and lipid raft association of the components of the ciliary neurotrophic factor receptor complex in Madin-Darby canine kidney cells, *J Cell Sci, 117*, 2063, 2004.
24. Dittrich, E., et al., Identification of a region within the cytoplasmic domain of the interleukin-6 (IL-6) signal transducer gp130 important for ligand-induced endocytosis of the IL-6 receptor, *J. Biol. Chem., 269*, 19014, 1994.
25. Mui, A., et al., Function of the common beta subunit of the GM-CSF/IL-3/IL-5 receptors, *Adv Exp Med Biol, 365*, 217, 1994.
26. Demoulin, J.B. and Renauld, J.C., Signalling by cytokines interacting with the interleukin-2 receptor gamma chain, *Cytokines Cell Mol Ther, 4*, 243, 1998.
27. Schuster, B., et al., Signalling of human CNTF revisited: the interleukin–6 (IL-6) receptor can serve as an a-receptor for ciliary neurotrophic factor (CNTF), *J. Biol. Chem., 278*, 9528–9535, 2003.
28. Elson, G.C., et al., CLF associates with CLC to form a functional heteromeric ligand for the CNTF receptor complex, *Nat. Neurosci., 3*, 867, 2000.
29. Derouet, D., et al., Neuropoietin, a new IL-6-related cytokine signaling through the ciliary neurotrophic factor receptor, *Proc Natl Acad Sci U.S.A., 101*, 4827, 2004.
30. Smith, C.A., Farrah, T., and Goodwin, R.G., The TNF receptor superfamily of cellular and viral proteins: activation costimulation and death, *Cell, 76*, 959, 1994.
31. Colotta, F., et al., Interleukin-1 type II receptor: a decoy target for IL-1 that is regulated by IL-4, *Science, 261*, 472, 1993.
32. Lang, D., et al., The type II IL-1 receptor interacts with the IL-1 receptor accessory protein: a novel mechanism of regulation of IL-1 responsiveness., *J. Immunol, 161*, 6871, 1998.
33. Müllberg, J., et al., The importance of shedding of membrane proteins for cytokine biology, *Eur. Cyt. Netw., 11*, 27, 2000.
34. Fernandez-Botran, R., Soluble cytokine receptors: their role in immunoregulation, *FASEB J., 5*, 2567, 1991.
35. Rose-John, S. and Heinrich, P.C., Soluble receptors for cytokines and growth factors: generation and biological function, *Biochem. J., 300*, 281, 1994.
36. Peters, M., Müller, A., and Rose-John, S., Interleukin-6 and soluble Interleukin-6 receptor: direct stimulation of gp130 and hematopoiesis, *Blood, 92*, 3495, 1998.
37. Rose-John, S. and Neurath, M., IL-6 trans-signaling: the heat is on, *Immunity, 20*, 2, 2004.
38. Symons, J.A., Young, P.R., and Duff, G.W., Soluble type II interleukin 1 (IL-1) receptor binds and blocks processing of IL-1 beta precursor and loses affinity for IL-1 receptor antagonist., *Proc Natl Acad Sci U.S.A., 92*, 1714, 1995.
39. Pflanz, S., et al., IL-27, a heterodimeric cytokine composed of EBI3 and p28 protein, induces proliferation of naive CD4(+) T cells, *Immunity, 16*, 779, 2002.
40. Parham, C., et al., A receptor for the heterodimeric cytokine IL-23 is composed of IL-12Rbeta1 and a novel cytokine receptor subunit, IL-23R, *J. Immunol., 168*, 5699, 2002.
41. Spriggs, M.K., Virus-encoded modulators of cytokines and growth factors, *Cytokine Growth Factor Rev, 10*, 1, 1999.

42. Pickup, D.J., Poxviral modifiers of cytokine responses to infection, *Infect Agents Dis, 3*, 116, 1994.

43. Chow, D.-C., et al., Structure of an extracellular gp130-cytokine receptor signalling complex., *Science, 291*, 2150, 2001.

44. Müllberg, J., et al., IL-6-receptor independent stimulation of human gp130 by viral IL-6, *J. Immunol., 164*, 4672, 2000.

45. Brown, S., Hu, N., and Hombria, J.C., Identification of the first invertebrate interleukin JAK/STAT receptor, the Drosophila gene domeless, *Curr Biol, 11*, 1700, 2001.

46. Kauppila, S., et al., Eiger and its receptor, Wengen, comprise a TNF-like system in Drosophila, *Oncogene, 22*, 4860, 2003.

47. Kiger, A.A., et al., Stem cell self-renewal specified by JAK-STAT activation in response to a support cell cue, *Science, 294*, 2542, 2001.

48. Schooltink, H. and Rose-John, S., Cytokines as therapeutic drugs, *J Interferon Cytokine Res, 22*, 505, 2002.

49. Reimold, A.M., TNFalpha as therapeutic target: new drugs, more applications, *Curr Drug Targets Inflamm Allergy, 1*, 377, 2002.

50. Kallen, K.-J., The role of transsignalling via the agonistic soluble IL-6 receptor, *Biochim Biophys Acta, 1592*, 323, 2002.

51. Atreya, R., et al., Blockade of IL-6 transsignaling abrogates established experimental colitis in mice by suppression of the antiapoptotic resistance of lamina propria T cells., *Nature Med., 6*, 583, 2000.

52. Hurst, S.M., et al., Control of leukocyte infiltration during inflammation: IL-6 and its soluble receptor orchestrate a temporal switch in the pattern of leukocyte recruitment., *Immunity, 14*, 705, 2001.

53. Nowell, M.A., et al., Soluble IL-6 receptor governs IL-6 activity in experimental arthritis: blockade of arthritis severity by soluble glycoprotein 130, *J. Immunol., 171*, 3202–3209, 2003.

4 Cytokines and Signaling

Michael David

CONTENTS

I. INTRODUCTION

Cytokines represent a group of transiently expressed, low molecular weight molecules that fulfill a vital role in intercellular communication, particularly in the immune system. Cytokines regulate numerous physiological and pathological events such as lymphocyte development and activation, programmed cell death, and inflammatory processes. Research over the past decade has lead to the discovery of many new cytokines and their respective cell surface receptors and has elucidated the signaling pathways that lead to cytokine receptor mediated gene induction. Equal to their complex biological effects, cytokine-induced transcriptional activation is the product of an intricate network of multiple signaling cascades. This chapter presents an overview of cytokine-induced signaling pathway with emphasis on signal transduction through STAT proteins and NFκB, as these two signaling pathways represent the most widely utilized signal transduction cascades activated by cytokine receptors.

II. THE JAK/STAT PATHWAY

The discovery of the Jak/STAT pathway is owed to the quest for understanding the mechanism by which interferons activate immediate early response genes.

The first STAT proteins were identified as the regulatory components of the type I interferon induced transcription factor complex, ISGF3 [1]. STAT1 and STAT2, in conjunction with the DNA-binding subunit p48, were found to bind the ISRE (Interferon Stimulated Response Element) enhancer as a heterotrimer in response to IFNα/β [2]. Alternatively, STAT1 homodimers activated via the IFN receptor can bind a distinct element termed GAS (IFN-activated sequence). Shortly after cloning of STAT1 and STAT2, numerous other cytokines and growth factors were reported to activate DNA-binding proteins with STAT-like characteristics (summarized in Table 4.I) [3,4]. Efforts to identify these proteins resulted in the cloning of five additional mammalian STAT genes (STAT3, STAT4, STAT5A, STAT5B, STAT6), which, in analogy to the interferon signaling model, bind to GAS-like elements [5]. Several of the STAT proteins are represented by various splice variants; however, STAT5A and STAT5B are encoded by two distinct genes. In contrast to the more restricted expression of STAT4 in spleen, heart, brain, peripheral blood cells, and testis, most STAT proteins are found rather ubiquitously. All seven STAT proteins share several conserved structural and functional domains (Figure 4.1). A carboxy-terminal tyrosine residue, which serves as the target for ligand-induced phosphorylation, is crucial for dimerization, nuclear translocation, and DNA-binding of STAT molecules. The src-homology 2 (SH2) domain of STAT proteins facilitates homo- or heterodimerization via reciprocal interaction with the phosphorylated tyrosine residue. In addition, the SH2 domain accounts for the specific interaction of STAT molecules with activated, tyrosine phosphorylated cytokine and growth factor receptors, as well as Jak tyrosine kinases [6]. Dimerization of STAT proteins allows the centrally located DNA-binding domains to interact with consensus palindromic sequences that differ only in their core nucleotide sequence [7]. Interestingly, STAT2 is unable to bind DNA in such a manner; rather, its function as a transcriptional activator appears to be restricted to its participation in the multimeric ISGF3 complex. The more divergent region carboxy-terminal to the tyrosine phosphorylation site harbors the transactivation domain (TAD). Posttranslational modification of several STAT members through serine phosphorylation [8], presumably by members of the ERK/SAPK kinase family [9], accounts for the regulated interaction of the STAT-TAD domains with transcriptional coactivators such as CBP/p300 or Minichromosome Maintenance proteins 5 (MCM5) [10,11].

It was again the paradigm of the interferon signaling cascade that lead to the discovery of the participation of the Janus tyrosine kinases in STAT-mediated gene transcription [12]. The demonstration that mutagenically ablated expression of Tyk2 or Jak1 results in IFNα/β unresponsive cells, whereas similar abrogation of Jak1 and Jak2 expression resulted in the loss of IFN responsiveness, provided the basis for the function of these tyrosine kinases and the identity of this novel signaling pathway [13,14]. Contrary to the ubiquitous expression of these kinases, a fourth family member, Jak3, was identified whose expression is restricted to cells of hematopoietic origin [15]. The most striking feature in the structure of Jak family members is the presence of a kinase-like domain of thus far unclear function in addition to the true kinase domain. Also noteworthy is the absence of any SH2- or SH3-like structures that are commonly found in cytoplasmic tyrosine kinases. The relatively large (120–140 kDa) Jak kinases are constitutively associated with many cytokine and growth factor receptors; nevertheless, this association can be increased after cytokine stimulation.

The general outline of the sequential events that lead to STAT-mediated gene transcription has been derived mostly from evidence gathered from the interferon

TABLE 4.1

Jak and STAT Family Members Activated by a Variety of Cytokines and Growth Factors

Receptor Family	Tyk 2	Jak 1	Jak 2	Jak 3	STAT 1	STAT 2	STAT 3	STAT 4	STAT 5	STAT 6
IFN										
IFNα/β	+	+	−	−	+	+	+	+	+	+
IFNg	−	+	+	−	+	−	−	−	+	−
gp130										
LIF	+	+	+	−	+	−	+	−	−	−
G-CSF	−	+	−	−	+	−	+	−	+	−
IL-6	+	+	+	−	+	−	+	−	+	−
IL-11	−	+	−	−	+	−	+	−	−	−
IL-12	+	−	+	−	−	−	+	+	−	−
CNTF	+	+	+	−	+	−	+	−	−	−
OSM	+	+	+	−	+	−	+	−	−	−
Common γ Chain										
IL-2	−	+	−	+	+	−	+	−	+	−
IL-4	−	+	−	+	+	−	+	−	−	+
IL-7	−	+	−	+	−	−	−	−	+	−
IL-9	−	+	−	+	−	−	−	−	+	−
IL-13	−	+	−	−	?	?	?	?	?	?
IL-15	−	+	−	+	−	−	−	−	+	−
Common β Chain										
IL-3	−	−	+	−	−	−	−	−	+	−
IL-5	−	−	+	−	−	−	−	−	+	−
GM-CSF	−	−	+	−	−	−	−	−	+	−
Single Chain										
EPO	−	−	+	−	−	−	−	−	+	−
GH	−	+	+	−	+	−	+	−	+	−
PRL	+	+	+	−	+	−	−	−	+	−
TPO	+	+	−	−	+	−	+	−	+	−
Growth Factors										
EGF	−	+	+	−	+	−	+	−	−	−
PDGF	+	+	+	−	+	−	+	−	−	+
CSF-1	−	+	+	−	+	−	+	−	−	−

signaling paradigm. It is believed that engagement of ligands to their respective receptors results in an increased local concentration of Jak proteins due to receptor aggregation and increased affinity of the receptors for Jak kinases. Subsequent cross-phosphorylation of the Jaks results in the activation of their kinase activity, such that they can phosphorylate tyrosine residues in the receptor chains. These

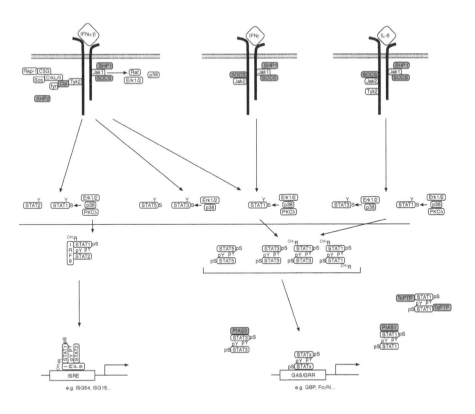

FIGURE 4.1 The Jak/STAT pathway.

phosphotyrosine moieties provide the docking sites for the STAT proteins via their SH2-domains, leading to their tyrosine phosphorylation, dimerization, and nuclear translocation. Parallel signaling events, presumably involving extracellular signaling regulated kinase/stress activated protein kinase (ERK/SAPK) family members, are responsible for further phosphorylation of the serine residues in the carboxy-terminus of the STAT proteins (Figure 4.1), which are required for efficient transcriptional response. In the case of growth factor receptors, not the presence of Jak kinases, but rather the intrinsic tyrosine kinase activity of the receptors is required for the tyrosine phosphorylation of STAT proteins.

As is the case with other signaling pathways, much of our understanding of the diverse functions of Jak and STAT proteins *in vivo* has been revealed through the generation of knockout mice [16]. As anticipated, STAT1-deficient mice display dramatically increased sensitivity toward viral and microbial pathogens, presumably due to their inability to respond to interferons [17,18]. Disruption of the STAT3 gene results in early embryonic lethality, and only the use of conditional gene targeting revealed an essential role for STAT3 in T cell and macrophage function [19,20]. As predicted by the restricted activation of STAT4 by IL-12, Th1-cell development is severely impaired by the disruption of the STAT4 gene [21,22]. Consistent with the activation of STAT5 by prolactin, STAT5a-deficient mice fail to lactate; in contrast, the absence of STAT5b causes failure to respond to growth hormone [23]. The role of STAT6 in IL-4 signaling was confirmed by the impairment of Th2 development and lack of IgE class switching. Jak1-deficient mice fail to nurse, display severely impaired lymphocyte development, and die perinatally [24,25]. The function of Jak2 in erythropoietin signaling is evidenced by embryonic lethality due to defective erythropoiesis in its absence [26]. Of particular interest are results obtained from Jak3-deficient mice [27] because mutations in Jak3 have also been identified in humans. Jak3 interacts with the common cytokine receptor chain, a component of the receptors for IL-2, IL-4, IL-7, IL-9, and IL-15, and mutation of which is responsible for X-linked severe combined immune-deficiency (X-SCID). Similarly, Jak3 deficiency in mice or mutations of Jak3 in humans causes SCID-like phenotypes [28].

Of equal importance to the activation of a signaling pathway is its spatially and temporally coordinated attenuation. Several independent mechanisms are responsible for the negative regulatory control over the Jak/STAT pathway. The two SH2-domain containing tyrosine phosphatases SHP-1 and SHP-2 are at least in part responsible for modulating the IFNα/β activation of the Jak/Stat pathway. SHP1 is associated with IFNAR1 and suppresses activation of the IFNα/β-induced Jak1 kinase [29]. Contradicting reports exist on the role of SHP2 in IFNα/β-mediated STAT activation. Expression of dominant-negative SHP2 mutants inhibits IFNα-mediated gene induction [30], whereas the absence of SHP2 in gene-targeted animals leads to an enhanced interferon response [31].

The existence of a nuclear tyrosine phosphatase that inactivates STAT1 has been described by several labs [32,33]. This crucial attenuator of the interferon response was recently identified as the nuclear 45kDa isoform of T cell protein tyrosine phosphatase (TcPTP) [34].

The SH2-domain containing suppressor of cytokine signaling (SOCS, aka CIS) proteins have been identified as a family of seven cytokine-inducible inhibitors of the Jak/STAT pathway [35,36]. These proteins bind to either tyrosine phosphorylated receptors or activated Jak kinases and mediate their ubiquitin-dependent degradation [37], thereby inactivating these essential signaling components in a classical negative feedback loop. A ubiquitin-proteosomal degradation pathway for STAT1 has also been suggested as an attenuation mechanism of the IFN activated Jak/STAT pathway [38].

In contrast to the cytokine-inducible SOCS proteins, the constitutively expressed protein inhibitors of activated STATs (PIAS) proteins do not prevent the phosphorylation of STAT proteins, but exert their negative regulatory role by association with tyrosine phosphorylated STAT dimers and preventing them from binding DNA [39,40]. Recently, PIAS proteins have been shown to perform additional functions by acting as small ubiquitin-like modifiers (SUMO) E3-ligases for c-jun and p53 [41,42].

III. CYTOKINE ACTIVATION OF NFκB

Stress events such as infections or tissue damage cause the immune system to evoke an inflammatory response leading to a cascade of local and systemic counteracting effects. Components or products of infectious microorganisms, e.g., LPS, induce the expression of cytokines such as IL-1 and tumor necrosis factor (TNF). The binding of these proinflammatory cytokines to their cell surface receptors rapidly induces a genetic program to respond to cellular stress and to generate inflammatory mediators (e.g., chemokines, acute phase proteins, proteases, and prostaglandins). Chronic inflammation can result in extensive tissue destruction and disease, such as rheumatoid arthritis. Therefore, it is important that signaling molecules transducing IL-1 and TNF signals not only be triggered quickly but also signal transiently.

The transcription factor NFκB fulfills these characteristics. Although it is a key mediator in the inflammatory response, NFκB was first isolated in B cells as a factor necessary for immunoglobulin kappa light chain transcription. Later, NFκB was found to exist in most cell types [43]. NFκB binding sites have been identified in a number of genes encoding cytokines and chemokines, adhesion molecules, acute phase proteins, anti-apoptotic genes, or transcription factors [44]. Thus, NFκB was implicated in innate immunity as well as in the adaptive immune response. NFκB exists as a homo- or heterodimer of any of the five thus far isolated subunits—RelA (p65), RelB, c-Rel, p50, and p52. The p50 and p52 subunits are first generated as the longer p105 and p100 forms, respectively, which are then proteolytically processed into the shorter, active forms. All members contain a conserved Rel homology domain (RHD) within their N-terminus which contains the dimerization, IκB-binding and nuclear localization signal (NLS) regions. Additionally, RelA, RelB, and c-Rel contain a C-terminal transactivation domain that is absent in p50 and p52. Hence, p50 and p52 homodimers are transcriptionally repressive [43].

TNF and IL-1, and LPS itself, activate a pathway that results in the degradation of inhibitors of NFκB (IκB), freeing NFκB proteins from the cytosolic tether and exposing the NLS (see Figure 4.2). Similar to transcriptional induction via STAT proteins, the activation of NFκB does not require protein synthesis. Once in the nucleus, NFκB negatively regulates its own activity by inducing transcription of IκB, which enters the nucleus and chaperones NFκB molecules out through an active export pathway involving the IκB nuclear export sequence (NES) [45].

The IκB proteins bind to NFκB dimers and mask the NFκB NLS. The IκB family members (IκBα, IκBβ, IκBε, IκBγ, and Bcl-3) contain ankyrin repeats that bind the NFκB RHD, an N-terminal regulatory domain, and a C-terminal PEST motif for proteolytic degradation [43]. Activators of the NFκB pathway trigger serine phosphorylation of IκB proteins (Ser32 and Ser36 on IκBα), which targets IκB for ubiquitin-mediated destruction by the 26S proteasome [46]. Two closely related proteins with IκB kinase (IKK) activity were identified as IKKα and IKKβ. Additionally, a putative scaffolding subunit was isolated and called IKKγ (NFκB essential modulator = NEMO). IKKα and IKKβ are 52% similar and both contain an N-terminal serine/threonine kinase domain, a leucine zipper, and a helix-loop-helix (HLH) motif [47]. Upon stimulation by TNF and IL-1, the IKK subunits are phosphorylated on serine residues. Substitution of alanines for these serines abrogated IKKβ kinase activity and eliminated responses to TNFα and IL-10. However, mutations of the homologous serines within IKKα left TNF- and IL-1-induced NFκB activity intact, suggesting that IKKβ is the main IKK kinase subunit through which

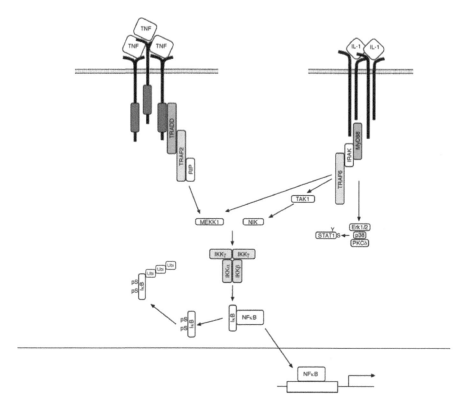

FIGURE 4.2 Cytokine activation of NFκB.

proinflammatory cytokines signal [48]. The IKKα and IKKβ knockout mice confirmed these earlier biochemical experiments, as IKKα-/- embryonic fibroblasts are normal in proinflammatory cytokine induced NFκB activation. In contrast, IKKβ-/- mice, which die between embryonic day 12.5 and 14.5 similar to RelA-/- embryos from to severe fetal liver destruction, lack the ability to suppress TNF induced apoptosis via NFκB. Indeed, when RelA-/- or IKKβ-/- mice are crossed to TNF receptor 1 or TNFα deficient mice, embryonic survival is rescued [45].

Analysis of the activation sites of the IKK kinases revealed that the mitogen activated protein kinase kinase kinase (MAPKKK) NFκB-Inducing kinase (NIK) activates the IKK complex and NFκB. NIK was originally isolated in a yeast two hybrid screen for TNF associated factor 2 (TRAF2) and was found to inhibit TNF- and IL-1-induced NFκB activation when expressed in a dominant negative form [49]. The natural mouse mutant *alymphoplasia* (*aly*) were identified to have a single amino acid mutation (G855R) within the C-terminal TRAF interacting region of NIK. Interestingly, embryonic fibroblasts from NIK-/- mice are only defective in lymphotoxin-induced but not IL-1- or TNF-induced NFκB activation, suggesting that NIK plays a critical role in lymphotoxin signaling but not in the TNF or IL-1 pathways, despite the results of the earlier overexpression studies [50].

In addition to NIK, MAP/ERK Kinase Kinase 1 (MEKK1) can also phosphorylate IKKα *in vitro* [51]. However, although MEKK1 activity is induced by TNF, MEKK1 is not as potent of a stimulator of IKK activity as NIK [50]. Furthermore, inactivation of MEKK1 does not impair NFκB activation by TNF or IL-1 [52]. Thus, additional kinases are likely to be involved in activating IKK in particular *in vivo* situations.

A number of adapter proteins link the IL-1 and TNF receptors to IKK activation and consequently NFκB activation. The TNFR1 is a death domain containing receptor. The death domains of TNFR1 interact with the death domain of FADD, initiating pro-apoptotic signals, but also bind the death domain of TRADD, thereby inducing the anti-apoptotic NFκB pathway. In turn, TRADD binds to TRAF2 [53]. A dominant negative form of TRAF2 inhibits TNF-induced NFκB activation, nevertheless, TRAF2-/- fibroblasts have no defect in TNF-induced NFκB activation [54]. However, TRAF2 recruits RIP1 (receptor interacting protein) to the TNFR complex, and RIP1-/- cells are deficient in NFκB activity [55].

The binding of IL-1 to the IL-1R also leads to the development of a signaling scaffold resulting in NFκB activation. First, the adapter protein MyD88 (myeloid differentiation factor 88) binds to the receptor. Interacting through its death domain, MyD88 binds the serine/threonine kinase IRAK (IL-1R associated kinase) [56], which in turn recruits TRAF6 to the complex. Both IRAK-/- and TRAF6-/- mice display defects in IL-1-induced NFκB activity. Additionally, the MAPKKK TAK1 (TGFβ activated kinase) associates with TRAF6 in an IL-1-dependent manner, leading to activation of NFκB in a NIK dependent manner [57].

IV. ADDITIONAL SIGNALING CASCADES
IN CYTOKINE RESPONSES

In addition to the STATs, numerous other proteins undergo rapid tyrosine phosphorylation following exposure of cells to IFNα/β and other cytokines. Two members of the insulin receptor substrate (IRS) family of adapter proteins, IRS-1 and IRS-2, become tyrosine phosphorylated in response to a number of cytokines such as IFNα/β, IL-4, or IL-9, allowing for the SH2-domain mediated binding of the p85 regulatory subunit of PI-3 kinase [58,59]. The resulting activation of PI-3 kinase promotes the activity of proto-oncogene Akt (PKB) [60,61] and PKCd [62], which facilitates the survival and proliferation of cytokine-stimulated cells. Interestingly, PI-3 kinase activity might also be essential for certain IFNβ-mediated transcriptional responses involving p65 [63], as well as the antiproliferative effects of IFNα/β [64], illustrating that the biological responses attributed to this signaling pathway can be altered due to modulation through other signaling events.

Two additional adapter proteins undergo rapid tyrosine phosphorylation after cytokine stimulation. The guanine nucleotide exchange factor and proto-oncogene Vav is activated by cytokines such as IL-3 or IL-6, and contribute to their mitogenic responses. In contrast, disrupted expression of the Vav abrogates the growth inhibitory effects of IFNα [65–67]. The proto-oncogene c-Cbl modulates cytokine signaling not only due to its function as an E3-ubiquitin ligase that promotes degradation of components of the cytokine signaling cascade, but also by providing docking sites for the src family kinase fyn [68] as well as the adapters CrkL and CrkII [69–71]. These SH2 and SH3 domain containing proteins link c-Cbl to Sos and C3G, a guanine nucleotide exchange factor for Rap-1 [72,73]. CrkL can associate with STAT5 and form a GAS-binding complex [74], and expression of CrkL and CrkII is essential for the antiproliferative effects of IFNα/β [75]. Another signaling cascade regulated by cytokines is the MAP kinase pathway [76,77]. IFNα/β activates Raf-1 as well as B-Raf, two serine kinases ultimately responsible for the activation of p42MAP kinase, in a Jak1-kinase and STAT1-dependent manner [78–80]. Likewise, it has also been reported that the p38 SAPK is activated by IFNα/β [81,82]. Ser 727 of Stat1, which is conserved in Stat3 and Stat4, is positioned within a consensus phosphorylation site for proline-directed serine kinases such as MAP kinases [8]. However, as other cytokine-activated kinases, such as PKCd [62] or CamKII [83], also have the ability to phosphorylate Ser727 of STAT1 *in vitro*, it remains unclear which kinase is ultimately responsible for the phosphorylation of STAT1 Ser727 *in vivo*.

V. CONCLUSIONS

In this chapter we have attempted to outline the events leading to cytokine mediated gene transcription by using that Jak/STAT and the NFκB pathways as two examples. It is important to remember that cytokine activation of gene transcription is not the result of a single, linear signaling cascade, but involves a complex network of interacting signal transduction pathways. Integration of numerous, often contradicting instructions received by the cell through a variety of cytokine and growth factor receptors, ultimately determines the qualitative and quantitative transcriptional response.

REFERENCES

1. Fu, X.-Y. et al., ISGF-3, the transcriptional activator induced by IFN-a, consists of multiple interacting polypeptide chains. *PNAS*, 87, 8555, 1990.
2. Fu, X.-Y., A transcription factor with SH2 and SH3 domains is directly activated by an interferon-a induced cytoplasmic protein tyrosine kinase(s). *Cell*, 70, 323, 1992.
3. Larner, A.C. and Finbloom, D.S., Protein tyrosine phosphorylation as a mechanism which regulates cytokine activation of early response genes. *Biochimica et Biophysica Acta*, 1266, 278, 1995.
4. Larner, A.C. et al., Tyrosine phosphorylation of DNA binding proteins by multiple cytokines. *Science*, 261, 1730, 1993.
5. Schindler, C. and Darnell, J.E., Jr., Transcriptional responses to polypeptide ligands: the JAK-STAT pathway. *Annu Rev Biochem*, 64, 621, 1995.
6. Heim, M.H. et al., Contribution of STAT SH2 groups to specific interferon signaling by the Jak-STAT pathway. *Science*, 267, 1347, 1995.
7. Seidel, H.M. et al., Spacing of palindromic half sites as a determinant of selective STAT (signal transducers and activators of transcription) DNA binding and transcriptional activity. *PNAS*, 92, 3041, 1995.
8. Wen, Z., Zhong, Z., and Darnell, J.E., Jr., Maximal activation of transcription by Stat1 and Stat3 requires both tyrosine and serine phosphorylation. *Cell*, 82, 241, 1995.

9. David, M. et al., Requirement for MAP kinase (ERK2) activity in interferon a/b-stimulated gene expression through stat proteins. *Science*, 269, 1721, 1995.

10. Korzus, E. et al., Transcription factor-specific requirements for coactivators and their acetyltransferase functions. *Science*, 279(5351), 703, 1998.

11. Zhang, J.J. et al., Ser727-dependent recruitment of MCM5 by Stat1alpha in IFN-gamma-induced transcriptional activation. *Embo J*, 17(23), 6963, 1998.

12. Velazquez, L. et al., A protein tyrosine kinase in the interferon a/b signaling pathway. *Cell*, 70, 313, 1992.

13. Muller, M. et al., The protein tyrosine kinase JAK1 complements defects in the interferon-a/b and -g signal transduction. *Nature*, 366, 129, 1993.

14. Watling, D. et al., Complementation by the protein tyrosine kinase JAK2 of a mutant cell line defective in the interferon-g signal transduction pathway. *Nature*, 366, 166, 1993.

15. Kawamura, M. et al., Molecular cloning of L-JAK, a Janus family protein-tyrosine kinase expressed in natural killer cells and activated leukocytes. *PNAS*, 91, 6374, 1994.

16. Akira, S., Functional roles of STAT family proteins: lessons from knockout mice. *Stem Cells*, 17(3), 138, 1999.

17. Durbin, J.E. et al., Targeted disruption of the mouse *Stat1* gene results in compromised innate immunity to viral disease. *Cell*, 84, 443, 1996.

18. Meraz, M.A. et al., Targeted disruption of the *Stat1* gene in mice reveals unexpected physiologic specificity in the JAK-STAT signaling pathway. *Cell*, 84, 431, 1996.

19. Takeda, K. et al., Stat3 activation is responsible for IL-6-dependent T cell proliferation through preventing apoptosis: generation and characterization of T cell-specific Stat3-deficient mice. *J Immunol*, 161(9), 4652, 1998.

20. Takeda, K. et al., Targeted disruption of the mouse Stat3 gene leads to early embryonic lethality. *Proc Natl Acad Sci U.S.A.*, 94(8), 3801, 1997.

21. Kaplan, M.H. et al., Impaired IL-12 responses and enhanced development of Th2 cells in Stat4-deficient mice. *Nature*, 382(6587), 174, 1996.

22. Thierfelder, W.E. et al., Requirement for Stat4 in interleukin-12-mediated responses of natural killer and T cells. *Nature*, 382(6587), 171, 1996.

23. Teglund, S. et al., Stat5a and Stat5b proteins have essential and nonessential, or redundant, roles in cytokine responses. *Cell*, 93(5), 841, 1998.

24. Takeda, K. et al., Essential role of Stat6 in IL-4 signalling. *Nature*, 380, 627, 1996.

25. Shimoda, K. et al., Lack of IL-4-induced Th2 response and IgE class switching in mice with dirupted Stat6 gene. *Nature*, 380, 630, 1996.

26. Neubauer, H. et al., Jak2 deficiency defines an essential developmental checkpoint in definitive hematopoiesis. *Cell*, 93(3), 397, 1998.

27. Nosaka, T. et al., Defective lymphoid development in mice lacking Jak3. *Science*, 270, 800, 1995.

28. Russell, S.M. et al., Mutation of Jak3 in a patient with SCID: essential role of Jak3 in lymphoid development. *Science*, 270, 797, 1995.

29. David, M. et al., Differential regulation of the IFNa/b-stimulated Jak/Stat pathway by the SH2-domain containing tyrosine phosphatase SHPTP1. *MCB*, 15(12), 7050, 1995.

30. David, M. et al., The SH2-domain containing tyrosine phosphatase PTP1D is required for IFNa/b-induced gene expression. *JBC*, 271(27), 15862, 1996.

31. You, M., Yu, D.H., and Feng, G.S., Shp-2 tyrosine phosphatase functions as a negative regulator of the interferon-stimulated Jak/STAT pathway. *Mol Cell Biol*, 19(3), 2416, 1999.

32. David, M. et al., A nuclear tyrosine phosphatase downregulates interferon-induced gene expression. *MCB*, 13(12), 7515, 1993.

33. Haspel, R.L. and Darnell, J.E., Jr., A nuclear protein tyrosine phosphatase is required for the inactivation of Stat1. *Proc Natl Acad Sci U.S.A.*, 96(18), 10188, 1999.

34. ten Hoeve, J. et al., Identification of a nuclear Stat1 protein tyrosine phosphatase. *Mol Cell Biol*, 22(16): p. 5662 2002.

35. Hilton, D.J., Negative regulators of cytokine signal transduction. *Cell Mol Life Sci*, 55(12), 1568, 1999.

36. Chen, X.P., Losman, J.A., and Rothman, P., SOCS proteins, regulators of intracellular signaling. *Immunity*, 13(3), 287, 2000.

37. Kile, B.T. et al., The SOCS box: a tale of destruction and degradation. *Trends Biochem Sci*, 27(5), 235, 2002.

38. Kim, T.K. and Maniatis, T., Regulation of Interferon-g-Activated STAT1 by the ubiquitin-proteasome pathway. *Science*, 273, 1717, 1996.

39. Liu, B. et al., Inhibition of Stat1-mediated gene activation by PIAS1. *Proc Natl Acad Sci U.S.A.*, 95(18), 10626, 1998.

40. Chung, C.D. et al., Specific inhibition of Stat3 signal transduction by PIAS3. *Science*, 278(5344), 1803, 1997.

41. Schmidt, D. and Muller, S., Members of the PIAS family act as SUMO ligases for c-Jun and p53 and repress p53 activity. *Proc Natl Acad Sci U.S.A.*, 99(5), 2872, 2002.

42. Kotaja, N. et al., PIAS proteins modulate transcription factors by functioning as SUMO-1 ligases. *Mol Cell Biol*, 22(14), 5222. 2002.

43. Ghosh, S., May, M.J., and Kopp, E.B., NF-kappa B and Rel proteins: evolutionarily conserved mediators of immune responses. *Annu Rev Immunol*, 16, 225, 1998.

44. Pahl, H.L., Activators and target genes of Rel/NF-kappaB transcription factors. *Oncogene*, 18(49), 6853, 1999.

45. Karin, M., The beginning of the end: IkappaB kinase (IKK) and NF-kappaB activation. *J Biol Chem*, 274(39), 27339, 1999.

46. Zandi, E. and Karin, M., Bridging the gap: composition, regulation, and physiological function of the IkappaB kinase complex. *Mol Cell Biol*, 19(7), 4547, 1999.

47. Karin, M., How NF-kappaB is activated: the role of the IkappaB kinase (IKK) complex. *Oncogene*, 18(49), 6867, 1999.

48. May, M.J. and Ghosh, S., Signal transduction through NF-kappa B. *Immunol Today*, 19(2), 80, 1998.

49. Mercurio, F. and Manning, A.M., Multiple signals converging on NF-kappaB. *Curr Opin Cell Biol*, 11(2), 226, 1999.

50. Karin, M. and Delhase, M., The I kappa B kinase (IKK) and NF-kappa B: key elements of proinflammatory signalling. *Semin Immunol*, 12(1), 85, 2000.

51. Lee, F.S. et al., Activation of the IkappaB alpha kinase complex by MEKK1, a kinase of the JNK pathway. *Cell*, 88(2), 213, 1997.

52. Israel, A., The IKK complex: an integrator of all signals that activate NF-kappaB? *Trends Cell Biol*, 10(4), 129, 2000.

53. Wallach, D. et al., Tumor necrosis factor receptor and Fas signaling mechanisms. *Annu Rev Immunol*, 17, 331, 1999.

54. Arch, R.H., Gedrich, R.W., and Thompson, C.B., Tumor necrosis factor receptor-associated factors (TRAFs)—a family of adapter proteins that regulates life and death. *Genes Dev*, 12(18), 2821, 1998.

55. Barkett, M. and Gilmore, T.D., Control of apoptosis by Rel/NF-kappaB transcription factors. *Oncogene*, 18(49), 6910, 1999.

56. Adachi, O. et al., Targeted disruption of the MyD88 gene results in loss of IL-1- and IL-18-mediated function. *Immunity*, 9(1), 143, 1998.

57. Ninomiya-Tsuji, J. et al., The kinase TAK1 can activate the NIK-I kappaB as well as the MAP kinase cascade in the IL-1 signalling pathway. *Nature*, 398(6724), 252, 1999.
58. Uddin, S. et al., Interferon-a engages the insulin receptor substrate-1 to associate with the phosphatidylinositol 3'-kinase. *JBC*, 270(27), 15938, 1995.
59. Platanias, L.C. et al., The type I interferon receptor mediates tyrosine phosphorylation of insulin receptor substrate 2. *J Biol Chem*, 271(1), 278, 1996.
60. Nguyen, H. et al., Roles of phosphatidylinositol 3-kinase in interferon-gamma-dependent phosphorylation of STAT1 on serine 727 and activation of gene expression. *J Biol Chem*, 276(36), 33361, 2001.
61. Uddin, S. et al., Interferon-dependent activation of the serine kinase PI 3'-kinase requires engagement of the IRS pathway but not the Stat pathway. *Biochem Biophys Res Commun*, 270(1), 158, 2000.
62. Uddin, S. et al., Protein kinase C-delta (PKC-delta) is activated by type I interferons and mediates phosphorylation of Stat1 on serine 727. *J Biol Chem*, 277(17), 14408, 2002.
63. Rani, M.R. et al., Requirement of phosphoinositide 3-kinase and Akt for interferon-beta-mediated induction of the beta-R1 (SCYB11) gene. *J Biol Chem*, 277(41), 38456, 2002.
64. Yang, C.H., Murti, A., and Pfeffer, L.M., STAT3 complements defects in an interferon-resistant cell line: evidence for an essential role for STAT3 in interferon signaling and biological activities. *Proc Natl Acad Sci U.S.A.*, 95(10), 5568, 1998.
65. Micouin, A. et al., p95 (vav) associates with the type I interferon (IFN) receptor and contributes to the antiproliferative effect of IFN-alpha in megakaryocytic cell lines. *Oncogene*, 19(3), 387, 2000.
66. Uddin, S. et al., The vav proto-oncogene product (p95vav) interacts with the Tyk-2 protein tyrosine kinase. *FEBS Lett*, 403(1), 31, 1997.
67. Platanias, L.C. and Sweet, M.E., Interferon A induces rapid tyrosine phosphorylation of the *vav* proto-oncogene product in hematopoietic cells. *JBC*, 269(5), 3143, 1994.
68. Uddin, S. et al., Interaction of p59fyn with interferon-activated Jak kinases. *Biochem Biophys Res Commun*, 235(1), 83, 1997.
69. de Jong, R. et al., Crkl is complexed with tyrosine-phosphorylated Cbl in Ph-positive leukemia. *J Biol Chem*, 270(37), 21468, 1995.
70. Barber, D.L. et al., Erythropoietin and interleukin-3 activate tyrosine phosphorylation of CBL and association with CRK adaptor proteins. *Blood*, 89(9), 3166, 1997.
71. Reedquist, K.A. et al., Stimulation through the T cell receptor induces Cbl association with Crk proteins and the guanine nucleotide exchange protein C3G. *J Biol Chem*, 271(14), 8435, 1996.
72. Tanaka, S. et al., C3G, a guanine nucleotide-releasing protein expressed ubiquitously, binds to the Src homology 3 domains of CRK and GRB2/ASH proteins. *Proc Natl Acad Sci U.S.A.*, 91(8), 3443, 1994.
73. Smit, L., van der Horst, G., and Borst, J., Sos, Vav, and C3G participate in B cell receptor-induced signaling pathways and differentially associate with Shc-Grb2, Crk, and Crk-L adaptors. *J Biol Chem*, 271(15), 8564, 1996.
74. Fish, E.N. et al., Activation of a CrkL-stat5 signaling complex by type I interferons. *J Biol Chem*, 274(2), 571, 1999.
75. Platanias, L.C. et al., CrkL and CrkII participate in the generation of the growth inhibitory effects of interferons on primary hematopoietic progenitors. *Exp Hematol*, 27(8), 1315, 1999.
76. Chang, F. et al., Signal transduction mediated by the Ras/Raf/MEK/ERK pathway from cytokine receptors to transcription factors: potential targeting for therapeutic intervention. *Leukemia*, 17(7), 1263, 2003.

77. Chang, F. et al., Regulation of cell cycle progression and apoptosis by the Ras/Raf/MEK/ERK pathway (Review). *Int J Oncol*, 22(3), 469, 2003.

78. Stancato, L. et al., Beta interferon and Oncostatin M activate Raf-1 and mitogen activated protein kinase through a JAK1-dependent pathway. *Mol. Cell. Biology*, 17(7), 3833, 1997.

79. Sakatsume, M. et al., Interferon gamma activation of Raf-1 is Jak1-dependent and p21ras-independent. *J Biol Chem*, 273(5), 3021, 1998.

80. Stancato, L.F. et al., Activation of Raf-1 by interferon gamma and oncostatin M requires expression of the Stat1 transcription factor. *J Biol Chem*, 273(30), 18701, 1998.

81. Uddin, S. et al., Activation of the p38 mitogen-activated protein kinase by type I interferons. *J Biol Chem*, 274(42), 30127, 1999.

82. Goh, K.C., Haque, S.J., and Williams, B.R., p38 MAP kinase is required for STAT1 serine phosphorylation and transcriptional activation induced by interferons. *Embo J*, 18(20), 5601, 1999.

83. Nair, J.S. et al., Requirement of Ca2+ and CaMKII for Stat1 Ser-727 phosphorylation in response to IFN-gamma. *Proc Natl Acad Sci U.S.A.*, 99(9), 5971, 2002.

5 Inflammatory Complexities in the CNS: New Insights into the Effects of Intracellular Redox State and Viral Infection in Modulating the Biology of Oligodendrocytes and Their Precursor Cells

Mark Noble, Margot Mayer-Pröschel, David Mock, and Chris Pröschel

CONTENTS

When it comes to considering the biology of inflammation in the central nervous system (CNS), it is clear that the famous maxim of Gertrude Stein that "a rose is a rose is a rose" simply is not true. Inflammation can cause damage, can exacerbate damage, and, paradoxically, can enhance the repair of damage. What appears to be the same insult may have different effects in different regions of the CNS and may also have different effects in different genetic backgrounds. Such complexity makes it very difficult to know best how to limit or, as some have suggested (e.g., [1]), promote inflammation so as to achieve reliable clinical benefit in the heterogeneous milieu of human injury.

Given such a varied response to inflammation in the CNS, how can this complexity be effectively analyzed? One aspect of this analysis is going to have to be the continued dissection of the inflammatory response at the cellular level, with particular attention to understanding the multiple components of this response that extend beyond the mere production of inflammatory cytokines. Such additional components will include understanding the response of the range of cell types affected by inflammation, the effects of combinations of agents on cells (which may be very different from the effects of single agents), and, not to be forgotten, the direct effects of those physiological insults associated with the mounting of an inflammatory response in the first place.

In this chapter we will discuss several components of the inflammatory response that are currently under study in our laboratories. Of critical importance are the various roles of redox modulation in altering cellular function, including modifying the response of cells to inflammatory cytokines. In addition, in further dissection of the individual components of an inflammatory action, we consider the effects of viral infection of CNS precursor cells, as exemplified by common members of the herpes virus family.

I. OLIGODENDROCYTES AND THEIR ANCESTRY

As this chapter will be concerned not only with oligodendrocytes, but also with the precursor cells that give rise to them, we first briefly summarize current views on the oligodendrocyte lineage and at least some of the signaling molecules that modulate their differentiation. We will use the term *precursor cell* generically to include both neuroepithelial stem cells (that can give rise to all of the cell types of the CNS) and lineage-restricted progenitor cells (that only give rise to restricted subsets of CNS cell types).

A. O-2A/OPCs

The cell that is believed to be the direct ancestor of the oligodendrocyte is a precursor cell that, at least *in vitro*, has the capacity to generate both oligodendrocytes and a particular subset of astrocytes (known as type-2 astrocytes). For this reason, these precursor cells originally were named oligodendrocyte-type-2 astrocyte (O-2A) progenitor cells [2]. These cells constitutively differentiate into oligodendrocytes but require exposure to specific environmental signals in order to give rise to astrocytes [3,4]. Whether these progenitor cells generate type-2 astrocytes during normal development has been a matter of considerable controversy and still remains unclear [5,6]. Difficulties in identifying type-2 astrocytes *in vivo* as a derivative of these progenitor cells has led multiple laboratories to refer to this identical population of cells as oligodendrocyte precursor cells (OPCs). In recognition of this still unresolved controversy, the abbreviation O-2A/OPC will be used throughout this chapter.

Although O-2A/OPCs do not require the action of cell-extrinsic signaling molecules to generate oligodendrocytes, they do require the action of such molecules to undergo division. The most extensively characterized mitogen for these precursor cells is the platelet derived growth factor AA homodimer PDGF-A [7,8]. In the intact CNS, PDGF-A is ubiquitously distributed, being synthesized by both astrocyte and neuronal populations [9,10]. Overexpression of PDGF-A results in a dramatic increase in the number of spinal cord O-2A/OPCs [11], whereas in PDGF-A knockouts, the number of these precursor cells is dramatically reduced [12].

In at least some cases, coexposure to PDGF with other cytokines alters the balance between self-renewal and differentiation in dividing O-2A/OPCs. Coexposure to fibroblast growth factor-2 (FGF-2), for example, causes these precursor cells to become trapped in a continuous program of self-renewal and appears to almost completely block the generation of oligodendrocytes [13]. Coexposure to NT-3 and PDGF also greatly enhances the extent of self-renewal [14]. Such studies have revealed what has proven to be an important general principle in precursor cell biology, which is that the mitotic capacity of a lineage-restricted progenitor cell is regulated in part by the specific combination of signaling molecules to which the cell is exposed. The number of divisions such progenitor cells can undergo can be greatly altered by exposure to different combinations and concentrations of cell-extrinsic signaling molecules.

The ability of certain combinations of growth factors to enhance progenitor cell self-renewal is counterbalanced by the action of other signaling molecules that promote oligodendrocyte generation. The most extensively studied of such signaling molecules is thyroid hormone (TH), which has been of particular interest due to the severe deficiencies in myelination that occur in children or experimental animals that are hypothyroid for genetic, nutritional, or experimental reasons [14–19]. Conditioned medium from cultured oligodendrocytes also inhibits the proliferation of O-2A/OPCs and promotes oligodendrocyte generation, and this effect may be mediated in part by transforming growth factor-ß (TGF-ß [20,21]). Still other cytokines, such as leukemia inhibitory effector (LIF) and ciliary neurotrophic factor (CNTF), also have been found to enhance the generation of oligodendrocytes from O-2A/OPCs [22–24]. It is important to note that these same factors, when applied to other precursor cells, appear to promote the generation of astrocytes [25–30]. Moreover, if applied to O-2A/OPCs together with extracellular matrix produced by endothelial cells, CNTF enhances astrocyte induction by the as yet unknown inducing factors present in the matrix [23,31].

B. GRP Cells

The only cell that arises earlier in development than O-2A/OPCs and has been shown to be able to give rise to O-2A/OPCs is a tripotential precursor cell that can be isolated from the embryonic spinal cord [27] and also can be generated directly from neuroepithelial stem cells. This cell, which has been named as the glial-restricted precursor (GRP) cell, is restricted to the generation of glia and can generate oligo-dendrocytes, type-1 astrocytes, and type-2 astrocytes. GRP cells do not give rise to neurons, either when transplanted into neurogenic zones of the CNS or when grown *in vitro* in conditions that promote generation of neurons from neuroepithelial stem cells or from neuron-restricted precursor (NRP) cells [27,32]. GRP cells represent one of the first two lineage-restricted populations to arise during differentiation of neuroepithelial stem cells (NSCs) *in vitro* (the other lineage-restricted population being neuron-restricted precursor cells [33]). GRP cells arise early *in vivo* and can be isolated directly from the E12.5 rat spinal cord, a stage of development that precedes the appearance of any differentiated glia [27,34]. In contrast, in the rat spinal cord, O-2A/OPCs (defined as cells that undergo bipotential differentiation into oligodendrocytes and type-2 astrocytes when examined in clonal culture) cannot be isolated from the rat spinal cord until at least E16 [35].

GRP cells differ from O-2A/OPCs in a variety of ways [27,35]. Freshly isolated GRP cells are dependent upon exposure to FGF-2 for both their survival and their division, whereas division and survival of O-2A/OPCs can be promoted by PDGF and other growth factors. Consistent with this difference in growth factor-response patterns, GRP cells freshly isolated from the E13.5 spinal cord do not express receptors for PDGF, although they do express such receptors with continued growth *in vitro* or *in vivo*. These populations also differ in their response to inducers of differentiation. For example, exposure of GRP cells to the combination of FGF-2 and CNTF induces these cells to differentiate into astrocytes (primarily express-ing the antigenic phenotype of type-2 astrocytes [27]). In contrast, exposure of O-2A/OPCs to FGF-2 + CNTF promotes the generation of oligodendrocytes [23].

A further striking difference between GRP cells and O-2A/OPCs is that GRP cells readily generate astrocytes following their transplantation into the adult CNS [32]. This is in striking contrast to primary O-2A/OPCs, which thus far only generate oligo-dendrocytes in such transplantations [6] although it has been reported that O-2A/OPC cell lines will generate astrocytes if transplanted in similar circumstances [36].

Antigenic and *in situ* analysis of development *in vivo* has confirmed that cells with the antigenic phenotype of GRP cells arise in spinal development several days prior to the appearance of astrocytes (as defined by expression of glial fibrillary acid protein; GFAP) and prior to the appearance of cells expressing markers of radial glia [34]. Thus, these cells can be isolated directly from the developing spinal cord, and cells with the appropriate antigenic phenotype have been found to exist *in vivo* at appropriate ages to play important roles in gliogenesis.

II. THE OLIGODENDROCYTE AS A TARGET
OF INFLAMMATORY DAMAGE

The analysis of inflammatory damage to oligodendrocytes has mostly been carried out in the context of multiple sclerosis (MS), but it is clear that inflammatory processes damage white matter in multiple other contexts.

Causation of white matter damage in MS appears to be quite complex. One component thought to contribute to MS-associated damage is an autoimmune reaction against myelin antigens. In most MS plaques, it is possible to visualize immunoglobulins and deposits of complement at the lesion site [37–39]. It has even been suggested that it is possible to observe deposition of antibodies against such specific antigens as myelin oligodendrocyte glycoprotein on dissolving myelin in active lesions [40], although it is clear that MS patients produce antibodies against a variety of myelin antigens. Indeed, it seems clear that as this disease progresses, the continued destruction of myelin causes an autovaccination process that is associated with a phenomenon called epitope spreading, in which the number of antigens recognized continues to increase [41–44]. The immune reaction leading to myelin destruction also appears to include a T-cell mediated immune reaction, which secondarily leads to macrophage activation. Indeed the range of possible immune-mediated destructive mechanisms that can lead to myelin destruction, and the substantial heterogeneity of the disease process itself, makes it seem likely that MS is more correctly viewed as a constellation of diseases that share certain characteristic features (see, e.g. [45,46] for review). For recent reviews on a variety of aspects of MS, the reader is referred to, e.g. [47–52].

One of the best studied contributors to death of oligodendrocytes and the precursor cells that give rise to them is exposure to supranormal levels of glutamate, one of the major contributors to CNS damage following traumatic injury. Glutamate toxicity for oligodendrocytes and their precursor cells has been demonstrated *in vitro* and has been shown to occur in isolated spinal dorsal columns [53] and *in vivo* following infusion of AMPA/kainate agonists into the optic nerve [54,55] or subcortical white matter [56]. The glutamate receptors expressed by oligodendrocytes and their precursors are of the AMPA-binding subclass, and it appears that AMPA antagonists can protect oligodendrocytes against ischemic damage, at least *in vitro* [57].

Along with having the potential to be cytotoxic in its own right, glutamate also may enhance the toxicity of other physiological insults. For example, ischemic injury is characterized by excessive release of glutamate into the extrasynaptic space [58,59]. Ischemia is also characterized by transient deprivation of oxygen and glucose, a physiological insult that is toxic for oligodendrocytes. Strikingly, the toxicity associated with deprivation of oxygen and glucose is further enhanced by coexposure to glutamate [56,57,60].

Glutamate-mediated damage of oligodendrocytes could be of physiological importance in a variety of settings. One dramatic example of oligodendrocyte death in which these pathways have been invoked is that of the ischemic injury occurring in birth trauma, which can be associated with periventricular leukomalacia and cerebral palsy [61]. It also must be considered whether glutamate contributes to the demyelination seen in multiple sclerosis and other inflammatory disorders of the CNS, particularly as it has been observed that glutamate levels are increased in the CNS of patients with demyelinating disorders, with levels correlating with disease severity [62,63]. In this context, it is of potential interest that chronic infusion of kainate (an AMPA receptor agonist) into white matter tracts is associated with the generation of lesions that have many of the characteristics of multiple sclerosis lesions, including extensive regions of demyelination with plaque formation, massive oligodendrocyte death, axonal damage, and inflammation [55]. Although acute infusion of kainate produces lesions that are repaired by endogenous cells, lesions induced by chronic kainate infusion are not spontaneously repaired.

Other potential contributors to oligodendrocyte death are the inflammatory cytokine tumor necrosis factor-α (TNF-α) and, surprisingly, the pro-form of nerve growth factor (ProNGF). It is known from *in vitro* and *in vivo* experiments that TNF-α induces cell death of oligodendrocytes [22,64,65]. TNF-α, along with glutamate, is released by activated microglia. As the release of glutamate can then reactivate microglia, it is possible that inflammation elicits a set of feedback responses that build upon each other with the eventual result of tissue destruction [66,67].

The neurotrophin receptor p75 also is induced by various injuries to the nervous system. Recent studies have shown that p75 is required for the death of oligodendrocytes following spinal cord injury, and its action is mediated mainly by proNGF [68]. Oligodendrocytes undergoing apoptosis express p75 and the absence of p75 results in a decrease in the number of apoptotic oligodendrocytes and increased survival of oligodendrocytes. ProNGF is likely responsible for activating p75 *in vivo* because the proNGF from the injured spinal cord induced apoptosis among p75(+/+), but not among p75(−/−), oligodendrocytes in culture, and its action was blocked by proNGF-specific antibody.

In vivo, it is unlikely to ever be the case that single factors act alone, and in this regard the interplay between glutamate and tumor necrosis factor-α (TNF-α) is of particular interest in regard to induction of demyelination. The combination of glutamate and TNF-α shows a highly lethal synergy when applied together in the thoracic grey matter of the spinal cord [69]. It is not yet known if similar synergies occur in respect to the killing of oligodendrocytes, either by TNF-α or by ProNGF, but such combinatorial effects seem likely.

III. INTRACELLULAR REDOX STATE AND OLIGODENDROCYTE SURVIVAL

One of the goals of studying oligodendrocyte death has been to identify means of protecting these cells, of which one of the most broadly effective appears to be through application of agents able to protect cells from oxidative stress. Such protective substances can be of many different sorts, ranging from the most common dietary antioxidants to novel chemicals that do not occur in nature.

One of our particular interests has been in using glutathione pro-drugs as a means of modulating intracellular redox state, with N-acetyl-L-cysteine (NAC) being of particular utility. NAC is one of several compounds that can be used to augment intracellular levels of glutathione, the major scavenger of reactive oxidative intermediates present in all eukaryotic forms of life [70]. Glutathione is generally required to protect cells against damage by oxidants and is able to reduce and thereby detoxify these potentially damaging chemical species. NAC enters cells readily and replenishes intracellular cysteine required to produce glutathione, thus leading to an increase in glutathione levels. NAC itself also may react with reactive oxidative intermediates (ROIs), thus directly protecting cells against these toxic compounds. This twofold action of NAC places this compound in a wholly different class from other ROI scavengers, such as superoxide dismutase, catalase, ascorbate, and α-tocopherol, none of which enhance intracellular production of glutathione. As a third potential mode of action relevant to its mucolytic activity, NAC is thought to act directly to reduce mucus glycoproteins to smaller subunits [71].

One of the utilities of NAC that has received considerable attention over the last decade is as an inhibitor of a variety of inducers on cell death, including both TNF-α [72,73] and glutamate. Application of NAC, and also of Vitamin C or Trolox (a water-soluble Vitamin E analog) to cultures of oligodendrocytes is effective at blocking cell death induced by exposure to TNF-α or glutamate [65]. The efficacy of NAC *in vivo* is considerable, ranging from preventing the death of rats injected with lethal doses of TNF-α (74) to enhancing neuronal survival following a moderate transient ischemic episode in the forebrain [75]. NAC is also efficacious at blocking the action of other inflammatory mediators, such as lipopolysaccharide [76], and is even able to block induction of autoimmune allergic encephalomyelitis, the most widely studied animal model of MS [77].

Along with being able to protect cells against "death by mugging" (e.g., exposure to cytotoxic agents), NAC is also able to help protect against "death by starvation," such as exposure of cells to suboptimal levels of trophic factors (65). For example, concentrations of CNTF or insulin-like growth factor-I (IGF-I) that have little or no effect on oligodendrocyte survival are associated with significant levels of survival if these cells are coexposed with 1mM NAC (which has no effect on cell survival when applied on its own [65]). Similar results are seen when suboptimal concentrations of nerve growth factor (NGF) are used to rescue spinal ganglion neurons, with coexposure to 1mM NAC being associated with a 13-fold increase in the number of surviving neurons. Such effects of NAC can also be seen *in vivo*. NAC administration has been found to rescue motor neurons from normal developmental cell death *in vivo* [78] and can also reduce acute retinal ganglion cell death following a lesion to the optic tectum [79] or superior colliculus [80].

IV. REDOX MODULATION OF THE FUNCTION OF OLIGODENDROCYTE PRECURSOR CELLS

Cytotoxic agents such as TNF-α cause oxidative stress as an important component of their activity. As we have discussed, pharmacological manipulation of redox state can be used to modify the action of these molecules and protect cells. Such analysis, however, treats cells as though redox state was a homogeneous property, with all cells being equal in this regard.

In the following sections, we will consider evidence demonstrating that it is overly simplistic to view intracellular redox state as a homogeneous state that affects all cells equally, at least during development. As we shall see, different precursor cell populations may differ considerably in their intracellular redox state, with profound implications for the way in which they respond to cell-extrinsic signaling molecules. Such findings will then be considered in the context of the heterogeneous response of CNS tissues to inflammatory cytokines.

Studies on the effects of redox modulation on the oligodendrocyte lineage revealed that intracellular redox state is a potent regulator of the critical precursor cell functions of division and differentiation [81]. In our initial studies on the relationship between intracellular redox state and precursor cell function, we found that there is a direct correlation between the redox state of precursor cells at the time of their isolation from the organism and their tendency to either divide or differentiate when induced to divide *in vitro* [81–83]. O-2A/OPCs that are more reduced *in vivo* undergo a great deal more self-renewal *in vitro* than those that are oxidized at the time of their isolation [81].

The ability of intracellular redox state to modulate the balance between self-renewal and differentiation in dividing progenitor cells appears to be utilized by the organism to modulate development in different regions of the CNS, as indicated by our comparative studies on O-2A/OPCs isolated from different CNS regions of the developing CNS (82). We initiated studies on O-2A/OPCs from different CNS regions in an attempt to understand the great variations seen in the timing of both neurogenesis and gliogenesis in different parts of the brain and spinal cord. For example, neuron production in the rat spinal cord is largely complete by the time of birth, is still ongoing in the rat cerebellum for at least several days after birth, and continues in the olfactory system and in some regions of the hippocampus of multiple species throughout life. Similarly, myelination has long been known to progress in a rostral–caudal direction, beginning in the spinal cord significantly earlier than in the brain (e.g., [84,86]). Even within a single CNS region, myelination is not synchronous. In the rat optic nerve, for example, myelinogenesis occurs with a retinal-to-chiasmal gradient, with regions of the nerve nearest the retina becoming myelinated first [84–87]. The cortex itself shows the widest range of timing for myelination, both initiating later than many other CNS regions (e.g., [84–86]) and exhibiting an ongoing myelinogenesis that can extend over long periods of time. This latter characteristic is seen perhaps most dramatically in the human brain, for which it has been suggested that myelination may not be complete until after several decades of life [88,89].

We now believe that differences in intracellular redox regulation, and associated regulation of the balance between self-renewal and differentiation, in O-2A/OPCs isolated from different CNS regions may play an important role in modulating the differing timing of oligoendrocyte generation and myelination in these regions [82]. In particular, cells isolated from optic nerve, optic chiasm, and cortex of identically aged rats show marked differences in their tendency to undergo self-renewing division and in their sensitivity to known inducers of oligodendrocyte generation. Precursor cells isolated from the cortex, a CNS region where myelination is a more protracted process than in the optic nerve, appear to be intrinsically more likely to begin generating oligodendrocytes at a later stage and over a longer time period than cells isolated from the optic nerve. For example, in conditions where optic nerve-derived O-2A/OPCs generated oligodendrocytes within 2 days, oligodendrocytes arose from chiasm-derived cells after 5 days and from cortical O-2A/OPCs only after 7–10 days. These differences, which appear to be cell-intrinsic, were manifested both in reduced percentages of clones producing oligodendrocytes and in a lesser representation of oligodendrocytes in individual clones. In addition, responsiveness of optic nerve-, chiasm-, and cortex-derived O-2A/OPCs to TH and ciliary neurotrophic factor (CNTF), well-characterized inducers of oligodendrocyte generation, was inversely related to the extent of self-renewal observed in basal division conditions. In precise accordance with our first studies on the effects of redox modulation on the balance between self-renewal and differentiation [81], we found that cortical O-2A/OPCs examined at the time of isolation from the developing rat are more reduced than those derived from the optic nerve [82].

The above results indicate that the O-2A/OPC population may be more complex than initially envisaged, with the properties of the precursor cells resident in any particular region being reflective of differing physiological requirements of the tissues to which these cell contribute. For example, as discussed earlier, a variety of experiments have indicated that the O-2A/OPC population of the optic nerve arises from a germinal zone located in or near the optic chiasm and enters the nerve by migration [90,91]. Thus, it would not be surprising if the progenitor cells of the optic chiasm expressed properties expected of cells at a potentially earlier developmental stage than those cells that are isolated from optic nerve of the same physiological age. Such properties would be expected to include the capacity to undergo a greater extent of self-renewal, much as has been seen when the properties of O-2A/OPCs from optic nerves of embryonic rats and postnatal rats have been compared [92]. In respect to the properties of cortical progenitor cells, physiological considerations also appear to be consistent with our observations. The cortex is one of the last regions of the CNS in which myelination is initiated, and the process of myelination also can continue for extended periods in this region [84–86]. If the biology of a precursor cell population is reflective of the developmental characteristics of the tissue in which it resides, then one might expect that O-2A/OPCs isolated from this tissue would not initiate oligodendrocyte generation until a later time than occurs with O-2A/OPCs isolated from structures in which myelination occurs earlier. In addition, cortical O-2A/OPCs might be physiologically required to make oligodendrocytes for a longer time due to the long period of continued development in this tissue, at least as this has been defined in the human CNS (e.g., [88,89]).

V. REDOX MEMORY

The above studies are discussed in a developmental context, as it is this context that currently provides our clearest understanding of a likely *in vivo* relevance, but possible relevance for the analysis of inflammation also needs to be considered. The idea that different precursor cells of the same lineage may have profound differences in their biological properties is a new concept that needs to be examined more broadly in respect to other lineages, to precursor cells of the adult CNS, and also in respect to differentiated cells. Those cells that are themselves more reduced would be expected to be more resistant to damage by inflammatory cytokines. As we shall discuss next, differences in redox state in different precursor cell populations may also be associated with differences in the vulnerability to damage of the differentiated cells generated from them.

Our analysis of the relationship between intracellular redox state and cellular function has recently begun to shed light on the puzzling heterogeneity of cellular responses that occurs in association with exposure to cytotoxic stimuli. Both *in vitro* and in chronic degenerative conditions, such as Parkinson's disease, pathological stimuli most frequently kill only a subset of putatively identical cells. Although virtually all members of a cellular population may succumb over prolonged periods of stress, or in association with catastrophic levels of exposure or injury, the most frequently observed situation is one in which only a proportion of a particular cell type is killed. Virtually nothing is known, however, about the biological underpinnings of differential vulnerability within a population of putatively identical cells. For example, it is not even known if vulnerability is randomly distributed in a population or whether, in contrast, clonally related cells exhibit similar levels of vulnerability. It has been suggested that there may be an association between vulnerability and the stage of the cell cycle a cell is in at the time of exposure. As differential vulnerability is also observed in populations of nondividing cells this cannot, however, be the sole explanation for the occurrence of differential levels of vulnerability.

One well-studied example of the heterogeneity of cellular vulnerability is provided by the response of oligodendrocytes, the myelin-forming cells of the central nervous system, to tumor necrosis factor-α (TNF-α). Multiple studies have revealed that oligodendrocytes are killed by TNF-α, a vulnerability thought to be of possible importance in the destruction of myelin that occurs in multiple sclerosis. Studies have indicated a correlation between the levels of TNF-α mRNA in acute MS lesions and the extent of demyelination and oligodendrocyte pathology (93). Yet, several studies reveal that a plateau of killing is observed at levels of TNF-α that kill only a subset of ~50% of oligodendrocytes and increases in TNF-α beyond this level are not associated with elimination of increasing numbers of cells (e.g., [65]). As this inability of increasing TNF levels to kill more than a subset of cells is seen in pure oligodendrocyte cultures [65], the heterogeneous response cannot be a consequence of a protective action of contaminating cells like astrocytes.

We have recently found that the vulnerability of oligodendrocytes to TNF-α-mediated killing is not randomly distributed but instead is similar among clonally related cells. When clonal cultures are exposed to TNF-α, there is a preferential vulnerability of oligodendrocytes found in small sized clones. Moreover, when we compared vulnerability of oligodendrocytes derived from tissues that generate clones of different sizes, we found that those derived from tissue that yielded smaller sized clones also generated oligodendrocytes that were more vulnerable to TNF-α-mediated killing.

Our earlier studies [82] demonstrated that clonal size is related to intracellular redox state, suggesting that a key determinant of the vulnerability of oligodendrocytes to TNF-α might be the redox state of the O2A-OPCs from which they are derived. That such a possibility is correct is indicated by findings that (1) oligodendrocytes generated from O-2A/OPCs isolated from corpus callosum are more resistant to TNF-α than those generated from optic nerve-derived O-2A/OPCs. The corpus callosum-derived O-2A/OPCs are more reduced than their optic nerve counterparts. (2) Purification of O-2A/OPCs on the basis of their intracellular redox state (as in [81]), followed by differentiation of these cells into oligodendrocytes, shows that the oligodendrocytes derived from the more oxidized progenitor cells are more vulnerable to killing by TNF-α. (3) Moreover, transient pretreatment of O-2A/OPCs with NAC prior to the induction of oligodendrocyte generation yields cells that are less vulnerable to TNF-α-mediated killing. (4) Transient pharmacological manipulation of O-2A/OPC intracellular redox state prior to oligodendrocyte generation was also associated with differences in the glutamate vulnerability of the oligodendrocytes generated from these cells, with oligodendrocytes derived from more reduced progenitors being less vulnerable to glutamate-mediated killing.

The above results provide the surprising new concept that the vulnerability of oligodendrocytes to killing by TNF-α and glutamate is modulated by the intracellular redox state of the progenitor cells from which they are derived. If the progenitor cell is relatively reduced, then the oligodendrocytes generated are resistant to killing by TNF-α. In contrast, relatively more oxidized progenitors give rise to TNF-α—sensitive oligodendrocytes. Such results were obtained with oligodendrocytes generated from cells that are intrinsically more reduced or oxidized *in vivo*, with oligodendrocytes generated from O-2A/OPCs in which intracellular redox state was altered pharmacologically prior to the induction of differentiation. This differential vulnerability was not restricted to TNF-α–sensitivity and seems to modulate vulnerability to glutamate.

To discuss these results, we introduce the concept of a heritable redox memory, by which we specifically mean a cellular behavior that is correlated with the intracellular redox state of the ancestor of the cell being analyzed. Such a heritable redox memory may be seen in progenitor cells and in differentiated cells. In previous studies we found that the intracellular redox state of a progenitor cell at the time it is isolated from the animal has predictable consequences regarding the behavior of the progenitors derived from that founder cell [81,82]. We found that those progenitors more reduced at the time of their isolation from the animal give rise to cells that continued to exhibit an enhanced tendency to undergo self-renewing division *in vitro* for several days to several weeks *in vitro*. This correlation was observed when we separated optic nerve-derived O-2A/OPCs into relatively more reduced or oxidized populations and when we compared cells from different regions of the CNS. Particularly striking results were seen when we compared the behavior of O-2A/OPCs derived from cortex and optic nerve of P7 rats. Cortex-derived progenitors exhibited a much greater tendency to undergo self-renewal than optic nerve-derived progenitors grown in identical conditions and were also more reduced than optic nerve-derived progenitors at the time of their isolation from the animal.

It is of particular interest to consider the potential relevance of the finding that cell populations that appear seemingly homogenous differ in their intracellular redox state to the heterogeneity of cellular responses to physiological stressors. Degenerative disorders routinely show a heterogeneous pattern of cell loss, and the basis for this is unknown. Could this be due to redox-related differences in the fundamental physiological properties of cells?

In addition to the heterogeneity of response within particular cellular populations, another problem to which the findings of a heritable redox memory may be relevant is that of delayed toxicity reactions. Exposure to such physiological stressors as inflammation during early development may be associated with a delayed loss of particular cell types, as seen for example in the association between fetal CNS inflammation and delayed loss of dopaminergic neurons in the sustantia nigra. As many environmental toxicants (including heavy metals and alcohol) are potent pro-oxidants, the present results raise the concern as to whether early exposure to such agents creates a heritage of increased vulnerability to other insults. Although it is admittedly difficult to imagine that the intracellular redox state of a precursor cell could have effects that are separated in time by long intervals, we nonetheless have been impressed by the remarkable persistence of redox-based memory. For example, cortical progenitors isolated from P7 rats continue to preferentially undergo self-renewing divisions for at least 6 weeks *in vitro* when grown in the presence of PDGF, whereas self-renewal of progenitors derived from optic nerves of the same animals rarely continues for more than 5–7 days in these conditions. This maintenance of a self-renewing phenotype was seen even though cells were grown in atmospheric oxygen concentrations, which are several times the oxygen concentration these cells would normally encounter *in vivo*. Despite growth in such a pro-oxidizing environment, however, the behavior of cells was correlated with the intracellular redox state of the founder cell of any particular clonal family. Maintenance of a heritable redox memory in such conditions raises the possibility that such a memory may be surprisingly stable. A prediction, to be tested in future experiments, is that oxidative stress during early development is associated with a subsequent increase in vulnerability to different physiological stressors.

VI. INTRACELLULAR REDOX STATE MODIFIES THE ACTION OF CELL SIGNALING PATHWAYS

The remarkable effects of intracellular redox states on precursor cell division and differentiation, and on oligodendrocyte survival, appear to be explained by a relatively simple hypothesis by which the redox state of a cell is able to selectively amplify the activities of some signaling pathways and dampen the activities of others. Such a hypothesis appears to be well supported by at least some data. Our own studies demonstrated that increasing intracellular glutathione content by as little as 10–15% is associated with a greatly enhanced response of oligodendrocytes and neurons to survival factors [65]. Studies by others have yielded complementary results. Reducing the glutathione content of lymphocytes (for example, by growth in the presence of buthionine sulfoximine, which inhibits glutathione production) reduces their response to mitogens and other activators [94–96]. Indeed, as little as a 10% decrease in average glutathione levels significantly decreases calcium influx in peripheral blood lymphocytes stimulated with anti-CD3 antibody [96]. When we analyzed intracellular redox state with a dye that appears to offer a broad indicator of redox state, we also found that changes in the range of 10–15% in average fluorescence values were associated with marked changes in responsiveness to inducers of division or differentiation.

For the cell types that we have studied, there is a remarkable predictability to the effects of redox modulation on signaling pathway function. Rendering cells more reduced appears to increase the response to inducers of cell survival or cell division, while dampening the response to inducers of death or differentiation. In contrast, cells that are more oxidized exhibit a reduced response to survival factors or mitogens, while exhibiting an enhanced response to inducers of differentiation or death.

That redox modulation plays a fundamental role in controlling normal cell behavior is indicated by the effects of cell signaling molecules on this physiological parameter, as well as by our findings that the redox state of cells at the time of their isolation from the animal allows prediction of their behavior in multiple functional assay. Many different signaling pathways appear to converge on regulation of redox state, and redox alterations can in turn modulate several different pathways of possible relevance in modulation of self-renewal and differentiation. Multiple components of the redox regulatory network can be altered by exposure of a cell to such cell-extrinsic signaling molecules as neurotrophins [97,98], type 1 interferon [99], stem cell factor [100], TGF-ß [101], inflammatory cytokines (e.g., [102]), and TH [103], and also by ras activation [104]. Our research thus far indicates that the direction of the redox changes induced by various signaling molecules is highly consistent with our other studies. We found that growth factors that enhance self-renewal (i.e., NT-3, FGF-2) made O-2A/OPCs more reduced, wheras those that promote differentiation (i.e., TH, bone morphogenetic proteins) make cells more oxidized. Moreover these changes in intracellular redox state appear to be essential components of the means by which such factors alter the balance between self-renewal and differentiation [81], and pharmacological interference with the redox alterations induced by these signaling molecules effectively abrogated their effects.

The above simple hypothesis suggests that modulation of intracellular redox state by signaling molecules normally encountered in the environment may have profound effects for understanding the biology of inflammation. Multiple signals encountered in the inflammatory environment (e.g., TNF-α, glutamate) render cells more oxidized. Such an increased oxidative state may render cells less responsive to the very signals (mitogens and survival factors) that are produced as a component of regenerative processes.

As we discuss in the following section, however, it appears that the complex situation of inflammation is far from being understood, even at the level of interaction between two of the signaling molecules most commonly expressed in such situations.

VII. EFFECTS OF VIRAL INFECTION ON GLIAL PRECURSOR CELL FUNCTION

Inflammatory responses are mounted in response to many adverse events in the CNS, including infection with viruses or bacteria, acute injury, and many syndromes of chronic degeneration. Despite their diverse origins, these inflammatory responses share a number of common features, including the cells involved in mounting the inflammatory response and the types of cytokines that are expressed.

There is one inducer of an inflammatory response that merits special attention, however, due to its ability to directly alter the genetic machinery of the affected cells, this being viral infection. The ability of viruses to infect cells and subvert their normal functions to the benefit of the virus places this particular aspect of inflammation in a somewhat special class, as befits an interplay that must represent one of the most ancient, if not the most ancient, battles in all of biology.

Virus infections in the CNS are not uncommon and have been the subject of a great deal of attention. Such studies have revealed that a viral infection can mobilize an inflammatory response that actually enhances viral replication. Moreover, viral proteins can not only subvert functions of the infect cells, but can mimic naturally occurring proteins. In some cases, this mimicry extends to proteins involved in modulating the inflammatory response, while in other cases resemblance to host proteins can produce an autoimmune response to host tissues.

One of the most common inducers of an inflammatory response in the CNS is infection with one or more of the human herpesviruses. An increasingly varied and complex array of inflammatory and other mechanisms of viral pathogenesis have been recently described, many of which involve the production of chemokines, predominantly by infected microglia and astrocytes. For example, both TNF-α and IFN-γ, discussed earlier for their potent effects in the oligodendrocyte lineage, as well as IL-6 are produced in the mouse CNS after corneal inoculation with HSV-1 [105]. Similarly, infection of lymphocytes with the related human herpesvirus 6 (HHV-6) has been shown to induce expression of a variety of TNF-associated factors and receptors as well as the IFN-γ activating factor, IL-18 [106]. Other studies have demonstrated production of these same cytokines together with IL-1β and the β-chemokines RANTES, MIP-1α, MIP-1β, and MCP-1 in microglia or astrocytes infected with either human cytomegalovirus or HSV-1 [107]. In addition to the potentially toxic effects of proinflammatory cytokines discussed in previous sections, effects on cell fate and differentiation also seem likely; IL-6, in particular, appears to interfere with the production of new neurons from neural stem cells by biasing their differentiation towards astrocyte generation [26].

Along with inducing an inflammatory response, viral gene products themselves may mimic host cytokines, cytokine receptors, or cell-signaling molecules [108]. The HHV-6 genome, for example, encodes several chemokine mimics, some of which appear to promote mononuclear cell chemotaxis (i.e., the U83 gene product), while others (i.e., U51) down-modulate host RANTES expression and mononuclear cell responses. These seemingly paradoxical mechanisms are carefully orchestrated, however, to alternatively incite and evade host inflammatory responses permitting, respectively, viral dissemination or persistence.

Viral genes also cause inflammatory problems in other ways, by stimulating inappropriate autoimmune responses. For example, molecular mimicry herpesvirus peptide/MHC complexes can promote pathological cross-reactive inflammatory responses to normal neuroglial cells or their component proteins. This has been recently demonstrated in the cross-recognition, by a T-cell line from patients with multiple sclerosis, of an Epstein–Barr virus peptide and one derived from myelin basic protein (MBP) [109]. Similar cross-recognition has been demonstrated for HHV-6 and MBP peptides (110). Finally Herpesvirus DNA or dsRNA may rapidly trigger CNS Toll-like receptor innate inflammatory responses even in the absence of replicating virus or viral proteins.

It has long been known that viruses alter cellular function through genetic mechanisms, but it is becoming increasingly apparent that such physiological parameters as redox modulation are also altered over relatively short time courses. Herpesvirus infections have also been shown to alter the intrinsic redox state of the cell within 10–30 minutes after infection, an event both necessary for viral replication and with profound consequences for cell fate and differentiation decisions as discussed earlier in this chapter. Recent work with human CMV in arterial smooth muscle cells, for example, has demonstrated the rapid generation of intracellular ROS which activate NF-κB which, in turn, mediates both expression of the CMV IE gene promoter and host genes involved in immune and inflammatory responses such as COX-2. Pretreatment with NAC profoundly reduced viral replication and the inflammatory response. Similar studies examining HSV infection of macrophages have demonstrated virus-mediated triggering of NF-κB by a signaling pathway that is dependent upon mitochondrial generation of ROS [111–115].

An ability of viral infection to generate ROS in cells may itself amplify the inflammatory response. As many viruses have evolved to benefit from an inflammatory response, with control of expression of their early genes often being regulated by such components of the inflammatory response as NF-κB activation, this will itself contribute to viral perpetuation.

At the genetic level, some of the most profound modifications of cellular function may occur in the case of nonlytic infection. Although maintenance of nonlytic viral infection in the brain has been generally considered asymptomatic and without clinical consequence, it is clear that this view is not correct. It has long been known that viral infection can alter cell functions such as myelin formation without accompanying cell death [116,117]; more recently persistent, nonlytic infection with lymphocytic choriomeningitis virus was shown to render immature Schwann cells defective in myelin formation in the absence of accompanying cytopathology [118].

Although all of the herpesviruses can cause neuroglial cell death by lytic infection, as occurs in HSV-1 encephalitis or CMV encephalitis in the immunocompromised host, CNS latency and persistence is the more common outcome of infection with these agents. CNS infection with HHV-6 can indeed cause leukoencephalitis in both immunocompetent and immunocompromised patients accompanied by demyelinative lesions that fail to repair [119–121]. More typically, however, CNS infections by HHV-6 lead to a putatively benign viral latency. Indeed, the development of PCR-based detection techniques has lead to the recognition that members of the human herpesvirus family, in particular, are frequent residents of the human brain, occurring in up to 70% of human brain tissues at autopsy in the case of HHV-6 [122]. HHV-6 is known to infect both primary human oligodendrocytes *in vitro* and *in vivo* [123,124] as well as human astrocytes [125], but the effects of HHV-6 virus infection on glial precursor cell functions have only recently begun to be explored [126].

The degree to which latent infections are truly benign needs to be questioned, particularly in the context of repair of demyelinating damage. Studies of the interactions between common resident herpesviruses and oligodendrocyte precursors (OPCs) are of particular interest in the context of demyelinating diseases because of an accumulating consensus that most, if not all, remyelinating cells originate from this precursor cell pool [50,127,128]. Although normal numbers of OPCs are frequently present within early demyelinating lesions, cells within the precursor pool appear to remain quiescent and unable to appropriately proliferate or differentiate [129–133]. The exact mechanisms that contribute to this repair failure remain largely speculative although, as discussed below, viral infections may prove to be relevant in this regard.

Recent studies from our laboratories have demonstrated that HHV can cause latent infection in human glial precursor cells. Human glial precursors were found to express surface CD46, the major cell surface receptor for HHV-6 [126]. HHV-6 infection of glial precursors resulted in early productive infection as evidenced by the demonstration of late viral transcripts by RT-PCR, the expression of viral structural proteins by immunocytochemistry, and the identification of characteristic herpesvirus virions within infected glial precursors by electron microscopy. Interestingly, direct infection of the precursor population with virus was not accompanied by increased cell death compared to uninfected control cells. If anything, there was a weak trend toward enhanced survival in infected cells.

One of the most striking consequences of HHV-6 infection of human glial precursor cells is to suppress cell division, as measured by decreased uptake of BrdU and Ki-67 immunostaining and a suppression of cell growth over time compared to uninfected controls. Subsequent flow cytometric studies identified a specific G1/S-phase arrest with both A and B strains of HHV-6. FAC-sorted HHV-6A or B-infected precursors also displayed a highly significant loss of the self-renewing glial precursor cellular pool, as assayed by expression of the glial precursor cell marker A2B5. The loss of A2B5 was accompanied by a corresponding induction of galactocerebroside-expressing oligodendrocytes. Thus, virus-mediated suppression of human glial precursor cell proliferation and cell cycle arrest appears to be accompanied by loss of the self-renewing glial progenitor pool and an increase in differentiated oligodendroglial cell types. The inhibition of proliferation and cell-cycle arrest observed in our studies are consistent with the previously reported ability of human herpesviruses to cause cell cycle arrest in other cell types [134].

Other members of the herpesvirus family also appear to compromise precursor cell function. For example, studies using murine CMV (MCMV) have provided a valuable animal model for studying the effects of herpesvirus infections in the CNS and have also demonstrated profound effects on both proliferation and differentiation of CNS stem and progenitor cells [135]. In these studies single CNS stem cells prepared from the brains of fetal mice proliferated and formed floating neurospheres while infected cells developed productive infection, cytopathic effect, and the beginnings of lytic cell death by 5 days postinfection. Within hours of infection, at a time prior to notable cytopathic effects, a significant decrease in stem cell DNA synthesis was noted as measured by reduced uptake of BrDU. MCMV infection also caused a decrease in the number of neurospheres able to regenerate in culture and inhibited neuronal differentiation more profoundly than glial differentiation as assayed by staining for expression of markers of neuronal or astrocytic development, as compared with uninfected controls. In addition, MCMV infection inhibited the migration of CNS stem cells after transplantation into neonatal rat brain, providing still another means by which precursor cell function can be compromised by viral infection.

Studies on MCMV also indicate that infection with at least some members of the herpesvirus family may also target astrocytes and stem cells of the subventricular zone. Studies using transgenic mouse models expressing the MCMV immediate early (IE) enhancer/promoter connected to a LacZ reporter showed astrocyte-specific expression of the transgene in the brain and in primary glial cultures from these animals. No expression in neurons, oligodendroglia, microglia, or endothelial cells was observed [136]. In a subsequent study, activation of the MCMV-IE promoter was found to be restricted to endothelial cells early in gestation, whereas in late gestation it was found to be largely restricted to the subventricular zone in a pattern similar to expression of Musashi 1, a marker expressed in neural progenitor and stem cells. In cultures of progenitor cells differentiated from neural stem cells, expression of the transgene was detected in glial progenitor cells expressing GFAP, nestin, and Musashi 1 but not in neuronal progenitor cells expressing MAP2 [137].

The effects of viral infection on precursor cell function demonstrate that even in the absence of any inflammatory response, these potent inducers of inflammation are themselves capable of altering cellular function in ways that may severely compromise repair processes. Of particular concern to us are the effects of infection with HHV-6, which is the most abundant herpesvirus in the human brain. In particular, disruption of precursor cell proliferation due to HHV-6 infection might limit the availability of precursor cells for recruitment to the site of injury. An additional contribution to the failure of remyelination might also be inappropriate differentiation of precursor cells that are already present at the lesion site, thus also contributing to a limitation of cell expansion. In fact, our results suggest that HHV-6 infection of precursor cells leads to increased differentiation of precursor cells towards GalC$^+$ oligodendrocytes compared to noninfected controls. These prematurely differentiated oligodendrocytes may be more vulnerable to the inflammatory environment generated by viral infection than precursor cells. One of the major inflammatory cytokines, TNF-α, has been described in demyelinated lesions [93,138] and, as also discussed earlier, is highly toxic to oligodendrocytes *in vivo* and *in vitro* [139–142]. Such concerns may also extend to other members of the HHV family, as we also have observed that a significant number of glial precursor cells express CD21, a known receptor for the γ-human herpesvirus Epstein–Barr [143]. The presence of the viral receptor CD21on precursor cells raises the interesting possibility that effects on glial progenitors may extend to other CNS resident human herpesviruses besides HHV-6.

REFERENCES

1. Schwartz, M., 2001. Harnessing the immune system for neuroprotection: therapeutic vaccines for acute and chronic neurodegenerative disorders. *Cell. Mol. Neurobiol.* 21:617–627.
2. Raff, M.C., Miller, R.H., and Noble, M.,1983. A glial progenitor cell that develops *in vitro* into an astrocyte or an oligodendrocyte depending on the culture medium. *Nature* 303:390–396.
3. Hughes, S.M., Lillien, L.E., Raff, M.C., Rohrer, H., and Sendtner, M.,1988. Ciliary neurotrophic factor induces type-2 astrocyte differentiation in culture. *Nature* 335:70–73.

4. Lillien, L.E. and Raff, M.C., 1990. Differentiation signals in the CNS: type-2 astrocyte development *in vitro* as a model system. *Neuron* 5:5896–6273.

5. Knapp, P.E., 1991. Studies of glial lineage and proliferation *in vitro* using an early marker for committed oligodendrocytes. *J. Neurosci. Res.* 30:336–345.

6. Espinosa de los Monteros, A., Zhang, M., and De Vellis, J., 1993. O2A progenitor cells transplanted into the neonatal rat brain develop into oligodendrocytes but not astrocytes. *Proc Natl Acad Sci U.S.A.* 90:50–54.

7. Noble, M., Murray, K., Stroobant, P., Waterfield, M.D., and Riddle, P., 1988. Platelet-derived growth factor promotes division and motility and inhibits premature differentiation of the oligodendrocyte/type-2 astrocyte progenitor cell. *Nature* 333:560–562.

8. Richardson, W.D., Pringle, N., Mosley, M., Westermark, B., and Dubois-Dalcq, M., 1988. A role for platelet-derived growth factor in normal gliogenesis in the central nervous system. *Cell* 53:309–319.

9. Yeh, J.J., Ruit, K.G., Wang, Y.X., Parks, W.C., Snider, W.D., and Deuel, T.F., 1991. PDGF A-chain is expressed by mammalian neurons during development and in maturity. *Cell* 64:209–219.

10. Hutchins, J.B. and Jefferson, V.E., 1992. Developmental distribution of platelet-derived growth factor in the mouse central nervous system. *Brain Res. Dev. Brain Res.* 67:121–135.

11. Calver, A., Hall, A., Yu, W., Walsh, F., Heath, J., Betsholtz, C., and Richardson, W., 1998. Oligodendrocyte population dynamics and the role of PDGF *in vivo*. *Neuron* 20:869–882.

12. Fruttiger, M., Karlsson, L., Hall, A., Abramsson, A., Calver, A., Bostrom, H., Willetts, K., Bertold, C., Heath, J., Betsholtz, C., and Richardson, W., 1999. Defective oligodendrocyte development and severe hypomyelination in PDGF-A knockout mice. *Development* 126:457–467.

13. Bogler, O., Wren, D., Barnett, S.C., Land, H., and Noble, M., 1990. Cooperation between two growth factors promotes extended self-renewal and inhibits differentiation of oligodendrocyte-type-2 asytocytes (O-2A) progenitor cells. *Proc. Natl. Acad. Sci, U.S.A.* 87:6368–6372.

14. Ibarrola, N., Mayer-Proschel, M., Rodriguez-Pena, A., and Noble, M., 1996. Evidence for the existence of at least two timing mechanisms that contribute to oligodendrocyte generation *in vitro*. *Dev. Biol.* 180:1–21.

15. Walters, S.N. and Morell, P., 1981. Effects of altered thyroid states on myelinogenesis. *J. Neurochem.* 36:1792–1801.

16. Siragusa, V., Boffelli, S., Weber, G., Triulzi, F., Orezzi, S., Scotti, G., and Chiumello, G., 1997. Brain magnetic resonance imaging in congenital hypothyroid infants at diagnosis. *Thyroid*: 761–764.

17. Ibarrola, N. and Rodriguez-Pena, A., 1997. Hypothyroidism coordinately and transiently affects myelin protein gene expression in most rat brain regions during postnatal development. *Brain Res.* 752:285–293.

18. Rodriguez-Pena, A., 1999. Oligodendrocyte development and thyroid hormone. *J. Neurobiol.* 40:497–512.

19. Jagannathan, N.R., Tandon, N., Raghunathan, P., and Kochupillai, N., 1998. Reversal of abnormalities of myelination by thyroxine therapy in congenital hypothyroidism: localized *in vivo* proton magnetic resonance spectroscopy (MRS) study. *Brain Res. Dev. Brain Res.* 109:179–186.

20. McKinnon, R.D., Piras, G., Ida, J.A., Jr., and Dubois Dalcq, M., 1993. A role for TGF-beta in oligodendrocyte differentiation. *J. Cell Biol.* 121:1397–1407.

21. Louis, J.C., Muir, D., and Varon, S., 1992. Autocrine inhibition of mitotic activity in cultured oligodendrocyte-type-2 astrocyte (O-2A) precursor cells. *Glia* 6:30–38.

22. Louis, J.C., Magal, E., Takayama, S., and Varon, S., 1993. CNTF protection of oligodendrocytes against natural and tumor necrosis factor-induced death. *Science* 259:689–692.

23. Mayer, M., Bhakoo, K., and Noble, M., 1994. Ciliary neurotrophic factor and leukemia inhibitory factor promote the generation, maturation and survival of oligodendrocytes *in vitro*. *Development* 120:142–153.

24. Barres, B.A., Schmidt, R., Sendnter, M., and Raff, M.C., 1993. Multiple extracellular signals are required for long-term oligodendrocyte survival. *Development* 118:283–295.

25. Johe, K.K., Hazel, T.G., Muller, T., Dugich-Djordjevic, M.M., and McKay, R.D., 1996. Single factors direct the differentiation of stem cells from the fetal and adult central nervous system. *Genes Dev* 10:3129–3140.

26. Bonni, A., Sun, Y., Nadal-Vicens, M., Bhatt, A., Frank, D.A., Rozovsky, I., Stahl, N., Yancopoulos, G.D., and Greenberg, M.E., 1997. Regulation of gliogenesis in the central nervous system by the JAK-STAT signaling pathway. *Science* 278: 477–483.

27. Rao, M., Noble, M., and Mayer-Pröschel, M., 1998. A tripotential glial precursor cell is present in the developing spinal cord. *Proc. Natl. Acad. Sci. U.S.A.* 95:3996–4001.

28. Mi, H. and Barres, B.A., 1999. Purification and characterization of astrocyte precursor cells in the developing rat optic nerve. *J. Neurosci.* 19:1049–1061.

29. Park, J.K., Williams, B.P., Alberta, J.A., and Stiles, C.D., 1999. Bipotent cortical progenitor cells process conflicting cues for neurons and glia in a hierarchical manner. *J. Neurosci.* 19:10383–10389.

30. Aberg, M.A., Ryttsen, F., Hellgren, G., Lindell, K., Rosengren, L.E., MacLennan, A.J., Carlsson, B., Orwar, O., and Eriksson, P.S., 2001. Selective introduction of antisense oligonucleotides into single adult CNS progenitor cells using electroporation demonstrates the requirement of STAT3 activation for CNTF-induced gliogenesis. *Mol. Cell. Neurosci.* 17:426–443.

31. Lillien, L.E., Sendtner, M., and Raff, M.C., 1990. Extracellular matrix-associated molecules collaborate with ciliary neurotrophic factor to induce type-2 astrocyte development. *J. Cell. Biol.* 111:635–644.

32. Herrera, J., Yang, H., Zhang, S.C., Proschel, C., Tresco, P., Duncan, I.D., Luskin, M., and Mayer-Proschel, M., 2001. Embryonic-derived glial-restricted precursor cells (GRP cells) can differentiate into astrocytes and oligodendrocytes *in vivo*. *Exp. Neurol.* 171:11–21.

33. Mayer-Pröschel, M., Kalyani, A., Mujtaba, T., and Rao, M.S., 1997. Isolation of lineage-restricted neuronal precursors from multipotent neuroepithelial stem cells. *Neuron* 19:773–785.

34. Liu, Y., Wu, Y., Lee, J.C., Xue, H., Pevny, L.H., Kaprielian, Z., and Rao, M.S., 2002. Oligodendrocyte and astrocyte development in rodents: an *in situ* and immunohistological analysis during embryonic development. *Glia* 40:25–43.

35. Gregori, N., Proschel, C., Noble, M., and Mayer-Pröschel, M., 2002. The tripotential glial-restricted precursor (GRP) cell and glial development in the spinal cord: generation of bipotential oligodendrocyte-type-2 astrocyte progenitor cells and dorsal-ventral differences in GRP cell function. *J. Neurosci.* 22:248–256.

36. Franklin, R.J. and Blakemore, W.F., 1995. Glial-cell transplantation and plasticity in the O-2A lineage—implications for CNS repair. *Trends Neurosci.* 18:151–156.

37. Prineas, J.W. and Graham, J.S., 1981. Multiple sclerosis: capping of surface immunoglobulin G on macrophages involved in myelin breakdown. *Ann. Neurol.* 10:149–158.

38. Gay, F.W., Drye, T.J., Dick, G.W., and Esiri, M.M., 1997. The application of multifactorial cluster analysis in the staging of plaques in early multiple sclerosis. Identification and characterization of primary demyelinating lesions. *Brain* 120:1461–1483.

39. Barnum, S.R., 2002. Complement in central nervous system inflammation. *Immunol. Res.* 26:7–13.

40. Genain, C.P., Cannella, B., Hauser, S.L., and Raine, C.S., 1999. Autoantibodies to MOG mediate myelin damage in MS. *Nat. Med.* 5:170–175.

41. Tuohy, V.K., Yu, M., Yin, L., Kawczak, J.A., Johnson, J.M., Mathisen, P.M., Weinstock-Guttman, B., and Kinkel, R.P., 1998. The epitope spreading cascade during progression of experimental autoimmune encephalomyelitis and multiple sclerosis. *Immunol. Rev.* 164:93–100.

42. Goebels, N., Hofstetter, H., Schmidt, S., Brunner, C., Wekerle, H., and Hohlfeld, R., 2000. Repertoire dynamics of autoreactive T cells in multiple sclerosis patients and healthy subjects: epitope spreading versus clonal persistence. *Brain* 123:508–518.

43. Tuohy, V.K. and Kinkel, R.P., 2000. Epitope spreading: a mechanism for progression of autoimmune disease. *Arch. Immunol. Ther. Exp. (Warsz.)* 48:347–351.

44. Vanderlugt, C.L. and Miller, S.D., 2002. Epitope spreading in immune-mediated diseases: implications for immunotherapy. *Nat. Rev. Immunol.* 2:85–95.

45. Lassmann, H., 1999. The pathology of multiple sclerosis and its evolution. *Phil. Trans. R. Soc. Lond. B* 354:1635–1640.

46. Lassmann, H., Bruck, W., and Lucchinetti, C., 2001. Heterogeneity of multiple sclerosis pathogenesis: implications for diagnosis and therapy. *Trends Mol. Med.* 7:115–121.

47. Hemmer, B., Kieseier, B., Cepok, S., and Hartung, H.P., 2003. New immunopathologic insights into multiple sclerosis. *Curr. Neurol. Neurosci. Rep.* 3:246–255.

48. Neuhaus, O., Archelos, J.J., and Hartung, H.P., 2003. Immunomodulation in multiple sclerosis: from immunosuppression to neuroprotection. *Trends Pharmacol. Sci.* 24:131–138.

49. Noseworthy, J.H., 2003. Treatment of multiple sclerosis and related disorders: what's new in the past 2 years? *Clin. Neuropharmacol.* 26:28–37.

50. Bruck, W., Kuhlmann, T., and Stadelmann, C., 2003. Remyelination in multiple sclerosis. *J. Neurol. Sci.* 206:181–185.

51. Galetta, S.L., Markowitz, C., and Lee, A.G., 2002. Immunomodulatory agents for the treatment of relapsing multiple sclerosis: a systematic review. *Arch. Intern. Med.* 162:2161–2169.

52. Waxman, S.G., 2002. Ion channels and neuronal dysfunction in multiple sclerosis. *Arch. Neurol.* 59:1377–1380.

53. Li, S. and Stys, P.K., 2000. Mechanisms of ionotropic glutamate receptor-mediated excitotoxicity in isolated spinal cord white matter. *J. Neurosci.* 20:1190–1198.

54. Matute, C., Sanchez-Gomez, M.V., Martinez-Millan, L., and Miledi, R., 1997. Glutamate receptor-mediated toxicity in optic nerve oligodendrocytes. *Proc. Natl. Acad. Sci. U.S.A.* 94:8830–8835.

55. Matute, C., 1998. Properties of acute and chronic kainate excitotoxic damage to the optic nerve. *Proc. Natl. Acad. Sci. U.S.A.* 95:10229–10234.

56. McDonald, J.W., Althomsons, S.P., Hyrc, K.L., Choi, D.W., and Goldberg, M.P., 1998. Oligodendrocytes from forebrain are highly vulnerable to AMPA/kainate receptor-mediated excitotoxicity. *Nat. Med.* 4:291–297.

57. Fern, R. and Möller, T., 2000. Rapid ischemic cell death in immature oligodendrocytes: a fatal glutamate release feedback loop. *J. Neurosci.* 20:34–42.

58. Lee, J.-M. et al., 1999. The changing landscape of ischaemic brain injury mechanisms. *Nature* 399:A7–A14.

59. Choi, D.W., 1988. Calcium-mediated neurotoxicity: relationship to specific channel types and role in ischemic damage. *Trends Neurosci.* 11:465–469.

60. Lyons, S.A. and Kettenmann, H., 1998. Oligodendrocytes and microglia are selectively vulnerable to combined hypoxia and hypoglycemia injury *in vitro. J. Cereb. Blood Flow Metab. Brain Dis.* 18:521–530.

61. Kinney, H.C. and Armstrong, D.D., 1997. Perinatal neuropathology. In *Greenfield's neuropathology.* D.I. Graham and P.L. Lantos, editors. Arnold.

62. Stover, J.F. et al., 1997. Neurotransmitters in cerebrospinal fluid reflect pathological activity. *Eur. J. Clin. Invest.* 27:1038–1043.

63. Barkhatova, V.P. et al., 1998. Changes in neurotransmitters in multiple sclerosis. *Neurosci. Behav. Physiol.* 28:341–344.

64. Butt, A.M. and Jenkins, H.G., 1994. Morphological changes in oligodendrocytes in the intact mouse optic nerve following intravitreal injection of tumour necrosis factor. *J. Neuroimmunol.* 51:27–33.

65. Mayer, M. and Noble, M., 1994. N-acetyl-L-cysteine is a pluripotent protector against cell death and enhancer of trophic factor-mediated cell survival *in vitro. Proc. Natl. Acad. Sci. U.S.A.* 91:7496–7500.

66. Piani, D. et al., 1991. Murine macrophages induce NMDA receptor mediated neurotoxicity *in vitro* by secreting glutamate. *Neurosci. Lett.* 133:159–162.

67. Noda, M. et al., 1999. Glutamate release from microglia via glutamate transporter is enhanced by amyloid-£] peptide. *Neuroscience* 92:1465–1474.

68. Beattie, M.S., Harrington, A.W., Lee, R., Kim, J.Y., Boyce, S.L., Longo, F.M., Bresnahan, J.C., Hempstead, B.L., and Yoon, S.O., 2002. ProNGF induces p75-mediated death of oligodendrocytes following spinal cord injury. *Neuron* 36:375–386.

69. Hermann, G.E., Rogers, R.C., Bresnahan, J.C., and Beattie, M.S., 2001. Tumor necrosis factor-α induces cFOS and strongly potentiates glutamate-mediated cell death in the rat spinal cord. *Neurobiol. Dis.* 8:590–599.

70. Meister, A., Anderson, M.E., and Hwang, O., 1986. Intracellular cysteine and glutathione delivery systems. *J. Am. Coll. Nutr.* 5:137–151.

71. Livingstone, C.R., Andrews, M.A., Jenkins, S.M., and Marriott, C., 1990. Model systems for the evaluation of mucolytic drugs: acetylcysteine and S-carboxymethyl-cysteine. *J. Pharm. Pharmacol.* 42:73–78.

72. Klebanoff, S.J., Vadas, M.A., Harlan, J.M., Sparks, L.H., Gamble, J.R., Agosti, J.M., and Waltersdorph, A.M., 1986. Stimulation of neutrophils by tumor necrosis factor. *J. Immunol.* 136:4220.

73. Kapp, A., Zeck-Kapp, G., and Blohm, D., 1989. Human tumor necrosis factor is a potent activator of the oxidative metabolism in human polymorphonuclear neutrophilic granulocytes: comparison with human lymphotoxin. *J. Invest. Dermatol.* 92:348.

74. Zimmerman, R.J., Marafino, B.J., Chan, Jr., A., Landre, P., and Winkelhake, J.L., 1989. The role of oxidant injury in tumor cell sensitivity to recombinant human tumor necrosis factor *in vivo.* Implications for mechanisms of action. *J. Immunol.* 142:1405.

75. Knuckey, N.W., Palm, D., Primiano, M., Epstein, M.H., and Johanson, C.E., 1995. N-acetylcysteine enhances hippocampal neuronal survival after transient forebrain ischemia in rats. *Stroke* 26:305–311.

76. Schreck, R. and Baeuerle., 1991. A role for oxygen radicals as second messengers. *Trends Cell Biol.* 1:39–42.
77. Lehmann, D., Karussis, D., Misrachi-Koll, R., Shezen, E., Ovadia, H., and Abramsky, O., 1994. Oral administration of the oxidant-scavenger N-acetyl-L-cysteine inhibits acute experimental autoimmune encephalomyelitis. *J. Neuroimmunol.* 50:35–42.
78. Caldero, J., Prevette, D., Mei, X., Oakley, R.A., Li, L., Milligan, C., Houenou, L., Burek, M., and Oppenheim, R.W., 1998. Peripheral target regulation of the development and survival of spinal sensory and motor neurons in the chick embryo. *J. Neurosci.* 18:356–370.
79. Castagne, V. and Clarke, P.G., 1996. Axotomy-induced retinal ganglion cell death in development: its time-course and its diminution by antioxidants. *Proc. R. Soc. Lond. B. Biol. Sci.* 263:1193–1197.
80. Cui, Q. and Harvey, A.R., 1995. At least two mechanisms are involved in the death of retinal ganglion cells following target ablation in neonatal rats. *J. Neurosci.* 15:8143–8155.
81. Smith, J., Ladi, E., Mayer-Pröschel, M., and Noble, M., 2000. Redox state is a central modulator of the balance between self-renewal and differentiation in a dividing glial precursor cell. *Proc. Natl. Acad. Sci. U.S.A.* 97:10032–10037.
82. Power, J., Mayer-Proschel, M., Smith, J., and Noble, M., 2002. Oligodendrocyte precursor cells from different brain regions express divergent properties consistent with the differing time courses of myelination in these regions. *Dev. Biol.* 245:362–375.
83. Noble, M., Smith, J., Power, J., and Mayer-Pröschel, M., 2003. Redox state as a central modulator of precursor cell function. *Ann. N.Y. Acad. Sci.* 991:251–257.
84. Foran, D.R. and Peterson, A.C., 1992. Myelin acquisition in the central nervous system of the mouse revealed by an MBP-LacZ transgene. *J. Neurosci.* 12:4890–4897.
85. Kinney, H.C., Brody, B.A., Kloman, A.S., and Gilles, F.H., 1988. Sequence of central nervous system myelination in human infancy. II. Patterns of myelination in autopsied infants. *J. Neuropath. Exp. Neurol.* 47:217–234.
86. Macklin, W.B. and Weill, C.L., 1985. Appearance of myelin proteins during development in the chick central nervous system. *Dev. Neurosci.* 7:170–178.
87. Skoff, R.P., Toland, D., and Nast, E., 1980. Pattern of myelination and distribution of neuroglial cells along the developing optic system of the rat and rabbit. *J. Comp. Neurol.*, 191:237–253.
88. Benes, F.M., Turtle, M., Khan, Y., and Farol, P., 1994. Myelination of a key relay zone in the hippocampal formation occurs in the human brain during childhood, adolescence and adulthood. *Arch. Gen. Psychiat.* 51:477–484.
89. Yakovlev, P.L. and Lecours, A.R., 1967. The myelogenetic cycles of regional maturation of the brain. In *Regional development of the brain in early life.* A. Minkowski et al., editors. Blackwell, Oxford. 3–70.
90. Small, R.K., Riddle, P., and Noble, M., 1987. Evidence for migration of oligodendrocyte-type-2 astrocyte progenitor cells into the developing rat optic nerve. *Nature* 328:155–157.
91. Ono, K., Bansal, R., Payne, J., Rutishauser, U., and Miller, R.H., 1995. Early development and dispersal of oligodendrocyte precursors in the embryonic chick spinal cord. *Development* 121:1743–1754.
92. Gao, F. and Raff, M., 1997. Cell size control and a cell-intrinsic maturation program in proliferating oligodendrocyte precursor cells. *J. Cell Biol.* 138:1367–1377.
93. Bitsch, A., Kuhlman, T., Da Costa, C., Bunkowski, S., Polak, T., and Bruck, W., 2000. Tumor necrosis factor alpha mRNA expression in early multiple sclerosis lesions: correlation with demyelinating activity and oligodendrocyte pathology. *Glia* 29:366–375.

94. Fidelus, R.K. and Tsan, M.F., 1986. Enhancement of intracellular glutathione promotes lymphocyte activation by mitogen. *Cell. Immunol.* 97:155–160.

95. Fidelus, R.K., Ginouves, P., Lawrence, D., and Tsan, M.F., 1987. Modulation of intracellular glutathione concentrations alters lymphocyte activation and proliferation. *Exp. Cell Res.* 170:269.

96. Staal, F., Anderson, M., Staal, G., Herzenberg, L., Gitler, C., and Herzenberg, L., 1994. Redox regulation of signal transduction: tyrosine phosphorylation and calcium influx. *Proc. Natl. Acad. Sci. U.S.A.* 91:3619–3622.

97. Sampath, D. and Perez-Polo, R., 1997. Regulation of antioxidant enzyme expression by NGF. *Neurochem. Res.* 22:351–362.

98. Gabaizadeh, R., Staecker, H., Liu, W., and Van De Water, T., 1997. BDNF protection of auditory neurons from cisplatin involves changes in intracellular levels of both reactive oxygen species and glutathione. *Brain Res. Mol. Brain Res.* 50:71–78.

99. Lewis, J., Huq, A., and Najarro, P., 1996. Inhibition of mitochondrial function by interferon. *J. Biol. Chem.* 271:184–190.

100. Lee, J., 1998. Inhibition of p53-dependent apoptosis by the KIT tyrosine kinase: regulation of mitochondrial permeability transition and reactive oxygen species generation. *Oncogene* 17:1653–1662.

101. Islam, K., Kayanoki, Y., Kaneto, H., Suzuki, K., Asahi, M., Fujii, J., and Taniguchi, N., 1997. TGF-beta1 triggers oxidative modifications and enhances apoptosis in HIT cells through accumulation of reactive oxygen species by suppression of catalase and glutathione peroxidase. *Free Radic. Biol. Med.* 22:1007–1017.

102. Hampton, M., Fadeel, B., and Orrenius, S., 1998. Redox regulation of the caspases during apoptosis. *Ann. N. Y. Acad. Sci.* 854:328–335.

103. Pillar, T. and Seitz, H., 1997. Thyroid hormone and gene expression in the regulation of mitochondrial respiratory function. *Eur. J. Endocrinol.* 136:231–239.

104. Lee, A., Fenster, B., Ito, H., Takeda, K., Bae, N., Hirai, T., Yu, Z., Ferrans, V., Howard, B., and Finkel, T., 1999. Ras proteins induce senescence by altering the intracellular levels of reactive oxygen species. *J. Biol. Chem.* 274:7936–7940.

105. Shimeld, C., Whiteland, J.L., Williams, N.A., Easty, D.L., and Hill, T.J., 1997. Cytokine production in the nervous system of mice during acute and latent infection with herpes simplex virus type 1. *J. Gen. Virol.* 78:3317–3325.

106. Mayne, M., Cheadle, C., Soldan, S.S., Cermelli, C., Yamano, Y., Akhyani, N., Nagel, J.E., Taub, D.D., Becker, K.G., and Jacobson, S., 2001. Gene expression profile of herpesvirus-infected T cells obtained using immunomicroarrays: induction of proinflammatory mechanisms. *J. Virol.* 75:11641–11650.

107. Lokensgard, J.R., Cheeran, M.C., Hu, S., Gekker, G., and Peterson, P.K., 2002. Glial cell responses to herpesvirus infections: role in defense and immunopathogenesis. *J. Infect. Dis.* 186:S171–S179.

108. Murphy, P.M., 2001. Viral exploitation and subversion of the immune system through chemokine mimicry. *Nat. Immunol.* 2:116–122.

109. Lang, H.L., Jacobsen, H., Ikemizu, S., Andersson, C., Harlos, K., Madsen, L., Hjorth, P., Sondergaard, L., Svejgaard, A., Wucherpfennig, K., Stuart, D.I., Bell, J.I., Jones, E.Y., and Fugger, L., 2002. A functional and structural basis for TCR cross-reactivity in multiple sclerosis. *Nat. Immunol.* 3:940–943.

110. Tejada-Simon, M.V., Zang, Y.C., Hong, J., Rivera, V.M., Killian, J.M., and Zhang, J.Z., 2002. Detection of viral DNA and immune responses to the human herpesvirus 6 101-kilodalton virion protein in patients with multiple sclerosis and in controls. *J. Virol.* 76:6147–6154.

111. Speir, E., 2000. Cytomegalovirus gene regulation by reactive oxygen species. Agents in atherosclerosis. *Ann. N. Y. Acad. Sci.* 899:363–374.

112. Mogensen, T.H., Melchjorsen, J., Hollsberg, P., and Paludan, S.R., 2003. Activation of NF-kappa B in virus-infected macrophages is dependent on mitochondrial oxidative stress and intracellular calcium: downstream involvement of the kinases TGF-beta-activated kinase 1, mitogen-activated kinase/extracellular signal-regulated kinase kinase 1, and I kappa B kinase. *J. Immunol.* 170:6224–6233.

113. Paludan, S.R., Ellermann-Eriksen, S., Lovmand, J., and Mogensen, S.C., 1999. Interleukin-4-mediated inhibition of nitric oxide production in interferon-gamma-treated and virus-infected macrophages. *Scand. J. Immunol.* 49:169–176.

114. Paludan, S.R., Ellermann-Eriksen, S., Malmgaard, L., and Mogensen, S.C., 2000. Inhibition of NO production in macrophages by IL-13 is counteracted by Herpes simplex virus infection through tumor necrosis factor-alpha-induced activation of NK-kappa B. *Eur. Cytokine Netw.* 11:275–282.

115. Paludan, S.R., Ellermann-Eriksen, S., Kruys, V., and Mogensen, S.C., 2001. Expression of TNF-alpha by herpes simplex virus-infected macrophages is regulated by a dual mechanism: transcriptional regulation by NF-kappa B and activating transcription factor 2/Jun and translational regulation through the AU-rich region of the 3′untranslated region. *J. Immunol.* 167:2202–2208.

116. Oldstone, M.B., Holmstoen, J., and Welsh, R.M., Jr., 1977. Alterations of acetylcholine enzymes in neuroblastoma cells persistently infected with lymphocytic choriomeningitis virus. *J. Cell. Physiol.* 91:459–472.

117. Oldstone, M.B., Sinha, Y.N., Blount, P., Tishon, A., Rodriguez, M., von Wedel, R., and Lampert, P.W., 1982. Virus-induced alterations in homeostasis: alteration in differentiated functions of infected cells *in vivo. Science* 218:1125–1127.

118. Rambukkana, A., Kunz, S., Min, J., Campbell, K.P., and Oldstone, M.B., 2003. Targeting Schwann cells by nonlytic arenaviral infection selectively inhibits myelination. *Proc. Natl. Acad. Sci. U.S.A.* 100:16071–16076.

119. Ito, M., Baker, J.V., Mock, D.J., Goodman, A.D., Blumberg, B.M., Shrier, D.A., and Powers, J.M., 2000. Human herpesvirus 6-meningoencephalitis in an HIV patient with progressive multifocal leukoencephalopathy. *Acta Neuropathol. (Berl.)* 100:337–341.

120. Carrigan, D.R., Harrington, D., and Knox, K.K., 1996. Subacute leukoencephalitis caused by CNS infection with human herpesvirus-6 manifesting as acute multiple sclerosis. *Neurology* 47:145–148.

121. Novoa, L.J., Nagra, R.M., Nakawatase, T., Edwards-Lee, T., Tourtellotte, W.W., and Cornford, M.E., 1997. Fulminant demyelinating encephalomyelitis associated with productive HHV-6 infection in an immunocompetent adult. *J. Med. Virol.* 52:301–308.

122. Sanders, V.J., Felisan, S., Waddell, A., and Tourtellotte, W.W., 1996. Detection of herpesviridae in postmortem multiple sclerosis brain tissue and controls by polymerase chain reaction. *J. Neurovirol.* 2:249–258.

123. Albright, A.V., Lavi, E., Black, J.B., Goldberg, S., O'Connor, M.J., and Gonzalez-Scarano, F., 1998. The effect of human herpesvirus-6 (HHV-6) on cultured human neural cells: oligodendrocytes and microglia. *J. Neurovirol.* 4:486–494.

124. Goodman, A.D., Mock, D.J., Powers, J.M., Baker, J.V., and Blumberg, B.M., 2003. Human herpesvirus 6 genome and antigen in acute multiple sclerosis lesions. *J. Infect. Dis.* 187:1365–1376.

125. He, J., McCarthy, M., Zhou, Y., Chandran, B., and Wood, C., 1996. Infection of primary human fetal astrocytes by human herpesvirus 6. *J. Virol.* 70:1296–1300.

126. Dietrich, J., Blumberg, B.M., Roshal, M., Baker, J.V., Hurley, S.D., Mayer-Proschel, M., and Mock, D.J., 2004. Infection with an endemic human herpes virus disrupts critical glial precursor cell functions. *J. Neurosci.* 24:4875–4883.

127. Blakemore, W.F. and Keirstead, H.S., 1999. The origin of remyelinating cells in the central nervous system. *J. Neuroimmunol.* 98:69–76.

128. Franklin, R.J., 2002. Why does remyelination fail in multiple sclerosis? *Nat. Rev. Neurosci.* 3:705–714.

129. Chang, A., Nishiyama, A., Peterson, J., Prineas, J.W., and Trapp, B.D., 2000. NG2+ oligodendrocyte progenitor cells in adult human brain and multiple sclerosis lesions. *J. Neurosci.* 20:6404–6412.

130. Chang, A., Tourtellotte, W.W., Rudick, R., and Trapp, B.D., 2002. Premyelinating oligodendrocytes in chronic lesions of multiple sclerosis. *N. Engl. J. Med.* 346:165–173.

131. Wolswijk, G., 1998. Chronic stage multiple sclerosis lesions contain a relatively quiescent population of oligodendrocyte precursor cells. *J. Neurosci.* 18:601–609.

132. Wolswijk, G., 2000. Oligodendrocyte survival, loss and birth in lesions of chronic-stage multiple sclerosis. *Brain* 123:105–115.

133. Wolswijk, G., 2002. Oligodendrocyte precursor cells in the demyelinated multiple sclerosis spinal cord. *Brain* 125:338–349.

134. Flemington, E.K., 2001. Herpesvirus lytic replication and the cell cycle: arresting new developments. *J. Virol.* 75:4475–4481.

135. Kosugi, I., Shinmura, Y., Kawasaki, H., Arai, Y., Baba, S., and Tsutsui, Y., 2000. Cytomegalovirus infection of the central nervous system stem cells from mouse embryo: a model for developmental brain disorders induced by cytomegalovirus. *Lab. Invest.* 80:1373–1383.

136. Aiba-Masago, S., Baba, S., Li, R.Y., Shinmura, Y., Kosugi, I., Arai, Y., Nishimura, M., and Tsutsui, Y., 1999. Murine cytomegalovirus immediate-early promoter directs astrocyte-specific expression in transgenic mice. *Am. J. Pathol.* 154:735–743.

137. Li, R.Y., Baba, S., Kosugi, I., Arai, Y., Kawasaki, H., Shinmura, Y., Sakakibara, S.I., Okano, H., and Tsutsui, Y., 2001. Activation of murine cytomegalovirus immediate-early promoter in cerebral ventricular zone and glial progenitor cells in transgenic mice. *Glia* 35:41–52.

138. Raine, C.S., Bonetti, B., and Cannella, B., 1998. Multiple sclerosis: expression of molecules of the tumor necrosis factor ligand and receptor families in relationship to the demyelinated plaque. *Rev. Neurol. (Paris)* 154:577–585.

139. Selmaj, K.W. and Raine, C.S., 1988. Tumor necrosis factor mediates myelin and oligodendrocyte damage *in vitro. Ann. Neurol.* 23:339–346.

140. Selmaj, K., Raine, C.S., Farooq, M., Norton, W.T., and Brosnan, C.F., 1991. Cytokine cytotoxicity against oligodendrocytes. Apoptosis induced by lymphotoxin. *J. Immunol.* 147:1522–1529.

141. McLarnon, J.G., Michikawa, M., and Kim, S.U., 1993. Effects of tumor necrosis factor on inward potassium current and cell morphology in cultured human oligodendrocytes. *Glia* 9:120–126.

142. McLaurin, J., D'Souza, S., Stewart, J., Blain, M., Beaudet, A., Nalbantoglu, J., and Antel, J.P., 1995. Effect of tumor necrosis factor alpha and beta on human oligodendrocytes and neurons in culture. *Int. J. Dev. Neurosci.* 13:369–381.

143. Tsoukas, C.D. and Lambris, J.D., 1993. Expression of EBV/C3d receptors on T cells: biological significance. *Immunol. Today* 14:556–559.

6 Cytokines and Development of the Nervous System

Limin Gao and Robert H. Miller

CONTENTS

I. OVERVIEW

The elaboration of the vertebrate central nervous system (CNS) results from the complex interplay of cellular interactions that establish positional information, development of cell type specificity, and precise connectivity between distinct cell populations. These diverse responses are orchestrated through the localized expression of a broad range of cytokines and the subsequent responses of target cells. As the molecular nature of specific cellular events in neural development become clear, several general themes have emerged. First, specification of the entire nervous system and cellular specification within the nervous system depend critically on both time and space such that mislocalization frequently leads to fundamental changes in cell fate. Second, signaling by individual cytokines is frequently used in different environmental settings to generate quite distinct cellular results. Third, integration of competing or antagonistic signals frequently determine the outcome of signaling. And finally, regulated cell death is frequently used to select distinct cell types from a palate of potential cell fates. In this chapter, we will briefly review the initial specification of the neural tube and discuss in more detail the role of a specific subset of cytokines in the regulation of patterning of the emerging nervous system and determination of neural cell fate. Ultimately, the identification of critical signaling mechanisms utilized during development will provide insights into neurological dysfunction of the mature tissue as well as provide candidates for novel therapeutic interventions in a wide range of neurological diseases.

II. EARLY SPECIFICATION OF THE NEURAL TUBE

The vertebrate central nervous system (CNS) is initially derived from a cytologically homogeneous sheet of epithelial cells termed the *neural plate,* which arises from dorsal ectoderm during the cytological reorganization that occurs during gastrulation. Somewhat remarkably, the specification of the neural plate appears to result not from the actions of "active inducer molecules" but rather through local suppression or the avoidance of the molecular signals responsible for the induction of nonneural cells in the surrounding ectoderm. As embryogenesis proceeds, the neural plate subsequently undergoes a series of morphogenetic movements to form the neural tube, a process that involves establishment of the dorsal–ventral axis, differential cell proliferation, infolding, segregation from the overlying ectoderm, and fusion of the lateral margins of the neural tube. Anterior–posterior patterning in the emerging CNS is established very early, and during neural tube closure the rostral or anterior end undergoes dramatic expansion and regionalization. Such regionalization is manifest in the portioning into three primary vesicles of the brain: the forebrain, the midbrain, and the hindbrain. The long, uniformly narrow caudal neural tube undergoes further differentiation to form the spinal cord. Each of theses regions is characterized by unique cytoarchitecture and specification of distinct cell types. This early regionalization of the neural plate along its anteroposterior axis depends on signals from the local environment. Here we will focus on the cytokines that play important roles in controlling neural induction, dorsalventral and rostrocaudal patterning of the neural tube.

A. INDUCTION AND ANTEROPOSTERIOR PATTERNING
OF THE NEURAL PLATE

Environmental signals contribute to the establishment of the anterioposterior axis of the CNS at the time of gastrulation. Classical tissue grafting experiments in amphibian embryos [1] suggested that the neural-inducing signals produced at gasturation by a distinct population of mesodermal cells located at the dorsal lip of the blastpore (Spemann's organizer) were critically involved in establishing the rostrol–caudal axis. Subsequent studies have begun to define the molecular bases of this signaling.

For example, inhibition of the antineurogenic signal transmitted between ectodermal cells is mediated by secreted proteins of the transforming growth factor- (TGF-) family, specifically bone morphogenic protein4 (BMP4) and leads to neural induction. Three lines of evidence support the hypothesis that BMP4 signaling is involved in the neural induction. First, BMP4 is expressed in the early nonneural ectoderm, and its expression is absent in neural plate cells during neural induction [2,3,4]. Second, BMP4 inhibits the expression of neural markers and contributes to epidermal differentiation in dissociated ectodermal cells [5]. Third, and most important, three secreted proteins expressed by Spemann's organizer or the dorsal blastpore lip called *follistatin, noggin,* and *chordin* can bind directly to and antagonize the action of BMPs, and these inhibitors are able to induce the expression of marker of anterior neural plate cells in naïve ectodermal cells [6]. Together, such studies suggest that the inhibition of BMP4 signaling may be a critical component that contributes to the induction of anterior neural plate.

Specification of posterior neural tissue depends on an alternative series of environmental cues. For these regions of the CNS, studies on the signaling pathway identified secreted proteins of the fibroblast growth factor (FGF) family as candidate posterior neural inducers. For example, when BMP signaling is reduced or eliminated, FGFs have been shown to induce neural tissue with the characteristics of spinal cord [7,8,9]. In addition, exposure of ectoderm to TGF inhibitor–noggin and FGF can induce neural tissue that is characteristic of the midbrain and hindbrain, regions that are at intermediate levels of the neuraxis [7]. Such observations have led to the hypothesis that the mechanisms that control anteroposterior patterning of the neural tube may involve the coordinated actions of inhibitors of BMP signaling and FGFs [10]. Not all the observations fit well into this model, however, and it is likely that other signaling systems influence rostral caudal patterning. For example, retinoids have been implicated in the generation of posterior neural tissue [11,12,13], whereas the role of other neural inducers such as hepatocyte nuclear factor (HNF) [14] and Cerberus protein [15] is currently less well defined. It may also be that the amphibian CNS is somewhat less complex than its mammalian homologue because mutations in the murine BMP4 gene do not result in an obvious expansion in neural tissue at the expense of epidermal ectoderm [16], which is different from studies in Xenopus. This lack of predicted phonotype suggests either the presence of other BMPs that function in a manner similar to that of BMP4, or the presence of other positive inductive signals of neural induction. What remains unresolved is the connection between neural induction per se and the acquisition of anteroposterior regional identity. It seems most likely that neural induction and the early regional fate of neural cells are linked processes rather than independent events [17].

B. DORSOVENTRAL PATTERNING OF THE NEURAL PLATE

The second major axis in the emerging CNS is the dorso-ventral axis that has critical influences on neural cell fate. Both axes form during similar developmental epochs such that as the antero-posterior axis is forming, dorso-ventral segregation is also underway and, like rostral-caudal axis, is influenced by extrinsic signals. One of the major ventralizing signals is sonic hedgehog (Shh), the vertebrate homologue of the *Drosophila hedgehog* gene. Secreted from the notochord and floor plate, Shh controls

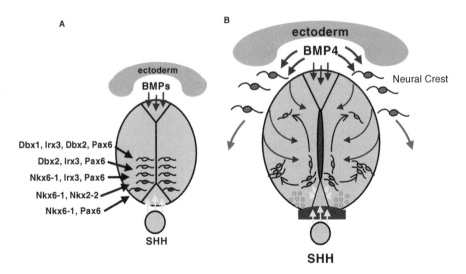

FIGURE 6.1 (A) Early in the development of the spinal cord, ventral cell types are generated in response to inducing signals, mainly Shh, initially released from the adjacent notochord that induces the formation of the floor plate. Generation of dorsal cell types is controlled by signals (BMPs) from the overlying epidermal ectoderm, which direct the induction of neural crest and roof plate cells. After closure of the neural tube, Shh expression is maintained in the floor plate and BMPs are expressed in the roof plate and dorsal neural tube. The concentration gradient of Shh regulates the ventral expression of a series of combinations of transcription factors in ventral progenitor cells. Similarly, different dorsal cell types are generated at different BMP concentration thresholds. (B) Neural crest cells initially originate from progenitors located at the border of the epidermal ectoderm and neural plate, and constitute a population of migratory progenitor cells that differentiate into the majority of neurons and nonneuronal cells in the peripheral nervous system. BMP4 signaling plays an important role in the specification of neural crest cell fate. Another population of highly migratory progenitor cells is oligodendrocyte progenitors cells (OPCs) which originate from the ventral ventricular zone of the spinal cord. These cells or their progeny subsequently migrate dorsally and laterally to populate the white matter of the whole spinal cord. Shh released from notochord and floor plate contributes to the induction of OPCs in the ventral ventricular domain which also gives rise to the motor neurons.

the identity and patterning of cell types generated in the ventral half of the neural tube through actions at multiple concentration thresholds [17]. At high concentrations, Shh induces the most ventral cell types including floor plate and motor neurons, whereas at lower concentrations Shh induces distinct populations of interneurons. The ventralizing influence of Shh is antagonized by signals from dorsal regions. Dorsalizing signals, including TGF-s, FGF, and Wnt families, maintain dorsal cell identity including neural crest and sensory interneurons by antagonizing signals from the overlying ectoderm. Here we will focus on the roles of the TGF family in patterning the dorsal neural tube (Figure 6.1A).

One early characteristic of dorsally derived neural cell populations is the neural crest. Neural crest cells initially emerge from dorsal neural tube around the time of tube closure and subsequently migrate throughout the embryo where they give rise to the majority of the peripheral nervous system (Figure 6.1B). The most dorsally localized cells that remain in the neural tube develop into a specific population of cells in the dorsal midline termed the *roof plate*. In avian embryos, BMP4 and BMP7 are not only expressed in epidermal ectoderm at early neural plate stages, but also have the ability to mimic the neural crest inducing activity of the epidermal ectoderm *in vitro* [18]. Additionally, these BMPs are capable of inducing roof plate cells from neural plate explants [19], and this induction could be blocked by antagonists of BMP signaling such as noggin. The inhibitors noggin, chordin, and follistatin are secreted proteins that bind BMPs and the related family member activin to prevent the activation of specific BMP and activin receptors. *In vitro* studies found noggin but not follistatin can block the neural crest-inducing activity of BMP4, whereas follistatin but not noggin can block the inductive activity of BMP7. These data strongly support BMPs as a dorsalizing signal controlling dorsal neural cell identity and pattern. Both noggin and follistatin can block the signaling of a number of different BMPs as well as activins; however, the precise identity of the relevant TGFß-related signaling factors is still not clear. Direct *in vivo* evidence for BMPs in controlling the induction of neural crest cells and dorsal spinal cord development has not been provided by murine genetics studies, although BMP2, BMP4, and BMP7 are the most likely mediators of the dorsalizing signals from epidermal ectoderm according to their pattern and timing of expression. The potential for multiple molecular cues suggests functional redundancy between BMP family members in the early specification of dorsal cell fate [20,21].

Several pathways by which cells in the neural plate respond to BMP signaling from the adjacent epidermal ectoderm have been identified. Expression of genes encoding the transcription factors Pax3, Pax7, Msx1, and Msx2 in neural plate cells are elevated in response to BMPs treatment or exposure to epidermal ectoderm [18,22]. Although there is evidence that Msx genes represent a general component of the cellular response to BMP signaling in a variety of tissues [23] and loss of Pax3 or Pax7 function can disrupt the differentiation of certain neural crest derivatives at later stages of neural crest development [24,25], the roles of these genes in cellular response to dorsalizing signals has not been resolved [18]. Zinc finger class transcription factors are also involved in the specification of neural crest cells by BMP signaling. Among them, the gene encoding SLUG is induced by BMPs and epidermal ectoderm [18,26], yet genetic studies in mice demonstrated that the *Slug* gene is not essential for neural crest development [27]. Overexpression of Slug in Xenopus ectodermal tissue is also insufficient for neural crest induction. In the presence of Wnt- or FGF-like signals, however, Slug-expressing ectoderm will generate neural crest, thereby bypassing the need for BMP in this process [28].

Not only are BMPs involved in the initial specification of neural crest cells, but they are also important in the specification of premigratory neural crest cells and in their subsequent emigration from the neural tube (Figure 6.1B). These events reflect differential adhesion/de-adhesion events and activation of distinct signaling pathways. The cell adhesion molecule cadherin 6B is upregulated in premigratory neural crest cells [29] and can be induced by BMPs in neural plate tissue [30]. In the subsequent step of the delamination of premigratory neural crest cells from the neural epithelium prior to their migration to the periphery, *rhoB*, a gene encoding a member of the rho family, GTP-binding proteins appears to be important. The expression of rhoB is induced by BMPs in chick neural plate cells and is expressed in the dorsal neural tube, and this expression persists transiently in migrating neural crest cells. Inhibition of rhoB activity prevents the delamination of neural crest cells from neural tube explants, suggesting that rhoB has a role in the delamination of neural crest cells from the dorsal neural tube [30].

The inductive effects of BMPs are not restricted to the neural crest. For example, BMP-mediated signals derived from overlying ectoderm also induce dorsal midline cells of the roof plate [19]. After the closure of the neural tube, the ectodermal expression of BMP4 and BMP7 is downregulated [18,31] and roof plate cells become a secondary source of TGF-related signals that are required for the generation of different classes of dorsal horn interneurons that are critical in processing of sensory stimuli [19]. During the period of interneuron generation, at least six members of the TGFß family (BMP4, BMP5, BMP7, DSL1, GDF6/7, and activinB) are expressed in nested domains by cells near or in the roof plate [19,32]. Like ventralizing signals, BMPs are thought to be expressed in a gradient that is important for the generation of distinct kinds of interneurons. Neural progenitors close to the roof plate that are exposed to a high level of BMPs preferentially generate D1 neurons, whereas progenitors further from the roof plate generate D2 interneurons in response to a low level of BMPs [19]. Genetic studies in mice identified a BMP family member, GDF7, expressed selectively by roof plate cells after neural tube closure, which plays a critical role in regulating the formation of a specific population of dorsal neural progenitors fated to generate D1A interneurons [32]. Two members of BMP family, BMP6 and 7, that are expressed by the roof plate cells are not able to compensate for the loss of GDF7 function, which indicates that multiple BMPs expressed by roof plate cells have distinct and nonredundant functions in dorsal patterning of the dorsal spinal cord [32].

In conclusion, spatial patterning in the emerging nervous system and the determination of dorsal neuronal fates seems to involve the differential responses of neural cells to TGFß signaling at different developmental stages. At the stage of neural crest induction, neural progenitor cells respond to the epidermal ectoderm-derived BMP signaling to direct the generation of neural crest and roof plate cells, whereas at later stages, BMPs secreted by the roof plate cells direct the dorsal interneuron fates in cooperation with other TGFß family members such as activin.

III. REGIONAL SPECIFICATION OF THE DEVELOPING CNS

A. FOREBRAIN AND MIDBRAIN

Similar environmental signals are important for the appropriate development of other regions of the CNS. Studies using molecular markers for BMP gene expression demonstrated that BMPs are expressed from the caudal spinal cord to the lamina terminalis at the base of the telencephalon [20,32,33,34]. Furthermore, during early development, BMP7 is expressed by the prechordal mesoderm that underlies the prospective ventral midline of the forebrain [35], and in these regions appears to cooperate with Shh to induce the specialized ventral midline cells of the rostral diencephalon. Later in development, BMP2, BMP4, BMP5, BMP6, and BMP7 are coexpressed in a region of dorsomedial telencephalon [34], where high-level expression of Msx is also found. These descriptive studies suggest that BMPs may also regulate local cell growth and regionalization in the dorsal forebrain.

B. HINDBRAIN

Dorsal cell patterning in the hindbrain is also controlled by inductive signals mediated by BMPs in a similar manner to that in spinal cord [20,33,36,37,38,39,40]. At the level of the hindbrain, the neural tube is subdivided by a series of transient constrictions into domains termed *rhombomeres* [41]. The elimination of neural crest cells from r3 and r5 in response to repressive signals from even numbered rhombomeres appears to depend on induction of BMP4 expression in the dorsal neural tube in r3 and r5 [42]. *In vitro* studies in isolated explants found that BMP4 can stimulate apoptosis that subsequently deletes neural crest cells from r3 and r5 but does not have a similar effect in r4 explants [43,44], suggesting localized responses to these environmental cues.

C. CEREBELLUM

Patterning of the developing cerebellum is considerably more complex than in other CNS regions. In addition to inductive signaling involving Wnt and FGF proteins [45], cerebellar development also depends on patterning signals that specify dorsoventral identity. The presumptive dorsalizing signals Bmp6, Bmp7, and Gdf7 are expressed by dorsal midline cells adjacent to the rhombic lip, and these BMPs can induce the expression of granule neuron markers in explants of ventral metencephalic neural plates. In addition, neural cells exposed to BMPs can form mature granule neurons after transplantation into the early postnatal cerebellum, suggesting initiation of the granule cell specification depends on BMP signaling. The subsequent development of granule cells is influenced by Shh, suggesting that in this region as in more rostral regions, these two signaling systems combine to generate the appropriate cellular pool.

In conclusion, TGF family of secreted proteins, especially BMP subfamily members, play important roles in early patterning of the CNS and a variety of different regions. The precise role of these proteins and the consequences of their signaling depend in large part on the cellular context in which the signals are received.

IV. DETERMINATION OF PRECISE CELLULAR IDENTITY IN THE CNS

The precise identity of individual neural cells in the developing CNS is a consequence of both the overall patterning of the neural tube, as well as specific cell interactions that occur in distinct locations of the tissue. How these various inputs are integrated into specific cellular readouts is an area of intense investigation. Substantial evidence suggests that the neuroepithelial cells of the early neural tube are multipotent stem cells that can self-renew and differentiate into neurons, astrocytes, and oligodendrocytes. Regional specification begins as early as the stem cell stage, and stem cells isolated from different regions of the developing CNS, while retaining multipotent potential, appear to have different differentiation biases [46,47,48]. These observations imply that regional identity can affect stem cell fate determination and define the functional identity of the neurons generated. What is less clear is whether similar spatial restrictions influence the appearance of the other major classes of neural cells, astrocytes and oligodendrocytes. It is broadly believed that the coordination of extrinsic signals and intrinsic property of stem cells conferred by their temporal and regional identity will control the probability of self-renewal or differentiation and cell fate determination of stem cells. Identifying the extrinsic signals (growth factors) and defining their respective activities are important for understanding the development of CNS and provide critical information for implication of stem cells in tissue regeneration and repair. Here the role of some major extracellular signals, especially cytokines that influence properties of neural stem cells (NSCs), will be considered (Figure 6.2).

A. INSTRUCTIVE ASTROCYTE DIFFERENTIATION OF NSCS BY SIGNALING OF BMPS AND IL-6 TYPE CYTOKINES

Astrocytes represent a major population of cells in the vertebrate CNS whose development is relatively poorly understood. Astrocytes are a heterogeneous population of cells that are thought to share the characteristic of expression of glial fibrillary acid protein (GFAP) at some stage in their development. Classical studies suggest that the precursors of at least some astrocytes are radial glial cells that directly convert to an astrocyte fate. More recently, however, it has been proposed that GFAP+ cells may represent multipotent stem cells capable of giving rise to neurons, astrocytes, and oligodendrocytes. The molecular cues responsible for the development of an astrocyte phenotype from stem cells are beginning to be elucidated and include members of the interleukin 6 family. The interleukin (IL)-6 family of cytokines includes IL-6, IL-11, LIF, ciliary neurotrophic factor (CNTF), oncostatin M (OSM), and cardiotrophin-1 (CT-1) and cardiotrophin-like cytokine (CLC). Mice lacking the LIF receptor show a phenotype of decreased numbers of astrocytes [49], whereas *in vitro* studies showed that CNTF is a potent inducer of astrocyte differentiation. The signaling mechanisms underlying this process are controversial. CNTF stimulated both the Janus kinase-signal transducer and activator of transcription (JAK-STAT) and Ras-mitogen-activated protein kinase (MAPK) signaling pathways in cortical precursor cells. Some evidence supports the notion that the JAK-STAT signaling pathway alone is responsible for the enhanced differentiation of these precursors along a glial lineage [50], whereas other studies implicate CNTF activation of both JAK-STAT and MAPK pathways with differential kinetics. In this model, activation of the MAPK pathway is required early in astrocyte differentiation, whereas activation of STAT proteins is required for commitment to an astrocytic

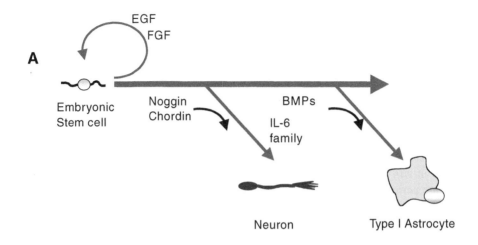

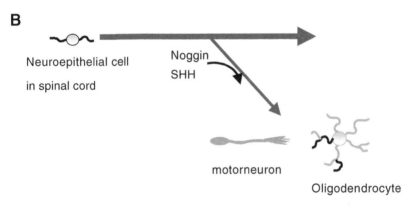

FIGURE 6.2 (A) EGF and FGF are important in regulating neural stem cell proliferation. Differentiation of embryonic stem (ES) cells in medium supplemented with the BMP4 antagonist–noggin resulted in neurons comprising >91% of surviving cells after 72 hours. Similarly, addition of chordin, another antagonist of BMP4, increases neural differentiation with lower (55%) efficiency. When cultured *in vitro* for 3–7 days, addition of BMPs (BMP2,4,5,6,7) induces the elaboration and irreversible differentiation of astrocytes in a dose and time-dependent manner. When cultured for a shorter period of time (2 days), addition of BMP2, 4, 7 in combination with IL-6 family members resulted in astrocyte differentiation. (B) In spinal cord, coordinated expression of SHH and BMP antagonist–noggin in ventral regions contributes to the induction of motoneurons and oligodendrocytes in the same domain from neuroepithelial cells.

fate, and thus activation of both the MAPK and JAK-STAT pathways are positively coupled to the differentiation of astroglial cells *in vitro* [50,51].

Isolation of neural stem cells and the ability to propagate these cells *in vitro* as neurospheres has facilitated further identification of astrocyte induction signals. Studies with murine embryonic subventricular zone (SVZ) multipotent progenitor cells demonstrated the expression of selective BMP ligands and transcripts for the

BMP-selective type I and II receptor subunits. Addition of BMPs (Dpp class-BMP2 and 4, BMP5 and 6, and OP1/BMP7) can induce the elaboration and irreversible differentiation of GFAP-immunoreactive astrocytes in a dose and time-dependent way, whereas other TGF family members, such as activin, TGF1 and glial cell line derived neurotrophic factor (GDNF) cannot, suggesting that in the TGF family there is distinct ligand selectivity. In some cases, the induction of astroglia fate is accompanied with the concurrent suppression of neuronal and oligodendroglial cell fates [52]. One concern with these studies is how good the neurospheres model is for normal astrocyte development. Clearly, culture interval and conditions can influence the responsiveness of neurospheres derived cells. In the above studies, cells were cultured *in vitro* for 3 to 7 days. When challenged with similar ligands after shorter culture periods (2 days), neither IL-6 family members nor BMPs (BMP2, BMP4, or BMP7) alone have astrocyte inducing activity, whereas a simultaneous treatment of the neuroepithelial cells with these two cytokines results in synergy and astrocyte differentiation [53,54,55]. The signaling underlying this response is complex. Studies with LIF and BMP2 found that STAT3 and Smads form a complex bridged by p300, which is a member of the CREB-binding protein (CBP)/P300 family of transcriptional coactivators. The CBP/P300 family members are ubiquitously expressed [56] and have been shown to interact with various transcription factors such as AP-1, Myb, and nuclear receptors [57,58,59]. P300 protein preferentially interacts with STAT3 and Smad 1 at its amino and carboxyl terminus, respectively, and P300 acts as an adaptor molecule linking these transcriptions factors. The resultant complex is believed to be involved in the cooperative transcriptional activation of the GFAP promoter mediated by LIF and BMP2. The reason IL-6 family members or BMPs alone can induce astrocytes after culture for 3 to 7 days may reflect the accumulation of endogenous IL-6 family cytokines or BMPs in long-term culture that cooperate with the exogenous administration of cytokines to contribute to the induction of astrocytes. This hypothesis is supported by *in vitro* and *in vivo* evidence that all members of the IL-6 cytokine family, BMP2, BMP4, BMP7, and their receptor components are expressed in neuroepithelial cells [53,54,55]. It seems likely that there are as yet undefined antagonistic signals that ensure that not all neuroepithelial cells assume an astrocyte fate early in development.

B. CYTOKINES AND CELL-FATE ALTERATION FROM NEUROGENESIS TO ASTROCYTOGENESIS

The canonical cells of the nervous system are neurons, and these cells arise from neural stem cells and detailed understanding of the genesis of neurons, and how this is influenced by glial-inducing factors is critical to developing strategies for neural repair. In vertebrate nervous systems, neurogenesis is regulated by basic helix-loop-helix (bHLH) transcription factors—Mash1, Neurogenin and Neuro D. Exposure of telencephalic neural progenitor cultures to BMP2 significantly reduced the number of undifferentiated neural precursors and neurons but increased the number of S100-beta-positive astrocytes. BMP2 up-regulated the expression of negative HLH factors Id1, Id3 (Id: inhibitor of differentiation), and Hes-5 that inhibited the transcriptional activity of Mash1 and neurogenin. Neurogenesis in neuroepithelial cells was greatly inhibited by ectopic expression of either Id1 or Id3, suggesting these HLH proteins are important in the BMP2-mediated alteration in the neurogenic fate of these cells [60]. BMP4 and BMP7 are found to have the same antineurogenic activity [61]. Although the antineurogenic effect of BMP2 is in part mediated via induction of gene expression for Id1, Id3, and HES-5 in the neuroepithelial cells, this is not the only regulatory mechanism. Notch signaling also plays a role in inhibiting neuronal differentiation, and HES-5 is known to be induced by Notch activation, raising the possibility of cross-talk between Notch and BMP signaling. Recent studies with mouse neuroepithelial cells showed that BMP2 can enhance Notch-induced transcriptional activation of Hes-5 and Hes-1. Smad1 and the intracellular domain of Notch are able to form a complex containing P/CAF and p300. These results suggest a novel cooperation between Notch signaling and BMP signaling [62].

Regulated cell death is also important in the final development of neural cell fate, and this may depend on the age of the responsive cells. In early rat cerebral cortex cells, BMP2 and BMP4 promotes cell death and inhibits proliferation, whereas at E16, 1–10ng/ml of BMP2 enhances neuronal and astroglial development and inhibits oligodendroglial differentiation. Increasing the concentration of BMP2 to 100ng/ml promoted cell death [63]. Such studies suggest that temporal regulation of BMPs expression is important for the development of cerebral cortex and that the cellular readout of BMP stimulation depends on the stage of development.

C. OLIGODENDROGENESIS IN THE SPINAL CORD

During development of the neural tube, the acquisition of ventral cellular fate depends on the secreted molecule Shh [64,65,66]. In caudal regions of the CNS, Shh is derived from the notochord and floor plate, and the ventral to dorsal diffusion of Shh has been proposed to direct the patterns of neurogenesis [67, 68]. As discussed above, this ventralization is antagonized by BMPs secreted by epidermal ectoderm and the roof plate that play critical roles in induction of neural crest cells and dorsal interneurons [18,19,69,70,71]. Both Shh and BMPs act as gradient morphogens and are thought to have antagonistic activities in controlling cell fate along the dorsoventral axis of the neural tube [72]. The effectiveness of these potent signaling systems is more complex, however. The prospective ventral cells are also exposed to BMP activities because ventral regions of the spinal cord also express BMPs although their effects are inhibited by the local expression of the BMP antagonist noggin [73]. An in vitro study found that the induction of motor neurons from uncommitted neural progenitor cells is increased with Shh and attenuated by BMPs [74]. Somewhat surprisingly, oligodendrocytes, the myelinating cells of the CNS, arise from the same domain of the ventral neural tube that give rise to motor neurons [75] and are also influenced by the effects of Shh and BMPs.

Shh is critical for the normal development of spinal cord oligodendrocytes. Many studies demonstrate that local expression of Shh contributes to the induction of oligodendrocytes from the ventral ventricular zone [76,77,78] and this is mediated through induction of the Olig transcription factors. More recently it has become clear that BMPs also negatively regulate oligodendrocyte appearance. Implantation of noggin-secreting cells adjacent to the dorsal chick spinal cord or *in vivo* ablation of the dorsal part of the neural tube early in development promoted the oligodendrogenesis dorsal to their normal domain of appearance, indicating the dorsally derived endogenous BMPs inhibits the appearance of oligodendrocytes [79]. Cells that express BMP4 include midline dorsal cells [80], overlying skin, and some dorsal root ganglion neurons. The suggested mechanism is that inhibition of oligodendrogenesis early in the specification period by BMPs is mediated by repression of the transcription factor, olig2. In this model, the localized origin of oligodendrocyte precursors is controlled by precisely balanced activities of Shh and BMPs. Consistent with this hypothesis studies in *Xenopus* using BMP-coated beads implanted into the developing neural tube demonstrated that BMP is sufficient to inhibit the appearance of oligodendrocytes from the ventral spinal cord, whereas implantation of Shh-coated beads or antiBMP coated beads is able to induce oligodendrocytes from the dorsal spinal cord. Similar effects of BMPs are found in rat spinal cord by using the function-blocking antibody specific for BMP4 [81]. The dorsal expression of BMPs is retained during early postnatal development, and it may be this continued expression that mediates the timing of local differentiation of oligodendrocyte precursors [81].

D. Cellular Fate Choice Between Astrocyte and Oligodendrocyte

Plasticity in cell fate is retained in some neural precursors throughout much of the development. In the oligodendrocyte lineage, for example, oligodendrocyte precursors retain the potential to generate a subset of astrocytes and possibly neurons. These cells constitutively differentiate into oligodendrocytes, but in the presence of serum they express both GFAP and A2B5 antigens and are called *type-2-astrocytes* [82] and the precursors are therefore called *O2A progenitors*. Several other factors in addition to serum, including CNTF and LIF, can also convert oligodendrocyte precursors into astrocytes (Figure 6.3). Both molecules promote oligodendrogenesis in cultures of dividing O2A progenitors isolated from brain [83], but when combined with unknown factors present in the extracellular matrix, LIF and CNTF both promoted the differentiation of O2A progenitors into type-2 astrocytes [84,85,86]. BMPs can also promote differentiation of O2As isolated from postnatal rat cortex into astrocytes and concurrently suppress oligodendrocyte differentiation in a dose-dependent fashion. BMP type I and type II serine/threonine kinase receptor subunits are expressed by O2A progenitor cells [87]. The responses of oligodendrocyte precursor cells to the BMPs are stage specific during the developmental of the oligodendrocyte lineage cells [88]. *In vitro* studies of postnatal rat cerebellum found Smad1 and BMP4 proteins are present in the cerebellum, and BMP4 can promote the differentiation of astrocytes [89]. Cultures of multipotent neural progenitor cells derived from the embryonic mammalian SVZ showed an enhanced elaboration and maturation of astrocytes when treated with BMPs, whereas both neuronal and oligodendroglial differentiation are suppressed [52,90]. In addition to BMP ligands, the multipotent SVZ progenitor cells also express BMP-selective type I and type II receptor subunits [52]. This suggests that BMPs have a role in astroglial lineage commitment of multipotent progenitor cells. *In vivo* studies of transgenic mice

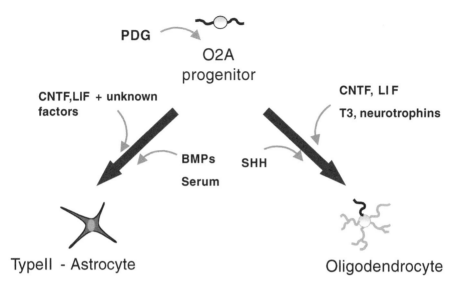

FIGURE 6.3 O2A progenitors are bipotential cells that are highly migratory and proliferate in response to PDGF. These cells constitutively differentiate into oligodendrocytes although high concentration of serum promotes type-2 astrocyte differentiation. In cultures of proliferating O2A progenitors isolated from brain, addition of CNTF and LIF promote differentiation of oligodendrocytes; but when combined with unknown factors derived from the extracellular matrix, CNTF and LIF promote the differentiation into type-2 astrocytes. *In vitro* culture of optic nerve studies demonstrated that several factors (T3, neurotrophins, bFGF, and PDGF) can increase oligodendrocyte numbers. BMPs are found to promote differentiation of O2As isolated from postnatal rat cortex into type-2 astrocytes and suppress oligodendrogenesis in a dose-dependent fashion.

overexpressing BMP4 found a remarkable increase of the number of astrocyte in various brain regions and a reduction of oligodendrocytes but no changes in the number of neurons. Further studies found that BMP4 released from the transgenic neurons drive the astroglial lineage commitment that could be inhibited by the BMP inhibitor noggin. Together these observations support the notion that BMPs are a critical regulator of astroglial development. They appear to act on different precursor cell populations, including stem cells and O2A progenitors. Because a major response of the adult nervous system to a variety of insults is to increase astrocyte number and differentiation, it becomes critical to define the roles of BMP and other astrocyte-inducing molecules in regulating injury responses.

V. CHEMOKINES AND NEURONAL DEVELOPMENT

In many other tissues, injury responses are orchestrated by the local expression of chemokines. Chemokines (chemotactic cytokines) are well known for directing the trafficking of various types of leukocytes and are important for normal immune responses as well as in inflammatory disease. Recently, an increasing number of studies have begun to reveal important roles for chemokines in the development of the nervous system. Chemokines are classified into four classes based on the positions of key cysteine residues—C, CC, CXC, and CX3C—and their actions are mediated through high- and low-affinity G-protein-coupled receptors [91,92]. Four of five CXCRs (CXCR2–CXCR5) are expressed in the CNS. Subsets of projection neurons in various regions of the spinal cord and brain express high levels of CXCR2 [93]. One CXCR2 ligand, IL-8, is expressed by activated astrocytes, and *in vitro* studies found it can enhance the survival of hippocampal neurons [93]. In the spinal cord, chemokine growth-related oncogene (GRO; CXCL1 in the systematic nomenclature) is produced by subsets of astrocytes and significantly enhances the proliferative response of immature oligodendrocyte precursors to platelet-derived growth factor (PDGF) [94]. Increased levels of GRO are related to enhanced oligodendrocyte progenitor proliferation in the dysmyelinating mutant *jimpy* (*jp*) mice [95]. In addition, signaling through CXCL1 and its receptor CXCR2 inhibits oligodendrocyte precursor migration in the developing spinal cord. CXCL1 expression in the white matter is temporo-spatially regulated and the inhibition of migration is rapid, reversible, and dose dependent. In the developing CXCR2 knock-out mouse spinal cord, the number of oligodendrocytes is reduced and abnormally located at the periphery of the white matter [96]. The control of CNS patterning by chemokines appears to be relatively common. CXCR4 is highly expressed by astrocytes, microglia, and neurons from both the CNS and PNS [97,98]. The only known CXCR4 ligand, SDF-1, was found to trigger migration of neuronal progenitors in a dose-dependent manner *in vitro* [99,100]. *In vivo* studies of CXCR4 and SDF-1 null mice demonstrate dramatically affected development of fetal cerebellum, including the absence of foliation and the premature migration of the external granular layer (EGL) cells into the internal granular layer [101,102,103]. This regulation of migration may also extend to control of axon guidance. For example, CXCR4/SDF-1 signaling was reported to act as chemorepellant or chemoattractant of the CXCR4-expressing growth cones of several types of developing neurons depending on the environmental conditions [104]. Although functional studies clearly demonstrate that chemokines are important in the development of nervous system by influencing cell migration, proliferation, and differentiation, chemokine receptor knockout mice do not normally have dramatic abnormalities in the brain (CXCR4 null mice are exceptional). Therefore, it seems likely that other chemokines and their receptors also contribute to the organization of the developing nervous system. Detailed analyses of the role of different chemokines in CNS development and in injury responses in the CNS will likely reveal profound effects of these molecules in the developing and mature nervous system.

REFERENCES

1. Spemann, H.; Mangold, H., Induction of embryonic primordia by implantation of organizers from a different species. *Int J Dev Biol*, 45, 13, 2001.
2. Hemmati-Brivanlou, A., Thomsen, G.H., Ventral mesodermal patterning in Xenopus embryos: expression patterns and activities of BMP-2 and BMP-4, *Dev Genet*, 17, 78–89, 1995.
3. Schmidt, J.E., Suzuki, A., Ueno, N., Kimelman, D., Localized BMP-4 mediates dorsal/ventral patterning in the early Xenopus embryo, *Dev Biol*, 169, 37, 1995.

4. Fainsod, A., Steinbeisser, H., De Robertis, E.M., On the function of BMP-4 in patterning the marginal zone of the Xenopus embryo, *EMBO J*, 13, 5015, 1994.

5. Wilson, P.A., Hemmati-Brivanlou, A., Induction of epidermis and inhibition of neural fate by Bmp-4, *Nature,* 376, 331, 1995.

6. Hemmati-Brivanlou, A., Kelly, O.G., Melton, D.A., Follistatin, an antagonist of activin, is expressed in the Spemann organizer and displays direct neuralizing activity, *Cell,* 77, 283, 1994.

7. Lamb, T.M., Harland, R.M., Fibroblast growth factor is a direct neural inducer, which combined with noggin generates anterior-posterior neural pattern, *Development*, 121, 3627, 1995.

8. Isaacs, H.V., Tannahill, D., Slack, J.M., Expression of a novel FGF in the Xenopus embryo: a new candidate inducing factor for mesoderm formation and anteroposterior specification, *Development* 114, 711, 1992.

9. Kengaku, M., Okamoto, H., bFGF as a possible morphogen for the anteroposterior axis of the central nervous system in Xenopus, *Development,* 121, 3121, 1995.

10. Tanabe, Y. and Jessell, T.M., Diversity and pattern in the developing spinal cord, *Science*, 274, 1115, 1996.

11. Durston, A.J., Timmermans, J.P., Hage, W.J., Hendriks, H.F., de Vries, N.J., Heideveld, M., Nieuwkoop, P.D., Retinoic acid causes an anteroposterior transformation in the developing central nervous system, *Nature,* 340, 140, 1989.

12. Sive, H.L., Draper, B.W., Harland, R.M., Weintraub, H., Identification of a retinoic acid-sensitive period during primary axis formation in Xenopus laevis, *Genes Dev*, 4, 932, 1990.

13. Hill, J., Clarke, J.D., Vargesson, N., Jowett, T., Holder, N., Exogenous retinoic acid causes specific alterations in the development of the midbrain and hindbrain of the zebrafish embryo including positional respecification of the Mauthner neuron, *Mech Dev*, 50, 3, 1995.

14. Ang, S.L., Rossant, J., HNF-3 beta is essential for node and notochord formation in mouse development, *Cell*, 78, 561, 1994.

15. Bouwmeester, T., Kim, S., Sasai, Y., Lu, B., De Robertis, E.M., Cerberus is a head-inducing secreted factor expressed in the anterior endoderm of Spemann's organizer, *Nature,* 382, 595, 1996.

16. Winnier, G., Blessing, M., Labosky, P.A., Hogan, B.L., Bone morphogenetic protein-4 is required for mesoderm formation and patterning in the mouse, *Genes Dev*, 9, 2105, 1995.

17. Tanabe, Y., Jessell, T.M., Diversity and pattern in the developing spinal cord, *Science*, 274, 1115, 1996.

18. Liem, K.F. Jr., Tremml, G., Roelink, H., Jessell, T.M., Dorsal differentiation of neural plate cells induced by BMP-mediated signals from epidermal ectoderm, *Cell*, 82, 969, 1995.

19. Liem, K.F. Jr., Tremml, G., Jessell, T.M., A role for the roof plate and its resident TGF beta-related proteins in neuronal patterning in the dorsal spinal cord, *Cell*, 91, 127, 1997.

20. Dudley, A.T., Robertson, E.J., Overlapping expression domains of bone morphogenetic protein family members potentially account for limited tissue defects in BMP7 deficient embryos, *Dev Dynamics,* 208, 349, 1997.

21. Schulte-Merker, S., Lee, K.J., McMahon, A.P., Hammerschmidt, M., The zebrafish organizer requires chordino, *Nature,* 387, 862, 1997.

22. Monsoro-Burq, A.H., Duprez, D., Watanabe, Y., Bontoux, M., Vincent, C., et al., The role of bone morphogenetic proteins in vertebral development. *Development,* 122, 3607, 1996.

23. Davidson, D., The function and evolution of Msx genes: pointers and paradoxes, *Trends Genet,* 11, 405, 1995.

24. Stuart, E.T., Kioussi, C., Gruss, P., Mammalian Pax genes, *Annu Rev Genet,* 28, 219, 1994.

25. Mansouri, A., Stoykova, A., Torres, M., Gruss, P., Dysgenesis of cephalic neural crest derivatives in Pax7-/- mutant mice, *Development,* 122, 831, 1996.

26. Dickinson, M.E., Selleck, M.A., McMahon, A.P., Bronner-Fraser, M., Dorsalization of the neural tube by the non-neural ectoderm, *Development,* 121, 2099, 1995.

27. Jiang, R., Lan, Y., Norton, C.R., Sundberg, J.P., Gridley, T., The slug gene is not essential for mesoderm or neural crest development in mice, *Dev Biol,* 198, 277, 1998.

28. LaBonne, C., Bronner-Fraser, M., Neural crest induction in Xenopus: evidence for a two signal model, *Development,* 125, 2403, 1998.

29. Nakagawa, S., Takeichi, M., Neural crest cell–cell adhesion controlled by sequential and subpopulation-specific expression of novel cadherins, *Development,* 121, 1321, 1995.

30. Liu, J.P., Jessell, T.M., A role for rhoB in the delamination of neural crest cells from the dorsal neural tube, *Development,* 125, 5055, 1998.

31. Watanabe, Y. et al., Lateral and axial signals involved in avian somite patterning: a role for BMP4, *Cell,* 84, 461, 1996.

32. Lee, K.J., Mendelsohn, M., Jessell, T.M., Neuronal patterning by BMPs: a requirement for GDF7 in the generation of a discrete class of commissural interneurons in the mouse spinal cord, *Genes Dev,* 12, 3394, 1998.

33. Lyons, K.M., Hogan, B.L., Robertson, E.J., Colocalization of BMP 7 and BMP 2 RNAs suggests that these factors cooperatively mediate tissue interactions during murine development, *Mech Dev,* 50, 71, 1995.

34. Furuta, Y., Piston, D.W., Hogan, B.L.M., Bone morphogenetic proteins (BMPs) as regulators of dorsal forebrain development, *Development,* 124, 2203, 1997.

35. Dale, J.K., Vesque, C., Lints, T.J., Sampath, T.K., Furley, A., et al., Cooperation of BMP7 and SHH in the induction of forebrain ventral midline cells by prechordal mesoderm, *Cell,* 90, 257, 1997.

36. Lee, K.J., Jessell, T.M., The specification of dorsal cell fates in the vertebrate central nervous system, *Annu Rev Neurosci,* 22, 261, 1999.

37. Pourquie, O., Fan, C.M., Coltey, M., Hirsinger, E., Watanabe, Y., et al., Lateral and axial signals involved in avian somite patterning: a role for BMP4, *Cell,* 84, 461, 1996.

38. Muhr, J., Jessell, T.M., Edlund, T., Assignment of early caudal identity to neural plate cells by a signal from caudal paraxial mesoderm, *Neuron,* 19, 487, 1997.

39. Schultheiss, T.M., Burch, J.B., Lassar, A.B., A role for bone morphogenetic proteins in the induction of cardiac myogenesis, *Genes Dev,* 11, 451, 1997.

40. Arkell, R., Beddington, R.S.P., BMP7 influences pattern and growth of the developing hindbrain of mouse embryos, *Development,* 124, 1, 1997.

41. Fraser, S., Keynes, R., Lumsden, A., Segmentation in the chick embryo hindbrain is defined by cell lineage restrictions, *Nature,* 344, 431, 1990.

42. Graham, A., Francis-West, P., Brickell, P., Lumsden, A., The signalling molecule BMP4 mediates apoptosis in the rhombencephalic neural crest, *Nature,* 372, 684, 1994.

43. Graham, A., Heyman, I., Lumsden, A., Even-numbered rhombomeres control the apoptotic elimination of neural crest cells from odd-numbered rhombomeres in the chick hindbrain, *Development,* 119, 233, 1993.

44. Graham, A., Francis, W.P., Brickell, P., Lumsden, A., The signalling molecule BMP4 mediates apoptosis in the rhombencephalic neural crest, *Nature,* 372, 684, 1994.

45. Joyner, A.L., Engrailed, wnt and pax genes regulate midbrain–hindbrain development, *Trends Genet*, 12, 15, 1996.
46. Shetty, A.K., Turner, D.A., *In vitro* survival and differentiation of neurons derived from epidermal growth factor-responsive postnatal hippocampal stem cells: inducing effects of brain-derived neurotrophic factor, *J Neurobiol*, 35, 395, 1998.
47. Ling, Z.D., Tong, C.W., Carvey, P.M., Partial purification of a pramipexole-induced trophic activity directed at dopamine neurons in ventral mesencephalic cultures, *Brain Res*, 791, 137, 1998.
48. Wichterle, H., Garcia-Verdugo, J.M., Herrera, D.G., Alvarez-Buylla, A., Young neurons from medial ganglionic eminence disperse in adult and embryonic brain, *Nat Neurosci*, 2, 461, 1999.
49. Ware, C.B., Horowitz, M.C., Renshaw, B.R., Hunt, J.S., Liggitt, D., Koblar, S.A., Gliniak, B.C., McKenna, H.J., Papayannopoulou, T., Thoma, B., et al., Targeted disruption of the low-affinity leukemia inhibitory factor receptor gene causes placental, skeletal, neural and metabolic defects and results in perinatal death, *Development*, 121, 1283, 1995.
50. Bonni, A., Sun, Y., Nadal-Vicens, M., Bhatt, A., Frank, D.A., Rozovsky, I., Stahl, N., Yancopoulos, G.D., Greenberg, M.E., Regulation of gliogenesis in the central nervous system by the JAK-STAT signaling pathway, *Science*, 278, 477, 1997.
51. Rajan, P., McKay, R.D., Multiple routes to astrocytic differentiation in the CNS, *J Neurosci*, 18, 3620, 1998.
52. Gross, R.E., Mehler, M.F., Mabie, P.C., Zang, Z., Santschi, L., Kessler, J.A., Bone, morphogenetic proteins promote astroglial lineage commitment by mammalian subventricular zone progenitor cells, *Neuron*, 17, 595, 1996.
53. Ochiai, W., Yanagisawa, M., Takizawa, T., Nakashima, K., Taga, T., Astrocyte differentiation of fetal neuroepithelial cells involving cardiotrophin-1-induced activation of STAT3, *Cytokine*, 14, 264, 2001.
54. Nakashima, K., Yanagisawa, M., Arakawa, H., Taga, T., Astrocyte differentiation mediated by LIF in cooperation with BMP2, *FEBS Lett*, 457, 43, 1999.
55. Yanagisawa, M., Nakashima, K., Takizawa, T., Ochiai, W., Arakawa, H., Taga, T., Signaling crosstalk underlying synergistic induction of astrocyte differentiation by BMPs and IL-6 family of cytokines, *FEBS Lett*, 489, 139, 2001.
56. Goodman, R.H., Smolik, S., CBP/p300 in cell growth, transformation, and development, *Genes Dev*, 14, 1553, 2000.
57. Chrivia, J.C., Kwok, R.P., Lamb, N., Hagiwara, M., Montminy, M.R., Goodman, R.H., Phosphorylated CREB binds specifically to the nuclear protein CBP, *Nature*, 365, 855, 1993.
58. Kamei, Y., Xu, L., Heinzel, T., Torchia, J., Kurokawa, R., Gloss, B., Lin, S.C., Heyman, R.A., Rose, D.W., Glass, C.K., Rosenfeld, M.G., A CBP integrator complex mediates transcriptional activation and AP-1 inhibition by nuclear receptors, *Cell*, 85, 403, 1996.
59. Kurokawa, R., Kalafus, D., Ogliastro, M.H., Kioussi, C., Xu, L., Torchia, J., Rosenfeld, M.G., Glass, C.K., Differential use of CREB binding protein-coactivator complexes, *Science*, 279, 700, 1998.
60. Nakashima, K., Takizawa, T., Ochiai, W., Yanagisawa, M., Hisatsune, T., Nakafuku, M., Miyazono, K., Kishimoto, T., Kageyama, R., Taga, T., BMP2-mediated alteration in the developmental pathway of fetal mouse brain cells from neurogenesis to astrocytogenesis, *Proc Natl Acad Sci*, 98, 5868, 2001.
61. Yanagisawa, M., Takizawa, T., Ochiai, W., Uemura, A., Nakashima, K., Taga, T., Fate alteration of neuroepithelial cells from neurogenesis to astrocytogenesis by bone morphogenetic proteins, *Neurosci Res*, 41, 391, 2001.

62. Takizawa, T., Ochiai, W., Nakashima, K., Taga, T., Enhanced gene activation by Notch and BMP signaling cross-talk, *Nucleic Acids Res*, 31, 5723, 2003.

63. Mabie, P.C., Mehler, M.F., Kessler, J.A., Multiple roles of bone morphogenetic protein signaling in the regulation of cortical cell number and phenotype, *J Neurosci*, 19, 7077, 1999.

64. Marti, E., Bumcrot, D.A., Takada, R., McMahon, A.P., Requirement of 19K form of sonic hedgehog for induction of distinct ventral cell types in CNS explants, *Nature*, 375, 322, 1995.

65. Chiang, C., Litingtung, Y., Lee, E., Young, K.E., Corden, J.L., Westphal, H., Beachy, P.A., Cyclopia and defective axial patterning in mice lacking sonic hedgehog gene function, *Nature*, 383, 407, 1996.

66. Hammerschmidt, M., Brook, A., McMahon, A.P., The world according to hedgehog, *Trends Genet*, 13, 14, 1997.

67. Ericson, J., Rashbass, P., Schedl, A., Brenner-Morton, S., Kawakami, A., van Heyningen, V., Jessell, T.M., Briscoe, J., Pax6 controls progenitor cell identity and neuronal fate in response to graded Shh signaling, *Cell*, 90, 169, 1997.

68. Briscoe, J., Pierani, A., Jessell, T.M., Ericson, J., A homeodomain protein code specifies progenitor cell identity and neuronal fate in the ventral neural tube, *Cell*, 101, 435, 2000.

69. Basler, K., Edlund, T., Jessell, T.M., Yamada, T., Control of cell pattern in the neural tube: regulation of cell differentiation by dorsalin-1, a novel TGF beta family member, *Cell*, 73, 687, 1993.

70. Barth, K.A., Kishimoto, Y., Rohr, K.B., Seydler, C., Schulte-Merker, S., Wilson, S.W., Bmp activity establishes a gradient of positional information throughout the entire neural plate, *Development*, 126, 4977, 1999.

71. Nguyen, V.H., Trout, J., Connors, S.A., Andermann, P., Weinberg, E., Mullins, M.C., Dorsal and intermediate neuronal cell types of the spinal cord are established by a BMP signaling pathway, *Development*, 127, 1209, 2000.

72. Hall, A.K., Miller, R.H., Emerging roles for bone morphogenetic proteins in central nervous system glial biology, *J Neurosci Res*, 76, 1, 2004.

73. McMahon, J.A., Takada, S., Zimmerman, L.B., Fan, C.M., Harland, R.M., McMahon, A.P., Noggin-mediated antagonism of BMP signaling is required for growth and patterning of the neural tube and somite, *Genes Dev*, 12, 1438, 1998.

74. Liem, K.F. Jr., Jessell, T.M., Briscoe, J., Regulation of the neural patterning activity of sonic hedgehog by secreted BMP inhibitors expressed by notochord and somites, *Development*, 127, 4855, 2000.

75. Ono, K., Bansal, R., Payne, J., Rutishauser, U., Miller, R.H., Early development and dispersal of oligodendrocyte precursors in the embryonic chick spinal cord, *Development*, 121, 1743, 1995.

76. Pringle, N.P., Yu, W.P., Guthrie, S., Roelink, H., Lumsden, A., Peterson, A.C., Richardson, W.D., Determination of neuroepithelial cell fate: induction of the oligodendrocyte lineage by ventral midline cells and sonic hedgehog, *Dev Biol*, 177, 30, 1996.

77. Orentas, D.M., Miller, R.H., The origin of spinal cord oligodendrocytes is dependent on local influences from the notochord, *Dev Biol*, 177, 43, 1996.

78. Poncet, C., Soula, C., Trousse, F., Kan, P., Hirsinger, E., Pourquie, O., Duprat, A.M., Cochard, P., Induction of oligodendrocyte progenitors in the trunk neural tube by ventralizing signals: effects of notochord and floor plate grafts, and of sonic hedgehog, *Mech Dev*, 60, 13, 1996.

79. Mekki-Dauriac, S., Agius, E., Kan, P., Cochard, P., Bone morphogenetic proteins negatively control oligodendrocyte precursor specification in the chick spinal cord, *Development*, 129, 5117, 2002.

80. Basler, K., Edlund, T., Jessell, T.M., Yamada, T., Control of cell pattern in the neural tube: regulation of cell differentiation by dorsalin-1, a novel TGF beta family member, *Cell*, 73, 687, 1993.

81. Miller, R.H., Dinsio, K., Wang, R., Geertman, R., Maier, C.E., Hall, A.K., Patterning of spinal cord oligodendrocyte development by dorsally derived BMP4, *J Neurosci Res*, 76, 9, 2004.

82. Raff, M.C., Miller, R.H., Noble, M., 1983. A glial progenitor cell that develops *in vitro* into an astrocyte or an oligodendrocyte depending on culture medium, *Nature*, 303, 390, 1983.

83. Mayer, M., Bhakoo, K., Noble, M., Ciliary neurotrophic factor and leukemia inhibitory factor promote the generation, maturation and survival of oligodendrocytes *in vitro*, *Development*, 120, 143, 1994.

84. Hughes, S.M., Lillien, L.E., Raff, M.C., Rohrer, H., Sendtner, M., Ciliary neurotrophic factor induces type-2 astrocyte differentiation in culture, *Nature*, 335, 70, 1988.

85. Lillien, L.E., Raff, M.C., Differentiation signals in the CNS: type-2 astrocyte development *in vitro* as a model system, *Neuron*, 5, 111, 1990.

86. Gard, A.L., Williams, W.C., Burrell, M.R., Oligodendroblasts distinguished from O-2A glial progenitors by surface phenotype (O4+GalC-) and response to cytokines using signal transducer LIFR beta, *Dev Biol.*, 167, 596, 1995.

87. Mabie, P.C., Mehler, M.F., Marmur, R., Papavasiliou, A., Song, Q., Kessler, J.A., Bone morphogenetic proteins induce astroglial differentiation of oligodendroglial-astroglial progenitor cells, *J Neurosci*, 17, 4112, 1997.

88. Grinspan, J.B., Edell, E., Carpio, D.F., Beesley, J.S., Lavy, L., Pleasure, D., Golden, J.A., Stage-specific effects of bone morphogenetic proteins on the oligodendrocyte lineage, *J Neurobiol*, 43, 1, 2000.

89. Angley, C., Kumar, M., Dinsio, K.J., Hall, A.K., Siegel, R.E., Signaling by bone morphogenetic proteins and Smad1 modulates the postnatal differentiation of cerebellar cells, *J Neurosci*, 23, 260, 2003.

90. Mehler, M.F., Mabie, P.C., Zhu, G., Gokhan, S., Kessler, J.A., Developmental changes in progenitor cell responsiveness to bone morphogenetic proteins differentially modulate progressive CNS lineage fate, *Dev Neurosci*, 22, 74, 2000.

91. Baggiolini, M., Dewald, B., Moser, B., Interleukin-8 and related chemotactic cytokines—CXC and CC chemokines, *Adv Immunol*, 55, 97, 1994.

92. Murphy, P., The molecular biology of leukocyte chemoattractant receptors, *Annu Rev Immunol*, 12, 593, 1994.

93. Horuk, R., Martin, A.W., Wang, Z., Schweitzer, L., Gerassimides, A., Guo, H., Lu, Z., Hesselgesser, J., Perez, H.D., Kim, J., Parker, J., Hadley, T.J., Peiper, S.C., Expression of chemokine receptors by subsets of neurons in the central nervous system, *J Immunol*, 158, 2882, 1997.

94. Robinson, S., Tani, M., Strieter, R.M., Ransohoff, R.M., Miller, R.H., The chemokine growth-related oncogene-a promotes spinal cord oligodendrocyte precursor proliferation, *J Neurosci*, 18, 10457, 1998.

95. Wu, Q., Miller, R.H., Ranshoff, R.M., Robinson, S., Bu, J., Nishiyama, A., Elevated levels of the chemokine GRO-1 correlate with elevated oligodendrocyte progenitors in the Jimp mutant, *J Neurosci*, 20, 2609, 2000.

96. Tsai, H.H., Frost, E., To V., Robinson, S., French-Constant, C., Geertman, R., Ransohoff, R.M., Miller, R.H., The chemokine receptor CXCR2 controls positioning of oligodendrocyte precursors in developing spinal cord by arresting their migration, *Cell*, 110, 373, 2002.

97. Ohtani, Y., Minami, M., Kawaguchi, N., Nishiyori, A., Yamamoto, J., Takami, S., Satoh, M., Expression of stromal cell-derived factor-1 and CXCR4 chemokine receptor mRNAs in cultured rat glial and neuronal cells, *Neurosci Lett*, 249, 163, 1997.

98. Bajetto, A., Bonavia, R., Barbero, S., Piccioli, P., Costa, A., Florio, T., Schettini, G., Glial and neuronal cells express functional chemokine receptor CXCR4 and its natural ligand stromal cell-derived factor1, *J Neurochem*, 73, 2349, 1999.

99. Hesselgesser, J., Taub, D., Baskar, P., Greenberg, M., Hoxie, J., Kolson, D.L., Horuk, R., Neuronal apoptosis induced by HIV-1 gp120 and the chemokine SDF-1ais medi-ated by the chemokine receptor CXCR4, *Curr Biol*, 8, 595, 1998.

100. Lazarini, F., Casanova, P., Tham, T.N., De Clercq, E., Arenzana-Seisdedos F., Baleux F., Dubois-Dalcq M., Differential signaling of the chemokine receptor CXCR4 by stro-mal cell-derived factor 1 and the HIV glycoprotein in rat neurons and astrocytes, *Eur J Neurosci*, 12, 117, 2000.

101. Ma, Q., Jones, D., Borghesani, P.R., Segal, R.A., Nagasawa, T., Kishimoto, T., Bronson, R.T., Springer, T.A., Impaired B-lymphopoiesis, myelopoiesis, and derailed cerebellar neuron migration in CXCR4- and SDF-1-deficient mice, *Proc Natl Acad Sci*, 95, 9448, 1998.

102. Tachibana, K., Hirota, S., Lisaza, H., Yoshida, H., Kawabata, K., Kataoka, Y., Kitamura, Y., Matsushima, K., Yoshida, N., Nishikawa, S., Kishimoto, T., Nagasawa, T., The chemokine receptor CXCR4 is essential for vascularization of the gastrointestinal tract, *Nature*, 393, 591, 1998.

103. Zou, Y.R., Kottman, A.H., Kuroda, M., Taniuchi, I., Littman, D.R., Function of the chemokine receptor CXCR4 in haematopoiesis and in cerebella development, *Nature*, 393, 595, 1998.

104. Xiang, Y., Li, Y., Zhang, Z., Cui, K., Wang, S., Yuan, X., Wu, C.P., Poo, M.M., Duan, S., Nerve growth cone guidance mediated by Gprotein coupled receptors, *Nature Neu-rosci*, 5, 843–848, 2002.

7 Cytokines in CNS Inflammation

Trevor Owens, Rachel D. Wheeler, and Simone P. Zehntner

CONTENTS

I. INTRODUCTION

Until quite recently, the prevailing logic underpinning discussion of cytokines in CNS inflammation was that cytokines were of immune origin and had a negative impact on the target organ. However, cytokines (we include chemokines under the general heading of cytokines) are quasi-universal mediators of biological processes and so are involved at all levels in the interaction between a tissue or organ and the immune inflammatory response. The cytokine 'storm' that erupts during an inflammatory response in the CNS includes cytokines produced by glial cells as well as by infiltrating leukocytes. Even without extending the definition of cytokines to embrace neurotrophins, evidence for both glial and neuronal sources of these mediators has accumulated to the extent that CNS-resident cells can be seen as almost equal partners in inflammatory processes in the CNS. Factoring in neurotrophins, which are themselves produced by infiltrating leukocytes as well as by neural cells, only strengthens the impression of an equal partnership.

Leukocyte infiltration is a major event in CNS inflammation. The process of cellular entry may be discriminated from overt immune inflammation and, more importantly, from triggering by antigen recognition of T cells within the CNS. What follows from these initiating events includes glial cell activation and the activation of myeloid cells that coinfiltrate with T cells. Every stage of this cascade of cellular activation involves cytokine production. The purpose of this chapter is to provide an update to the many excellent reviews on the role of cytokines in the CNS, in asking whether and to what extent cytokines contribute to inflammatory pathology in the CNS and the subsequent repair and regeneration processes that define its outcome.

II. INFLAMMATORY RESPONSES IN THE CNS

A. INNATE VS. IMMUNE INFLAMMATION

There is an emerging sense that CNS inflammation is not an all-or-none process, but that different levels of inflammation may be defined. For instance, reactive microglia, one of the hallmarks of inflammation in the CNS, differ in their cytokine profile depending on whether they are associated with a prion disease or with infection [1]. Thus, the activation state of glia may differ, depending on the specific insult to which they are responding, e.g., infection, injury, autoimmune infiltration. Although pure glial activation may be considered as one end of a spectrum of inflammatory pathologies, there are reasons to be cautious about excluding the possibility of leukocyte infiltration in even such cases. Glial activation and even leukocyte entry to the CNS can occur under circumstances that do not involve antigen triggering of effector lymphocytes. As one specific illustration, the microglial and astroglial response that accompanies chemical-induced demyelination or experimental axotomy is also associated with entry of leukocytes, in the absence of blood–brain barrier breach [2–4]. What distinguishes these apparently innocuous (and, to date, functionally mysterious) infiltrates from inflammatory infiltration seems to be the triggering of immune effectors through recognition of antigen. The role of antigen-specific T cells therefore defines whether inflammation in the CNS should be considered as innate vs. immune. However, despite their critical role in initiation of inflammation, the contribution of T cells to the subsequent progression of an inflammatory response may ultimately be equaled if not surpassed by myeloid and glial cells.

B. LEUKOCYTE ENTRY TO THE CNS

The question of whether the CNS is truly 'immune-privileged' continues to be discussed. There has been a tendency to extend observations that CNS and peripheral circulatory systems are isolated from one another, as evidenced by orders of magnitude differences in immunoglobulin concentrations, to the movement of immune cells across the blood–brain barrier. There appears to be a consensus now that leukocytes can access the CNS across an intact blood–brain barrier, though this has yet to be observed directly by intravital microscopy (which allows observation of lymphocyte movement in pial blood vessels). Studies to date describe very little leukocyte adhesion in pial blood vessels of healthy mouse brain [5,6]. The presence of leukocytes in healthy brain is felt to reflect the immune surveillance role of T lymphocytes, though there is a major intellectual divide between those who support this theory and those whose work would suggest that any entry of activated T cells into the CNS has deleterious pathological consequences. Throughout the body, T cells have a propensity to infiltrate tissues, particularly when activated through TCR triggering, which induces selective adhesion ligand and chemokine receptor expression. However, some contentious issues remain, such as whether and to what extent a T cell or other leukocyte needs to be activated for entry into the CNS and the location of any non-CNS-specific immune cells within the CNS.

Conventional histological examination of the healthy CNS does not reveal the presence of immune cells, whereas flow cytometric analysis of CNS homogenates prepared from perfused animals using density gradient fractionation suggests that T cells with naïve status are present [7]. Where are the T cells that flow cytometry finds? A few round cells can be observed in meninges, by careful histological examination, and it is probable that the cells identified in flow cytometric analyses of healthy CNS are in fact extraparenchymal, although outside the perfusable blood compartment. In the inflamed CNS, it is clear that T cells do infiltrate the parenchyma and so cells identified from tissue homogenates by flow cytometry must include these. Studies that have addressed the location of non-TCR-triggered T cells, that may be considered as 'bystanders [8]. have shown that ovalbumin-specific T cells also infiltrate the parenchyma in rats [9], though the functional significance of these bystander T cells remains unclear.

Recent studies using chimeric mice reconstituted with fluorescent protein transgenic bone marrow have revealed that blood-derived cells enter the CNS parenchyma and in some cases take on the location and morphology of resident microglia [10,11]. Although these experiments identify cellular exchange between blood-derived and CNS-resident cells, such exchange takes place on a very small scale and is not sufficient to significantly affect the majority of parenchymal populations in the context of most experimental paradigms involving irradiation and reconstitution [12]. Nevertheless, such experiments do show the capacity (albeit limited) of leukocytes that have not been experimentally activated to enter CNS parenchyma and for cellular exchange to occur between blood and brain. One caveat is that such experiments depend on prior lethal irradiation, and it is not clear to what extent tissue damage caused by irradiation provokes either leukocyte infiltration or microglial turnover. It is also not clear whether leukocyte activation through homeostatic expansion in irradiated animals contributes to propensity for entry to the CNS. The impetus for cellular entry in all cases likely includes action of cytokines and chemokines.

C.   Chemokines Direct Leukocyte Entry to the CNS

Since their definition as a family and the explosion of interest in identifying and characterizing them, it has been accepted that chemokines are critical mediators directing immune cell entry into the CNS [13]. A major conundrum has been that production of chemokines, like many cytokines, is not constitutive but needs to be induced. In particular, the entry of adoptively transferred cells into the uninflamed CNS of a transfer recipient has been difficult to explain by a chemokine model, given that the CNS is not intrinsically inflamed. Recent work using the animal model, experimental autoimmune encephalomyelitis (EAE), describes the induction of the chemokines CCL1/TCA-3, CCL2/MCP-1 and CXCL10/IP-10 at the blood–brain barrier by the cytokine tumor necrosis factor-α (TNFα), produced by immune cells in blood [14]. Mice lacking the CCR8 receptor for chemokine CCL1/TCA-3 were resistant to EAE, as are CCL2/MCP-1-deficient mice, demonstrating the importance of chemokines in immune cell infiltration of the CNS [14,15]. Mice deficient in TNFα expression show a delay in onset of EAE, with accumulation of leukocytes in the perivascular space [16]. This chemokine-inducing role of cytokines produced by peripheral immune cells may be considered a fundamental mechanism for the extravasation of immune cells throughout the body. There may also be additional stimulus from antigen presentation at the blood–brain barrier, an issue that is somewhat contentious [17,18]. Chemokines also regulate immune cell entry into the CNS in innate inflammation, for instance in response to injury. In injury-reactive hippocampus or spinal cord, there is a lack of macrophage entry in mice that lack the CCR2 chemokine receptor [4,19].

Non-antigen-directed interactions between T cells and blood–brain barrier endothelia involving CD40L/CD154, whose expression is increased upon TCR triggering, have potential for chemokine induction [20]. This identifies a mechanism whereby activated T cells of any specificity can access the CNS. Whether a passive attribute of the peripherally activated T cell, or a more active interaction between the immune and nervous systems, chemokine induction is among the first steps required for immune cell entry to the CNS. Other signals that contribute to immune cell entry to the CNS, such as adhesion ligand/receptor interactions and action of matrix metalloproteinase enzymes, are beyond the scope of this review and the reader is referred to two among many fine reviews [21,22].

Cellular sources of chemokines in the CNS include astrocytes, endothelial cells, and microglia and likely include most if not all infiltrating leukocytes, though each chemokine has a distinct pattern of expression, e.g., in EAE, TCA-3/CCL1 is expressed specifically by microglia [14]. Overexpression of chemokines in the CNS through transgenesis has led to leukocyte infiltration in the case of CXCL1/KC/Groα [23], CCL21 [24], and CXCL10/IP-10 [25]. Different chemokines act on distinct immune cell populations, shaping the infiltrate composition. For instance, MBP promoter-driven CXCL1/KC or CCL21 expression induced neutrophil but not lymphocyte infiltration [23,24], whereas CD3$^+$ CXCR3$^+$ cells were recruited to the CNS of GFAP promoter-driven CXCL10/IP-10 transgenics, along with neutrophils that were CXCR3$^+$ [25]. Interestingly, when expressed under control of a GFAP promoter, CCL2/MCP-1 induced parenchymal leukocyte infiltration [26], whereas MBP promoter-driven expression resulted only in perivascular accumulation of leukocytes [27], reminiscent of the effects of immunization of mice lacking TNFα, described above. Whereas mice lacking expression of CCL2/MCP-1 or its principal receptor CCR2 were resistant to EAE [15,28], the same was not true for mice lacking the chemokine CCL5/RANTES or its receptor [29]. Thus, not all chemokines are critical for immune cell infiltration of the CNS. However, precise assignation of specific pathological effects to one chemokine or receptor is difficult because of redundancy and promiscuity in chemokine-receptor interactions and because of the multiplicity of chemokines and their sources in an inflammatory response.

D. T-CELL TRIGGERING IN THE CNS

The fact that T cells specific for myelin and for antigens that are not expressed in the CNS can be found in CNS isolates [7] reinforces the view that antigen recognition is not essential for T-cell entry to the CNS. Similarly, ovalbumin-specific T cells migrate to the CNS under certain conditions [8,9]. However, secondary reactivation or triggering of T cells is required for cytokine production post-entry to CNS, and this is likely the single major determinant of whether there will be an immune response. The antigen presenting cells that present antigenic peptides to T cells likely include resident microglia and coinfiltrating macrophages [30,31]. There is evidence from transgenic models that microglia can initiate autoimmune inflammation in the CNS [32], and it is known that macrophages are required for induction of EAE [33,34].

A recent study described the ability of T cells to induce apoptotic cell death of human neurons in primary or slice cultures by perforin or glutamate mechanisms [35]. Curiously, this cytotoxicity appeared to be mediated through simple cell–cell contact, regardless of the antigen specificity of the T cell, and might also be expected to affect cells outside the CNS with devastating effects. It would seem very important for maintenance of CNS integrity to control either access or activity of such cytotoxic T cells. The experimental systems reported may lack some biological regulatory control that normally prevents widespread degenerative effects of the activation of these T cells. Indeed, observations of direct cytotoxicity against neurons resulting from simple non-MHC-restricted contact with activated T cells challenge the notion that activated T cells have access to the CNS and perform an immune surveillance role. The two seem intellectually incompatible, and this issue requires resolution.

III. THE ROLE OF CYTOKINES IN CNS INFLAMMATION

Many inflammatory cytokines and chemokines are not expressed at detectable levels in the uninflamed CNS, but are rapidly induced following insult, injury, or immune

TABLE 7.1
Cytokine Actions in CNS Inflammation

Critical Points in Inflammation	Cytokines That Are Implicated
Leukocyte entry to CNS	TNFα, MCP-1/CCL2, TCA-3/CCL1, IP10/CXCL10, CXCL1/KC/Groα, CCL21, CCL5/RANTES, CXCL2/MIP2
Glial cell activation and mobilization	CXCL1/KC/Groα, TNFα
Reactivation of lymphocytes in CNS	IL-12 family, IL-18, IL-10, Type I interferons, osteopontin
Immune effector function	IFNγ (human), TNFα, lymphotoxin, TRAIL, LIGHT, NO, IL-17,
Immune regulation	IFNγ (rodent), TGFβ, IFNβ, IL-10, NO
Precursor mobilization	KC/Groα/CXCL1, PDGFα
Neuroprotection/remyelination	TGFβ, NT3, BDNF, LIF, CNTF, CLC/CLF

triggering. This has been shown by both selective [36] and unbiased screens. Cytokines and neurotrophic factors are prominent among the genes up-regulated following traumatic brain injury (for review see [37]). A microarray-based screen for gene expression in MS and EAE showed that among the genes selectively up-regulated were inflammatory cytokines, cytokine and chemokine receptors, and downstream signaling intermediates [38]. Cytokines have numerous functions within the inflammatory response, directing immune cell actions within the CNS, regulating ongoing immune cell entry, and mediating direct effects within the CNS. The eventual outcome of the CNS inflammatory response will depend on the circumstances within which it is occurring e.g., resolution of infection, autoimmune demyelination, neuronal injury. The rest of this chapter will focus on these aspects of cytokine action in CNS inflammation (see Table 7.1 for a summary).

IV. CYTOKINES THAT DIRECT IMMUNE RESPONSES IN THE CNS

A. THE TH1 VS. TH2 PARADIGM

Antigen-presenting cells (APCs) direct the quality of the emergent T cell response and have the capacity to subvert CNS immune responses toward inflammatory or noninflammatory outcomes, including whether CD4+ T cell activation generates a Th1 or a Th2 response. Although cell–surface interactions e.g., involving B7 family members or CD40, are known to play a role, the major influence on quality of T cell response is through release of specific regulatory cytokines by APCs. Macrophages that are induced with different colony-stimulating factors produce distinct cytokines and differentially elicit Th1 (APC1) vs. Th2 (APC2) responses [39]. Prominent among these are IL-12, IL-18, and IL-23, all of which are implicated in induction of Th1 immune responses and specifically interferon-gamma (IFNγ) production, and IL-10, which directs Th2 responses. These cytokines are discussed in more detail below. A question that arises is how are these APCs directed to induce Th1 or Th2 responses? The stimuli that regulate APC biology effectively represent the local microenvironment.

It has been proposed that the balance between Th1 and Th2 type responses controls the outcome of inflammation, such that a bias toward the IFNγ-producing Th1 phenotype is detrimental, whereas Th2-inducing conditions are beneficial (for review, see [40]). This may be exploited in MS and EAE to alleviate disease. For instance, glatiramer acetate, an immunotherapy for MS, has been shown to preferentially direct Th2 infiltration of the CNS in mice [41], and to direct dendritic cells to produce IL-10 and induce Th2 responses [42,43]. The importance of a Th1/Th2 balance in controlling the outcome of CNS inflammation has become less clear recently, as the detrimental role of Th1 responses, particularly IFNγ, has been questioned (see below). What is clear is that, like in any tissue inflammatory response, the APC in the CNS exerts enormous control over the immune response via cytokine production.

B. THE IL-12 FAMILY

Members of the IL-12 family of cytokines (IL-12, IL-23, IL-27) are heterodimers, consisting of a soluble cytokine receptor-like molecule that associates with a specific cytokine polypeptide, enabling the complex to be secreted from cells [44]. In the case of IL-12 and IL-23, both cytokines contain the p40 polypeptide chain, which has similarity to the soluble IL-6 receptor. In association with p19, p40 forms IL-23, and in association with p35, it generates IL-12. The IL-12 family of cytokines induces IFNγ production from T cells, but differs in the specific stage at which they mediate this action, as well as having additional individual functions.

Recent findings suggest that there may be critical differences between the periphery and the CNS in terms of the cytokines that are key in directing inflammatory responses. IL-12 is critical for generation of a Th1 response in peripheral lymphoid tissue. Interestingly, peripheral injection of IL-12 into mice that have received a spinal cord injury induces EAE-like symptoms, with inflammatory lesions present in the CNS that contain macrophages, CD4+ T cells and activated microglia [45]. However, within the CNS, IL-12 seems secondary to IL-23, as shown by relative susceptibility of mice lacking p19, p35 or p40 to induction of EAE, or its modulation by peripheral vs. CNS-directed anti-cytokine or cytokine-supplementing interventions [46–48]. In particular, mice lacking p19 only in the CNS were resistant to induction of EAE, indicating that expression of IL-12 is not sufficient, and expression of IL-23 in the CNS is critical. Both microglia and macrophages can produce IL-12 and IL-23. However, microglia specifically express receptors for IL-12 whereas macrophages can respond to both IL-12 and IL-23 [47]. Thus, macrophages represent at least one target for the critical actions of IL-23 in EAE.

Before the discovery of IL-23 in 2000, many studies had investigated the role of IL-12 in the CNS. Inflammatory disease develops spontaneously in GFAP promoter-driven IL-12 transgenics, identifying a potential pathological role for this cytokine in the CNS [49]. However, in light of the more recent report that IL-23 rather than IL-12 is the functionally relevant cytokine in the CNS in EAE, this observation may serve more as a demonstration of potential rather than true insight into physiological mechanism. There is evidence to suggest that pathology in the IL-12 overexpressing mouse may reflect induction of IFNγ (see Section V).

C. IL-18

Interleukin-18 is a member of the IL-1 family of cytokines, sharing several features with IL-1β such as protein folding structure and caspase-1 mediated processing. IL-18 mRNA is widely expressed, with prominent expression by cells of the myeloid and dendritic lineages [50]. Similar to the IL-1 family, IL-18 signals via a receptor that triggers an IRAK and MyD88 intracellular signaling pathway, leading to NFkB activation [51]. The principal regulation of IL-18 activity is via cleavage of an inactive 24kD pro-form by caspase-1, to generate an 18kD mature functional protein [52]. Caspase-1 mRNA is expressed as widely as IL-18, but caspase-1 itself is regulated by proteolysis. IL-18 activity is also regulated by a circulating natural inhibitor, IL-18 binding protein (IL-18BP), which exists as multiple isoforms, though only some of these block IL-18 activity [53]. To date, IL-18BP expression has been detected in macrophages and endothelial cells *in vivo* [54].

Like IL-12, IL-18 is also implicated in the induction of IFNγ production and was originally named the IFNγ inducing factor. IL-12 and IL-18 exert a powerful synergy in inducing IFNγ, by interregulation of receptor chains [55]. A negative feedback loop operates whereby IFNγ induces IL-18BP expression [56]. Caspase-1 expression is also regulated by IFNγ [57–60]. IL-18 can induce cytokines other than IFNγ, includ-ing IL-1β, TNFα, and IL-6 in peripheral blood cells, and also in microglia [61]. There is evidence that IL-18 plays an important role in CNS inflammation, at least in part by acting though NK cells to promote peripheral Th1 immune responses. Mice that lack IL-18 are resistant to EAE, and anti-IL-18 antibodies are protective [62,63]. Caspase-1 deficiency or inhibition prevents or inhibits EAE, associated with impaired development of encephalitogenic Th1 cells [64].

D. Other Potential Th1 Switch Mediators

From a microarray screen of genes expressed in MS, two cytokines were singled out for more detailed analysis, GM-CSF and osteopontin/Eta-1. Mice deficient in either of these cytokines exhibited attenuated or reduced susceptibility to EAE, and increased IL-10/IFNγ ratios in the case of osteopontin [38,65,66]. Osteopontin is implicated in induction of Th1 T cell responses, and is itself regulated by IFNγ [67]. It is expressed in neurons in the developing rat brain and may play a role in normal cerebellar and brainstem development [68]. It has also been shown to be synthesized by reactive microglia in kainic acid lesions in rat and in astrocytes and microglia following LPS administration, suggesting a role in innate CNS physiology for this cytokine-like mediator [69,70].

E. IL-10

IL-10 is a classical Th2 cytokine that inhibits production of IFNγ by Th1 cells. It acts as a homodimer and is made by Th2 cells, but other cellular sources include myeloid cells and B cells. Indeed, the production of IL-10 by B cells is regulated by contact with activated T cells [71]. Mice deficient in IL-10 are more susceptible to EAE (reviewed in [72]), and IL-10 is widely believed to be beneficial in inflammation by promoting Th2 responses. The frequency of IL-10-producing cells and expression levels have been inversely correlated to progression of MS [73–75], leading to the frequent use of IL-10 as a prognostic indicator.

F. Type I Interferons

IFNβ is in use as a clinical therapy for MS. This likely reflects multiple activities, including inhibition of leukocyte migration across the blood–brain barrier and promotion of a Th2 response [76,77]. In contrast to this beneficial action of IFNβ in MS patients, transgenic expression of IFNα in astrocytes induced a degenerative pathology in mice, reminiscent of human neurodegenerative diseases and viral encephalitis [78]. Surprisingly, crossing those transgenics to mice lacking signaling elements thought to be critical for the IFNα receptor, STAT-1 and STAT-2, resulted in exacerbated pathology and in the case of STAT-2 deficiency, mice died of medulloblastoma [79]. This was associated with an IFNγ-producing T cell infiltrate, whose induction by IFN-α was STAT-1 and STAT-2 independent. IFNγ was shown to be responsible for STAT-2 independent up-regulation of sonic hedgehog and Gli-1 by cerebellar granule neurons, which likely contributed to medulloblastoma. Taken together with therapeutic effect in MS, these findings do not suggest a simple picture of the potential role of type I interferons in CNS inflammation.

The primary immunological function of type I interferons (IFNα and IFNβ) is in innate immunity against viral infection, and they can be induced via Toll-like receptor (TLR) signaling [80,81]. It has been suggested that TLRs have functions beyond their original definition as pathogen receptors [82]. TLR expression has been detected on cultured microglia, astrocytes and oligodendrocytes, and their expression is up-regulated in MS and EAE [83,84]. These observations have led to a growing interest in TLRs as regulators of innate responses in the CNS.

V. EFFECTOR CYTOKINES—AS EACH FALLS, ANOTHER RISES TO TAKE ITS PLACE

A. The TNF Family

TNFα is an important cytokine in inflammation that is produced by many different cell types. The contribution of different cells to TNFα levels in the CNS has been demonstrated elegantly in EAE [85]. TNFα actions are mediated through two receptors (TNFR1 and TNFR2) and the role of TNFα in the CNS has become more complex as more actions have been identified. Originally, it was believed to be detrimental. Intense interest was generated by the fact that TNFα could mediate oligodendrocyte death *in vitro* [86], and this led to transgenic models being developed that supported a proinflammatory role. However, it was not always obvious whether phenotypes of these mice reflected direct demyelination or activity of TNFα on cells other than oligodendrocytes. TNFα plays an important role in induction of EAE, as mice lacking TNFα or TNFRI show a significant delay in onset of disease (reviewed in [72]). However, this role may have more to do with the induction of chemokines (as discussed above) than in mediating direct pathology [14], and the overall quality and progression of EAE once induced was not different in TNFα knockouts from wild-type mice. Studies of TNFα knockouts have been complicated by abnormal secondary lymphoid organ development, and it has been shown that soluble TNFα is preferentially required for induction of inflammation whereas membrane-associated TNFα is critical for most aspects of secondary lymphoid organ development [87]. Exacerbated EAE in mice lacking the regenerative neurotrophin ciliary neurotrophic factor (CNTF) was corrected by antibodies against TNFα, identifying a balance between propathologic TNFα and repair processes [88].

Trials that tested TNFα-blocking agents in MS have reported surprisingly discouraging results [89]. Despite such therapies being effective against other inflammatory diseases, particularly arthritis, these agents induced more relapses when administered to MS patients. Recent findings suggest that whereas fetal rodent oligodendrocytes are quite sensitive to cytotoxic effects of TNFα *in vitro*, human adult oligodendrocytes are not [151]. This may suggest that a reevaluation of the role of TNFα in MS is necessary. TNFα can act through its two receptors with distinct biological outcomes, adding to the complexity of action of this cytokine. More needs to be known about the relative expression of these receptors in the CNS and how TNFα signals are directed in the context of inflammation. It is possible that actions of TNFα on cells involved in regeneration or repair (see Section V1) counteract its proinflammatory effects.

B. Lymphotoxins, Light and Trail

The elucidation of TNFα function in inflammation has been complicated by the presence of several other related ligands. The TNFα-related lymphotoxin molecule LTαβ, which binds to both TNF receptors as well as to distinct LTβR receptors, and the LTααβ homologue LIGHT (homologous to lymphotoxins, exhibits inducible expression, and competes with herpes simplex virus glycoprotein D for HVEM, a receptor expressed by T lymphocytes) have also been implicated in CNS inflammation. It has been difficult to discriminate between effects of TNFα, LT and LIGHT due to their close genetic linkage within the MHC. The fact that lack of TNFRI and LTβR signaling both lead to variably defective lymphoid development also has complicated interpretations [90,91], and complex chimeric models have been used to try to resolve these issues [92]. A recent study using selective inhibitors showed that LT and not LIGHT is required for induction of EAE [93]. Interestingly, this study discriminated between EAE models that did or did not use *Bordetella pertussis* as a coadjuvant. Only those models not dependent on pertussis were blocked by the LT inhibitor. This reflects a proposed role for pertussis in facilitating blood–brain barrier permeability, which itself relates to its activity in biogenic amine generation [94,95]. The role of biogenic amines in controlling autoimmune susceptibility continues to receive attention [96]. The fact that inhibition of EAE depended on whether or not pertussis was used introduces a cautionary note to all studies of the role of cytokines in EAE and their relevance to MS, especially as EAE studies increasingly use 129/J or C57BL/6 mouse strains that are dependent on pertussis.

Another member of the TNFR family, TNF-related apoptosis-inducing ligand (TRAIL), has been implicated in the cytotoxic effects of CNS inflammation. TRAIL, acting in a membrane bound or soluble trimeric form, induces apoptosis by binding to specific receptors, but most normal cells also express decoy TRAIL receptors that protect them from TRAIL-mediated apoptosis. However, it has been reported that TRAIL induced significant cell death of neurons, astrocytes, oligodendrocytes and microglia when applied to human brain slices, compared to vehicle-treated slices, at doses reported to be effective against tumor cells [97]. Activated T cells can express TRAIL, so this study raises concerns about the potential damaging effect of TRAIL+ T cells entering the CNS. TRAIL inhibitors exacerbated EAE rather than inhibiting disease [98], so the net effect of TRAIL in the CNS remains to be clarified.

C. IFNγ

Despite the fact that IFNγ is taken to define the Th1 immune response, investigations on the role of this cytokine in CNS inflammation have generated confounding and controversial findings.

Notably, IFNγ appears to play a regulatory role in EAE but a proinflammatory role in MS. Levels of IFNγ mRNA and protein in the CNS of rodents and MS patients correlate to disease severity (reviewed in [72]). This initially led to suggestions that IFNγ might promote disease. Two studies in MS in particular support this. Intravenous administration of IFNγ to MS patients provoked an increase in attack rate, and the treatment was consequently stopped [99]—increased NK cell activity and MHC II-expressing monocytes were noted in the serum of some patients. A more recent study showed amelioration of secondary progressive MS in patients treated with antibodies against the IFNγ receptor [100]. Both studies point to a pathological role of IFNγ in MS, although neither could discriminate between effects of treatments in the periphery or within the CNS.

Studies in mice, however, have generated conflicting results. Systemic administration of anti-IFNγ antibodies in mice exacerbated EAE, to the extent of overcoming strain-dependent resistance in one study (reviewed in [72]). Furthermore, direct administration of IFNγ into the CNS of rats or mice ameliorated EAE [72,101]. A number of reports show that mice lacking IFNγ or its receptor develop a severe, nonremitting, often lethal EAE, with an inflammatory infiltrate that is dominated by neutrophils rather than macrophages (reviewed in [72]). In addition to suggesting a protective role for IFNγ in EAE, these knock out studies clearly demonstrate that IFNγ is not required for the induction of EAE in mice.

There is some evidence in rodents that IFNγ induces inflammation when introduced into the normal CNS. For instance, direct intra-CNS administration of IFNγ induced inflammation in rats. In addition, a number of transgenic studies in which IFNγ was expressed under the control of CNS-specific promoters support a proinflammatory role for IFNγ (reviewed in [72]). However, it is worth considering two points regarding these transgenic studies. One is that IFNγ is undetectable by any means in the healthy, uninflamed adult CNS, so that overexpression at any level constitutes a significant change to normal CNS physiology. Secondly, with the exception of one study using a GFAP promoter [102], only MBP promoters have been reported for overexpression of IFNγ in the CNS. Whether this introduces a bias, similar to that described for MCP-1 transgenics (discussed in an earlier section) is not known. Further support for a proinflammatory role for IFNγ comes from recent work from Pagenstecher and colleagues [103]. They showed that the Borna virus clearance-promoting effect of IL-12 in mice expressing IL-12 transgenically in the CNS under control of a GFAP promoter was dependent on IFNγ. This is consistent with previous data describing a nonlytic antiviral effect of IFNγ, but not IL-12, in organotypic slice cultures, in the absence of lymphocytes [104]. One hypothesis is that the primary protective role of IFNγ in the CNS is probably to mediate antiviral protection, and points to a physiological role for IFNγ in immune surveillance. However, this seems unlikely to contribute to its regulatory role in EAE. An alternative regulatory mechanism has been identified by Zehntner et al., in studies comparing wild-type and IFNγ knockout mice [152]. IFNγ was demonstrated to control nitric oxide (NO) production by neutrophils, which in turn regulates T cell effects. This NO-mediated regulatory effect likely extends to other myeloid cells that infiltrate the CNS, based on observations made in EAE and other inflammatory models [105–107]. These data strengthen the case for a regulatory role for IFNγ in the mouse, and it may be that the dogma of IFNγ being proinflammatory requires reevaluation, at least in rodents.

As discussed earlier, cytokines play a key role in regulating immune cell entry to the CNS through induction of chemokines. IFNγ is an important cytokine in inducing chemokine expression and therefore regulating immune cell entry. A characteristic chemokine profile dominated by CCL5/RANTES and CCL2/MCP-1 is induced in the presence of IFNγ and is associated with macrophage infiltration, but in the absence of IFNγ, expression of RANTES and MCP-1 is markedly reduced with concomitant increase in expression of CXCL2/MIP-2, associated with a shift toward neutrophil infiltration. Very similar correlations between chemokine profiles and infiltration patterns have been shown for SJL/J, BALB/c and C57BL/6 mice with EAE [108, 153, Zehntner et al., unpublished]. Abromson-Leeman et al. [153] showed that the CXCR2 receptor (which can bind MIP-2) was critical for directing neutrophil infiltration (Am J Pathol (in press)). Direct analysis of effects of viral vector-delivered IFNγ in the CNS confirms selective induction of the T cell and macrophage–chemotactic chemokines RANTES/CCL5 and IP-10/CCL10 (Millward and Owens, unpublished). The net effect of IFNγ in the mouse CNS therefore appears to be to suppress T-cell responses and to redirect inflammatory infiltrates.

There is some controversy as to whether IFNγ is produced by the adult nervous system. There have been reports that neuronal and glial cells cultured in vitro can be stained for IFNγ or express mRNA for IFNγ, suggesting a role for IFNγ in innate CNS physiology [109–111]. However, many other labs consistently fail to detect IFNγ mRNA or protein, or signaling molecules associated with an IFNγ response. These negative findings tend not to be reported, but contribute to a sense that if neural cells can express IFNγ, they usually don't. The capacity of myeloid-lineage cells to produce IFNγ is now well-documented, [112] and usually occurs in response to an inflammatory stimulus. This allows the possibility that microglia might contribute to IFNγ titers in the inflamed CNS. However, it should be noted that the levels of IFNγ produced by myeloid cells are usually far less than those produced by T or NK cells.

D.   Nitric Oxide and Inducible Nitric Oxide Synthase

Nitric oxide (NO) is a free radical that has a short half life but is capable of very potent local effects. It is produced predominantly by inducible nitric oxide synthase (iNOS) during inflammation, though neuronal and endothelial NOS enzymes also exist. IFNγ is a major regulator of iNOS expression, and iNOS levels are markedly diminished in IFNγ-deficient mice with EAE [108]. It is therefore not surprising that, similar to IFNγ, contradictory roles in inflammation have been described for nitric oxide. There are several mechanisms by which NO is proposed to mediate detrimental effects in inflammation e.g., disruption of the blood–brain barrier. NO may also be a mediator of oligodendrocyte cytopathology in MS [113]. In contrast, mice that lack iNOS show exacerbated EAE, and a number of recent studies have identified a T cell-suppressive role for NO (reviewed in [114]). Attempts to reconcile these disparities have included localizing inflammatory vs. suppressive NO production to CNS vs. peripheral compartments in bone marrow chimeras [107,115]. A disease-suppressive role dominated in these studies. Although NO may march in step with IFNγ, the mouse continues to be out of step with humans, in this regard at least.

E.   IL-17

Given mounting evidence that interferon-gamma may not be the critical proin-flammatory cytokine in rodent EAE, attention shifts to alternatives. The IL-17 family of cytokines represents one likely alternative. Several proinflammatory actions have been described for IL-17, including up-regulation of other proinflam-matory cytokines and chemokines that recruit macrophages and neutrophils, including MCP-1, and promotion of T cell activation. There are at least six IL-17 family members, which have distinct expression patterns. For instance, IL-17 is specifically expressed in some activated memory T cells, induced by IL-23 or IL-18, whereas IL-17D is expressed particularly in skeletal muscle and the nervous system. Activated T cells that produce IL-17 co-secrete TNFα and GM-CSF, and fall outside traditional Th1 or Th2 categories (reviewed in [116]). Given the actions of IL-17 and TNFα, described above, the potential for such T cells to induce inflammation is obvious. Consistent with this, IL-17 is up-regulated in blood and CSF in MS [38,117]. IL-23–driven IL-17–secreting, CD4+ T cells were recently shown to be encephalitogenic [154]. Activated T cells may not be the only source of IL-17, as TNFα or IL-1β induce up-regulation of IL-17 in astrocytes [118]. It remains to be determined whether IL-17 has any direct actions on CNS resident cells, but it is clear that IL-17 will emerge as a key proinflammatory molecule in CNS inflammation.

VI.   CYTOKINES IN REGENERATION
AND REMYELINATION

The influx of inflammatory immune cells into the CNS is conventionally regarded as deleterious with pathologic consequences. Indeed, many therapies directed against demyelinating disease are in fact directed against the inflammatory immune response, and their efficacy is testament to the pathological impact of the immune component of inflammation. However, at some point, an inflammatory response will resolve (though this may not occur successfully). Thus, some processes during inflammation will trigger anti-inflammatory mechanisms as well as any necessary repair processes. In the context of CNS inflammation, it is neuronal function that is vulnerable, either from direct toxicity on neurons, or damage to axons, myelin sheaths or oligodendrocytes, resulting in disruption of axon function. Thus, neuro-protection may reflect action on neurons or oligodendrocytes.

A. OLIGODENDROCYTE PRECURSOR CELLS

It is now apparent that adult oligodendrocytes are largely ineffective at remyelination, and that populations of oligodendrocyte precursor cells (OPCs) present in adult human and rodent CNS mediate this role [119–124]. In rodents, OPCs are identified by a combination of morphology, expression of surface markers, such as the glyco-proteins NG2, O4, PDGFαR and A2B5, and by their lack of expression of myelin proteins such as MBP and MOG [125–128]. OPCs are guided within the CNS by chemotactic cues. For instance, the chemokine CXCL1/KC/Groα, acting through the CXCR2 receptor, has been implicated in OPC guidance in the developing spinal cord, and may also regulate OPC proliferation [129,130]. Interestingly, CXCR2 also mediates CXCL2/MIP-2-stimulated neutrophil infiltration into the CNS during inflammation [131], demonstrating that very different functions can be mediated by the same receptor, depending on the context. The fate of OPCs is determined in large part by their access to growth and survival factors. The factors that control static maintenance of OPCs in uninflamed tissue are not defined, although their phenotypic definition by expression of PDGFαR speaks to a role for (likely astrocyte-derived) PDGFα. Neurotrophins such as neurotrophin-3 (NT3) and BDNF, acting through p75 neurotrophin receptors, also play a critical role in oligodendrocyte development [132–134].

B. NEUROPROTECTIVE ROLES FOR CYTOKINES AND CHEMOKINES

Cells whose activation is associated with inflammation, both infiltrating immune cells and resident glial cells, may at the same time contribute to repair and regen-eration [135,136], particularly by secretion of neuroprotective factors. These may include cytokines (e.g., leukemia inhibitory factor, LIF; transforming growth factor beta, TGFβ), neurotrophins (e.g., BDNF; CNTF), or chemokines (e.g., CXCL1/KC). There has been a dramatic increase recently in studies investigating the function and regulation of factors that promote neuroregeneration and remyelination, particularly following CNS injury or inflammation. Although the functions of many of these molecules have been known for some time, their expression and regulation by immune cells is just beginning to be investigated.

Neurotrophins promote axonal survival, oligodendrocyte proliferation and remy-elination. This has been shown for NT3 and BDNF by grafting cells expressing these neurotrophins into injured rat spinal cord [137]. Infiltrating T cells and macrophages that express neurotrophins such as BDNF have been identified in MS or EAE lesions [138–140]. Glial cells also produce neurotrophins, and are induced to express neurotrophin receptors in MS [141–143]. Subsets of T cells, for instance those immunized in the gut, can express neurotrophic cytokines such as TGFβ [144], which has both immunoprotective and neuroregenerative effects. Importantly, glatiramer acetate-specific T cells, which are induced during MS therapy, can produce TGFβ and BDNF [41,140].

Neuroprotective molecules are important in limiting the damage caused by inflammation. For instance, demyelination was exacerbated in CNTF-deficient mice with EAE [145].

CNTF is protective for mature oligodendrocytes *in vitro* [146] and has been implicated in control of myelination [147]. The fact that the enhanced demyelination in these mice was counteracted by anti-TNFα antibodies adds to evidence for TNFα as a mediator of demyelinating pathology, although there remain contradictory findings (see the next paragraph) [145]. CNTF acts by binding to its receptor, CNTFR, and signaling through a complex with the LIF receptor and gp130 signaling chain. Several molecules that act through these signaling molecules have neuroprotective actions on either neurons or oligodendrocytes, including LIF, which enhances oligodendrocyte survival in the face of inflammatory cytotoxicity [148]. An alternative ligand for CNTFR is the CNTF homologue cardiotrophin-like cytokine (CLC), which is secreted as a complex with cytokine-like factor (CLF) [149,150]. CLC/CLF has neuroprotective functions, and its actions on oligodendrocytes have yet to be determined.

As described above for IFNγ, proinflammatory cytokines can also have regulatory functions. There is evidence that TNFα can play a regenerative role, in addition to its more well-defined proinflammatory functions. Mice that lacked either TNFα or TNFR2 had impaired remyelination in a toxin-induced model of demyelination [126]. TNFα is proposed to act directly on OPCs via TNFR2 to promote remyelination. This may help explain why therapies directed against TNFα were ineffective in MS, although there are likely to be other possible explanations.

VII. CONCLUSIONS

This review has sought to examine current understanding of the role of cytokines in CNS inflammation, looking at various mediators in some detail but attempting to understand their diverse roles as a functional unit. Cytokines are pleiotropic molecules that fulfill a wide range of roles in every step of the inflammatory process (Figure 7.1). They are involved from the early stages, regulating immune cell entry into the CNS through induction of specific chemokines. Cytokines signal the recognition of antigen, control the development of CNS immune responses toward a Th1 or Th2 profile, and drive further immune cell infiltration. Cytokines feed back to amplify inflammatory processes but also feed forward to initiate antiinflammatory mechanisms that will ultimately lead to resolution of the inflammatory response, as well as triggering neuroprotective and repair mechanisms, e.g., mobilization of OPCs. The role of an individual cytokine is often complex, as its actions may differ depending on the stage of inflammation at which it is acting. For instance, the

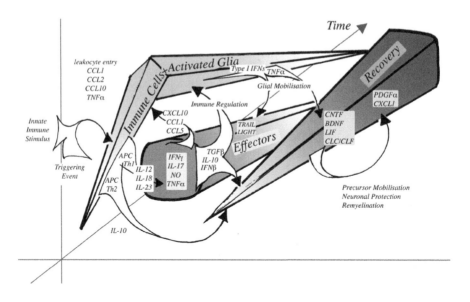

FIGURE 7.1 Cytokines in CNS inflammation. Over time, the intricate milieu of cytokines within the CNS direct a diverse array of events, contributing to cellular entry, the effector phase, and resolution of inflammation, as well as tissue recovery. Both infiltrating immune cells and activated resident glia produce and respond to individual cytokines. These cytokines play multiple roles.

expression levels of a cytokine's receptors may vary at different stages of the process, dictating what action the cytokine has at any one point. Alternatively, different ligands for an individual chemokine receptor may be present at different stages, such that very different effects can be mediated through the same receptor.

In light of their potency in regulating immune and glial responses, are cytokines sufficient to drive inflammatory pathology in the CNS? The transgenic models in which putative inflammatory cytokines have been expressed under control of CNS-specific promoters and which show spontaneous evolution of inflammatory, demyelinating, or degenerative phenotypes suggest that individual cytokines may be sufficient. However, the caveats in that interpretation are significant, and there are as many transgenic and other modes of cytokine expression in the CNS that did not result in pathology to disallow firm conclusions. The fact that individual cytokine knockout mice do not show a lack or diminution of glial responses once they develop EAE argues at least for redundancy among proinflammatory cytokines.

Inflammatory immune responses in the CNS have traditionally been considered detrimental, but these responses involve the production of cytokines and chemokines that not only promote adaptive immune responses, but can also potentially trigger protective or regenerative processes. Furthermore, these mediators are produced by activated glial cells as well as by infiltrating leukocytes. When interpreting the action of a cytokine, it is essential to consider the context, both in terms of the receptor through which actions are being mediated (where multiple receptors exist) and the stage of inflammation. The study of inflammation has taught us much about cytokines and will continue to do so.

ACKNOWLEDGMENTS

Research in the Owens laboratory is supported by the Canadian Institutes of Health Research, the Multiple Sclerosis Society of Canada, The National Multiple Sclerosis Society (NY), and the Wadsworth Foundation.

REFERENCES

1. Perry, V.H. et al., Atypical inflammation in the central nervous system in prion disease, *Curr Opin Neurol* 15, 349, 2002.
2. Bechmann, I. et al., Presence of B7—2 (CD86) and lack of B7—1 (CD80) on myelin phagocytosing MHC-II-positive rat microglia is associated with nondestructive immunity *in vivo*, *Faseb J* 15, 1086, 2001.
3. Matsushima, G.K. and Morell, P., The neurotoxicant, cuprizone, as a model to study demyelination and remyelination in the central nervous system, *Brain Pathol* 11, 107, 2001.
4. Babcock, A.A. et al., Chemokine expression by glial cells directs leukocytes to sites of axonal injury in the CNS, *J Neurosci* 23, 7922, 2003.
5. Kerfoot, S.M. and Kubes, P., Overlapping roles of P-selectin and alpha 4 integrin to recruit leukocytes to the central nervous system in experimental autoimmune encephalomyelitis, *J Immunol* 169, 1000, 2002.
6. Piccio, L. et al., Molecular mechanisms involved in lymphocyte recruitment in inflamed brain microvessels: critical roles for P-selectin glycoprotein ligand-1 and heterotrimeric G(i)-linked receptors, *J Immunol* 168, 1940, 2002.
7. Brabb, T. et al., *In situ* tolerance within the central nervous system as a mechanism for preventing autoimmunity, *J Exp Med* 192, 871, 2000.
8. Krakowski, M.L. and Owens, T., Naive T lymphocytes traffic to inflamed central nervous system, but require antigen recognition for activation, *Eur J Immunol* 30, 1002, 2000.
9. Bauer, J. et al., T-cell apoptosis in inflammatory brain lesions: destruction of T cells does not depend on antigen recognition, *Am J Pathol* 153, 715, 1998.
10. Vallieres, L. and Sawchenko, P.E., Bone marrow-derived cells that populate the adult mouse brain preserve their hematopoietic identity, *J Neurosci* 23, 5197, 2003.
11. Priller, J. et al., Targeting gene-modified hematopoietic cells to the central nervous system: use of green fluorescent protein uncovers microglial engraftment, *Nat Med* 7, 1356, 2001.
12. Hickey, W.F. and Kimura, H., Perivascular microglial cells of the CNS are bone marrow-derived and present antigen *in vivo*, *Science* 239, 290, 1988.
13. Rossi, D. and Zlotnik, A., The biology of chemokines and their receptors, *Annu Rev Immunol* 18, 217, 2000.
14. Murphy, C.A. et al., Interactions between hemopoietically derived TNF and central nervous system-resident glial chemokines underlie initiation of autoimmune inflammation in the brain, *J Immunol* 169, 7054, 2002.
15. Huang, D. et al., Absence of monocyte chemoattractant protein 1 in mice leads to decreased local macrophage recruitment and antigen-specific T helper cell type 1 immune response in experimental autoimmune encephalomyelitis, *J Exp Med* 193, 713, 2001.
16. Korner, H. et al., Critical points of tumor necrosis factor action in central nervous system autoimmune inflammation defined by gene targeting, *J Exp Med* 186, 1585, 1997.

17. Antel, J.P. and Owens, T., Immune regulation and CNS autoimmune disease, *J Neuroimmunol* 100, 181, 1999.

18. Prat, A. et al., B7 expression and antigen presentation by human brain endothelial cells: requirement for proinflammatory cytokines, *J Neuropathol Exp Neurol* 59, 129, 2000.

19. Ma, M. et al., Monocyte recruitment and myelin removal are delayed following spinal cord injury in mice with CCR2 chemokine receptor deletion, *J Neurosci Res* 68, 691, 2002.

20. Omari, K.M. et al., Induction of beta-chemokine secretion by human brain microvessel endothelial cells via CD40/CD40L interactions, *J Neuroimmunol* 146, 203, 2004.

21. Ransohoff, R.M., Mechanisms of inflammation in MS tissue: adhesion molecules and chemokines, *J Neuroimmunol* 98, 57, 1999.

22. Yong, V.W. et al., Metalloproteinases in biology and pathology of the nervous system, *Nat Rev Neurosci* 2, 502, 2001.

23. Tani, M. et al., Neutrophil infiltration, glial reaction, and neurological disease in transgenic mice expressing the chemokine N51/KC in oligodendrocytes, *J Clin Invest* 98, 529, 1996.

24. Chen, S.C. et al., Central nervous system inflammation and neurological disease in transgenic mice expressing the CC chemokine CCL21 in oligodendrocytes, *J Immunol* 168, 1009, 2002.

25. Boztug, K. et al., Leukocyte infiltration, but not neurodegeneration, in the CNS of transgenic mice with astrocyte production of the CXC chemokine ligand 10, *J Immunol* 169, 1505, 2002.

26. Bennett, J.L. et al., CCL2 transgene expression in the central nervous system directs diffuse infiltration of CD45(high)CD11b(+) monocytes and enhanced Theiler's murine encephalomyelitis virus-induced demyelinating disease, *J Neurovirol* 9, 623, 2003.

27. Fuentes, M.E. et al., Controlled recruitment of monocytes and macrophages to specific organs through transgenic expression of monocyte chemoattractant protein-1, *J Immunol* 155, 5769, 1995.

28. Izikson, L. et al., Resistance to experimental autoimmune encephalomyelitis in mice lacking the CC chemokine receptor (CCR)2, *J Exp Med* 192, 1075, 2000.

29. Tran, E.H. et al., Induction of experimental autoimmune encephalomyelitis in C57BL/6 mice deficient in either the chemokine macrophage inflammatory protein-1 alpha or its CCR5 receptor, *Eur J Immunol* 30, 1410, 2000.

30. Ford, A.L. et al., Normal adult ramified microglia separated from other central nervous system macrophages by flow cytometric sorting. Phenotypic differences defined and direct *ex vivo* antigen presentation to myelin basic protein-reactive CD4+ T cells compared, *J Immunol* 154, 4309, 1995.

31. Carson, M.J., Microglia as liaisons between the immune and central nervous systems: functional implications for multiple sclerosis, *Glia* 40, 218, 2002.

32. Zehntner, S.P. et al., Constitutive expression of a costimulatory ligand on antigen-presenting cells in the nervous system drives demyelinating disease, *Faseb J* 17, 1910, 2003.

33. Brosnan, C.F. et al., The effects of macrophage depletion on the clinical and pathologic expression of experimental allergic encephalomyelitis, *J Immunol* 126, 614, 1981.

34. Tran, E.H. et al., Immune invasion of the central nervous system parenchyma and experimental allergic encephalomyelitis, but not leukocyte extravasation from blood, are prevented in macrophage-depleted mice, *J Immunol* 161, 3767, 1998.

35. Nitsch, R. et al., Direct impact of T cells on neurons revealed by two-photon microscopy in living brain tissue, *J Neurosci* 24, 2458, 2004.

36. Hofman, F.M. et al., Tumor necrosis factor identified in multiple sclerosis brain, *J Exp Med* 170, 607, 1989.
37. Marciano, P.G. et al., Expression profiling following traumatic brain injury: a review, *Neurochem Res* 27, 1147, 2002.
38. Lock, C. et al., Gene-microarray analysis of multiple sclerosis lesions yields new targets validated in autoimmune encephalomyelitis, *Nat Med* 8, 500, 2002.
39. Kapsenberg, M.L. and Kalinski, P., The concept of type 1 and type 2 antigen-presenting cells, *Immunol Lett* 69, 5, 1999.
40. Becher, B. et al., Brain-immune connection: immuno-regulatory properties of CNS-resident cells, *Glia* 29, 293, 2000.
41. Aharoni, R. et al., Glatiramer acetate-specific T cells in the brain express T helper 2/3 cytokines and brain-derived neurotrophic factor *in situ*, *Proc Natl Acad Sci U.S.A* 100, 14157, 2003.
42. Vieira, P.L. et al., Glatiramer acetate (copolymer-1, copaxone) promotes Th2 cell development and increased IL-10 production through modulation of dendritic cells, *J Immunol* 170, 4483, 2003.
43. Kim, H.J. et al., Inflammatory potential and migratory capacities across human brain endothelial cells of distinct glatiramer acetate-reactive T cells generated in treated multiple sclerosis patients, *Clin Immunol* 111, 38, 2004.
44. Trinchieri, G. et al., The IL-12 family of heterodimeric cytokines: new players in the regulation of T cell responses, *Immunity* 19, 641, 2003.
45. Ahmed, Z. et al., Interleukin-12 induces mild experimental allergic encephalomyelitis following local central nervous system injury in the Lewis rat, *J Neuroimmunol* 140, 109, 2003.
46. Segal, B.M. et al., An interleukin (IL)-10/IL-12 immunoregulatory circuit controls susceptibility to autoimmune disease, *J Exp Med* 187, 537, 1998.
47. Cua, D.J. et al., Interleukin-23 rather than interleukin-12 is the critical cytokine for autoimmune inflammation of the brain, *Nature* 421, 744, 2003.
48. Becher, B. et al., Experimental autoimmune encephalitis and inflammation in the absence of interleukin-12, *J Clin Invest* 110, 493, 2002.
49. Pagenstecher, A. et al., Astrocyte-targeted expression of IL-12 induces active cellular immune responses in the central nervous system and modulates experimental allergic encephalomyelitis, *J Immunol* 164, 4481, 2000.
50. Nakanishi, K. et al., Interleukin-18 regulates both Th1 and Th2 responses, *Annu Rev Immunol* 19, 423, 2001.
51. Adachi, O. et al., Targeted disruption of the MyD88 gene results in loss of IL-1- and IL-18-mediated function, *Immunity* 9, 143, 1998.
52. Fantuzzi, G. and Dinarello, C. A., Interleukin-18 and interleukin-1 beta: two cytokine substrates for ICE (caspase-1), *J Clin Immunol* 19, 1, 1999.
53. Kim, S.H. et al., Structural requirements of six naturally occurring isoforms of the IL-18 binding protein to inhibit IL-18, *Proc Natl Acad Sci U.S.A* 97, 1190, 2000.
54. Corbaz, A. et al., IL-18-binding protein expression by endothelial cells and macrophages is up-regulated during active Crohn's disease, *J Immunol* 168, 3608, 2002.
55. Chang, J.T. et al., The costimulatory effect of IL-18 on the induction of antigen-specific IFN-gamma production by resting T cells is IL-12 dependent and is mediated by up-regulation of the IL-12 receptor beta2 subunit, *Eur J Immunol* 30, 1113, 2000.
56. Hurgin, V. et al., The promoter of IL-18 binding protein: activation by an IFN-gamma-induced complex of IFN regulatory factor 1 and CCAAT/enhancer binding protein beta, *Proc Natl Acad Sci U.S.A* 99, 16957, 2002.

57. Karlsen, A.E. et al., Interferon-gamma induces interleukin-1 converting enzyme expression in pancreatic islets by an interferon regulatory factor-1-dependent mechanism, *J Clin Endocrinol Metab* 85, 830, 2000.
58. Dai, C. and Krantz, S.B., Interferon gamma induces upregulation and activation of caspases 1, 3, and 8 to produce apoptosis in human erythroid progenitor cells, *Blood* 93, 3309, 1999.
59. Okamoto, T. et al., Chronic hepatitis in interferon-gamma transgenic mice is associated with elevated CPP32-like activity and interleukin-1beta-converting enzyme activity suppression, *Jpn J Pharmacol* 79, 289, 1999.
60. Kanzaki, M. and Morris, P.L., Identification and regulation of testicular interferon-gamma (IFNgamma) receptor subunits: IFNgamma enhances interferon regulatory factor-1 and interleukin-1beta converting enzyme expression, *Endocrinology* 139, 2636, 1998.
61. Wheeler, R.D. et al., Interleukin-18 induces expression and release of cytokines from murine glial cells: interactions with interleukin-1 beta, *J Neurochem* 85, 1412, 2003.
62. Shi, F.D. et al., IL-18 directs autoreactive T cells and promotes autodestruction in the central nervous system via induction of IFN-gamma by NK cells [In Process Citation], *J Immunol* 165, 3099, 2000.
63. Wildbaum, G. et al., Neutralizing antibodies to IFN-gamma-inducing factor prevent experimental autoimmune encephalomyelitis, *J Immunol* 161, 6368, 1998.
64. Furlan, R. et al., Caspase-1 regulates the inflammatory process leading to autoimmune demyelination, *J Immunol* 163, 2403, 1999.
65. Chabas, D. et al., The influence of the proinflammatory cytokine, osteopontin, on autoimmune demyelinating disease, *Science* 294, 1731, 2001.
66. Jansson, M. et al., Cutting edge: attenuated experimental autoimmune encephalomyelitis in eta-1/osteopontin-deficient mice, *J Immunol* 168, 2096, 2002.
67. O'Regan, A.W. et al., Osteopontin augments CD3-mediated interferon-gamma and CD40 ligand expression by T cells, which results in IL-12 production from peripheral blood mononuclear cells, *J Leukoc Biol* 68, 495, 2000.
68. Lee, M.Y. et al., Expression of osteopontin mRNA in developing rat brainstem and cerebellum, *Cell Tissue Res* 306, 179, 2001.
69. Choi, J.S. et al., Induction and temporal changes of osteopontin mRNA and protein in the brain following systemic lipopolysaccharide injection, *J Neuroimmunol* 141, 65, 2003.
70. Kim, S.Y. et al., Osteopontin in kainic acid-induced microglial reactions in the rat brain, *Mol Cells* 13, 429, 2002.
71. Duddy, M.E. et al., Distinct profiles of human B cell effector cytokines: a role in immune regulation? *J Immunol* 172, 3422, 2004.
72. Owens, T. et al., Genetic models for CNS inflammation, *Nat Med* 7, 161, 2001.
73. Huan, J. et al., Monomeric recombinant TCR ligand reduces relapse rate and severity of experimental autoimmune encephalomyelitis in SJL/J mice through cytokine switch, *J Immunol* 172, 4556, 2004.
74. Vandenbark, A.A. et al., Diminished frequency of interleukin-10-secreting, T-cell receptor peptide-reactive T cells in multiple sclerosis patients might allow expansion of activated memory T cells bearing the cognate BV gene, *J Neurosci Res* 66, 171, 2001.
75. van Boxel-Dezaire, A.H. et al., Decreased interleukin-10 and increased interleukin-12p40 mRNA are associated with disease activity and characterize different disease stages in multiple sclerosis, *Ann Neurol* 45, 695, 1999.
76. Arnason, B.G., Immunologic therapy of multiple sclerosis, *Annu Rev Med* 50, 291, 1999.

77. Becher, B. et al., Interferon-gamma secretion by peripheral blood T-cell subsets in multiple sclerosis: correlation with disease phase and interferon-beta therapy, *Ann Neurol* 45, 247, 1999.
78. Akwa, Y. et al., Transgenic expression of IFN-alpha in the central nervous system of mice protects against lethal neurotropic viral infection but induces inflammation and neurodegeneration, *J Immunol* 161, 5016, 1998.
79. Wang, J. et al., Dysregulated Sonic hedgehog signaling and medulloblastoma consequent to IFN-alpha-stimulated STAT2-independent production of IFN-gamma in the brain, *J Clin Invest* 112, 535, 2003.
80. Hertzog, P.J. et al., The interferon in TLR signaling: more than just antiviral, *Trends Immunol* 24, 534, 2003.
81. Yeh, W.C. and Chen, N.J., Immunology: another toll road, *Nature* 424, 736, 2003.
82. Medzhitov, R., Toll-like receptors and innate immunity, *Nat Rev Immunol* 1, 135, 2001.
83. Bsibsi, M. et al., Broad expression of Toll-like receptors in the human central nervous system, *J Neuropathol Exp Neurol* 61, 1013, 2002.
84. Zekki, H. et al., The clinical course of experimental autoimmune encephalomyelitis is associated with a profound and sustained transcriptional activation of the genes encoding toll-like receptor 2 and CD14 in the mouse CNS, *Brain Pathol* 12, 308, 2002.
85. Juedes, A.E. et al., Kinetics and cellular origin of cytokines in the central nervous system: insight into mechanisms of myelin oligodendrocyte glycoprotein-induced experimental autoimmune encephalomyelitis, *J Immunol* 164, 419, 2000.
86. Selmaj, K.W. and Raine, C.S., Tumor necrosis factor mediates myelin and oligodendrocyte damage *in vitro*, *Ann Neurol* 23, 339, 1988.
87. Ruuls, S.R. et al., Membrane-bound tnf supports secondary lymphoid organ structure but is subservient to secreted tnf in driving autoimmune inflammation, *Immunity* 15, 533, 2001.
88. Linker, R.A. et al., CNTF is a major protective factor in demyelinating CNS disease: a neurotrophic cytokine as modulator in neuroinflammation, *Nat Med* 8, 620, 2002.
89. TNF neutralization in MS: results of a randomized, placebo-controlled multicenter study. The Lenercept Multiple Sclerosis Study Group and The University of British Columbia MS/MRI Analysis Group, *Neurology* 53, 457, 1999.
90. Suen, W.E. et al., A critical role for lymphotoxin in experimental allergic encephalomyelitis, *J Exp Med* 186, 1233, 1997.
91. Frei, K. et al., Tumor necrosis factor alpha and lymphotoxin alpha are not required for induction of acute experimental autoimmune encephalomyelitis, *J Exp Med* 185, 2177, 1997.
92. Riminton, S.D. et al., Challenging cytokine redundancy: inflammatory cell movement and clinical course of experimental autoimmune encephalomyelitis are normal in lymphotoxin-deficient, but not tumor necrosis factor-deficient, mice, *J Exp Med* 187, 1517, 1998.
93. Gommerman, J.L. et al., A role for surface lymphotoxin in experimental autoimmune encephalomyelitis independent of LIGHT, *J Clin Invest* 112, 755, 2003.
94. Adamson, P. et al., Lymphocyte trafficking through the blood-brain barrier is dependent on endothelial cell heterotrimeric G-protein signaling, *Faseb J* 16, 1185, 2002.
95. Linthicum, D.S. et al., Acute experimental autoimmune encephalomyelitis in mice. I. Adjuvant action of Bordetella pertussis is due to vasoactive amine sensitization and increased vascular permeability of the central nervous system, *Cell Immunol* 73, 299, 1982.

96. Ma, R.Z. et al., Identification of Bphs, an autoimmune disease locus, as histamine receptor H1, *Science* 297, 620, 2002.
97. Nitsch, R. et al., Human brain-cell death induced by tumour-necrosis-factor-related apoptosis-inducing ligand (TRAIL), *Lancet* 356, 827, 2000.
98. Hilliard, B. et al., Roles of TNF-related apoptosis-inducing ligand in experimental autoimmune encephalomyelitis, *J Immunol* 166, 1314, 2001.
99. Panitch, H.S. et al., Exacerbations of multiple sclerosis in patients treated with gamma interferon, *Lancet* 1, 893, 1987.
100. Skurkovich, S. et al., Randomized study of antibodies to IFN-gamma and TNF-alpha in secondary progressive multiple sclerosis, *Mult Scler* 7, 277, 2001.
101. Furlan, R. et al., Intrathecal delivery of IFN-gamma protects C57BL/6 mice from chronic-progressive experimental autoimmune encephalomyelitis by increasing apoptosis of central nervous system-infiltrating lymphocytes, *J Immunol* 167, 1821, 2001.
102. LaFerla, F.M. et al., Regional hypomyelination and dysplasia in transgenic mice with astrocyte-directed expression of interferon-gamma, *J Mol Neurosci* 15, 45, 2000.
103. Hofer, M. et al., Cerebral expression of interleukin-12 induces neurological disease via differential pathways and recruits antigen-specific T cells in virus-infected mice, *Am J Pathol* 165, 949, 2004.
104. Friedl, G. et al., Borna disease virus multiplication in mouse organotypic slice cultures is site-specifically inhibited by gamma interferon but not by interleukin-12, *J Virol* 78, 1212, 2004.
105. Chu, C.Q. et al., Failure to suppress the expansion of the activated CD4 T cell population in interferon gamma-deficient mice leads to exacerbation of experimental autoimmune encephalomyelitis, *J Exp Med* 192, 123, 2000.
106. Kusmartsev, S.A. et al., Gr-1+ myeloid cells derived from tumor-bearing mice inhibit primary T cell activation induced through CD3/CD28 costimulation, *J Immunol* 165, 779, 2000.
107. Willenborg, D.O. et al., IFN-gamma is critical to the control of murine autoimmune encephalomyelitis and regulates both in the periphery and in the target tissue: a possible role for nitric oxide, *J Immunol* 163, 5278, 1999.
108. Tran, E.H. et al., IFN-gamma shapes immune invasion of the central nervous system via regulation of chemokines, *J Immunol* 164, 2759, 2000.
109. Olsson, T. et al., Gamma-interferon-like immunoreactivity in axotomized rat motor neurons, *J Neurosci* 9, 3870, 1989.
110. Neumann, H. et al., Interferon gamma gene expression in sensory neurons: evidence for autocrine gene regulation, *J Exp Med* 186, 2023, 1997.
111. De Simone, R. et al., Interferon gamma gene expression in rat central nervous system glial cells, *Cytokine* 10, 418, 1998.
112. Frucht, D.M. et al., IFN-gamma production by antigen-presenting cells: mechanisms emerge, *Trends Immunol* 22, 556, 2001.
113. Smith, K.J. and Lassmann, H., The role of nitric oxide in multiple sclerosis, *Lancet Neurol* 1, 232, 2002.
114. Willenborg, D.O. et al., Our shifting understanding of the role of nitric oxide in autoimmune encephalomyelitis: a review, *J Neuroimmunol* 100, 21, 1999.
115. Zehntner, S.P. et al., Bone marrow-derived vs. parenchymal sources of inducible nitric oxide synthase in experimental autoimmune encephalomyelitis, *J Neuroimmunol* 150, 70, 2004.
116. Aggarwal, S. and Gurney, A.L., IL-17: prototype member of an emerging cytokine family, *J Leukoc Biol* 71, 1, 2002.

117. Matusevicius, D. et al., Interleukin-17 mRNA expression in blood and CSF mono-nuclear cells is augmented in multiple sclerosis, *Mult Scler* 5, 101, 1999.

118. Meeuwsen, S. et al., Cytokine, chemokine and growth factor gene profiling of cultured human astrocytes after exposure to proinflammatory stimuli, *Glia* 43, 243, 2003.

119. Carroll, W.M. et al., Identification of the adult resting progenitor cell by autoradio-graphic tracking of oligodendrocyte precursors in experimental CNS demyelination, *Brain* 121 (Pt 2), 293, 1998.

120. Di Bello, I.C. et al., Generation of oligodendroglial progenitors in acute inflammatory demyelinating lesions of the rat brain stem is associated with demyelination rather than inflammation, *J Neurocytol* 28, 365, 1999.

121. Blakemore, W. F. et al., Remyelinating the demyelinated CNS, *Novartis Found Symp* 231, 289, 2000.

122. Zhang, S.C. et al., Adult brain retains the potential to generate oligodendroglial progenitors with extensive myelination capacity, *Proc Natl Acad Sci U.S.A* 96, 4089, 1999.

123. Levine, J.M. et al., The oligodendrocyte precursor cell in health and disease, *Trends Neurosci* 24, 39, 2001.

124. Windrem, M.S. et al., Progenitor cells derived from the adult human subcortical white matter disperse and differentiate as oligodendrocytes within demyelinated lesions of the rat brain, *J Neurosci Res* 69, 966, 2002.

125. Tsai, H.H. and Miller, R.H., Glial cell migration directed by axon guidance cues, *Trends Neurosci* 25, 173, 2002.

126. Arnett, H.A. et al., TNFalpha promotes proliferation of oligodendrocyte progenitors and remyelination, *Nat Neurosci* 4, 1116, 2001.

127. Keirstead, H.S. et al., Response of the oligodendrocyte progenitor cell population (defined by NG2 labelling) to demyelination of the adult spinal cord, *Glia* 22, 161, 1998.

128. Ye, P. et al., Mouse NG2+ oligodendrocyte precursors express mRNA for proteolipid protein but not its DM-20 variant: a study of laser microdissection-captured NG2+ cells, *J Neurosci* 23, 4401, 2003.

129. Robinson, S. et al., The chemokine growth-regulated oncogene-alpha promotes spinal cord oligodendrocyte precursor proliferation, *J Neurosci* 18, 10457, 1998.

130. Tsai, H.H. et al., The chemokine receptor CXCR2 controls positioning of oligoden-drocyte precursors in developing spinal cord by arresting their migration, *Cell* 110, 373, 2002.

131. Bell, M.D. et al., Recombinant human adenovirus with rat MIP-2 gene insertion causes prolonged PMN recruitment to the murine brain, *Eur J Neurosci* 8, 1803, 1996.

132. Barres, B.A. et al., A crucial role for neurotrophin-3 in oligodendrocyte development, *Nature* 367, 371, 1994.

133. Cosgaya, J.M. et al., The neurotrophin receptor p75NTR as a positive modulator of myelination, *Science* 298, 1245, 2002.

134. Chan, J.R. et al., Neurotrophins are key mediators of the myelination program in the peripheral nervous system, *Proc Natl Acad Sci U.S.A* 98, 14661, 2001.

135. David, S. et al., Macrophages can modify the nonpermissive nature of the adult mammalian central nervous system, *Neuron* 5, 463, 1990.

136. Schwartz, M. et al., Innate and adaptive immune responses can be beneficial for CNS repair, *Trends Neurosci* 22, 295, 1999.

137. McTigue, D.M. et al., Neurotrophin-3 and brain-derived neurotrophic factor induce oligodendrocyte proliferation and myelination of regenerating axons in the contused adult rat spinal cord, *J Neurosci* 18, 5354, 1998.

138. Kerschensteiner, M. et al., Activated human T cells, B cells, and monocytes produce brain-derived neurotrophic factor *in vitro* and in inflammatory brain lesions: a neuroprotective role of inflammation?, *J Exp Med* 189, 865, 1999.

139. Gielen, A. et al., Increased brain-derived neurotrophic factor expression in white blood cells of relapsing-remitting multiple sclerosis patients, *Scand J Immunol* 57, 493, 2003.

140. Chen, M. et al., Glatiramer acetate-reactive T cells produce brain-derived neurotrophic factor, *J Neurol Sci* 215, 37, 2003.

141. Elkabes, S. et al., Brain microglia/macrophages express neurotrophins that selectively regulate microglial proliferation and function, *J Neurosci* 16, 2508, 1996.

142. Ladiwala, U. et al., p75 neurotrophin receptor expression on adult human oligodendrocytes: signaling without cell death in response to NGF, *J Neurosci* 18, 1297, 1998.

143. Dowling, P. et al., Up-regulated p75NTR neurotrophin receptor on glial cells in MS plaques, *Neurology* 53, 1676, 1999.

144. Khoury, S.J. et al., Oral tolerance to myelin basic protein and natural recovery from experimental autoimmune encephalomyelitis are associated with downregulation of inflammatory cytokines and differential upregulation of transforming growth factor beta, interleukin 4, and prostaglandin E expression in the brain, *J Exp Med* 176, 1355, 1992.

145. Linker, R.A. et al., CNTF is a major protective factor in demyelinating CNS disease: a neurotrophic cytokine as modulator in neuroinflammation, *Nat Med* 8, 620, 2002.

146. D'Souza, S.D. et al., Ciliary neurotrophic factor selectively protects human oligodendrocytes from tumor necrosis factor-mediated injury, *J Neurosci Res* 43, 289, 1996.

147. Stankoff, B. et al., Ciliary neurotrophic factor (CNTF) enhances myelin formation: a novel role for CNTF and CNTF-related molecules, *J Neurosci* 22, 9221, 2002.

148. Butzkueven, H. et al., LIF receptor signaling limits immune-mediated demyelination by enhancing oligodendrocyte survival, *Nat Med* 8, 613, 2002.

149. DeChiara, T.M. et al., Mice lacking the CNTF receptor, unlike mice lacking CNTF, exhibit profound motor neuron deficits at birth, *Cell* 83, 313, 1995.

150. Elson, G.C. et al., CLF associates with CLC to form a functional heteromeric ligand for the CNTF receptor complex, *Nat Neurosci* 3, 867, 2000.

151. Wosik, K. et al., Resistance of human adult oligodendrocytes to AMPA/kainate receptor-mediated glutamate injury, *Brain* 127, 2636, 2004.

152. Zehntner, S.P. et al., Neutrophils that infiltrate the central nervous system regulate T cell responses, *J Immunol* 174, 5124, 2005.

153. Abromson-Leeman, S. et al., T-cell properties determine disease site, clinical presentation, and cellular pathology of experimental autoimmune encephalomyelitis, *Am J Pathol* 165, 1519, 2004.

154. Langrish C.L. et al., IL-23 drives a pathogenic T cell population that includes autoimmune inflammation. *J. Exp. Med.* 201, 233, 2005.

8 Cytokines and Immune Regulation in the Nervous System

Wendy Smith Begolka, Eileen J. McMahon, and Stephen D. Miller

CONTENTS

I. INTRODUCTION

Experimental autoimmune encephalomyelitis (EAE) is a CD4+ T cell-mediated inflammatory demyelinating disease of the central nervous system (CNS) that serves as an animal model of multiple sclerosis (MS) with which it shares numerous clinical, immunological, and histological parallels. EAE may be induced by active immunization with myelin proteins or peptides or through the adoptive transfer of myelin-specific CD4+ T cells and leads to a relapsing–remitting, chronic, or monophasic clinical presentation based on the strain of rodent and the inducing antigen. These differences in disease manifestation have led to numerous studies investigating the potential intrinsic pathogenic and regulatory mechanisms of CNS autoimmune disease and have often focused on the roles of pro- and anti-inflammatory cytokines in these processes.

The implication of cytokines as critical to resistance/susceptibility to EAE as well as disease initiation, progression, and modulation of CNS autoimmunity stems from several lines of evidence including (i) detection of cytokines in the CNS or from encephalitogenic T cells and regulatory cells, (ii) antibody depletion/inhibition studies, (iii) *in vivo* administration of cytokines, (iv) use of transgenic/knockout animals, (v) gene therapy studies, and (vi) various "tolerance"-inducing protocols. This review, therefore, will focus on our current understanding of the central role that pro- and anti-inflammatory cytokines play during the natural course of EAE and discuss the potential for cytokine-based therapeutic strategies to impact the balance between pathogenesis and recovery.

II. PRO-INFLAMMATORY CYTOKINE PROFILES IN EAE

A. INTERLEUKIN-6 (IL-6)

IL-6 is a pleiotropic cytokine known to regulate hematopoiesis and contribute significantly to inflammation and immune responses [1]. Although it has been implicated in numerous autoimmune diseases, its role in MS was first suggested by several studies describing its upregulation in the blood, cerebrospinal fluid (CSF), and brain tissue of MS patients [2]. More recently, microarray analysis from MS lesions revealed that the nuclear factor that mediates IL-6 transcription (NF-IL6) was highly upregulated in all the MS samples examined [3]. In EAE, IL-6 gene expression is also markedly upregulated in CNS tissue and correlates well with clinical severity [4].

Its important role in the initiation of EAE has been demonstrated by numerous studies showing that IL-6$^{-/-}$ mice are resistant to MOG-induced EAE [5–8]. Although the studies agree on the dramatic effect of the IL-6 deficiency on clinical disease, inflammation, and infiltration of cells into the CNS, they differ on the proposed mechanisms of this effect. One report found peripheral T cell proliferation and Th1/Th2 cytokine production was impaired in the IL-6$^{-/-}$ mice, suggesting the deficiency is in preventing T helper cell differentiation during priming [8]. These observations are in contrast to another study where no defect was noted in T cell responses, yet a dramatic lack of upregulated VCAM-1 and ICAM-1 on CNS endothelial cells from immunized IL-6$^{-/-}$ mice was found, suggesting the lack of disease is potentially due to the inability of the encephalitogenic T cells to gain access to the CNS [6]. Further complicating the effects of IL-6 deficiency in EAE are studies showing that lymph node (LN) cells from immunized IL-6$^{-/-}$ mice produce less IFN-γ but more IL-4 and IL-10 on MOG$_{35-55}$ stimulation, suggesting that immune deviation may account for the resistance in EAE [5,9].

Although it remains unclear how IL-6 deficiency is conferring resistance to EAE, its importance is largely restricted to the initiation phase of disease, with little influence on the effector phase, as IL-6$^{-/-}$ mice are fully susceptible to passive transfer EAE if donor cells are wild type [5,7]. One study did report that IL-6$^{-/-}$ recipient mice had slightly decreased disease severity with less CNS infiltration, suggesting that local production of IL-6 may be influencing cells in the target organ [5,7]. In support of this effect, *in vivo* administration of neutralizing antibodies to IL-6 has also been shown to decrease severity of passive transfer EAE [10].

B. TUMOR NECROSIS FACTOR-α (TNF-α)

TNF-α is a prototypic inflammatory cytokine long implicated in autoimmune diseases. It is strongly produced by activated macrophages but is also expressed by T cells, NK cells, mast cells, endothelial cells, fibroblasts, microglia, and astrocytes [11,12]. TNF-α is produced in two biologically active forms, a transmembrane protein and a cleaved soluble factor, that are able to bind two receptors, TNF-R1 (p55) or TNF-R2 (p75). In addition to its role in apoptosis, TNF-α can also potently promote inflammation by increasing antigen presentation, inducing dendritic cell (DCs) maturation, promoting astrocyte proliferation, and regulating cell trafficking though its dramatic effects on chemokine and adhesion molecule expression [13,14]. Although TNF-neutralizing strategies have proven useful in the treatment and prevention of inflammation and autoimmunity, the exact role of TNF-α in MS and EAE has proven difficult to fully elucidate [15]. Many initial studies pointed to its importance in promoting disease.

TNF-α and its receptors are upregulated in the peripheral immune organs and the CNS during EAE as well as in active MS lesions and antigen presenting cells (APCs) and T cells isolated from the CNS during EAE produce TNF-α [16–19]. *In vivo* administration of TNF-α increases disease and promotes relapse, whereas neutralizing it delays or decreases disease [20–24]. Finally, because TNF-α can also induce the apoptosis of oligodendrocytes *in vitro*, it is possible that it can directly mediate demyelination [25].

Conflicting evidence began with the studies of knockout mice with reports that TNF-α deficiency either made no difference or even enhanced disease [26,27]. Nevertheless, because the TNF-α gene resides within the MHC complex — a chromosomal region known to contain genes that influence EAE susceptibility — and because mixed background or partially backcrossed mice were used in these studies, interpreting their results is difficult [28]. When the TNF-α gene was knocked out directly in C57BL/6 (B6) mice, the deficiency caused a significant delay in disease onset and duration but no difference in mean peak disease [29,30]. T cell priming in the periphery was also unaffected, and the numbers of infiltrating cells into the CNS were similar [29,30]. There were however dramatic differences in the pattern of infiltration, as knockout mice had leukocytes, particularly macrophages, clustered in perivascular areas that failed to extend deep into the parenchyma, suggesting a critical role for TNF-α in regulating trafficking within the target organ itself [29,31]. Bone marrow chimeras in which wild-type mice were reconstituted with TNF-α$^{-/-}$ bone marrow also confirmed the delay in disease onset but further suggested that hematopoietically derived cells, such as T cells and macrophages, are more critical sources of TNF-α than CNS resident glial cells [32].

Recent studies propose that this altered trafficking, and therefore late onset disease, is likely due to the failure of chemokine upregulation [14,32]. On day 14 postimmunization, when wild-type mice are sick and TNF-α$^{-/-}$ are not, numerous chemokines including IP-10, MCP-1, and TCA-3 have failed to be upregulated in the CNS of TNF-α$^{-/-}$ mice. However at onset of disease in the knockout mice, these chemokines do increase, suggesting that lack of TNF-α can ultimately be overcome by other redundant factors [32]. Even in naïve mice, deficiency in TNF-α has dramatic effects on constitutive chemokine expression in peripheral lymph organs, which likely contributes to the altered microarchitecture seen in TNF-α$^{-/-}$ animals [14,31,33].

Knockout mouse systems have demonstrated that TNF-α likely uses TNF-R1, and not TNF-R2, to mediate its pro-inflammatory effects [34–36]. During acute disease, TNF-R1$^{-/-}$ mice had either significantly reduced disease severity or were completely resistant. Demyelination and infiltration were also diminished, and some reports showed a concomitant decrease in T cell proliferation and IFN-γ production. There was even evidence of immune deviation as stimulated T cells produced significantly higher quantities of IL-5 in the absence of TNF-R1 [36]. Also noted was increased clinical disease with greater IL-2 production and increased infiltration in TNF-R2-deficient mice, suggesting TNF-α, or other TNF-R2 ligands, also had anti-inflammatory or protective functions [35,36].

Some early studies with TNF-α$^{-/-}$ mice also suggest an anti-inflammatory role for TNF-α, and studies blocking TNF in human MS patients support this protective role. Neutralizing antibody treatment unexpectedly exacerbated immune activation and MRI activity in two rapidly progressive MS patients, and a larger study using TNFR-Ig fusion protein also noted an increase in the clinical exacerbation rate in the TNFR-Ig treated group [37, 38]. These contrary results were difficult to reconcile until it was suggested that whereas TNF-α promotes inflammation at disease initiation and in the acute phase, it may have a role in downregulating myelin reactivity late in EAE [34]. This report found that although MBP-specific T cell reactivity had largely regressed in the periphery of wild-type mice by 50 days postimmunization, TNF-α$^{-/-}$ mice maintained reactivity for greater than 70 days. Furthermore, clinical disease in wild-type mice decreased and stabilized at lower severity after the acute phase, whereas TNF-α$^{-/-}$ mice or TNF-R1/2 double knockout mice chronically progressed. Interestingly, evidence from the cuprizone model of demyelination suggests that TNF-α, through TNF-R2, has a role in promoting remyelination, and deficiency in TNF-α or TNF-R2 may prevent attempts by oligodendrocyte progenitors to repair damage [39]. Therefore, as with many cytokines, the role of TNF-α in EAE and MS appears to be multifaceted, and therapies to regulate its function in human patients should be developed and employed with caution.

C. Lymphotoxin-α (LT-α)

LT-α shares many structural and functional properties with TNF-α and was even originally called TNF-β [40]. LT-α is secreted and forms a homotrimer, which binds TNF-R1 and TNF-R2. In fact, its proinflammatory and proapoptotic functions appeared to be largely redundant with TNF-α until a related membrane-bound molecule, LT-β, was discovered. LT-β cannot function alone but forms a heterotrimer with LT-α, remaining at the cell surface and binding a distinct receptor, LTβR. Activated T cells primarily produce LT-α, although NK cells and a subset of B cells are also potential sources. LT-α is critical for the development of secondary lymph organs as mice deficient in LT-α, LT-β, or LTβR all lack peripheral lymph nodes, Peyer's patches, and organized splenic germinal centers [41–43]. Due to this gross immune alteration, results from studies using these mice must be cautiously interpreted. LT-α also resides within the MHC, making the use of backcrossed knockout mice problematic. Similar to TNF-α, LT-α is found expressed in the CNS and CSF of MS patients and in rodents with EAE, and it has also been shown to be toxic to oligodendrocytes in culture, suggesting a direct role in the demyelinating process [44–46]. Furthermore, treatments that blocked LT-α or TNF-R1/2 ameliorated active and passive EAE, although none of these studies distinguished between LT-α and TNF-α [21,23,47].

Also similar to TNF-α, initial studies with knockout mice found conflicting results, as one report indicated LT-α was not required for EAE because LT-α/TNF-α double knockout mice were fully susceptible and another reported decreased EAE in both LT-α$^{-/-}$ and LT-β$^{-/-}$ mice [27,48]. Bone marrow chimeras studies using RAG-1$^{-/-}$ mice reconstituted with LT-α$^{-/-}$ or LT-α$^{+/+}$ bone marrow were also equally susceptible to EAE, suggesting no overt functional, nonredundant role for LT-α [24]. This was further supported by experiments showing TNF-α$^{-/-}$ mice reconstituted with TNF-α$^{-/-}$ or TNF-α$^{-/-}$LT-α$^{-/-}$ bone marrow showed equally delayed EAE disease course, suggesting no additional effect of the LT-α deficiency.

More recently, the role of LT-α was revisited using a LTβR-Ig fusion protein that blocks surface LT-α/LT-β heterotrimer and demonstrated that LTβR-Ig treatment had no effect on MOG$_{35-55}$-induced EAE [49]. However, in models that do not rely on pertussis toxin, disruption of LT-α/LT-β yielded disparate results. Blocking LT-α/LT-β in a rat model of EAE significantly ameliorated disease and decreased T cell recall responses. In PLP$_{139-151}$-induced EAE, the initial acute disease was unaffected yet relapse was prevented. Lastly, studies in the DO11.10 TCR transgenic system suggested that LT-α/LT-β had a role in the reactivation and not the initial priming of antigen-specific cells, leading the authors to speculate surface LT-α/LT-β may be necessary to orchestrate the proper timing and positioning of cells in the lymph nodes for optimal sustenance and maturation of T cell responses [49]. Although more research is required to elucidate an exact mechanism of LTβR-Ig treatment, these studies highlight a previously underappreciated role for LT-α in EAE.

D. INTERFERON-γ (IFN-γ)

IFN-γ is the prototypic Th1 cytokine. Produced by activated T cells and NK cells, it primarily affects macrophage and dendritic cell (DCs) function by upregulating antigen presentation capabilities through increasing MHC Class II, costimulatory and adhesion molecules. It can also induce the production of proinflammatory cytokines and nitric oxide (NO), a molecule long believed to be directly involved in the destruction of myelin in both MS and EAE [50]. IFN-γ has similar effects on microglia and can also upregulate the antigen presentation abilities and cytokine and chemokine production in astrocytes [51–54]. Finally, IFN-γR is also present on oligodendrocytes and may contribute directly to cell death [55].

In MS and EAE, IFN-γ production has long been associated with disease pathogenesis due to its central role in Th1 responses. IFN-γ is present in MS lesions and MS patients have increased numbers of myelin-reactive cells secreting IFN-γ in their blood and CSF [56,57]. In EAE, IFN-γ mRNA levels in the CNS correlate well with disease, with the highest level shortly before peak of acute disease and at the height of subsequent relapses [58]. Myelin-specific IFN-γ-producing cells can also be found in the CNS at peak of disease, and it has been repeatedly demonstrated that myelin-specific Th1 clones producing IFN-γ but not Th2 clones are encephalitogenic [59–62].

Given such evidence, it was widely hypothesized that blocking IFN-γ would be a useful therapy in EAE and possibly MS, however the opposite has proven true. In repeated studies, *in vivo* administration of neutralizing antibody to IFN-γ increased the severity of EAE, whereas recombinant IFN-γ administration was protective [63–65]. Similar results were also obtained in a study localizing IFN-γ production to the CNS by intrathecal injection of a nonreplicating viral vector encoding IFN-γ, wherein mice receiving the IFN-γ-encoding virus showed earlier onset of MOG_{35-55} EAE, and the majority of mice recovered completely from disease [66]. This recovery correlated with increased apoptosis of CD4+ T cells within the CNS, suggesting IFN-γ was involved in turning off pathogenic T cell responses in the target organ.

Studies with knockout mice have corroborated the paradoxically protective role of IFN-γ. A deficiency in IFN-γ made normally resistant BALB/c mice susceptible to EAE [67]. Likewise, IFN-γ$^{-/-}$ mice on susceptible backgrounds developed equal to more severe EAE compared to wild-type controls [67,68]. Although increased expression of MHC Class II was absent in the CNS, there were significant increases in infiltration, including increased numbers of activated CD4+ T cells [69,70]. IFN-γR$^{-/-}$ mice were also highly susceptible to disease induction with a high degree of mortality, and cells from IFN-γR$^{-/-}$ knockout mice transfer an exacerbated form of disease to either knockout or wild-type recipients. Interestingly, however, if the recipients were wild-type and could respond to IFN-γ, they fully recovered, suggesting that host cells are capable of downregulating and controlling disease [71]. CD4+ T cells appear to be responsible for this exacerbated disease in both IFN-γ$^{-/-}$ and IFN-γR$^{-/-}$ mice, as T cells from IFN-γ$^{-/-}$ mice also show increased proliferation and decreased apoptosis in response to antigen *in vitro*, whereas addition of exogenous IFN-γ reversed this effect [70].

Interestingly, this exacerbated disease was similar to that observed in mice deficient in iNOS, the enzyme responsible for NO production in myeloid cells [72]. Although dogma held that NO was likely detrimental in EAE by mediating destruction of myelin, recent evidence points to its important role in inhibiting T cell proliferation and downregulating immune responses [73,74]. It is of interest then that passive transfer of EAE with IFN-γ$^{-/-}$ donor cells into wild-type recipients, although causing severe disease, does not cause CNS upregulation of iNOS [75]. It was also found that wild-type macrophages could inhibit the hyperproliferation of IFN-γR$^{-/-}$ splenocytes during *in vitro* stimulation and that this inhibition was prevented by NO blockers, proving the IFN-γ-induced NO production from macrophages was responsible for downregulating T cell proliferation [76]. Finally, passive transfer of EAE with IFN-γR$^{-/-}$ donor cells into bone marrow chimeras with IFN-γR selectively expressed in either the CNS or periphery showed that cells from both locations could mediate this downregulation, suggesting both microglia and peripheral macrophages or DCs are important sources of NO. Therefore, although the effect of IFN-γ in promoting Th1 immune responses through its upregulation of antigen presentation cannot be refuted, the studies in EAE suggest this effect may be superceded by its role in inducing iNOS and subsequently downregulating T cell proliferation.

E. Interleukin-18 (IL-18)

IL-18 is a cytokine produced by DCs, macrophages, and microglia on innate stimuli activation [53,77,78]. IL-18 has been shown to be a potent inducer of IFN-γ and cytotoxicity function in NK cells and can synergize with IL-12 in driving differentiation of Th1 cells and IFN-γ production through upregulation of the IL-12Rβ2 chain [79–81]. IL-18R is expressed constitutively or inducibly on a wide variety of cells including T, NK and B cells, DCs, microglia, astrocytes, and even oligodendrocytes, suggesting that the effects IL-18 may be broad [82–84].

Several studies describe elevated IL-18 in serum or CSF of patients with MS, with the highest levels occurring in MS patients with active lesions and in patients with secondary chronic progressive versus relapsing–remitting MS [85,86]. Peripheral blood mononuclear cells (PBMC) from MS patients also secrete elevated levels of IL-18 on anti-CD3/CD28 activation in a CD40-CD40L dependent manner [87].

A direct role for IL-18 in CNS autoimmunity was first shown in the rat EAE system where *in vivo* administration of neutralizing antibodies to IL-18 significantly diminished severity of passive and active EAE [88]. This treatment also decreased IFN-γ and TNF-α production and increased IL-4 production on peptide stimulation, suggesting immune deviation contributed to the decrease in disease. Studies with knockout mice further substantiated the role of IL-18, as IL18$^{-/-}$ mice were largely resistant to MOG$_{35-55}$ EAE, although no IL-4 increases were noted [89]. Furthermore, the defect appeared to be in peripheral immune responses and priming events because IL-18$^{-/-}$ mice were equally susceptible to adoptive EAE if wild-type donor cells were used.

Surprisingly, the defect appeared to originate in the NK cell compartment. NK cells from immunized IL-18$^{-/-}$ mice had decreased cytotoxicity function and secreted less IFN-γ [89]. More conclusively, transfer of wild-type NK cells into IL-18$^{-/-}$ mice before immunization increased both incidence and severity of disease, although the same effect was not observed with IFN-γ$^{-/-}$ NK cells. Overall, the results would support a role for IL-18 in promoting IFN-γ production from NK cells that then presumably aid in driving the Th1 responses responsible for clinical disease. IL-18, therefore, appears to be another molecule linking the innate and adaptive immune responses and contributing to autoimmune responses.

F. INTERLEUKIN-12 (IL-12)

IL-12 is a heterodimeric molecule (p70) made up of 2 subunits, p40 and p35, and is produced by activated antigen-presenting cells, such as DCs, B cells, macrophages and microglia, in response to Toll-like receptor stimulation and immune signals such as CD40 engagement [90,91]. It is thought to be critical for the differentiation of Th1 cells as IL-12$^{-/-}$ mice were defective (although not completely) in their production of IFN-γ [92,93]. In addition, IL-12 can upregulate CCR5 expression on T cells in a IFN-γ-independent manner, suggesting an additional role in modulating T cell migration [94]. IL-12 mediates its effects through the IL-12R complex made up of the IL-12Rβ1 and β2 chains, which become upregulated on differentiating Th1 T cells and subsequently signals through STAT-4 to stimulate IFN-γ production [95,96]. Although IL-12R is expressed on T cells and NK cells, it can also be found on peripheral and CNS APCs, reenforcing IL-12 production in a positive feedback loop [97–99].

In MS, the p40 subunit of IL-12 has been localized to lesions, and numerous studies have established a correlation between increased CNS inflammation and disease severity and elevated levels of IL-12 in CSF or PBMCs [100–102]. In adoptive EAE, IL-12 p40 mRNA levels peak at the preclinical stage of disease and subsequently decrease through peak of acute and recovery [103]. In relapsing–remitting EAE, CNS IL-12 expression is also detected at preclinical stages of disease (preceding IFN-γ expression) and then decreases during remission and increases again at relapse, suggesting a direct role in mediating pathogenic autoimmune responses [104]. Infiltrating macrophages in particular appear to be an important source of IL-12 in the EAE CNS *in vivo*. Although both macrophages and microglia express mRNA for both chains of the IL-12R after EAE onset, only macrophages express both protein subunits for IL-12, possibly suggesting a preferential ability to drive IFN-γ production [105].

Direct evidence for IL-12 in EAE pathogenesis was established when preincubation of donor cells with recombinant IL-12 increased transferred EAE disease severity and production of IFN-γ and TNF-α [106]. *In vivo* administration of IL-12 to actively primed mice had similar effects. The role of IL-12 in promoting EAE was further substantiated by numerous studies that found treatment with neutralizing antibodies to IL-12 decreased the incidence and severity of EAE and decreased T cell proliferation and IFN-γ production [106–109]. Surprisingly, anti-IL-12 could modulate disease in both IFN-γ knockout or wild-type animals, suggesting IL-12 capable of mediating additional effects independent of IFN-γ [110]. Anti-IL-12 treatment did increase IL-10 production in naïve and primed animals, suggesting IL-12-mediated suppression of IL-10 is a significant contribution to disease pathogenesis [110]. In support of this, it was found that whereas T cells from anti-IL-12 treated mice induced a mild disease on adoptive transfer, cotreatment with anti-IL-10 restored normal disease, indicating anti-IL-12 was working through IL-10 [110]. Studies from knockout mice also corroborate with these observations, as p40$^{-/-}$ and IL-12Rβ1$^{-/-}$ mice are resistant to actively induced EAE [110–112].

Conflicting data exists, however, on whether p40 is involved only in disease initiation or also in its maintenance and expansion. A MOG-specific encephalitogenic

T cell line could induce disease in p40$^{-/-}$ mice, suggesting IL-12 was not required to maintain effector function of an already differentiated Th1 line, yet p40$^{-/-}$ mice had reduced incidence and clinical severity of passive EAE if they received splenocytes from a MOG-immunized wild-type mouse [111,113]. Moreover, after active priming, bone marrow chimeric mice lacking the p40 subunit in the CNS but reconstituted with IL-12-secreting capabilities in the periphery, still had decreased disease incidence with a concomitant delay of onset and decreased severity, suggesting local production of p40 was important in promoting and maintaining disease [111].

Although collectively these data argue strongly for the critical role of IL-12 in CNS autoimmunity, the importance was questioned with the unexpected discovery that p35$^{-/-}$ or IL-12Rβ2$^{-/-}$ mice were fully susceptible to EAE and in some cases, exhibited a more severe clinical disease [114,115]. Splenocytes and LN cells from immunized p35$^{-/-}$ or IL-12Rβ2$^{-/-}$ mice were able to secrete IFN-γ in response to MOG$_{35-55}$, and intracellular cytokine staining revealed that T cells, microglia, and macrophages from the CNS of immunized p35$^{-/-}$ mice were making TNF-α [114,115]. Therefore, in contrast to the p40 subunit of IL-12 and the β1 chain of the IL-12R, it appeared Th1 T cell priming and disease induction/progression remained intact in the absence of the p35 subunit and the β2 chain. These inconsistencies were explained by the discovery of a new cytokine, IL-23.

G. INTERLEUKIN-23 (IL-23)

IL-23 shares the p40 subunit with IL-12 but also has a unique subunit, p19. Similarly, its heterodimeric receptor also shares the IL-12Rβ1 subunit with IL-12, but additionally uses a unique subunit known as IL-23R [116]. It is predominantly secreted by APCs such as DCs, macrophages, and microglia, and IL-23R is expressed on many cell types including T cells (particularly of the memory phenotype), NK cells, DCs, and some macrophages [97,105,116]. IL-23 also shares overlapping functional characteristics with IL-12 such as promoting proliferation and IFN-γ production in T cells and upregulating CD40 and iNOS expression in APCs [105,117]. However, distinct functions do exist, as IL-23 can also induce strong proliferation in memory CD4$^+$ cells, cause DCs to immunogenically activate T cells with a normally tolerogenic tumor peptide and induce IL-1β and TNFα expression in peritoneal macrophages [105,117,118].

Although expression of IL-23 and IL-23R has not yet been extensively examined in MS, their expression on macrophages and microglia has been examined in EAE. Using real-time PCR, a 30-fold increase in the expression of p19 was detected in a mixed population of CNS-derived macrophages and microglia at early stages of disease [97]. Surprisingly though, only macrophages acquired IL-23R expression after EAE onset, whereas microglia did not, implying that each population differed in their ability to respond [105].

Although the initial studies showing resistance to EAE in p40$^{-/-}$ and IL-12Rβ1$^{-/-}$ mice were attributed to a lack of IL-12 responsiveness, the discovery of IL-23 question these conclusions. As discussed previously, p35$^{-/-}$ mice that lack IL-12 but have IL-23, are fully susceptible to EAE, indicating IL-23 and not IL-12 is critical in the induction of EAE [111,114]. Conversely, p19$^{-/-}$ mice lacking IL-23 but not IL-12, were completely resistant to MOG-induced EAE [105]. Likewise, injection of a viral vector encoding IL-23 into the brains of immunized p19$^{-/-}$ mice reconstituted disease, albeit a muted version. In contrast to MOG-immunized p35$^{-/-}$ and p40$^{-/-}$ mice, p19$^{-/-}$ mice did mount strong Th1 responses *in vitro*, maintaining proliferation and IFN-γ production while avoiding Th2 bias. p19$^{-/-}$ T cells were also capable of initially migrating into the CNS, as increased CD4$^+$ T cells and infiltrating macrophages are initially evident in knockout mice. However, these infiltrates then failed to expand, microglia remain unactivated, and disease never occured [105]. Collectively, these data suggest that although IL-12 is integral to disease initiation and priming of CD4 responses, IL-23 is involved in promoting and maintaining the chronic inflammation within the CNS, an action necessary for the development of disease.

III. ANTI-INFLAMMATORY CYTOKINE PROFILES IN EAE

A. INTERLEUKIN-10 (IL-10)

During the course of CNS autoimmunity, IL-10 can be produced by macrophages, Th2 cells, B cells and DCs, in addition to astrocytes and microglia [19,53,119,120]. IL-10 has well-documented inhibitory effects on Th1 proliferation and effector cytokine production and is also a potent inhibitor of APC function through its ability to decrease MHC Class II and B7 expression and proinflammatory cytokine and chemokine production [121–123]. In contrast, IL-10 is also known to promote B cell differentiation, survival, and immunoglobulin production/class switching and induce CD8$^+$ T cell proliferation and cytotoxic activity. Based on these potent immunomodulatory functions, IL-10 has been investigated as an anti-inflammatory mediator in both EAE and MS.

Some of the earliest studies examining the role of endogenous IL-10 during autoimmunity reported decreased IL-10 mRNA expression coincident with disease remission in both the mouse and rat EAE systems, and a similar decrease in IL-10 mRNA expression in the blood of MS patients prior to a clinical relapse supports these findings [4,101,124]. There have however been reports in other EAE models that demonstrate no such correlation [58,125].

Follow-up studies designed to modulate active EAE *in vivo* using systemic IL-10 and anti-IL10 therapies have demonstrated differential effects depending on the timing of treatment. IL-10 administration close to immunization or during the pre-clinical phase of EAE led to either complete disease inhibition or significant disease attenuation, whereas delay of treatment until disease onset had no effect [126–128]. In contrast, in a MBP-specific adoptive transfer system, there was no effect to a modest exacerbation in clinical severity following either IL-10 or anti-IL-10 treatment across all times of administration relative to day of transfer [129].

The use of IL-10 knockout and transgenic mice has corroborated these observations. IL-10 transgenic mice under control of the MHC Class II or CD2 promoters are highly resistant to disease induction by active immunization despite the generation of myelin-specific Th1 cells, whereas IL-10-deficient mice have a heightened susceptibility to EAE induction and develop a disease of significantly increased severity without a clinical remission [110,130–132]. The absence of spontaneous recovery was associated with a lack of IL-10 production by B cells. It was also demonstrated that IFN-γ production by myelin-specific cells was increased substantially in the absence of IL-10 and decreased in the presence of transgenic IL-10, again supporting the integral role for this cytokine in the generation of pathogenic Th1 responses. Taken together, these data are consistent with an early kinetic requirement for IL-10 to exert its immunoregulatory function in the suppression of myelin-specific Th1 cell generation and effector function.

To avoid the paradox and complications of systemic approaches, several targeted therapies have been employed to address the effects of CNS-localized and in some instances antigen-specific, production of IL-10. Transfer of a $PLP_{139-151}$-specific T cell clone overexpressing IL-10 was able to significantly inhibit the severity of acute EAE without a concomitant delay in disease onset, yet the transfer of a retrovirally transduced MBP-specific hybridoma expressing high levels of IL-10 was unable to alter the development or pathogenesis of EAE [133,134]. Similar contrary observations have been reported in an adenovirus system where in one report adenoviral delivery of mouse IL-10 did not prevent EAE progression, whereas in another using human IL-10, disease onset was completely inhibited, and the therapy was used successfully to down-modulate ongoing pathogenesis and prevent disease relapse [126,135]. Likewise, neither intracranial (i.c.) infection with a herpes simplex virus (HSV) containing IL-10, nor the i.c. transfer of IL-10-DNA-cationic liposome complexes (CLC) postdisease had any effect on disease progression, whereas i.c. transfer of retrovirally transfected fibroblasts expressing IL-10 postimmunization was able to effectively decrease disease severity, but did not affect disease onset [135–137]. These contrary observations may be due to the different efficacies of each treatment in the amount of IL-10 actually produced in the CNS (or the number of cells that reached the CNS), in addition to the cell producing IL-10, as it is known that the various approaches will target infection of different cell types and vary widely in the amount of transgenic protein production.

The effects of IL-10 treatment in other models of autoimmunity have underscored observations from the EAE system and illustrate the importance of the cell producing IL-10 and the timing of its administration for beneficial effects on disease progression. In the non-obese–diabetic (NOD) mouse model of autoimmunity, systemic and prolonged treatment with IL-10 resulted in reduced severity of insulitis and the occurrence of diabetes, as did the early neutralization of IL-10 [138–140]. Similarly, whereas islet-specific Th1 cells overexpressing IL-10 could prevent the adoptive transfer of diabetes, localized transgenic production of IL-10 by islet cells accelerated destruction of the insulin-producing beta cells [141,142]. Altogether, it appears that there is potential for downregulation of autoimmunity using IL-10, however the effects on clinical disease will be most beneficial when issues related to kinetics of production, and the localization of IL-10 producing cells within the inflammatory environment are more rigorously controlled.

B. INTERLEUKIN-4 (IL-4)

IL-4 is another anti-inflammatory cytokine believed to play a critical role in the intrinsic regulation of autoimmunity and, in contrast to IL-10, is limited in its production to only Th2 cells, mast cells, and NK1.1 T cells. The effects of IL-4 in the differentiation of naïve T cells toward a Th2 phenotype is well documented, as is the ability of IL-4 to modulate effector function of Th1 cells via immune deviation [143,144]. IL-4 has also been shown to promote the maturation of DCs and activate B cells while inducing class switching to the Th2-associated IgG1 isotype. Interestingly, a few studies have also indicated that microglia can respond to IL-4 with increased activation and proliferation, yet a corresponding suppression of MHC Class II expression and TNF-α, IL-10, and IL-1β production, which could be a critical regulation pathway during CNS inflammation [145,146].

The expression of IL-4 has been reported in MS lesions and blood from MS patients [147], yet no significant association between clinical episode onset or remission has been demonstrated to date [19,120,148]. Much like IL-10, the correlation between spontaneous EAE remission and endogenous IL-4 production has also been contradictory with some reports demonstrating increased levels of IL-4 mRNA in the remitted CNS after both the primary acute phase as well as subsequent relapsing episodes and others reporting no such connection [58,125,149]. In addition, production of IL-4 may confer some level of initial resistance to EAE induction, as T cells from resistant BALB/c and male SJL/J mice preferentially make IL-4 rather than IFNγ following active immunization with spinal cord homogenate [108,150].

Further evidence for the modulatory role of IL-4 stems from the repeated observation that PLP/MBP-specific Th2 clones secreting IL-4 are ineffective at adoptively transferring EAE [60,61,151–153]. Likewise, addition of IL-4 to *in vitro* cultures of encephalitogenic Th1 cells decreases IFN-γ production and reduces their pathogenic potential, yet attempts to ameliorate disease induction or progression by cotransfer of Th2 populations has produced mixed results [144,152–154]. Interestingly however, it has been reported that preimmunization with nonencephalitogenic peptides/ proteins in IFA led to the IL-4 production that protected mice from EAE and promoted immune deviation of potential pathogenic Th1 cells [155,156]. These observations lend support for the regulatory capability of IL-4 during the inductive phase of an immune response, but only when present in high amounts in a microenvironment that can modulate the development, effector function or migration of Th1 cells.

To that end, systemic administration of IL-4 completely prevents EAE onset if provided frequently (every 8 hours) in large concentrations from the time of immunization through day 11 as well as reduces the severity of ongoing EAE and demyelination within the CNS and results in decreased peripheral and CNS proinflammatory Th1 cytokines [127,144,157,158]. Similarly, CNS-targeting of IL-4 by retrovirally transduced MBP-specific hybridomas or by HSV expressing IL-4 led to diminished EAE severity and inhibited autoimmune progression [134,136,159]. IL-4 gene therapy has also proven beneficial in other models of autoimmunity such as collagen-induced arthritis and NOD systems where production of IL-4 within the pancreas protects the animals from both insulitis and diabetes [160–162].

These disease inhibiting observations are at odds however with the several studies using IL-4 knockout and transgenic mice. Contrary to expectations, IL-4-deficient PL/J and B6 mice and IL-4 T cell transgenic B6 mice have repeatedly been shown to develop normal disease pathology with equal frequency and severity following active or adoptive disease induction [131,132,163,164]. This is in contrast to other reports demonstrating that IL-4-deficient B6 and BALB/c display a modestly exacerbated clinical course and express increased CNS proinflammatory cytokines compared to wild-type controls [131,165]. This discrepancy may be due in part to the use of pertussis toxin in the EAE induction protocol, as its absence leads to a less severe disease.

Despite these differences, and in studies where animals were followed until recovery, IL4$^{-/-}$ mice underwent normal disease remission indicating possible redundancy in the activity of anti-inflammatory Th2 cytokines. In fact, increased CNS IL-10 expression was evident in IL-4$^{-/-}$ mice coincident with clinical recovery, indicating IL-4 is not necessary for the onset of remission [163]. However, given that Th2 responses are absent in IL-4$^{-/-}$ mice, the source of IL-10 would appear to be dissociated from the Th2 pathway, and the mechanism underlying disease remission is possibly distinct from immune deviation. It would appear from these studies that IL-4 is less critical for regulation of MOG-induced EAE in B6 mice and more central to modulation of PLP/MBP-induced EAE in SJL mice, although disease induction in IL4$^{-/-}$ mice in traditionally susceptible strains has yet to be completed. In light of these data, there remains compelling evidence in support of the ability of IL-4 to downregulate Th1 inflammatory responses, although its exact role in the regulation of EAE is complex.

C. TRANSFORMING GROWTH FACTOR-β (TGF-β)

The TGF-β cytokine family, composed of three closely related isoforms, TGF-β_{1-3}, are produced by many cells of both the lymphoid and nonlymphoid lineages. TGF-β_1 has widespread pleiotropic effects that result in both stimulatory and inhibitory outcomes on cell proliferation, differentiation and migration depending on the cell type, location and kinetics of production, in addition to its potent immunosuppressive capabilities [166]. TGF-β_1 has been shown to mediate most of its suppressive effects in the lymphoid system and decreases Th1 production of the pro-inflammatory cytokines IFN-γ and TNF-α, induces apoptosis and inhibits APC function through decreases in MHC Class II and costimulatory molecules [167–170]. In addition, TGF-β_1 inhibits iNOS activity resulting in suppressed production of NO radicals from macrophages, decreases mRNA expression of chemokine receptors on microglia, and decreases the ability of B cells to isotype switch and produce IgM and IgG [171–174]. Lastly, TGF-β_1 has been shown to diminish cytolytic activity and interferon responsiveness of both NK cells and CD8$^+$ T cells [175,176]. Despite these potent immunosuppressive capabilities, TGF-β_1 is also known to provide and stimulate production of a chemotactic gradient for leukocytes during the induction of an immune response and promote the development of Th2-type responses with increased IL-4 production. Therefore the biphasic nature of TGF-β_1 effects places this cytokine in a class of its own.

There is some evidence that TGF-β_1 may play a regulatory role during MS, as increased amounts of TGF-β_1 mRNA or protein is detectable in PBMC and CSF of MS patients during periods of clinical remission and clinical stability [147,177,178]. This regulatory role is also supported by data from MS patients in interferon-β1a/b therapy wherein multiple studies have indicated an increased serum expression of TGF-β_1 after 3–6 months of treatment [179,180]. During EAE, endogenous TGF-β_1 mRNA and protein production in the CNS has also been reported to correlate with the onset of remission, or with increased resistance to disease induction [45,181–183]. Similarly, during the course of actively induced EAE in the Lewis rat, spontaneous recovery is associated with the proliferation of an antigen-specific, CD4+, Vβ8-bearing T cell population capable of secreting TGF-β [184].

There is considerable support for the disease-limiting role of TGF-β during autoimmunity that has come from reports wherein (1) incubation of TGF-β1 with encephalitogenic T cells reduced their adoptive transfer potential [170,185]; (2) retrovirally transduced encephalitogenic T cells expressing TGF-β1 lose the ability to transfer EAE and afford protection from active disease induction [186,187]; (3) treatment of EAE with anti-TGF-β exacerbated clinical EAE [181,188,189]; and (4) early/preclinical systemic or CNS-localized administration of TGF-β protected animals from disease onset or reduced acute phase severity when provided at the time of immunization [137,158,181,190,191]. Disappointingly however, a Phase I clinical trial examining treatment of MS with TGF-β_2 resulted in no significant change in the expanded disability status scale (EDSS) score or MRI lesion activity during the treatment period, and the trial was ultimately stopped due to significant side effects associated with kidney toxicity [192].

Collectively, these results suggest that the primary role of TGF-β as a mediator during EAE is to limit the extent of inflammation and as a result, the severity of disease. The molecular and cellular mechanisms by which TGF-β exerts these effects have not been conclusively determined and are likely dependent on TGF-β's composite interactions within both the innate and adaptive immune system and possible synergy with other immunoregulatory cytokines to modulate CNS inflammation.

D. TOLERANCE-INDUCING STRATEGIES AND ANTI-INFLAMMATORY CYTOKINES

Perhaps the most convincing data to implicate anti-inflammatory cytokines as key mediators in the regulation of EAE, and pathogenic Th1 responses, stems from effective tolerance-inducing therapies that promote the endogenous expression of these cytokines as a result of the tolerogenic approach, rather than those that have attempted to directly manipulate their expression and involves several mechanisms including bystander suppression, immune deviation, and active suppression by regulatory T cells.

IL-10, IL-4, and TGF-β have been shown to be produced in an antigen-specific manner following intranasal and low/continuous oral administration of myelin peptides/proteins, with a concomitant decrease in effector encephalitogenic Th1 proliferation and production of IFN-γ and IL-2, resulting in protection from active EAE disease induction [193–196]. Another therapeutic regimen known to ameliorate clinical EAE, the use of altered peptide ligands (APL), is mediated to a significant extent through the production of IL-4 by responding T cells that inhibits the generation of encephalitogenic Th1 responses to both the native, as well as intra- and intermolecular myelin epitopes [197,198]. In this regard, it has been shown that a Th2 clone specific for a $PLP_{139-151}$ APL can prevent adoptive EAE when co-cultured with encephalitogenic splenocytes, an effect dependent on the secretion of IL4, IL-10, IL-13, and TGF-β by the APL clone [199]. Similarly, induction of nonencephalitogenic Th2 cells through immunization with IFA has been shown to inhibit the generation of myelin-specific T cells via IL-10 or IL-4-mediated bystander suppression and prevent the development of EAE [155,156]. Lastly, our lab has found that i.v. injection of ethylene carbodiimide (ECDI)-fixed splenocytes coupled with myelin peptides/proteins is quite effective at preventing induction of EAE both pre- and post-priming in addition to postadoptive transfer, as well as effective in modulating disease progression through the inhibition of relapses [200,201]. Although the mechanisms involved in this method of tolerance are complex, we have previously determined that ECDI-coupled cell tolerance results in the anergy of Th1 specific proliferation and cytokine production with a concomitant stimulation of Th2 cells [202–204].

Other tolerizing therapies are also mediated in part by the production of anti-inflammatory cytokines. Injection of an aggregated form of an immunoglobulin chimera carrying the $PLP_{139-151}$ peptide induces endogenous IL-10 production from macrophages and DCs, dramatically reduces the severity of EAE and prevents epitope spreading [205]. SJL mice co-vaccinated with genes for both $PLP_{139-151}$ and IL-4 were protected from severe active EAE induction by the same peptide and resulted in a functional Th2 shift in autoreactive cells (as measured by increased IL-4 and IL-10 recall production) that was able to confer protection to naïve recipients [206]. Likewise, co-vaccination of B6 mice with MOG and IL-4 genes at peak of the acute phase resulted in a dramatic reduction in disease severity as compared to control mice and a reduction in subsequent relapses [206].

IV. CONCLUSIONS

Although we have discussed each cytokine individually relative to their unique roles in potentiating or modulating CNS inflammation, the situation is certainly more complex when considering the dynamic interactions that occur in an inflammatory microenvironment containing multiple cell types producing multiple cytokines. Considering the existing evidence, it would appear that balance between Th1 and Th2 responses and the innate and adaptive immune system is pivotal in controlling inflammation and the potential for ensuing autoimmunity, and the cytokines discussed here are integral to this complex and highly regulated process.

REFERENCES

1. Ishihara, K. and Hirano, T., IL-6 in autoimmune disease and chronic inflammatory proliferative disease, *Cytokine Growth Factor Rev.*, 13, 357, 2002.
2. Maimone, D. et al., Cytokine levels in the cerebrospinal fluid and serum of patients with multiple sclerosis, *J. Neuroimmunol.*, 32, 67, 1991.

3. Lock, C. et al., Gene-microarray analysis of multiple sclerosis lesions yields new targets validated in autoimmune encephalomyelitis, *Nat. Med.*, 8, 500, 2002.

4. Kennedy, M.K. et al., Analysis of cytokine mRNA expression in the central nervous system of mice with experimental autoimmune encephalomyelitis reveals that IL-10 mRNA expression correlates with recovery, *J. Immunol.*, 149, 2496, 1992.

5. Okuda, Y. et al., IL-6 plays a crucial role in the induction phase of myelin oligodendrocyte glucoprotein 35–55 induced experimental autoimmune encephalomyelitis, *J. Neuroimmunol.*, 101, 188, 1999.

6. Eugster, H.P. et al., IL-6-deficient mice resist myelin oligodendrocyte glycoprotein-induced autoimmune encephalomyelitis, *Eur. J. Immunol.*, 28, 2178, 1998.

7. Mendel, I. et al., Interleukin-6 functions in autoimmune encephalomyelitis: a study in gene-targeted mice, *Eur. J. Immunol.*, 28, 1727, 1998.

8. Samoilova, E.B. et al., IL-6-deficient mice are resistant to experimental autoimmune encephalomyelitis: roles of IL-6 in the activation and differentiation of autoreactive T cells, *J. Immunol.*, 161, 6480, 1998.

9. Okuda, Y. et al., Enhancement of Th2 response in IL-6-deficient mice immunized with myelin oligodendrocyte glycoprotein, *J. Neuroimmunol.*, 105, 120, 2000.

10. Gijbels, K. et al., Administration of neutralizing antibodies to interleukin-6 (IL- 6) reduces experimental autoimmune encephalomyelitis and is associated with elevated levels of IL-6 bioactivity in central nervous system and circulation, *Molecular Medicine*, 1, 795, 1995.

11. Chung, I.Y. and Benveniste, E.N., Tumor necrosis factor-α production by astrocytes. Induction by lipopolysaccharide, IFN-gamma, and IL-1β, *J. Immunol.*, 144, 2999, 1990.

12. Vassalli, P., The pathophysiology of tumor necrosis factors, *Ann. Rev. Immunol.*, 10, 411, 1992.

13. Selmaj, K.W., Tumour necrosis factor and anti-tumour necrosis factor approach to inflammatory demyelinating diseases of the central nervous system, *Ann. Rheum. Dis.*, 59 Suppl 1, i94, 2000.

14. Sedgwick, J.D. et al., Tumor necrosis factor: a master-regulator of leukocyte movement, *Immunol. Today*, 21, 110, 2000.

15. Feldmann, M. et al., Biological insights from clinical trials with anti-TNF therapy, *Springer Semin. Immunopathol.*, 20, 211, 1998.

16. Juedes, A.E. et al., Kinetics and cellular origin of cytokines in the central nervous system: insight into mechanisms of myelin oligodendrocyte glycoprotein-induced experimental autoimmune encephalomyelitis, *J. Immunol.*, 164, 419, 2000.

17. Renno, T. et al., TNF-alpha expression by resident microglia and infiltrating leukocytes in the central nervous system of mice with experimental allergic encephalomyelitis. Regulation by Th1 cytokines, *J. Immunol.*, 154, 944, 1995.

18. Spuler, S. et al., Multiple sclerosis: prospective analysis of TNF-alpha and 55 kDa TNF receptor in CSF and serum in correlation with clinical and MRI activity, *J. Neuroimmunol.*, 66, 57, 1996.

19. Cannella, B. and Raine, C.S., The adhesion molecule and cytokine profile of multiple sclerosis lesions, *Ann. Neurol.*, 37, 424, 1995.

20. Crisi, G.M. et al., Staphylococcal enterotoxin B and tumor-necrosis factor-alpha-induced relapses of experimental allergic encephalomyelitis: protection by transforming growth factor-beta and interleukin-10, *Eur. J. Immunol.*, 25, 3035, 1995.

21. Ruddle, N.H. et al., An antibody to lymphotoxin and tumor necrosis factor prevents transfer of experimental allergic encephalomyelitis, *J. Exp. Med.*, 172, 1193, 1990.

22. Korner, H. et al., Unimpaired autoreactive T-cell traffic within the central nervous system during tumor necrosis factor receptor-mediated inhibition of experimental autoimmune encephalomyelitis, *Proc. Natl. Acad. Sci. USA*, 92, 11066, 1995.

23. Selmaj, K. et al., Prevention of chronic relapsing experimental autoimmune encephalomyelitis by soluble tumor necrosis factor receptor I, *J. Neuroimmunol.*, 56, 135, 1995.

24. Riminton, D.S. et al., Challenging cytokine redundancy: inflammatory cell movement and clinical course of experimental autoimmune encephalomyelitis are normal in lymphotoxin-deficient, but not tumor necrosis factor-deficient mice, *J. Exp. Med.*, 187, 1517, 1998.

25. Selmaj, K.W. and Raine, C.S., Tumor necrosis factor mediates myelin and oligodendrocyte damage *in vitro*, *Ann. Neurol.*, 23, 339, 1988.

26. Liu, J. et al., TNF is a potent anti-inflammatory cytokine in autoimmune-mediated demyelination, *Nature Med.*, 4, 78, 1998.

27. Frei, K. et al., Tumor necrosis factor alpha and lymphotoxin alpha are not required for induction of acute experimental autoimmune encephalomyelitis, *J. Exp. Med.*, 185, 2177, 1998.

28. Ewing, C. and Bernard, C.C., Insights into the aetiology and pathogenesis of multiple sclerosis, *Immunol. Cell Biol.*, 76, 47, 1998.

29. Korner, H. et al., Critical points of tumor necrosis factor action in central nervous system autoimmune inflammation defined by gene targeting, *J. Exp. Med.*, 186, 1585, 1997.

30. Kassiotis, G., Kranidioti, K., and Kollias, G., Defective CD4 T cell priming and resistance to experimental autoimmune encephalomyelitis in TNF-deficient mice due to innate immune hypo-responsiveness, *J. Neuroimmunol.*, 119, 239, 2001.

31. Ruuls, S.R. et al., Membrane-bound TNF supports secondary lymphoid organ structure but is subservient to secreted TNF in driving autoimmune inflammation, *Immunity*, 15, 533, 2001.

32. Murphy, C.A. et al., Interactions between hemopoietically derived TNF and central nervous system-resident glial chemokines underlie initiation of autoimmune inflammation in the brain, *J. Immunol.*, 169, 7054, 2002.

33. Muller, G., Hopken, U.E., and Lipp, M., The impact of CCR7 and CXCR5 on lymphoid organ development and systemic immunity, *Immunol. Rev.*, 195, 117, 2003.

34. Kassiotis, G. and Kollias, G., Uncoupling the proinflammatory from the immunosuppressive properties of tumor necrosis factor (TNF) at the p55 TNF receptor level: implications for pathogenesis and therapy of autoimmune demyelination, *J. Exp. Med.*, 193, 427, 2001.

35. Eugster, H.P. et al., Severity of symptoms and demyelination in MOG-induced EAE depends on TNFR1, *Eur. J. Immunol.*, 29, 626, 1999.

36. Suvannavejh, G.C. et al., Divergent roles for p55 and p75 tumor necrosis factor receptors in the pathogenesis of MOG(35-55)-induced experimental autoimmune encephalomyelitis, *Cell. Immunol.*, 205, 24, 2000.

37. van Oosten, B.W. et al., Increased MRI activity and immune activation in two multiple sclerosis patients treated with the monoclonal anti-tumor necrosis factor antibody cA2, *Neurology*, 47, 1531, 1996.

38. TNF neutralization in MS: results of a randomized, placebo-controlled multicenter study. The Lenercept Multiple Sclerosis Study Group and The University of British Columbia MS/MRI Analysis Group, *Neurology*, 53, 457, 1999.

39. Arnett, H.A. et al., TNF alpha promotes proliferation of oligodendrocyte progenitors and remyelination, *Nat. Neurosci.*, 4, 1116, 2001.

40. Gommerman, J.L. and Browning, J.L., Lymphotoxin/light, lymphoid microenvironments and autoimmune disease, *Nat. Rev. Immunol.*, 3, 642, 2003.

41. Banks, T.A. et al., Lymphotoxin-alpha-deficient mice. Effects on secondary lymphoid organ development and humoral immune responsiveness, *J. Immunol.*, 155, 1685, 1995.

42. Futterer, A. et al., The lymphotoxin beta receptor controls organogenesis and affinity maturation in peripheral lymphoid tissues, *Immunity*, 9, 59, 1998.

43. Koni, P.A. et al., Distinct roles in lymphoid organogenesis for lymphotoxins alpha and beta revealed in lymphotoxin beta-deficient mice, *Immunity*, 6, 491, 1997.

44. Matusevicius, D. et al., Multiple sclerosis: the proinflammatory cytokines lymphotoxin-alpha and tumour necrosis factor-alpha are upregulated in cerebrospinal fluid mononuclear cells, *J. Neuroimmunol.*, 66, 115, 1996.

45. Issazadeh, S. et al., Cytokine production in the central nervous system of Lewis rats with experimental autoimmune encephalomyelitis: dynamics of mRNA expression for IL-10, interleukin-12, cytolysin, tumor necrosis factor alpha, and tumor necrosis beta, *J. Neuroimmunol.*, 61, 205, 1995.

46. Selmaj, K. et al., Cytokine cytotoxicity against oligodendrocytes. Apoptosis induced by lymphotoxin, *J. Immunol.*, 147, 1522, 1991.

47. Baker, D. et al., Control of established experimental allergic encephalomyelitis by inhibition of tumor necrosis factor (TNF) activity within the central nervous system using monoclonal antibodies and TNF receptor-immunoglobulin fusion proteins, *Eur. J. Immunol.*, 24, 2040, 1994.

48. Suen, W.E. et al., A critical role for lymphotoxin in experimental allergic encephalomyelitis, *J. Exp. Med.*, 186, 1233, 1997.

49. Gommerman, J.L. et al., A role for surface lymphotoxin in experimental autoimmune encephalomyelitis independent of LIGHT, *J. Clin. Invest*, 112, 755, 2003.

50. Sherman, M.P., Griscavage, J.M., and Ignarro, L.J., Nitric oxide-mediated neuronal injury in multiple sclerosis. [Review] [39 refs], *Medical Hypotheses*, 39, 143, 1992.

51. Aloisi, F. et al., Microglia are more efficient than astrocytes in antigen processing and in Th1 but not Th2 activation, *J. Immunol.*, 160, 4671, 1998.

52. Cornet, A. et al., Role of astrocytes in antigen presentation and naive T-cell activation, *J. Neuroimmunol.*, 106, 69, 2000.

53. Olson, J.K., Girvin, A.M., and Miller, S.D., Direct activation of innate and antigen presenting functions of microglia following infection with Theiler's virus, *J. Virol.*, 75, 9780, 2001.

54. Girvin, A.M. et al., Differential abilities of central nervous system resident endothelial cells and astrocytes to serve as inducible antigen-presenting cells, *Blood*, 99, 3692, 2002.

55. Vartanian, T. et al., Interferon-gamma-induced oligodendrocyte cell death: implications for the pathogenesis of multiple sclerosis, *Molecular Medicine*, 1, 732, 1995.

56. Olsson, T. et al., Increased numbers of T cells recognizing multiple myelin basic protein epitopes in multiple sclerosis, *Eur. J. Immunol.*, 22, 1083, 1992.

57. Link, H. et al., Virus-reactive and autoreactive T cells are accumulated in cerebrospinal fluid in multiple sclerosis, *J. Neuroimmunol.*, 38, 63, 1992.

58. Begolka, W.S. et al., Differential expression of inflammatory cytokines parallels progression of central nervous system pathology in two clinically distinct models of multiple sclerosis, *J. Immunol.*, 161, 4437, 1998.

59. Targoni, O.S. et al., Frequencies of neuroantigen-specific T cells in the central nervous system versus the immune periphery during the course of experimental allergic encephalomyelitis, *J. Immunol.*, 166, 4757, 2001.

60. Ando, D.G. et al., Encephalitogenic T cells in the B10.PL model of experimental allergic encephalomyelitis (EAE) are of the Th-1 lymphokine subtype, *Cell. Immunol.*, 124, 132, 1989.

61. Baron, J.L. et al., Surface expression of alpha 4 integrin by CD4 T cells is required for their entry into brain parenchyma, *J. Exp. Med.*, 177, 57, 1993.

62. Renno, T. et al., Selective enrichment of Th1 CD45RB low CD4+ T cells in autoimmune infiltrates in experimental allergic encephalomyelitis, *Int. Immunol.*, 6, 347, 1994.

63. Lublin, F.D. et al., Monoclonal anti-gamma interferon antibodies enhance experimental allergic encephalomyelitis, *Autoimmunity*, 16, 267, 1993.

64. Heremans, H. et al., Chronic relapsing experimental autoimmune encephalomyelitis (CREAE) in mice: enhancement by monoclonal antibodies against interferon-gamma, *Eur. J. Immunol.*, 26, 2393, 1996.

65. Duong, T.T. et al., Effect of anti-interferon-gamma monoclonal antibody treatment on the development of experimental allergic encephalomyelitis in resistant mouse strains, *J. Neuroimmunol.*, 53, 101, 1994.

66. Furlan, R. et al., Intrathecal delivery of IFN-gamma protects C57BL/6 mice from chronic-progressive experimental autoimmune encephalomyelitis by increasing apoptosis of central nervous system-infiltrating lymphocytes, *J. Immunol.*, 167, 1821, 2001.

67. Krakowski, M. and Owens, T., Interferon-gamma confers resistance to experimental allergic encephalomyelitis, *Eur. J. Immunol.*, 26, 1641, 1996.

68. Ferber, I.A. et al., Mice with a disrupted IFN-gamma gene are susceptible to the induction of experimental autoimmune encephalomyelitis (EAE)., *J. Immunol.*, 156, 5, 1996.

69. Tran, E.H., Prince, E.N., and Owens, T., IFN-gamma shapes immune invasion of the central nervous system via regulation of chemokines, *J. Immunol.*, 164, 2759, 2000.

70. Chu, C.Q., Wittmer, S., and Dalton, D.K., Failure to suppress the expansion of the activated CD4 T cell population in interferon gamma-deficient mice leads to exacerbation of experimental autoimmune encephalomyelitis, *J. Exp. Med.*, 192, 123, 2000.

71. Willenborg, D.O. et al., IFN-gamma plays a critical down-regulatory role in the induction and effector phase of myelin oligodendrocyte glycoprotein-induced autoimmune encephalomyelitis, *J. Immunol.*, 157, 3223, 1996.

72. Fenyk-Melody, J.E. et al., Experimental autoimmune encephalomyelitis is exacerbated in mice lacking the NOS2 gene, *J. Immunol.*, 160, 2940, 1998.

73. van der Veen, R.C. et al., Nitric oxide inhibits the proliferation of T-helper 1 and 2 lymphocytes without reduction in cytokine secretion, *Cell Immunol.*, 193, 194, 1999.

74. Juedes, A.E. and Ruddle, N.H., Resident and infiltrating central nervous system APCs regulate the emergence and resolution of experimental autoimmune encephalomyelitis, *J. Immunol.*, 166, 5168, 2001.

75. van der Veen, R.C., Dietlin, T.A., and Hofman, F.M., Tissue expression of inducible nitric oxide synthase requires IFN-gamma production by infiltrating splenic T cells: more evidence for immunosuppression by nitric oxide, *J. Neuroimmunol.*, 145, 86, 2003.

76. Willenborg, D.O. et al., IFN-gamma is critical to the control of murine autoimmune encephalomyelitis and regulates both in the periphery and in the target tissue: a possible role for nitric oxide, *J. Immunol.*, 163, 5278, 1999.

77. Okamura, H. et al., Cloning of a new cytokine that induces IFN-gamma production by T cells, *Nature*, 378, 88, 1995.

78. Prinz, M. and Hanisch, U.K., Murine microglial cells produce and respond to inter-leukin-18, *J. Neurochem.*, 72, 2215, 1999.
79. Takeda, K. et al., Defective NK cell activity and Th1 response in IL-18-deficient mice, *Immunity*, 8, 383, 1998.
80. Ahn, H.J. et al., A mechanism underlying synergy between IL-12 and IFN-gamma-inducing factor in enhanced production of IFN-gamma, *J. Immunol.*, 159, 2125, 1997.
81. Chang, J.T. et al., The costimulatory effect of IL-18 on the induction of antigen-specific IFN-gamma production by resting T cells is IL-12 dependent and is mediated by up-regulation of the IL-12 receptor beta2 subunit, *Eur. J. Immunol.*, 30, 1113, 2000.
82. Gutzmer, R. et al., Human dendritic cells express the IL-18R and are chemoattracted to IL-18, *J. Immunol.*, 171, 6363, 2003.
83. Cannella, B. and Raine, C.S., Multiple sclerosis: cytokine receptors on oligodendro-cytes predict innate regulation, *Ann. Neurol.*, 55, 46, 2004.
84. Andre, R. et al., Identification of a truncated IL-18R beta mRNA: a putative regulator of IL-18 expressed in rat brain, *J. Neuroimmunol.*, 145, 40, 2003.
85. Nicoletti, F. et al., Increased serum levels of interleukin-18 in patients with multiple sclerosis, *Neurology*, 57, 342, 2001.
86. Losy, J. and Niezgoda, A., IL-18 in patients with multiple sclerosis, *Acta Neurol. Scand.*, 104, 171, 2001.
87. Karni, A. et al., IL-18 is linked to raised IFN-gamma in multiple sclerosis and is induced by activated CD4(+) T cells via CD40-CD40 ligand interactions, *J. Neuroim-munol.*, 125, 134, 2002.
88. Wildbaum, G. et al., Neutralizing antibodies to IFN-gamma-inducing factor prevent experimental autoimmune encephalomyelitis, *J. Immunol.*, 161, 6368, 1998.
89. Shi, F.D. et al., IL-18 directs autoreactive T cells and promotes autodestruction in the central nervous system via induction of IFN-gamma by NK cells, *J. Immunol.*, 165, 3099, 2000.
90. Trinchieri, G., Proinflammatory and immunoregulatory functions of interleukin-12, *Int. Rev. Immunol.*, 16, 365, 1998.
91. Becher, B., Blain, M., and Antel, J.P., CD40 engagement stimulates IL-12 p70 production by human microglial cells: basis for Th1 polarization in the CNS, *J. Neuroimmunol.*, 102, 44, 2000.
92. Magram, J. et al., IL-12-deficient mice are defective but not devoid of type 1 cytokine responses, *Ann. N.Y. Acad. Sci.*, 795, 60, 1996.
93. Magram, J. et al., IL-12-deficient mice are defective in IFN gamma production and type 1 cytokine responses, *Immunity*, 4, 471, 1996.
94. Bagaeva, L.V., Williams, L.P., and Segal, B.M., IL-12 dependent/IFN gamma inde-pendent expression of CCR5 by myelin-reactive T cells correlates with encephalito-genicity, *J. Neuroimmunol.*, 137, 109, 2003.
95. Szabo, S.J. et al., Regulation of the interleukin (IL)-12R beta 2 subunit expression in developing T helper 1 (Th1) and Th2 cells, *J. Exp. Med.*, 185, 817, 1997.
96. Carter, L.L. and Murphy, K.M., Lineage-specific requirement for signal transducer and activator of transcription (Stat)4 in interferon gamma production from CD4(+) versus CD8(+) T cells, *J. Exp. Med.*, 189, 1355, 1999.
97. Li, J. et al., Differential expression and regulation of IL-23 and IL-12 subunits and receptors in adult mouse microglia, *J. Neurol. Sci.*, 215, 95, 2003.
98. Taoufik, Y. et al., Human microglial cells express a functional IL-12 receptor and produce IL-12 following IL-12 stimulation, *Eur. J. Immunol.*, 31, 3228, 2001.
99. Trinchieri, G., Interleukin-12 and the regulation of innate resistance and adaptive immunity, *Nat. Rev. Immunol.*, 3, 133, 2003.

100. Fassbender, K. et al., Increased release of interleukin-12p40 in MS: association with intracerebral inflammation, *Neurology*, 51, 753, 1998.

101. Boxel-Dezaire, A.H. et al., Decreased interleukin-10 and increased interleukin-12p40 mRNA are associated with disease activity and characterize different disease stages in multiple sclerosis, *Ann. Neurol.*, 45, 695, 1999.

102. Makhlouf, K., Weiner, H.L., and Khoury, S.J., Increased percentage of IL-12+ monocytes in the blood correlates with the presence of active MRI lesions in MS, *J. Neuroimmunol.*, 119, 145, 2001.

103. Bright, J.J. et al., Expression of IL-12 in CNS and lymphoid organs of mice with experimental allergic encephalitis, *J. Neuroimmunol.*, 82, 22, 1998.

104. Issazadeh, S. et al., Kinetics of expression of costimulatory molecules and their ligands in murine relapsing experimental autoimmune encephalomyelitis *in vivo*, *J. Immunol.*, 161, 1104, 1998.

105. Cua, D.J. et al., Interleukin-23 rather than interleukin-12 is the critical cytokine for autoimmune inflammation of the brain, *Nature*, 421, 744, 2003.

106. Leonard, J.P., Waldburger, K.E., and Goldman, S.J., Prevention of experimental autoimmune encephalomyelitis by antibodies against interleukin 12, *J. Exp. Med.*, 181, 381, 1995.

107. Ichikawa, M. et al., Anti-IL-12 antibody prevents the development and progression of multiple sclerosis-like relapsing–remitting demyelinating disease in NOD mice induced with myelin oligodendrocyte glycoprotein peptide, *J. Neuroimmunol.*, 102, 56, 2000.

108. Constantinescu, C.S. et al., Modulation of susceptibility and resistance to an autoimmune model of multiple sclerosis in prototypically susceptible and resistant strains by neutralization of interleukin-12 and interleukin-4, respectively, *Clin. Immunol.*, 98, 23, 2001.

109. Bright, J.J., Rodriguez, M., and Sriram, S., Differential influence of interleukin-12 in the pathogenesis of autoimmune and virus-induced central nervous system demyelination, *J. Virol.*, 73, 1637, 1999.

110. Segal, B.M., Dwyer, B.K., and Shevach, E.M., An Interleukin (IL)-10/IL-12 immunoregulatory circuit controls susceptibility to autoimmune disease, *J. Exp. Med.*, 187, 537, 1998.

111. Becher, B., Durell, B.G., and Noelle, R.J., Experimental autoimmune encephalitis and inflammation in the absence of interleukin-12, *J. Clin. Invest*, 110, 493, 2002.

112. Zhang, G.X. et al., Role of IL-12 receptor beta 1 in regulation of T cell response by APC in experimental autoimmune encephalomyelitis, *J. Immunol.*, 171, 4485, 2003.

113. Mendel, I. and Shevach, E.M., Differentiated Th1 autoreactive effector cells can induce experimental autoimmune encephalomyelitis in the absence of IL-12 and CD40/CD40L interactions, *J. Neuroimmunol.*, 122, 65, 2002.

114. Gran, B. et al., IL-12p35-deficient mice are susceptible to experimental autoimmune encephalomyelitis: evidence for redundancy in the IL-12 system in the induction of central nervous system autoimmune demyelination, *J. Immunol.*, 169, 7104, 2002.

115. Zhang, G.X. et al., Induction of experimental autoimmune encephalomyelitis in IL-12 receptor-beta 2-deficient mice: IL-12 responsiveness is not required in the pathogenesis of inflammatory demyelination in the central nervous system, *J. Immunol.*, 170, 2153, 2003.

116. Parham, C. et al., A receptor for the heterodimeric cytokine IL-23 is composed of IL-12Rbeta1 and a novel cytokine receptor subunit, IL-23R, *J. Immunol.*, 168, 5699, 2002.

117. Oppmann, B. et al., Novel p19 protein engages IL-12p40 to form a cytokine, IL-23, with biological activities similar as well as distinct from IL-12, *Immunity*, 13, 715, 2000.

118. Belladonna, M.L. et al., IL-23 and IL-12 have overlapping, but distinct, effects on murine dendritic cells, *J. Immunol.*, 168, 5448, 2002.

119. Fillatreau, S. et al., B cells regulate autoimmunity by provision of IL-10, *Nat. Immunol.*, 3, 944, 2002.

120. Hulshof, S. et al., Cellular localization and expression patterns of interleukin-10, interleukin-4, and their receptors in multiple sclerosis lesions, *Glia*, 38, 24, 2002.

121. Ding, L. et al., IL-10 inhibits macrophage costimulatory activity by selectively inhibiting the up-regulation of B7 expression, *J. Immunol.*, 151, 1224, 1993.

122. O'Keefe, G.M., Nguyen, V.T., and Benveniste, E. N., Class II transactivator and class II MHC gene expression in microglia: modulation by the cytokines TGF-beta, IL-4, IL-13 and IL-10, *Eur. J. Immunol.*, 29, 1275, 1999.

123. Oswald, I.P. et al., Interleukin-10 inhibits macrophage microbial activity by blocking the endogenous production of tumor necrosis factor alpha required as a costimulatory factor for interferon-gamma-induced activation, *Proc. Natl. Acad. Sci. U.S.A.*, 89, 8676, 1992.

124. Issazadeh, S. et al., Cytokines in relapsing experimental autoimmune encephalomyelitis in DA rats: persistent mRNA expression of proinflammatory cytokines and absent expression of interleukin-10 and transforming growth factor-beta, *J. Neuroimmunol.*, 69, 103, 1996.

125. Di Rosa, F. et al., Lack of Th2 cytokine increase during spontaneous remission of experimental allergic encephalomyelitis, *Eur. J. Immunol.*, 28, 3893, 1998.

126. Cua, D.J. et al., Central nervous system expression of IL-10 inhibits autoimmune encephalomyelitis, *J. Immunol.*, 166, 602, 2001.

127. Willenborg, D.O. et al., Cytokines and murine autoimmune encephalomyelitis: inhibition or enhancement of disease with antibodies to select cytokines, or by delivery of exogenous cytokines using a recombinant vaccinia virus system, *Scand. J. Immunol.*, 41, 31, 1995.

128. Nagelkerken, L., Blauw, B., and Tielemans, M., IL-4 abrogates the inhibitory effect of IL-10 on the development of experimental allergic encephalomyelitis in SJL mice, *International Immunology*, 9, 1243, 1997.

129. Cannella, B. et al., IL-10 fails to abrogate experimental autoimmune encephalomyelitis., *J. Neurosci. Res.*, 45, 735, 1996.

130. Cua, D.J. et al., Transgenic interleukin 10 prevents induction of experimental autoimmune encephalomyelitis, *J. Exp. Med.*, 189, 1005, 1999.

131. Bettelli, E. et al., IL-10 is critical in the regulation of autoimmune encephalomyelitis as demonstrated by studies of IL-10- and IL-4-deficient and transgenic mice, *J. Immunol.*, 161, 3299, 1998.

132. Samoilova, E.B., Horton, J.L., and Chen, Y., Acceleration of experimental autoimmune encephalomyelitis in interleukin-10-deficient mice: roles of interleukin-10 in disease progression and recovery, *Cell Immunol.*, 188, 118, 1998.

133. Mathisen, P.M. et al., Treatment of experimental autoimmune encephalomyelitis with genetically modified memory T cells, *J. Exp. Med.*, 186, 159, 1997.

134. Shaw, M.K. et al., Local delivery of interleukin 4 by retrovirus-transduced T lymphocytes ameliorates experimental autoimmune encephalomyelitis, *J. Exp. Med.*, 185, 1711, 1997.

135. Croxford, J.L. et al., Different therapeutic outcomes in experimental allergic encephalomyelitis dependent on the mode of delivery of IL-10: a comparison of the effects of protein, adenoviral or retroviral IL-10 delivery into the central nervous system, *J. Immunol.*, 166, 4124, 2001.

136. Broberg, E. et al., Expression of interleukin-4 but not of interleukin-10 from a replicative herpes simplex virus type 1 viral vector precludes experimental allergic encephalomyelitis, *Gene Ther.*, 8, 769, 2001.

137. Croxford, J.L. et al., Cytokine gene therapy in experimental allergic encephalomyelitis by injection of plasmid DNA-cationic liposome complex into the central nervous system, *J. Immunol.*, 160, 5181, 1998.

138. Pennline, K.J., Roque-Gaffney, E., and Monahan, M., Recombinant human IL-10 prevents the onset of diabetes in the nonobese diabetic mouse, *Clin. Immunol. Immunopathol.*, 71, 169, 1994.

139. Goudy, K.S. et al., Systemic overexpression of IL-10 induces CD4+CD25+ cell populations *in vivo* and ameliorates type 1 diabetes in nonobese diabetic mice in a dose-dependent fashion, *J. Immunol.*, 171, 2270, 2003.

140. Lee, M.S. et al., IL-10 is necessary and sufficient for autoimmune diabetes in conjunction with NOD MHC homozygosity, *J. Exp. Med.*, 183, 2663, 1996.

141. Moritani, M. et al., Prevention of adoptively transferred diabetes in nonobese diabetic mice with IL-10-transduced islet-specific Th1 lymphocytes. A gene therapy model for autoimmune diabetes, *J. Clin. Invest*, 98, 1851, 1996.

142. Wogensen, L., Lee, M.S., and Sarvetnick, N., Production of interleukin 10 by islet cells accelerates immune-mediated destruction of beta cells in nonobese diabetic mice, *J. Exp. Med.*, 179, 1379, 1994.

143. Seder, R.A. and Paul, W.E., Acquisition of lymphokine-producing phenotype by CD4+ T cells, *Annu. Rev. Immunol.*, 12, 635, 1994.

144. Racke, M.K. et al., Cytokine-induced immune deviation as a therapy for inflammatory autoimmune disease, *J. Exp. Med.*, 180, 1961, 1994.

145. Suzumura, A. et al., Interleukin-4 induces proliferation and activation of microglia but suppresses their induction of class II major histocompatibility complex antigen expression, *J. Neuroimmunol.*, 53, 209, 1994.

146. Kitamura, Y. et al., Interleukin-4-inhibited mRNA expression in mixed rat glial and in isolated microglial cultures, *J. Neuroimmunol.*, 106, 95, 2000.

147. Link, J., Interferon-gamma, interleukin-4 and transforming growth factor- beta mRNA expression in multiple sclerosis and myasthenia gravis, *Acta Neurologica Scandinavia*, Supplementum. 158, 1, 1994.

148. Boxel-Dezaire, A.H. et al., Cytokine and IL-12 receptor mRNA discriminate between different clinical subtypes in multiple sclerosis, *J. Neuroimmunol.*, 120, 152, 2001.

149. Merrill, J.E. et al., Inflammatory leukocytes and cytokines in the peptide-induced disease of experimental allergic encephalomyelitis in SJL and B10.PL mice, *Proc. Natl. Acad. Sci. U.S.A.*, 89, 574, 1992.

150. Bebo, B.F.J., Vandenbark, A.A., and Offner, H., Male SJL mice do not relapse after induction of EAE with PLP 139-151., *J. Neurosci. Res.*, 45, 680, 1996.

151. Kuchroo, V.K. et al., B7-1 and B7-2 costimulatory molecules differentially activate the Th1/Th2 developmental pathways: application to autoimmune disease therapy, *Cell*, 80, 707, 1995.

152. Fontana, A., Fierz, W., and Wekerle, H., Astrocytes present myelin basic protein to encephalitogenic T-cell lines., *Nature*, 307, 273, 1984.

153. Khoruts, A., Miller, S.D., and Jenkins, M.K., Neuroantigen-specific Th2 cells are inefficient suppressors of experimental autoimmune encephalomyelitis induced by effector Th1 cells, *J. Immunol.*, 155, 5011, 1995.

154. Katz, J.D., Benoist, C., and Mathis, D., T helper cell subsets in insulin-dependent diabetes, *Science*, 268, 1185, 1995.

155. Falcone, M. and Bloom, B.R., A T helper cell 2 (Th2) immune response against non-self antigens modifies the cytokine profile of autoimmune T cells and protects against experimental allergic encephalomyelitis., *J. Exp. Med.*, 185, 901, 1997.
156. Stohlman, S.A. et al., Activation of regulatory cells suppresses experimental allergic encephalomyelitis via secretion of IL-10, *J. Immunol.*, 163, 6338, 1999.
157. Inobe, J.I., Chen, Y., and Weiner, H.L., *In vivo* administration of IL-4 induces TGF-beta-producing cells and protects animals from experimental autoimmune encephalomyelitis, *Ann. NY Acad. Sci.*, 778, 390, 1996.
158. Piccirillo, C.A. and Prud'homme, G.J., Prevention of experimental allergic encephalomyelitis by intramuscular gene transfer with cytokine-encoding plasmid vectors, *Hum. Gene Ther.*, 10, 1915, 1999.
159. Furlan, R. et al., Central nervous system gene therapy with interleukin-4 inhibits progression of ongoing relapsing-remitting autoimmune encephalomyelitis in Biozzi AB/H mice, *Gene Ther.*, 8, 13, 2001.
160. Tarner, I.H. et al., Retroviral gene therapy of collagen-induced arthritis by local delivery of IL-4, *Clin. Immunol.*, 105, 304, 2002.
161. Kageyama, Y. et al., Plasmid encoding interleukin-4 in the amelioration of murine collagen-induced arthritis, *Arthritis Rheum.*, 50, 968, 2004.
162. Mueller, R., Krahl, T., and Sarvetnick, N., Pancreatic expression of interleukin-4 abrogates insulitis and autoimmune diabetes in nonobese diabetic (NOD) mice, *J. Exp. Med.*, 184, 1093, 1996.
163. Liblau, R., Steinman, L., and Brocke, S., Experimental autoimmune encephalomyelitis in IL-4-deficient mice, *Int. Immunol.*, 9, 799, 1997.
164. Zhao, M.L. and Fritz, R.B., Acute and relapsing experimental autoimmune encephalomyelitis in IL-4- and alpha/beta T cell-deficient C57BL/6 mice, *J. Neuroimmunol.*, 87, 171, 1998.
165. Falcone, M. et al., A critical role for IL-4 in regulating disease severity in experimental allergic encephalomyeltis as demonstrated in IL-4-deficient C57BL/6 and BALB/c mice, *J. Immunol.*, 160, 4822, 1998.
166. Prud'homme, G.J. and Piccirillo, C.A., The inhibitory effects of transforming growth factor-beta-1 (TGF-beta1) in autoimmune diseases, *J. Autoimmun.*, 14, 23, 2000.
167. Bright, J.J. and Sriram, S., TGF-beta inhibits IL-12-induced activation of JAK-STAT pathway in T lymphocytes, *J. Immunol.*, 161, 1772, 1998.
168. Gorham, J.D. et al., Low dose TGF-beta attenuates IL-12 responsiveness in murine Th cells, *J. Immunol.*, 161, 1664, 1998.
169. Czarniecki, C.W. et al., Transforming growth factor-beta 1 modulates the expression of class II histocompatibility antigens on human cells., *J Immunol*, 140, 4217, 1988.
170. Schluesener, H.J., Transforming growth factors type beta 1 and beta 2 suppress rat astrocyte autoantigen presentation and antagonize hyperinduction of class II major histocompatibility complex antigen expression by interferon-gamma and tumor necrosis factor-alpha, *J. Neuroimmunol.*, 27, 41, 1990.
171. Vodovotz, Y. and Bogdan, C., Control of nitric oxide synthase expression by transforming growth factor-beta: implications for homeostasis, *Prog. Growth Factor Res.*, 5, 341, 1994.
172. Tsunawaki, S. et al., Deactivation of macrophages by transforming growth factor-beta, *Nature*, 334, 260, 1988.
173. Kehrl, J.H. et al., Transforming growth factor-beta suppresses human B lymphocyte Ig production by inhibiting synthesis and the switch from the membrane form to the secreted form of Ig mRNA, *J. Immunol.*, 146, 4016, 1991.

174. Paglinawan, R. et al., TGFbeta directs gene expression of activated microglia to an anti-inflammatory phenotype strongly focusing on chemokine genes and cell migratory genes, *Glia*, 44, 219, 2003.

175. Rook, A.H. et al., Effects of transforming growth factor beta on the functions of natural killer cells: depressed cytolytic activity and blunting of interferon responsiveness, *J. Immunol.*, 136, 3916, 1986.

176. Kehrl, J.H. et al., Production of transforming growth factor beta by human T lymphocytes and its potential role in the regulation of T cell growth, *J. Exp. Med.*, 163, 1037, 1986.

177. Carrieri, P.B. et al., Profile of cerebrospinal fluid and serum cytokines in patients with relapsing-remitting multiple sclerosis: a correlation with clinical activity, *Immunopharmacol. Immunotoxicol.*, 20, 373, 1998.

178. Bertolotto, A. et al., Transforming growth factor beta1 (TGFbeta1) mRNA level correlates with magnetic resonance imaging disease activity in multiple sclerosis patients, *Neurosci. Lett.*, 263, 21, 1999.

179. Losy, J. and Michalowska-Wender, G., *In vivo* effect of interferon-beta 1a on interleukin-12 and TGF-beta(1) cytokines in patients with relapsing-remitting multiple sclerosis, *Acta Neurol. Scand.*, 106, 44, 2002.

180. Nicoletti, F. et al., Blood levels of transforming growth factor-beta 1 (TGF-beta1) are elevated in both relapsing remitting and chronic progressive multiple sclerosis (MS) patients and are further augmented by treatment with interferon-beta 1b (IFN-beta1b), *Clin. Exp. Immunol.*, 113, 96, 1998.

181. Racke, M.K. et al., Evidence of endogenous regulatory function of transforming growth factor-beta 1 in experimental allergic encephalomyelitis, *Int. Immunol.*, 4, 615, 1992.

182. Kjellen, P. et al., Genetic influence on disease course and cytokine response in relapsing experimental allergic encephalomyelitis, *Int. Immunol.*, 10, 333, 1998.

183. Mustafa, M. et al., The major histocompatibility complex influences myelin basic protein 63-88-induced T cell cytokine profile and experimental autoimmune encephalomyelitis, *Eur. J. Immunol.*, 23, 3089, 1993.

184. Karpus, W.J., Gould, K.E., and Swanborg, R.H., CD4+ suppressor cells of autoimmune encephalomyelitis respond to T cell receptor-associated determinants on effector cells by interleukin-4 secretion, *Eur. J. Immunol.*, 22, 1757, 1992.

185. Stevens, D.B., Gould, K.E., and Swanborg, R.H., Transforming growth factor-beta 1 inhibits tumor necrosis factor- alpha/lymphotoxin production and adoptive transfer of disease by effector cells of autoimmune encephalomyelitis, *J. Neuroimmunol.*, 51, 77, 1994.

186. Chen, L.Z. et al., Gene therapy in allergic encephalomyelitis using myelin basic protein-specific T cells engineered to express latent transforming growth factor-beta1, *Proc. Natl. Acad. Sci. U.S.A.*, 95, 12516, 1998.

187. Thorbecke, G.J. et al., When engineered to produce latent TGF-beta1, antigen specific T cells down regulate Th1 cell-mediated autoimmune and Th2 cell-mediated allergic inflammatory processes, *Cytokine Growth Factor Rev.*, 11, 89, 2000.

188. Johns, L.D. et al., Successful treatment of experimental allergic encephalomyelitis with transforming growth factor-beta 1, *J. Immunol.*, 147, 1792, 1991.

189. Santambrogio, L. et al., Studies on the mechanisms by which transforming growth factor- beta (TGF-beta) protects against allergic encephalomyelitis. Antagonism between TGF-beta and tumor necrosis factor, *J. Immunol.*, 151, 1116, 1993.

190. Schluesener, H.J. and Lider, O., Transforming growth factors beta 1 and beta 2: cytokines with identical immunosuppressive effects and a potential role in the regulation of autoimmune T cell function, *J. Neuroimmunol.*, 24, 249, 1989.

191. Kuruvilla, A.P. et al., Protective effect of transforming growth factor beta 1 on experimental autoimmune diseases in mice, *Proc. Natl. Acad. Sci. U.S.A.*, 88, 2918, 1991.
192. Calabresi, P.A. et al., Phase 1 trial of transforming growth factor beta 2 in chronic progressive MS, *Neurology*, 51, 289, 1998.
193. Faria, A.M. et al., Oral tolerance induced by continuous feeding: enhanced up-regulation of transforming growth factor-beta/interleukin-10 and suppression of experimental autoimmune encephalomyelitis, *J. Autoimmun.*, 20, 135, 2003.
194. Burkhart, C. et al., Peptide-induced T cell regulation of experimental autoimmune encephalomyelitis: a role for IL-10, *Int. Immunol.*, 11, 1625, 1999.
195. Khoury, S.J., Hancock, W.W., and Weiner, H.L., Oral tolerance to myelin basic protein and natural recovery from experimental autoimmune encephalomyelitis are associated with downregulation of inflammatory cytokines and differential upregulation of transforming growth factor beta, interleukin 4, and prostaglandin E expression in the brain, *J. Exp. Med.*, 176(5), 1355, 1992.
196. Chen, Y. et al., Mechanisms of recovery from experimental autoimmune encephalomyelitis: T cell deletion and immune deviation in myelin basic protein T cell receptor transgenic mice, *J. Neuroimmunol.*, 82, 149, 1999.
197. Brocke, S. et al., Treatment of experimental encephalomyelitis with a peptide analogue of myelin basic protein, *Nature*, 379, 343, 1996.
198. Nicholson, L.B. et al., An altered peptide ligand mediates immune deviation and prevents autoimmune encephalomyelitis, *Immunity*, 3, 397, 1995.
199. Young, D.A. et al., IL-4, IL-10, IL-13, and TGF-beta from an altered peptide ligand-specific Th2 cell clone down-regulate adoptive transfer of experimental autoimmune encephalomyelitis, *J. Immunol.*, 164, 3563, 2000.
200. McRae, B.L. et al., Functional evidence for epitope spreading in the relapsing pathology of experimental autoimmune encephalomyelitis, *J. Exp. Med.*, 182, 75, 1995.
201. Vanderlugt, C.L. et al., Pathologic role and temporal appearance of newly emerging autoepitopes in relapsing experimental autoimmune encephalomyelitis, *J. Immunol.*, 164, 670, 2000.
202. Peterson, J.D. et al., Split tolerance of Th1 and Th2 cells in tolerance to Theiler's murine encephalomyelitis virus, *Eur. J. Immunol.*, 23, 46, 1993.
203. Karpus, W.J., Peterson, J.D., and Miller, S.D., Anergy *in vivo*: down-regulation of antigen-specific CD4[+] Th1 but not Th2 cytokine responses, *Int. Immunol.*, 6, 721, 1994.
204. Kennedy, K.J. et al., Induction of antigen-specific tolerance for the treatment of ongoing, relapsing autoimmune encephalomyelitis—A comparison between oral and peripheral tolerance, *J. Immunol.*, 159, 1036, 1997.
205. Legge, K.L. et al., Coupling of peripheral tolerance to endogenous interleukin 10 promotes effective modulation of myelin-activated T cells and ameliorates experimental allergic encephalomyelitis, *J. Exp. Med.*, 191, 2039, 2000.
206. Garren, H. et al., Combination of gene delivery and DNA vaccination to protect from and reverse Th1 autoimmune disease via deviation to the Th2 pathway, *Immunity*, 15, 15, 2001.

9 Cytokines and Neurodegeneration

Greer M. Murphy, Jr., and
Parvathy Saravanapavan

CONTENTS

I. INTRODUCTION

Cytokines have been implicated in a variety of neurodegenerative conditions including Alzheimer's disease (AD), Parkinson's disease (PD), amyotrophic lateral sclerosis (ALS), prion diseases, and others. Many investigators have attributed a negative role to cytokines in neurodegenerative disease, hypothesizing that cytokine-induced inflammatory processes result in toxicity to nerve cells. Indeed, there is substantial evidence from cell culture and animal studies that cytokines can have direct or indirect neurotoxic effects. Recently, however, there has been increasing awareness that some processes mediated by cytokines may be potentially beneficial to the compromised neuron.

This chapter will review the evidence for cytokine-induced neurotoxicity as well as possible protective effects in neurodegeneration.

II. ALZHEIMER'S DISEASE

AD is the most common cause of dementia in the elderly. The neuropathological characteristics of AD include synaptic loss, neuronal cell death, reactive astrogliosis, microgliosis, neurofibrillary tangles, and extracellular neuritic plaques [1]. Plaques are composed of aggregated amyloid β (Aβ), which is thought to impair neuronal function as well as inducing an inflammatory response [2,3]. A variety of cytokines have been implicated in the pathophysiology of AD.

A. INTERLEUKIN-1

Interleukin-1 is the most widely studied of all cytokines in relation to AD. It has been known for many years that inflammatory cells are found in proximity to plaque lesions containing Aβ in AD [4,5]. However, it was not until the 1980s that investigation of the role of inflammation in AD attracted widespread attention. Interleukin 1 (IL-1; actually two closely related molecules, IL-1α and IL-1β with similar effects) is a ubiquitously expressed cytokine produced by microglia and astrocytes in the brain [6]. In a seminal paper, Griffin and coworkers showed that IL-1 antibodies decorated microglia and astrocytes in AD and Down syndrome brain [7] and that IL-1 levels were increased in brain homogenates from AD cases in comparison with controls. They hypothesized that IL-1 stimulated the astrogliosis characteristic of AD brain. Simultaneously, Goldgaber and colleagues demonstrated that IL-1 could regulate neuronal expression of the then newly identified amyloid beta precursor protein (APP) gene [8]. Later work has shown that APP cleavage products can in turn induce expression of IL-1 [9]. In another important early paper, Araujo and Cotman [10] showed that Aβ could induce glial expression of IL-1. These and other authors suggested that in AD a self-amplifying cascade might exist in which IL-1 stimulates formation of Aβ, which in turn stimulates additional production of IL-1. At the same time, McGeer and Rogers [11] suggested that cytokines and other mediators of inflammation, such as complement, might directly injure nerve cells in AD.

Although there is good evidence for increased levels of IL-1 in AD brain, the precise role of this cytokine in AD neuropathology is incompletely defined. Recent clinical data demonstrating efficacy of an NMDA-receptor antagonist in AD suggests that excitotoxicity could be important in neuronal injury in AD [12]. IL-1 induces iNOS expression by astrocytes, which in turn potentiates NMDA neurotoxicity [13]. In addition, IL-1 promotes intracellular calcium release after NMDA receptor stimulation [14]. Blocking NMDA receptors interferes with IL-1-induced toxicity in cell cultures [15]. IL-1 may also be involved in the process of neurofibrillary tangle formation in AD by promoting phosphorylation of tau through a p38-MAPK signaling pathway [16]. Cognitive impairment in AD may also be accelerated by IL-1 induction of acetylcholinesterase, which decreases levels of this key neurotransmitter [17].

Polymorphisms in the IL-1α and IL-1β genes have been associated with risk for AD. A promoter polymorphism (–889) upstream from IL-1a gene has been associated with AD in several studies [18–21], whereas one study found an IL-1β promoter variant (–511) was a risk factor for AD [21] and a coding region variant in the IL-β gene has been associated with risk for AD and age of onset [21,22]. None of these variants results in a structural change in the protein, so it is likely that they affect protein abundance or are in linkage disequilibrium with a functional variant. However, not all studies have confirmed the association of these IL-1 polymorphisms with AD [23,24]. Ethnic variation may affect the association between AD and these IL-1 variants [25,26]. One study found that an association between the IL-1α –889 variant and rate of decline in AD [27], but little else is known about how these polymorphisms modify phenotypes in AD. It is conceivable that IL-1 genetic variants could affect the degree of inflammatory neuropathology in AD, but so far this hypothesis has not been tested.

B. Neurotrophic Properties of IL-1 and Microglia

Although there is substantial evidence for IL-1-mediated neurotoxicity, under certain circumstances IL-1 may be protective. IL-1 induces the expression of the neurotrophin CNTF after traumatic brain injury [28] and potentiates the protective effects of NGF after excitotoxic injury to cultured neurons [29]. IL-1 also protects midbrain neurons from MPTP-induced injury [30]. Other microglial cytokines including IL-6 and TNF-α have neurotrophic effects under certain conditions [31–36]. Activated microglia likely injure nerve cells in a variety of disease states [37–39], but microglia can also be neuroprotective. Watanabe et al. [40] found that microglial supernatants protected dissociated neurons from excitotoxic injury. In another co-culture model, microglial–neuronal contact changed the microglial phenotype from toxic to protective [41]. Grafting of microglia into injured spinal cord results in neurite outgrowth and other regenerative neuronal responses [42]. Microglial activation was recently associated with better neuronal survival in an optic nerve crush injury model [43]. There is growing evidence that microglial expression of classic neurotrophins such as NGF may provide support for neurons in a variety of disease states [44–51]. Other potential positive functions of microglia in neurologic disease were recently reviewed [52]. There is a need for a balanced view of microglia, astrocytes, and their cytokine products in AD and other neurodegenerative disorders.

C. Interleukin-6

Interleukin-6 is a systemic cytokine that is increased during infection, inflammation, and tissue injury. IL-6 is induced by other inflammatory cytokines, by endotoxin, viral factors, and other proinflammatory agents. In the brain, IL-6 expression has been localized to neurons, astrocytes, and microglia [53–55]. Cellular localization of IL-6 expression likely varies with developmental stage and with disease pathology. Early work demonstrated IL-6 immunoreactivity in both diffuse and neuritic plaques in AD brain. The immunoreactivity was primarily extracellular, colocalizing with Aβ in the plaque, although a few neuronal processes around plaques were also reported as labeled [56]. The Tg2576 mouse model for AD overexpresses human APP with a double mutation designated K670N, M671L, found in a Swedish family with autosomal dominant AD [57]. These mice show progressive Aβ deposition, gliosis, and behavioral changes including memory deficits. IL-6 mRNA is elevated in the brains of Tg2576 mice before Aβ deposition, suggesting that increased expression of this cytokine could be an early neuropathologic change [58]. Because IL-6 is found extracellularly in AD brain, CSF IL-6 has also been examined. Results have been inconsistent [59,60]. Because IL-6 expression is upregulated in blood in a number of inflammatory states, it is reasonable to predict that this cytokine might be a peripheral marker for inflammation in AD, but there are conflicting reports regarding blood IL-6 levels in AD [61,62].

There is no consensus as to the role of elevated IL-6 in AD brain. Because IL-6 has a wide variety of effects on the expression of other cytokines and mediators of inflammation, it has been proposed that there is a cerebral acute phase response in AD driven by IL-6 that ultimately results in neuronal injury [63]. Transgenic mice overexpressing IL-6 in astrocytes show gliosis, increased expression of other cytokines, synaptic loss, and deficits in learning [64,65]. In some cell culture models, IL-6 enhances NMDA-induced neurotoxicity, possibly by promoting calcium influx [66]. When applied in combination with NMDA and Aβ, IL-6 potentiates neurotoxicity [67]. However, there is also evidence that IL-6 has neurotrophic properties. Intraventricular delivery of IL-6 can protect neurons after ischemic injury [68], and knockout of IL-6 expression increases neuronal vulnerability to apoptotic death [69]. IL-6 supports the proliferation of neuroblastoma cells [70], and in a neuronal–glia co-culture system, IL-6 protects neurons from methylmercury injury [71]. IL-6 also protects neuronal cultures from excitotoxic injury through an IL-1-mediated mechanism [72]. Further work is required to define the relative positive and negative effects of IL-6 in the brain.

D. TNF-α

Although TNF-α has a defined role in neurologic disorders such as multiple sclerosis, it is uncertain if this cytokine is important in AD. Levels of TNF-α in the CSF of AD patients were reported to be significantly higher than those in controls [73]. However, another study found no difference between AD patients and controls in CSF TNF-α [61] and actually reported significantly lower TNF-α levels in homogenates of brain regions with heavy neuropathologic change in AD. In Tg2576 mice modeling AD, TNF-α immunopositive microglia were localized with fibrillar Aβ deposits [74], but the cellular source of TNF-α in AD brain has not been determined. Following exposure of preaggregated Aβ(1-42) to AD and control human microglial cultures, an increase in the production of TNF-α was reported [75].

Potential mechanisms of action for TNF-α in AD have been proposed. Meda et al. showed a synergistic effect between Aβ and IFN-γ in triggering the production of NO and TNF-α from microglia [76]. The combination of TNF-α and IFN-γ in turn can trigger the production of Aβ and inhibit the secretion of sAPPα [77]. TNF-α in the presence of Aβ can enhance apoptosis in several cell types [78]. Carboxy-terminal APP in combination with IFN-α can increase TNF-α synthesis in human monocytic THP-1 cell line [79]. However, it is concerning that there is no evidence for increased IFN-γ in AD. Further, there is evidence that TNF-α can protect neurons against Aβ toxicity [80].

Genetic association studies have examined TNF-α gene polymorphisms in AD. A polymorphism in the regulatory region of the TNF-α gene (C-850T) was found to increase AD risk conferred by the APOE ε4 allele [81]. However, two other studies have failed to confirm this finding [82,83]. Another TNF-α regulatory region variant (G-308A) was shown to be associated with age of onset of AD [84]. Another group using a sib-pair design found an association of TNF-α regulatory region haplotypes with AD and a significant association between a TNF receptor 2 exon 6 polymorphism and late-onset AD families [85]. In summary, there are intriguing leads regarding a role for TNF-α in AD, but more study is required.

E. MACROPHAGE COLONY STIMULATING FACTOR

Macrophage colony stimulating factor (M-CSF, CSF-1) is a hematopoietic cytokine found in both the brain and the periphery. M-CSF is expressed by cultured neurons, astrocytes, endothelia, and microglia [86–91]. M-CSF induces proliferation, migration, and activation of microglia [92–96]. M-CSF is upregulated in AD brain [90] and in the PDAPP transgenic mouse model of AD [97]. PDAPP overexpress human APP with the familial V717F mutation [98]. They show increasing Aβ deposition, astroglioisis, and microgliosis with age. Microglia cultured at autopsy from patients with AD show increased expression of M-CSF when treated with Aβ, whereas microglia from elderly controls do not [99,100]. Further, we found that treatment of cultured microglia or organotypic hippocampal cultures with M-CSF results in a dramatic augmentation of Aβ-induced cytokine, chemokine, and nitric oxide production [101,102], and induces proliferation [103]. Together, these results suggest that in AD dysregulation of M-CSF could have important activating effects on microglia near Aβ deposits.

The receptor for M-CSF (M-CSFR) is encoded by the *c-fms* proto-oncogene and is expressed on microglia [97,104] and possibly on certain neurons [105,106]. The M-CSFR shows increased expression on microglia in AD brain [107] and in the PDAPP and Tg2576 transgenic mouse models for AD [97] (our unpublished results). The M-CSFR is also increased on microglia following ischemic and traumatic brain injury [104,106]. Overexpression of M-CSFR on cultured mouse and human microglia results in microglial activation and production of proinflammatory cytokines and nitric oxide [108]. Further, microglia overexpressing M-CSFR induce a paracrine inflammatory response in our microglia–hippocampal organotypic co-culture system [108]. These results suggest that increased M-CSF signaling via the M-CSFR is an essential step in activating microglia and promoting the microglial inflammatory response after a variety of neurologic insults.

Microglial phagocytosis may be important after therapeutic Aβ immunization in mouse models for AD [109]. We recently showed that increased expression of M-CSFR on microglia results in accelerated phagocytosis of Aβ via a macrophage scavenger receptor-dependent mechanism [110]. We also showed that increased M-CSFR expression by *c-fms* transfection of mouse and human microglia increases expression of Fcγ receptor and increases phagocytosis of opsonized Aβ via Fcγ receptor-dependant and independent mechanisms [111]. Likewise, M-CSF treatment of cultured microglia augments phagocytosis [112]. In AD and in nonimmunized PDAPP transgenic mice, increased M-CSF and M-CSFR expression may be a microglial response to increased Aβ deposition that promotes clearance, albeit too slowly to prevent disease progression. In PDAPP mice, in the setting of immunization, increased M-CSFR expression may be essential in Aβ clearance by enhancing expression of Fc-γ receptors on microglia and hence accelerating microglial Aβ phagocytosis. Microglia in AD and AD mouse models showing increased M-CSFR expression may be "primed" so that small increases in intracerebral anti-Aβ antibodies result in large increases in phagocytosis.

F. M-CSF: NEUROTOXIC OR NEUROPROTECTIVE?

The exact role of M-CSF in neuronal injury in AD is unclear. One hypothesis is that M-CSF acts as a synergistic cofactor for Aβ-induced microglial activation that in turn injures nerve cells via neurotoxic cytokines and NO derivatives such as peroxynitrite [101,113]. We found that M-CSF and Aβ in combination induce strong microglial activation in organotypic hippocampal co-cultures [102]. To our surprise, this inflammatory response did not injure neurons. Further, there are data that indicate M-CSF is neuroprotective. After mechanical or ischemic brain injury in adult op/op mice, which lack M-CSF expression, microglia fail to proliferate and hypertrophy [114,115]. Lack of microglial proliferation might be predicted to protect neurons from microglial-induced injury. Yet, after acute ischemic injury op/op mice show increased neuronal loss in comparison with controls [114]. Similarly, op/op mice are more vulnerable to the hippocampal neurotoxin TMT than are wild-type animals [116]. M-CSF also has trophic properties that support the growth of dissociated cerebellar and cortical neuronal cultures [105,106]. In an ischemia model with wild-type animals, treatment with M-CSF was neuroprotective [117]. We found that exogenously applied M-CSF protects neurons in organotypic cultures from NMDA-induced neuronal injury and apoptosis through inhibition of caspase-3 activity [118]. Excitotoxicity may be important in a wide variety of neurologic conditions, including AD [119–122]. Further work on the neuroprotective properties of M-CSF is warranted.

G. CHEMOKINES

Chemokines are small peptides that promote chemotaxis [123]. In the brain, chemokines may recruit astrocytes and microglia to sites of injury [124]. Xia and colleagues [125] first reported increased expression of the IL-8 B receptor on dystrophic neurites in AD brain. Increased expression of MCP-1 and MIP-1α on microglia, MIP-1β on astrocytes, and MIP-1α on neurons has also been observed in AD [126,127]. MCP-1 was also found to be elevated in microvessels from AD brain [128]. Another study reported increased astrocytic expression in AD of IP-10, whereas the receptor for this chemokine, CXCR3, was constitutively expressed on neurons. Recently, the chemokine receptor CCR1, which binds to a number of chemokines, was observed on dystrophic neurites around plaques containing the pathogenic Aβ 1-42 peptide at an early stage in AD, whereas it was generally undetectable in control brains [129]. Interestingly, a carefully performed study did not detect the CCR8 chemokine receptor on microglia in AD, although this receptor is present in other inflammatory conditions of the nervous system [130].

There are a number of possible functions for chemokines and their receptors in AD. A number of studies have found that Aβ can induce chemokine expression by cultured microglia and astrocytes [131–133] as well as in hippocampal organotypic cultures [102]. This suggests that in AD, Aβ deposition could stimulate chemokine production, which in turn could recruit glial cells to plaques, promoting a neurotoxic inflammatory response. Chemokine effects on neurons in AD may be complex. For example, activation of neuronal CXCR2 receptors may have neurotrophic effects and at the same time stimulate phosphorylation of tau, an important step in neurofibrillary tangle formation [134].

H. Transforming Growth Factors

TGF-β family cytokines are expressed by neurons, astrocytes, and microglia, depending on the developmental and neuropathological environment [135], and have anti-inflammatory and possibly neuroprotective properties [136,137]. Increased expression of TGF-β1 has been reported in AD in some studies [138–141]. One hypothesis is that TGF-β family cytokines are increased as a reaction to proinflammatory processes and hence could have a protective effect [142]. Mice lacking TGF-β1 expression show increased neuronal vulnerability to excitotoxic injury [143]. Further, although TGF-β cytokines may promote amyloidosis in the cerebral vasculature [141], in brain parenchyma this cytokine induces microglial clearance of Aβ plaques [144]. Taken together, these results indicate a positive role for TGFβ cytokines in AD.

III. PARKINSON'S DISEASE

PD is a common neurodegenerative disorder characterized by massive and selective loss of dopaminergic neurons in the substantia nigra pars compacta (SNc) and a decrease in the dopamine content in the striatum [145]. Intraneuronal inclusions known as Lewy bodies, containing the proteins α-synuclein, Parkin, ubiquitin, and tau, are found in the substantia nigra and elsewhere. Clinically, PD is characterized by a tremor at rest, gait abnormalities, bradykinesia, and often cognitive and behavioral changes. Although rare mutations in α-synuclein and Parkin have been identified that lead to familial PD [146,147], the etiology of most cases remains unknown. Further, the basis for the selective vulnerability of dopaminergic neurons within the SNc in PD is also unknown.

Both activated microglia and astrocytes have been implicated in the pathophysiology of PD. A large body of evidence indicates that glial-induced oxidative stress may be important in the progressive neuronal degeneration observed in PD [148].

The density of microglial cells may be higher in the substantia nigra compared to other midbrain areas [149], leading to the hypothesis that neurons in the substantia nigra area may be more susceptible to activated microglial-mediated injury. In mice, injection of bacterial LPS into the SNc caused a strong activation of microglia throughout the substantia nigra, followed by a marked degeneration of dopaminergic neurons [150]. Also, the pharmacological inhibition of microglial activation with naloxone prevented LPS induced SNc neuronal death [150]. Further, LPS results in killing of dopaminergic neurons in mixed glial cell cultures but not in pure neuronal cultures [151].

It has also been observed that there is a significant and specific reduction in the mitochondrial complex I activity in the SNc of patients with PD [152]. Thus mitochondrial complex I inhibitors have been widely used to mimic Parkinsonism in animal models. Exposure to the toxin 1-methyl-4-phenyl 1,2,23,6-tetrahydropyridine (MPTP), a mitochondrial complex I inhibitor, induces Parkinsonism in humans [153]. Autopsy studies of MPTP-induced Parkinsonism show gliosis and clustering of microglial cells around nerve cells, indicating an ongoing inflammatory response after the time-limited toxic insult and suggesting that activated microglial cells may contribute to neuronal degeneration [154]. In idiopathic PD, a preponderance of microglia over astrocytes has been reported [155]. In MPTP-treated mice modeling PD, iNOS-positive glial cells in the SNc were primarily microglia and not astrocytes [156]. Yet, astrocytosis in MPTP-treated mice may be more prominent than in idiopathic PD [157]. In the rotenone (another mitochondrial complex I inhibitor) mouse model for PD, the midbrain inflammatory response is primarily microglia [158]. Taken together, these data show that activated glial cells, particularly microglia, may promote death of dopaminergic neurons.

Activation of microglia in PD may produce neurotoxic compounds including reactive oxygen species, reactive nitrogen species, proinflammatory prostaglandins, and cytokines [159]. TNF-α, IL-1β, and IFN-γ have all been reported to be increased in SNc tissues and CSF of PD patients [160–164]. A microarray study also found increased expression of IL-1β and IL-6 in the substantia nigra and striatum of mice after MPTP treatment [165]. MPTP treatment of wild-type mice also resulted in a time-dependent expression of TNF-α in striatum, which preceded loss of dopaminergic markers [166].

TNF receptor type 1 and type 2 double knockout mice, but not mice with deletion of individual TNF receptors, are reported to be resistant to MPTP toxicity [166]. Dopaminergic neurons in the substantia nigra express type 1 TNF-α receptors (TNFR1) [167]. TNF-α induces trimerization of TNFR1 on binding that leads to autoproteolytic activation of caspase-8. Caspase-8 in turn can cleave effector caspases such as caspase-3 or amplify death signals through the translocation of BID, a proapoptotic member of the Bcl-family to the mitochondria [75,168]. Engagement of TNFR1 also activates the translocation of NFκB transcription factor to the nucleus, which can trigger cell death by apoptosis [169]. In fact, there is a 70-fold increase in nuclear translocation of NFκB in nigral dopaminergic neurons of PD patients [169]. Therefore, activation of the TNFR1 coupled transduction pathway may be associated with apoptotic dopaminergic cell death in PD.

MPTP-induced activation of microglia in mice can be inhibited by minocycline, which decreases formation of IL-1 and iNOS and slows SNc neuronal death [170]. In MPTP-treated mice administration of aspirin decreases the rate of dopamine depletion [171]. This may be due to inhibition of the COX-2 enzyme, which is induced by TNF-α and IL-1, and which plays an important role in inflammation [172,173]. Several genetic association studies have found an association that IL1β and TNF-α polymorphisms increase risk for PD [174,175]. These studies will require replication.

Lewy body dementias include dementia with Lewy bodies (DLB; "diffuse Lewy body disease") and the Lewy body variant of AD. The Lewy body variant of AD is characterized by cortical Lewy bodies as well as by the Aβ plaques and neurofibrillary tangles of AD. DLB, in contrast, is characterized primarily by cortical Lewy bodies without extensive plaque and neurofibrillary pathology. A few studies have examined cytokine expression in DLB. Increased expression of IL-1 and TNF-α has been associated with microglia and astrocytes found near Lewy body-containing neurons [176,177]. However, one study found no significant increase in IL-1 and IL-6 in CSF in DLB patients [178]. The role of cytokines in neuronal injury in DLB requires further investigation.

A. NEUROTROPHIC CYTOKINES IN PARKINSON'S DISEASE

Although there is substantial evidence for microglial neurotoxicity in PD, microglia may also have trophic properties. Glial derived neurotrophic factor (GDNF), which is released by activated microglia, induces dopaminergic nerve fiber sprouting [179–181]. Furthermore, in both mice and monkeys treated with MPTP, GDNF attenuates dopaminergic neuronal death and enhances dopaminergic function within injured neurons [182–184]. Cytokine-activated astrocytes may also protect dopaminergic neurons against degeneration by scavenging toxic compounds released by dying neurons in PD [185]. Thus, glial activation in PD may have both beneficial as well as harmful effects in PD.

IV. AMYOTROPHIC LATERAL SCLEROSIS

ALS ("Lou Gehrig Disease") is a predominantly sporadic neurodegenerative disorder characterized by the loss of cortical, brainstem, and spinal motor neurons that leads to a devastating progressive muscle weakness and atrophy, and eventual paralysis. Thus far, mutations in two genes have been associated with human familial ALS: Cu/Zn superoxide dismutase 1 (SOD1; mutations result in a toxic gain of function) [186], and alsin (ALS2; a guanine nucleotide exchange factor [187]). Transgenic mice expressing mutant human SOD1 have been developed that model many of the clinical and neuropathologic changes of ALS [188]. Microglial proliferation in ALS spinal cord is well documented in both humans and mouse models of ALS [137,189].

G93A-SOD1 mice overexpress a human SOD1 mutant transgene and demonstrate age-dependent motor symptoms reminiscent of ALS. There is good evidence for a progressive inflammatory response in the spinal cord in these animals. Hensley et al. (2002) found increases in IL1 and in TNF-α and its receptors prior to motor symptom onset (80 days) and further increases during the symptomatic stage of the disease (120 days) [190]. An early increase in TNF-α expression in these mice was confirmed in another study [191]. In the later stages of disease progression in transgenic mice, increases in other cytokines, including IL-6, M-CSF, TGF-β, and IFNγ, and the chemokines RANTES and KC have been reported [190,192]. In contrast, T-cell derived cytokines including IL-2, IL-3, and IL-4 were detected only at low levels.

The chemokine MCP-1, which may be involved in the recruitment of microglia to sites of injury, was reported as increased in CSF from ALS patients [193]. However, treating glial cell cultures with CSF from ALS patients did not result in increased NO and IL-6 production, suggesting that proinflammatory effects are likely to depend on cell–cell interactions [194]. In a recent microarray study of autopsied end-stage ALS spinal cord, none of the microglial cytokines found to be increased in mouse models for ALS were upregulated [195]. This may reflect biological differences from the mouse models, changes at the end stage of the disease, or an artifact of expression profiling on postmortem tissue.

Interestingly, despite evidence for a negative role of cytokines in ALS, a recent study demonstrated that neurons expressing mutant SOD1 in a transgenic model for ALS showed better survival in regions characterized by large numbers of activated astrocytes and microglia [196]. As with other neurodegenerative disorders, the negative and positive effects of cytokine-producing glial cells in ALS needs further evaluation.

Decreased levels of neurotrophic cytokines may contribute to ALS pathology. Ciliary neurotrophic factor (CNTF) is an IL-6 family survival factor that can prevent motor neuron degeneration and is expressed by astrocytes in the CNS and by Schwann cells [197–199]. Postmortem spinal cord samples from ALS patients show a decrease in CNTF levels [200]. Further, deletion of CNTF expression results in motor neuron disease in mice [201] and accelerates motor neuron degeneration in mice harboring mutant SOD1 transgenes [202]. Some studies have reported an association between null mutations in the CNTF gene and ALS [202], but this was not confirmed in a recent large sample study [203]. Treatment of ALS with CNTF has not been effective [204].

Cardiotropin-1 (CT-1), another member of the IL-6 cytokine family derived from muscle, may also be a neurotrophic factor for motoneurons [205]. Adenoviral gene delivery of CT-1 slowed the course of motoneuron degeneration in a mouse model for ALS [206]. Leukemia inhibitory factor (LIF) is another IL-6 family neurotrophic cytokine that is expressed by astrocytes [207,208]. An infrequent nonsynonymous single nucleotide polymorphism in the LIF gene that could affect neuroprotective properties of the cytokine was found in a sample of ALS patients but not controls [209]. However, in animal models for ALS, LIF has no clear neuroprotective properties [210].

V. PRION DISEASES

Human prion diseases include Creutzfeld-Jakob disease (CJD), variant CJD, Kuru, and Gerstmann-Straussler syndrome [211]. These disorders are characterized by a spongiform encephalopathy, deposition of prion protein (PrP) in plaques in the brain, and rapidly progressive dementia [212]. An important neuropathologic feature of prion disease is the accumulation of microglia surrounding prion amyloid plaques [213]. Astrogliosis is also present. The identity of the infectious agent that causes prion disease is unknown. Some believe PrP is responsible [214], whereas others believe that an agent with a nucleic acid genome is responsible [215]. In general, various animal models for CJD are produced by injecting brain homogenates from clinically ill animals that contain the infectious agent [216]. The cytokine profile of microglia in prion disease is controversial. IL-1β has been reported as increased in CJD CSF [217] as well as in the brain in certain mouse models for CJD [218,219], along with other microglial cytokines such as TNF-α [220]. However, studies using other mouse models have failed to detect increased microglial expression of classic proinflammatory cytokines [221,222]. This has led to the proposal that the glial reaction in CJD is part of an atypical inflammatory response [223], although some of the varying results may be due to differences among animal models. An interesting feature of CJD and animal models of CJD is increased expression of transforming growth factor-family cytokines [222,224–226]. TGF-β1 is increased in AD and has been associated with amyloid deposition and clearance in animal models for AD [141,144]. Whether TGF-β cytokines are involved in amyloidosis in prion diseases is at present unknown.

VI. HUNTINGTON'S DISEASE

Huntington's disease (HD) is an inherited autosomal-dominant disorder that results in neuronal loss and dysfunction in the striatum and cortex and is characterized clinically by a progressive deterioration of motor function and cognition [227]. The mutation responsible for HD is a CAG repeat encoding glutamine residues at the 5' end of the HD gene [228], which encodes the huntingtin protein. The huntingtin protein may be involved in signal transduction within cells [229], but the events leading to neuronal dysfunction in HD are not understood.

Ona and colleagues reported a significant increase in IL-1β in the cortex from a small sample of human HD patients in comparison with controls [230]. Caspase-1 processes pro-IL-1β to generate active mature IL-1β [231]. Mice expressing human huntingtin with an exon 1 expanded CAG repeat develop a phenotype similar to HD [232]. The expression of a dominant negative form of caspase-1 protein in a transgenic mouse model of HD resulted in decreased mature IL1β and delayed the onset of symptoms, neurotransmitter loss, and astrogliosis [230]. It is unclear if the therapeutic effect was due to decreased IL-1β levels or some other effect of caspase-1 inhibition. Minocycline decreases neuronal death in animal and cell culture models for HD [233,234]. Minocycline inhibits microglial activation [170], although the effect of this agent on specific cytokines in HD models is unknown.

As discussed above, IL-6 may have both neurotoxic as well as neuroprotective properties. IL-6 acts through a receptor complex composed of membrane bound IL-6 receptor (IL6R) and the gp130 protein (a transducing subunit) [235]. Bensadoun et al. [236] found a neuroprotective effect of IL-6 and an IL6-IL6R chimera in a quinolinic acid lesion mouse model of HD. Whether IL-6 plays a neuroprotective role in HD is unknown.

VII. VASCULAR DEMENTIA

Vascular dementia (VAD) may account for up to 20% of patients with dementia. VAD is a clinically heterogeneous disorder, but in general multiple infarcts in both cortical and subcortical structures and ischemic injury to white matter are thought to be the primary causes of cognitive impairment in VAD [237]. Reactive gliosis around infracted regions was noted by Alzheimer in his 1895 description of the pathology of VAD [238]. Glial activation, also noted in subsequent descriptions of VAD pathology [239], suggests that an inflammatory process could be involved in this disorder.

There is an extensive literature regarding the role of cytokines in ischemic brain injury [240]. However, relatively little work has been done specifically addressing the role of cytokines in VAD. Tarkowski and colleagues [73,241] showed a striking increase in the levels of TNF-α but not TNF-β, IL-1β, or IL-6 in the CSF of VAD patients compared to controls. TNF-α has been hypothesized to be a major factor in ischemic brain injury by promoting apoptosis and excitotoxicity and impairing perfusion of the infarcted region [242]. However, Cacabelos et al. [243] showed an increase in IL-1β levels in the cortex of VAD patients compared to that of the controls, indicating that other proinflammatory cytokines may also be important in VAD.

Granculocyte macrophage colony stimulating factor (GM-CSF) is a proinflammatory cytokine that induces proliferation of microglia [244]. Significantly higher levels of GM-CSF in the sera and CSF of patients with VAD compared to those in control subjects were observed by Tarkowski et al. [245]. In the same study cohort, there were no changes in the levels of IL-8, TGF-β, and IFN-γ in CSF of VAD patients compared to the controls. The cellular origin of the GM-CSF in VAD is unknown, but macrophages, endothelial cells, and astrocytes have the capacity to synthesize this cytokine, which could contribute to further glial activation.

A few genetic association studies of VAD have targeted cytokine genes. The C-850T polymorphism in the TNF-α gene promoter was found to be a risk factor for VAD in a sample of 81 subjects [81]. Although the functional relevance of this polymorphism is unknown, a polymorphism 13 bp upstream affects the expression of TNF-α message [246]. However, a subsequent study was unable to replicate an association between TNFα C-850T and VAD [83]. Another study found an association between a polymorphism in the IL-6 promoter and VAD [247]. Confirmation of these findings is necessary.

VIII. SUMMARY

The evidence reviewed here shows that all neurodegenerative disorders are characterized by changes in the expression of cytokines in the brain. It is likely that these changes are neither entirely harmful nor entirely helpful to neuronal function. For this reason, treatments aimed at inhibiting inflammatory processes in neurodegeneration may fail. For example, NSAID treatment in AD has proved ineffective thus far [248]. However, there may be ways to harness cytokine-driven inflammatory processes as therapeutics. For example, in AD, immunization with Aβ may promote clearance of this peptide by cytokine-activated microglia [109]. Yet, this positive process may have other inflammatory side events that complicate treatment [249].

Although many fine descriptive reports of cytokines in neurodegenerative disorders have been published, there is a need for a more comprehensive understanding of the interactions among these factors at different stages in the disease process, as well as more work on targeted rather than global cytokine inhibition or augmentation approaches. Studies of variation in cytokine genes as predictors of neurodegenerative disease risk or progression are in their infancy but may ultimately yield pharmaco-genetic markers that identify patients likely to benefit from therapeutics directed toward cytokines in the brain.

ACKNOWLEDGMENT

This work was supported by the National Institute of Mental Health (MH57833) and the National Institute on Aging (AG17824)

REFERENCES

1. Cummings, J.L., Alzheimer's disease, *N Engl J Med* 351 (1), 56–67, 2004.
2. Selkoe, D.J., Alzheimer disease: mechanistic understanding predicts novel therapies, *Ann Intern Med* 140 (8), 627–38, 2004.
3. Mattson, M.P., Pathways towards and away from Alzheimer's disease, *Nature* 430 (7000), 631–9, 2004.
4. Schechter, R., Yen, S.H., and Terry, R.D., Fibrous Astrocytes in senile dementia of the Alzheimer type, *J Neuropathol Exp Neurol* 40 (2), 95–101, 1981.
5. McGeer, P.L., Itagaki, S., Tago, H., and McGeer, E. G., Reactive microglia in patients with senile dementia of the Alzheimer type are positive for the histocompatibility glycoprotein HLA-DR, *Neurosci Lett* 79 (1-2), 195–200, 1987.
6. Rothwell, N., Interleukin-1 and neuronal injury: mechanisms, modification, and therapeutic potential, *Brain, Behavior, and Immunity* 17 (3), 152–157, 2003.
7. Griffin, W.S., Stanley, L.C., Ling, C., White, L., MacLeod, V., Perrot, L.J., White, C. L.d., and Araoz, C., Brain interleukin 1 and S-100 immunoreactivity are elevated in Down syndrome and Alzheimer disease, *Proc Natl Acad Sci U.S.A.* 86 (19), 7611–5, 1989.
8. Goldgaber, D., Harris, H.W., Hla, T., Maciag, T., Donnelly, R.J., Jacobsen, J.S., Vitek, M.P., and Gajdusek, D.C., Interleukin 1 regulates synthesis of amyloid beta-protein precursor mRNA in human endothelial cells, *Proc Natl Acad Sci U.S.A.* 86 (19), 7606–10, 1989.
9. Barger, S.W. and Harmon, A.D., Microglial activation by Alzheimer amyloid precursor protein and modulation by apolipoprotein E, *Nature* 388 (6645), 878–81, 1997.
10. Araujo, D.M. and Cotman, C.W., Beta-amyloid stimulates glial cells *in vitro* to produce growth factors that accumulate in senile plaques in Alzheimer's disease, *Brain Res* 569 (1), 141–5, 1992.
11. McGeer, P.L. and Rogers, J., Anti-inflammatory agents as a therapeutic approach to Alzheimer's disease, *Neurology* 42 (2), 447–9, 1992.
12. Reisberg, B., Doody, R., Stoffler, A., Schmitt, F., Ferris, S., and Mobius, H.J., Memantine in moderate-to-severe Alzheimer's disease, *N Engl J Med* 348 (14), 1333–41, 2003.

13. Hewett, S.J., Csernansky, C.A., and Choi, D.W., Selective potentiation of NMDA-induced neuronal injury following induction of astrocytic iNOS, *Neuron* 13 (2), 487–94, 1994.

14. Viviani, B., Bartesaghi, S., Gardoni, F., Vezzani, A., Behrens, M.M., Bartfai, T., Binaglia, M., Corsini, E., Di Luca, M., Galli, C.L., and Marinovich, M., Interleukin-1beta enhances NMDA receptor-mediated intracellular calcium increase through activation of the Src family of kinases, *J Neurosci* 23 (25), 8692–700, 2003.

15. Chao C.C., Hu S.X., Ehrlich L., and Peterson P.K., Interleukin-1 and Tumor Necrosis Factor-[alpha] Synergistically Mediate Neurotoxicity: involvement of Nitric Oxide and of N-Methyl-D-aspartate Receptors, *Brain, Behavior, and Immunity* 9 (4), 355–365, 1995.

16. Li, Y., Liu, L., Barger, S.W., and Griffin, W.S.T., Interleukin-1 Mediates Pathological Effects of Microglia on Tau Phosphorylation and on Synaptophysin Synthesis in Cortical Neurons through a p38-MAPK Pathway, *J. Neurosci* 23 (5), 1605–1611, 2003.

17. Li, Y., Liu, L., Kang, J., Sheng, J.G., Barger, S.W., Mrak, R.E., and Griffin, W.S., Neuronal-glial interactions mediated by interleukin-1 enhance neuronal acetylcholinesterase activity and mRNA expression, *J Neurosci* 20 (1), 149–55, 2000.

18. Du, Y., Dodel, R.C., Eastwood, B.J., Bales, K.R., Gao, F., Lohmuller, F., Muller, U., Kurz, A., Zimmer, R., Evans, R.M., Hake, A., Gasser, T., Oertel, W.H., Griffin, W.S., Paul, S.M., and Farlow, M.R., Association of an interleukin 1 alpha polymorphism with Alzheimer's disease, *Neurology* 55 (4), 480-3, 2000.

19. Nicoll, J.A., Mrak, R.E., Graham, D.I., Stewart, J., Wilcock, G., MacGowan, S., Esiri, M.M., Murray, L.S., Dewar, D., Love, S., Moss, T., and Griffin, W.S., Association of interleukin-1 gene polymorphisms with Alzheimer's disease [see comments], *Ann Neurol* 47 (3), 365–8, 2000.

20. Rebeck, G.W., Confirmation of the genetic association of interleukin-1A with early onset sporadic Alzheimer's disease, *Neurosci Lett* 293 (1), 75–7, 2000.

21. Grimaldi, L.M., Casadei, V.M., Ferri, C., Veglia, F., Licastro, F., Annoni, G., Biunno, I., De Bellis, G., Sorbi, S., Mariani, C., Canal, N., Griffin, W. S., and Franceschi, M., Association of early-onset Alzheimer's disease with an interleukin-1alpha gene polymorphism [see comments], *Ann Neurol* 47 (3), 361–5, 2000.

22. Sciacca, F.L., Ferri, C., Licastro, F., Veglia, F., Biunno, I., Gavazzi, A., Calabrese, E., Martinelli Boneschi, F., Sorbi, S., Mariani, C., Franceschi, M., and Grimaldi, L.M., Interleukin-1B polymorphism is associated with age at onset of Alzheimer's disease, *Neurobiol Aging* 24 (7), 927–31, 2003.

23. Green, E.K., Harris, J.M., Lemmon, H., Lambert, J.C., Chartier-Harlin, M. C., St. Clair, D., Mann, D.M.A., Iwatsubo, T., and Lendon, C.L., Are interleukin-1 gene polymorphisms risk factors or disease modifiers in AD?, *Neurology* 58 (10), 1566–1568, 2002.

24. Ehl, C., Kolsch, H., Ptok, U., Jessen, F., Schmitz, S., Frahnert, C., Schlosser, R., Rao, M.L., Maier, W., and Heun, R., Association of an interleukin-1beta gene polymorphism at position-511 with Alzheimer's disease, *Int J Mol Med* 11 (2), 235–8, 2003.

25. Tsai, S.J., Liu, H.C., Liu, T.Y., Wang, K.Y., and Hong, C.J., Lack of association between the interleukin-1alpha gene C(-889)T polymorphism and Alzheimer's disease in a Chinese population, *Neurosci Lett* 343 (2), 93–6, 2003.

26. Ma, S.L., Tang, N.L., Lam, L.C., and Chiu, H.F., Lack of association of the interleukin-1beta gene polymorphism with Alzheimer's disease in a Chinese population, *Dement Geriatr Cogn Disord* 16 (4), 265–8, 2003.

27. Murphy, G.M., Jr., Claassen, J.D., DeVoss, J.J., Pascoe, N., Taylor, J., Tinklenberg, J.R., and Yesavage, J.A., Rate of cognitive decline in AD is accelerated by the interleukin-1alpha-889 *1 allele, *Neurology* 56 (11), 1595–7, 2001.

28. Herx, L.M., Rivest, S., and Yong, V.W., Central Nervous system-initiated inflammation and neurotrophism in trauma: IL-1β is required for the production of ciliary neurotrophic factor, *J Immunol* 165 (4), 2232–2239, 2000.

29. Strijbos, P. and Rothwell, N., Interleukin-1 beta attenuates excitatory amino acid-induced neurodegeneration *in vitro*: involvement of nerve growth factor, *J. Neurosci.* 15 (5), 3468–3474, 1995.

30. Akaneya, Y., Takahashi, M., and Hatanaka, H., Interleukin-1 beta enhances survival and interleukin-6 protects against MPP+ neurotoxicity in cultures of fetal rat dopaminergic neurons, *Exp Neurol* 136 (1), 44–52, 1995.

31. Ali, C., Nicole, O., Docagne, F., Lesne, S., MacKenzie, E.T., Nouvelot, A., Buisson, A., and Vivien, D., Ischemia-induced interleukin-6 as a potential endogenous neuroprotective cytokine against NMDA receptor-mediated excitotoxicity in the brain, *J Cereb Blood Flow Metab* 20 (6), 956–66, 2000.

32. Liu, X.H., Xu, H., and Barks, J.D., Tumor necrosis factor-a attenuates N-methyl-D-aspartate-mediated neurotoxicity in neonatal rat hippocampus, *Brain Res* 851 (1-2), 94–104, 1999.

33. Knezevic-Cuca, J., Stansberry, K.B., Johnston, G., Zhang, J., Keller, E.T., Vinik, A. I., and Pittenger, G.L., Neurotrophic role of interleukin-6 and soluble interleukin-6 receptors in N1E-115 neuroblastoma cells, *J Neuroimmunol* 102 (1), 8–16, 2000.

34. Strijbos, P.J. and Rothwell, N.J., Interleukin-1 beta attenuates excitatory amino acid-induced neurodegeneration *in vitro*: involvement of nerve growth factor, *J Neurosci* 15 (5 Pt 1), 3468–74, 1995.

35. Thier, M., Marz, P., Otten, U., Weis, J., and Rose-John, S., Interleukin-6 (IL-6) and its soluble receptor support survival of sensory neurons, *J Neurosci Res* 55 (4), 411–22, 1999.

36. Yamada, M. and Hatanaka, H., Interleukin-6 protects cultured rat hippocampal neurons against glutamate-induced cell death, *Brain Res* 643 (1-2), 173–80, 1994.

37. Giese, A. and Kretzschmar, H.A., Prion-induced neuronal damage—the mechanisms of neuronal destruction in the subacute spongiform encephalopathies, *Curr Top Microbiol Immunol* 253, 203–17, 2001.

38. Glass, J.D. and Wesselingh, S.L., Microglia in HIV-associated neurological diseases, *Microsc Res Tech* 54 (2), 95–105, 2001.

39. McGeer, P.L. and McGeer, E.G., Mechanisms of cell death in Alzheimer disease—immunopathology, *J Neural Transm Suppl* 54, 159–66, 1998.

40. Watanabe, H., Abe, H., Takeuchi, S., and Tanaka, R., Protective effect of microglial conditioning medium on neuronal damage induced by glutamate, *Neurosci Lett* 289 (1), 53–6, 2000.

41. Zietlow, R., Dunnett, S.B., and Fawcett, J.W., The effect of microglia on embryonic dopaminergic neuronal survival *in vitro*: diffusible signals from neurons and glia change microglia from neurotoxic to neuroprotective, *Eur J Neurosci* 11 (5), 1657–67, 1999.

42. Rabchevsky, A.G. and Streit, W.J., Grafting of cultured microglial cells into the lesioned spinal cord of adult rats enhances neurite outgrowth, *J Neurosci Res* 47 (1), 34–48, 1997.

43. Shaked, I., Porat, Z., Gersner, R., Kipnis, J., and Schwartz, M., Early activation of microglia as antigen-presenting cells correlates with T cell-mediated protection and repair of the injured central nervous system, *J Neuroimmunol* 146 (1-2), 84–93, 2004.

44. Elkabes, S., DiCicco-Bloom, E.M., and Black, I.B., Brain microglia/macrophages express neurotrophins that selectively regulate microglial proliferation and function, *J Neurosci* 16 (8), 2508–21, 1996.

45. Nakajima, K., Kikuchi, Y., Ikoma, E., Honda, S., Ishikawa, M., Liu, Y., and Kohsaka, S., Neurotrophins regulate the function of cultured microglia, *Glia* 24 (3), 272–89, 1998.

46. Nakajima, K., Honda, S., Tohyama, Y., Imai, Y., Kohsaka, S., and Kurihara, T., Neurotrophin secretion from cultured microglia, *J Neurosci Res* 65 (4), 322–31, 2001.

47. Nakajima, K., Tohyama, Y., Kohsaka, S., and Kurihara, T., Ceramide activates microglia to enhance the production/secretion of brain-derived neurotrophic factor (BDNF) without induction of deleterious factors *in vitro*, *J Neurochem* 80 (4), 697–705, 2002.

48. Stadelmann, C., Kerschensteiner, M., Misgeld, T., Bruck, W., Hohlfeld, R., and Lassmann, H., BDNF and gp145trkB in multiple sclerosis brain lesions: neuroprotective interactions between immune and neuronal cells?, *Brain* 125 (Pt 1), 75–85, 2002.

49. Suzuki, H., Imai, F., Kanno, T., and Sawada, M., Preservation of neurotrophin expression in microglia that migrate into the gerbil's brain across the blood-brain barrier, *Neurosci Lett* 312 (2), 95–8, 2001.

50. Heese, K., Hock, C., and Otten, U., Inflammatory signals induce neurotrophin expression in human microglial cells, *J Neurochem* 70 (2), 699–707, 1998.

51. Liu, X., Mashour, G.A., Webster, H.F., and Kurtz, A., Basic FGF and FGF receptor 1 are expressed in microglia during experimental autoimmune encephalomyelitis: temporally distinct expression of midkine and pleiotrophin, *Glia* 24 (4), 390–7, 1998.

52. Schwartz, M., Macrophages and microglia in central nervous system injury: are they helpful or harmful?, *J Cereb Blood Flow Metab* 23 (4), 385–94, 2003.

53. Norris, J.G. and Benveniste, E.N., Interleukin-6 production by astrocytes: induction by the neurotransmitter norepinephrine, *J Neuroimmunol* 45 (1–2), 137–45, 1993.

54. Ye, S.M. and Johnson, R.W., Increased interleukin-6 expression by microglia from brain of aged mice, *J Neuroimmunol* 93 (1–2), 139–48, 1999.

55. Gadient, R.A. and Otten, U., Differential expression of interleukin-6 (IL-6) and interleukin-6 receptor (IL-6R) mRNAs in rat hypothalamus, *Neurosci Lett* 153 (1), 13–16, 1993.

56. Strauss, S., Bauer, J., Ganter, U., Jonas, U., Berger, M., and Volk, B., Detection of interleukin-6 and alpha 2-macroglobulin immunoreactivity in cortex and hippocampus of Alzheimer's disease patients, *Lab Invest* 66 (2), 223–30, 1992.

57. Hsiao, K., Chapman, P., Nilsen, S., Eckman, C., Harigaya, Y., Younkin, S., Yang, F., and Cole, G., Correlative memory deficits, Abeta elevation, and amyloid plaques in transgenic mice [see comments], *Science* 274 (5284), 99–102, 1996.

58. Tehranian, R., Hasanvan, H., Iverfeldt, K., Post, C., and Schultzberg, M., Early induction of interleukin-6 mRNA in the hippocampus and cortex of APPsw transgenic mice Tg2576, *Neurosci Lett* 301 (1), 54–8, 2001.

59. Blum-Degen, D., Muller, T., Kuhn, W., Gerlach, M., Przuntek, H., and Riederer, P., Interleukin-1 beta and interleukin-6 are elevated in the cerebrospinal fluid of Alzheimer's and *de novo* Parkinson's disease patients, *Neurosci Lett* 202 (1–2), 17–20, 1995.

60. Hampel, H., Schoen, D., Schwarz, M.J., Kotter, H.U., Schneider, C., Sunderland, T., Dukoff, R., Levy, J., Padberg, F., Stubner, S., Buch, K., Muller, N., and Moller, H. J., Interleukin-6 is not altered in cerebrospinal fluid of first-degree relatives and patients with Alzheimer's disease, *Neurosci Lett* 228 (3), 143–6, 1997.

61. Lanzrein, A.S., Johnston, C.M., Perry, V.H., Jobst, K.A., King, E.M., and Smith, A.D., Longitudinal study of inflammatory factors in serum, cerebrospinal fluid, and brain tissue in Alzheimer disease: interleukin-1beta, interleukin-6, interleukin-1 receptor antagonist, tumor necrosis factor-alpha, the soluble tumor necrosis factor receptors I and II, and alpha1-antichymotrypsin, *Alzheimer Dis Assoc Disord* 12 (3), 215–27, 1998.

62. Licastro, F., Pedrini, S., Caputo, L., Annoni, G., Davis, L.J., Ferri, C., Casadei, V., and Grimaldi, L.M.E., Increased plasma levels of interleukin-1, interleukin-6 and α-1-antichymotrypsin in patients with Alzheimer's disease: peripheral inflammation or signals from the brain?, *Journal of Neuroimmunology* 103 (1), 97–102, 2000.

63. Vandenabeele, P. and Fiers, W., Is amyloidogenesis during Alzheimer's disease due to an IL-1-/IL-6-mediated 'acute phase response' in the brain?, *Immunol Today* 12 (7), 217–9, 1991.

64. Heyser, C.J., Masliah, E., Samimi, A., Campbell, I.L., and Gold, L.H., Progressive decline in avoidance learning paralleled by inflammatory neurodegeneration in transgenic mice expressing interleukin 6 in the brain, *PNAS* 94 (4), 1500–1505, 1997.

65. Campbell, I.L., Abraham, C.R., Masliah, E., Kemper, P., Inglis, J.D., Oldstone, M.B., and Mucke, L., Neurologic disease induced in transgenic mice by cerebral overexpression of interleukin 6, *Proc Natl Acad Sci U.S.A.* 90 (21), 10061–5, 1993.

66. Qiu, Z., Sweeney, D.D., Netzeband, J.G., and Gruol, D.L., Chronic interleukin-6 alters NMDA receptor-mediated membrane responses and enhances neurotoxicity in developing CNS neurons, *J. Neurosci* 18 (24), 10445–10456, 1998.

67. Qiu, Z. and Gruol, D.L., Interleukin-6, beta-amyloid peptide and NMDA interactions in rat cortical neurons, *J Neuroimmunol* 139 (1–2), 51–7, 2003.

68. Loddick, S.A., Turnbull, A.V., and Rothwell, N.J., Cerebral interleukin-6 is neuroprotective during permanent focal cerebral ischemia in the rat, *J Cereb Blood Flow Metab* 18 (2), 176–9, 1998.

69. Penkowa, M., Giralt, M., Carrasco, J., Hadberg, H., and Hidalgo, J., Impaired inflammatory response and increased oxidative stress and neurodegeneration after brain injury in interleukin-6-deficient mice, *Glia* 32 (3), 271–85, 2000.

70. Knezevic-Cuca, J., Stansberry, K.B., Johnston, G., Zhang, J., Keller, E.T., Vinik, A.I., and Pittenger, G.L., Neurotrophic role of interleukin-6 and soluble interleukin-6 receptors in N1E-115 neuroblastoma cells, *Journal of Neuroimmunology* 102 (1), 8–16, 2000.

71. Eskes, C., Honegger, P., Juillerat-Jeanneret, L., and Monnet-Tschudi, F., Microglial reaction induced by noncytotoxic methylmercury treatment leads to neuroprotection via interactions with astrocytes and IL-6 release, *Glia* 37 (1), 43–52, 2002.

72. Carlson, N.G., Wieggel, W.A., Chen, J., Bacchi, A., Rogers, S.W., and Gahring, L.C., Inflammatory Cytokines IL-1α, IL-1β, IL-6, and TNF-α Impart Neuroprotection to an Excitotoxin Through Distinct Pathways, *J Immunol* 163 (7), 3963–3968, 1999.

73. Tarkowski, E., Blennow, K., Wallin, A., and Tarkowski, A., Intracerebral production of tumor necrosis factor-alpha, a local neuroprotective agent, in Alzheimer disease and vascular dementia, *J Clin Immunol* 19 (4), 223–30, 1999.

74. Benzing, W.C., Wujek, J.R., Ward, E.K., Shaffer, D., Ashe, K.H., Younkin, S.G., and Brunden, K.R., Evidence for glial-mediated inflammation in aged APP(SW) transgenic mice, *Neurobiol Aging* 20 (6), 581–9, 1999.

75. Hartmann, A., Michel, P.P., Troadec, J.D., Mouatt-Prigent, A., Faucheux, B.A., Ruberg, M., Agid, Y., and Hirsch, E.C., Is Bax a mitochondrial mediator in apoptotic death of dopaminergic neurons in Parkinson's disease?, *J Neurochem* 76 (6), 1785–93, 2001.

76. Meda, L., Cassatella, M.A., Szendrei, G.I., Otvos, L., Jr., Baron, P., Villalba, M., Ferrari, D., and Rossi, F., Activation of microglial cells by beta-amyloid protein and interferon-gamma, *Nature* 374 (6523), 647–50, 1995.

77. Blasko, I., Marx, F., Steiner, E., Hartmann, T., and Grubeck-Loebenstein, B., TNFalpha plus IFNgamma induce the production of Alzheimer beta-amyloid peptides and decrease the secretion of APPs, *Faseb J* 13 (1), 63–8, 1999.

78. Blasko, I., Schmitt, T.L., Steiner, E., Trieb, K., and Grubeck-Loebenstein, B., Tumor necrosis factor alpha augments amyloid beta protein (25–35) induced apoptosis in human cells, *Neurosci Lett* 238 (1–2), 17–20, 1997.

79. Chong, Y.H., Sung, J.H., Shin, S.A., Chung, J.H., and Suh, Y.H., Effects of the beta-amyloid and carboxyl-terminal fragment of Alzheimer's amyloid precursor protein on the production of the tumor necrosis factor-alpha and matrix metalloproteinase-9 by human monocytic THP-1, *J Biol Chem* 276 (26), 23511–7, 2001.

80. Barger, S.W., Horster, D., Furukawa, K., Goodman, Y., Krieglstein, J., and Mattson, M. P., Tumor necrosis factors alpha and beta protect neurons against amyloid beta-peptide toxicity: evidence for involvement of a kappa B-binding factor and attenuation of peroxide and Ca2+ accumulation, *Proc Natl Acad Sci U.S.A.* 92 (20), 9328–32, 1995.

81. McCusker, S.M., Curran, M.D., Dynan, K.B., McCullagh, C.D., Urquhart, D.D., Middleton, D., Patterson, C.C., McIlroy, S.P., and Passmore, A.P., Association between polymorphism in regulatory region of gene encoding tumour necrosis factor alpha and risk of Alzheimer's disease and vascular dementia: a case-control study, *Lancet* 357 (9254), 436–9, 2001.

82. Infante, J., Llorca, J., Berciano, J., and Combarros, O., No synergistic effect between 850 tumor necrosis factor-alpha promoter polymorphism and apolipoprotein E epsilon 4 allele in Alzheimer's disease, *Neurosci Lett* 328 (1), 71–3, 2002.

83. Terreni, L., Fogliarino, S., Quadri, P., Ruggieri, R.M., Piccoli, F., Tettamanti, M., Lucca, U., and Forloni, G., Tumor necrosis factor alpha polymorphism C-850T is not associated with Alzheimer's disease and vascular dementia in an Italian population, *Neurosci Lett* 344 (2), 135–7, 2003.

84. Alvarez, V., Mata, I.F., Gonzalez, P., Lahoz, C.H., Martinez, C., Pena, J., Guisasola, L.M., and Coto, E., Association between the TNFalpha-308 A/G polymorphism and the onset-age of Alzheimer disease, *Am J Med Genet* 114 (5), 574–7, 2002.

85. Perry, R.T., Collins, J.S., Wiener, H., Acton, R., and Go, R.C., The role of TNF and its receptors in Alzheimer's disease, *Neurobiol Aging* 22 (6), 873–83, 2001.

86. Hao, C., Guilbert, L.J., and Fedoroff, S., Production of colony-stimulating factor-1 (CSF-1) by mouse astroglia *in vitro*, *J Neurosci Res* 27 (3), 314–23, 1990.

87. Nohava, K., Malipiero, U., Frei, K., and Fontana, A., Neurons and neuroblastoma as a source of macrophage colony-stimulating factor, *Eur J Immunol* 22 (10), 2539–45, 1992.

88. Thery, C., Stanley, E.R., and Mallat, M., Interleukin 1 and tumor necrosis factor-alpha stimulate the production of colony-stimulating factor 1 by murine astrocytes, *J Neurochem* 59 (3), 1183–6, 1992.

89. Chung, Y., Albright, S., and Lee, F., Cytokines in the central nervous system: expression of macrophage colony stimulating factor and its receptor during development, *J Neuroimmunol* 52, 9–17, 1994.

90. Du Yan, S., Zhu, H., Fu, J., Yan, S.F., Roher, A., Tourtellotte, W.W., Rajavashisth, T., Chen, X., Godman, G.C., Stern, D., and Schmidt, A.M., Amyloid-beta peptide-receptor for advanced glycation endproduct interaction elicits neuronal expression of macrophage-colony stimulating factor: a proinflammatory pathway in Alzheimer disease, *Proc Natl Acad Sci U.S.A.* 94 (10), 5296–301, 1997.

91. Lee, S.C., Dickson, D.W., Liu, W., and Brosnan, C.F., Induction of nitric oxide synthase activity in human astrocytes by interleukin-1 beta and interferon-gamma, *J Neuroimmunol* 46 (1–2), 19–24, 1993.

92. Giulian, D. and Ingeman, J.E., Colony-stimulating factors as promoters of ameboid microglia, *J Neurosci* 8 (12), 4707–17, 1988.

93. Sawada, M., Suzumura, A., Yamamoto, H., and Marunouchi, T., Activation and proliferation of the isolated microglia by colony stimulating factor-1 and possible involvement of protein kinase C, *Brain Res* 509 (1), 119–24, 1990.

94. Ganter, S., Northoff, H., Mannel, D., and Gebicke-Harter, P. J., Growth control of cultured microglia, *J Neurosci Res* 33 (2), 218–30, 1992.

95. Kloss, C.U., Kreutzberg, G.W., and Raivich, G., Proliferation of ramified microglia on an astrocyte monolayer: characterization of stimulatory and inhibitory cytokines, *J Neurosci Res* 49 (2), 248–54, 1997.

96. Shafit-Zagardo, B., Sharma, N., Berman, J., Bornstein, M., and Brosnan, C., CSF-1 expression is unregulated in astorcyte cultures by IL-1 and TNF and affects microglial proliferation and morphology in organotypic cultures, *Int J Devl Neurosci* 22, 189–198, 1993.

97. Murphy, G.M., Jr., Zhao, F., Yang, L., and Cordell, B., Expression of macrophage colony-stimulating factor receptor is increased in the AbetaPP(V717F) transgenic mouse model of Alzheimer's disease, *Am J Pathol* 157 (3), 895–904, 2000.

98. Johnson-Wood, K., Lee, M., Motter, R., Hu, K., Gordon, G., Barbour, R., Khan, K., Gordon, M., Tan, H., Games, D., Lieberburg, I., Schenk, D., Seubert, P., and McConlogue, L., Amyloid precursor protein processing and A beta42 deposition in a transgenic mouse model of Alzheimer disease, *Proc Natl Acad Sci U.S.A.* 94 (4), 1550–5, 1997.

99. Lue, L.F., Walker, D.G., Brachova, L., Beach, T.G., Rogers, J., Schmidt, A.M., Stern, D.M., and Yan, S.D., Involvement of microglial receptor for advanced glycation endproducts (rage) in Alzheimer's disease: identification of a cellular activation mechanism, *Exp Neurol* 171 (1), 29–45, 2001.

100. Lue, L.F., Rydel, R., Brigham, E.F., Yang, L.B., Hampel, H., Murphy, G.M., Jr., Brachova, L., Yan, S.D., Walker, D.G., Shen, Y., and Rogers, J., Inflammatory repertoire of Alzheimer's disease and nondemented elderly microglia *in vitro*, *Glia* 35 (1), 72–9, 2001.

101. Murphy, G. M., Jr., Yang, L., and Cordell, B., Macrophage colony-stimulating factor augments beta-amyloid-induced interleukin-1, interleukin-6, and nitric oxide production by microglial cells, *J Biol Chem* 273 (33), 20967–71, 1998.

102. Vincent, V.A., Selwood, S.P., and Murphy, G.M., Proinflammatory effects of M-CSF and Abeta in hippocampal organotypic cultures, *Neurobiol Aging* 23 (3), 349–62, 2002.

103. Hamilton, J.A., Whitty, G., White, A.R., Jobling, M.F., Thompson, A., Barrow, C.J., Cappai, R., Beyreuther, K., and Masters, C.L., Alzheimer's disease amyloid beta and prion protein amyloidogenic peptides promote macrophage survival, DNA synthesis and enhanced proliferative response to CSF-1 (M-CSF), *Brain Res* 940 (1–2), 49–54, 2002.

104. Raivich, G., Haas, S., Werner, A., Klein, M.A., Kloss, C., and Kreutzberg, G.W., Regulation of MCSF receptors on microglia in the normal and injured mouse central nervous system: a quantitative immunofluorescence study using confocal laser microscopy, *J Comp Neurol* 395 (3), 342–58, 1998.

105. Murase, S. and Hayashi, Y., Expression pattern and neurotrophic role of the c-fms proto-oncogene M-CSF receptor in rodent Purkinje cells, *J Neurosci* 18 (24), 10481–92, 1998.

106. Wang, Y., Berezovska, O., and Fedoroff, S., Expression of colony stimulating factor-1 receptor (CSF-1R) by CNS neurons in mice, *J Neurosci Res* 57 (5), 616–632, 1999.

107. Akiyama, H., Nishimura, T., Kondo, H., Ikeda, K., Hayashi, Y., and McGeer, P. L., Expression of the receptor for macrophage colony stimulating factor by brain microglia and its upregulation in brains of patients with Alzheimer's disease and amyotrophic lateral sclerosis, *Brain Res* 639 (1), 171–4, 1994.

108. Mitrasinovic, O.M., Perez, G.V., Zhao, F., Lee, Y.L., Poon, C., and Murphy, G.M., Jr., Overexpression of macrophage colony-stimulating factor receptor on microglial cells induces an inflammatory response, *J Biol Chem* 276 (32), 30142–9, 2001.

109. Bard, F., Cannon, C., Barbour, R., Burke, R.L., Games, D., Grajeda, H., Guido, T., Hu, K., Huang, J., Johnson-Wood, K., Khan, K., Kholodenko, D., Lee, M., Lieberburg, I., Motter, R., Nguyen, M., Soriano, F., Vasquez, N., Weiss, K., Welch, B., Seubert, P., Schenk, D., and Yednock, T., Peripherally administered antibodies against amyloid beta-peptide enter the central nervous system and reduce pathology in a mouse model of Alzheimer disease, *Nat Med* 6 (8), 916–9, 2000.

110. Mitrasinovic, O.M. and Murphy, G.M., Jr., Accelerated phagocytosis of amyloid-beta by mouse and human microglia overexpressing the M-CSF receptor, *J Biol Chem* 277(33), 29889–96, 2002.

111. Mitrasinovic, O.M. and Murphy, G.M., Jr., Microglial overexpression of the M-CSF receptor augments phagocytosis of opsonized Abeta, *Neurobiol Aging* 24 (6), 807–15, 2003.

112. Mitrasinovic, O.M., Vincent, V.A., Simsek, D., and Murphy, G.M., Jr., Macrophage colony stimulating factor promotes phagocytosis by murine microglia, *Neurosci Lett* 344 (3), 185–8, 2003.

113. Yan, S.D., Roher, A., Chaney, M., Zlokovic, B., Schmidt, A.M., and Stern, D., Cellular cofactors potentiating induction of stress and cytotoxicity by amyloid beta-peptide, *Biochim Biophys Acta* 1502 (1), 145–57, 2000.

114. Berezovskaya, O., Maysinger, D., and Fedoroff, S., The hematopoietic cytokine, colony-stimulating factor 1, is also a growth factor in the CNS: congenital absence of CSF-1 in mice results in abnormal microglial response and increased neuron vulnerability to injury, *Int J Dev Neurosci* 13 (3–4), 285–99, 1995.

115. Raivich, G., Moreno-Flores, M.T., Moller, J.C., and Kreutzberg, G.W., Inhibition of posttraumatic microglial proliferation in a genetic model of macrophage colony-stimulating factor deficiency in the mouse, *Eur J Neurosci* 6 (10), 1615–8, 1994.

116. Bruccoleri, A. and Harry, G.J., Chemical-induced hippocampal neurodegeneration and elevations in TNFalpha, TNFbeta, IL-1alpha, IP-10, and MCP-1 mRNA in osteopetrotic (op/op) mice [In Process Citation], *J Neurosci Res* 62 (1), 146–55, 2000.

117. Berezovskaya, O., Maysinger, D., and Fedoroff, S., Colony stimulating factor-1 potentiates neuronal survival in cerebral cortex ischemic lesion, *Acta Neuropathol (Berl)* 92 (5), 479–86, 1996.

118. Vincent, V.A., Robinson, C.C., Simsek, D., and Murphy, G.M., Macrophage colony stimulating factor prevents NMDA-induced neuronal death in hippocampal organotypic cultures, *J Neurochem* 82 (6), 1388–97, 2002.

119. Olney, J.W., Wozniak, D.F., and Farber, N.B., Excitotoxic neurodegeneration in Alzheimer disease. New hypothesis and new therapeutic strategies, *Arch Neurol* 54 (10), 1234–40, 1997.

120. Sattler, R. and Tymianski, M., Molecular mechanisms of glutamate receptor-mediated excitotoxic neuronal cell death, *Mol Neurobiol* 24 (1–3), 107–29, 2001.

121. Lipton, S.A. and Rosenberg, P.A., Excitatory amino acids as a final common pathway for neurologic disorders, *N Engl J Med* 330 (9), 613–22, 1994.

122. Castellano, C., Cestari, V., and Ciamei, A., NMDA receptors and learning and memory processes, *Curr Drug Targets* 2 (3), 273–83, 2001.

123. Zlotnik, A. and Yoshie, O., Chemokines: a new classification system and their role in immunity, *Immunity* 12 (2), 121–7, 2000.

124. Mennicken, F., Maki, R., de Souza, E.B., and Quirion, R., Chemokines and chemokine receptors in the CNS: a possible role in neuroinflammation and patterning, *Trends Pharmacol Sci* 20 (2), 73–8, 1999.

125. Xia, M., Qin, S., McNamara, M., Mackay, C., and Hyman, B.T., Interleukin-8 receptor B immunoreactivity in brain and neuritic plaques of Alzheimer's disease, *Am J Pathol* 150 (4), 1267–74, 1997.

126. Ishizuka, K., Kimura, T., Igata-yi, R., Katsuragi, S., Takamatsu, J., and Miyakawa, T., Identification of monocyte chemoattractant protein-1 in senile plaques and reactive microglia of Alzheimer's disease, *Psychiatry Clin Neurosci* 51 (3), 135–8, 1997.

127. Xia, M.Q., Qin, S.X., Wu, L.J., Mackay, C.R., and Hyman, B.T., Immunohistochemical study of the beta-chemokine receptors CCR3 and CCR5 and their ligands in normal and Alzheimer's disease brains, *Am J Pathol* 153 (1), 31–7, 1998.

128. Grammas, P. and Ovase, R., Inflammatory factors are elevated in brain microvessels in Alzheimer's disease, *Neurobiology of Aging* 22 (6), 837–842, 2001.

129. Halks-Miller, M., Schroeder, M.L., Haroutunian, V., Moenning, U., Rossi, M., Achim, C., Purohit, D., Mahmoudi, M., and Horuk, R., CCR1 is an early and specific marker of Alzheimer's disease, *Ann Neurol* 54 (5), 638–46, 2003.

130. Trebst, C., Staugaitis, S.M., Kivisakk, P., Mahad, D., Cathcart, M.K., Tucky, B., Wei, T., Rani, M.R., Horuk, R., Aldape, K.D., Pardo, C.A., Lucchinetti, C.F., Lassmann, H., and Ransohoff, R.M., CC chemokine receptor 8 in the central nervous system is associated with phagocytic macrophages, *Am J Pathol* 162 (2), 427–38, 2003.

131. Gitter, B.D., Cox, L.M., Rydel, R.E., and May, P.C., Amyloid beta peptide potentiates cytokine secretion by interleukin-1 beta-activated human astrocytoma cells, *Proc Natl Acad Sci U.S.A.* 92 (23), 10738–41, 1995.

132. Meda, L., Bernasconi, S., Bonaiuto, C., Sozzani, S., Zhou, D., Otvos, L., Jr., Mantovani, A., Rossi, F., and Cassatella, M.A., Beta-amyloid (25–35) peptide and IFN-gamma synergistically induce the production of the chemotactic cytokine MCP-1/JE in monocytes and microglial cells, *J Immunol* 157 (3), 1213–8, 1996.

133. Szczepanik, A.M., Funes, S., Petko, W., and Ringheim, G.E., IL-4, IL-10 and IL-13 modulate A beta(1–42)-induced cytokine and chemokine production in primary murine microglia and a human monocyte cell line, *J Neuroimmunol* 113 (1), 49–62, 2001.

134. Xia, M. and Hyman, B.T., GROalpha/KC, a chemokine receptor CXCR2 ligand, can be a potent trigger for neuronal ERK1/2 and PI-3 kinase pathways and for tau hyperphosphorylation-a role in Alzheimer's disease?, *J Neuroimmunol* 122 (1–2), 55–64, 2002.

135. Finch, C.E., Laping, N.J., Morgan, T.E., Nichols, N.R., and Pasinetti, G.M., TGF-beta 1 is an organizer of responses to neurodegeneration, *J Cell Biochem* 53 (4), 314–22, 1993.

136. Zhu, Y., Ahlemeyer, B., Bauerbach, E., and Krieglstein, J., TGF-beta1 inhibits caspase-3 activation and neuronal apoptosis in rat hippocampal cultures, *Neurochem Int* 38 (3), 227–35, 2001.

137. Bruno, V., Battaglia, G., Casabona, G., Copani, A., Caciagli, F., and Nicoletti, F., Neuroprotection by glial metabotropic glutamate receptors is mediated by transforming growth factor-beta, *J Neurosci* 18 (23), 9594–600, 1998.

138. Peress, N.S. and Perillo, E., Differential expression of TGF-beta 1, 2 and 3 isotypes in Alzheimer's disease: a comparative immunohistochemical study with cerebral infarction, aged human and mouse control brains, *J Neuropathol Exp Neurol* 54 (6), 802–11, 1995.

139. van der Wal, E.A., Gomez-Pinilla, F., and Cotman, C.W., Transforming growth factor-beta 1 is in plaques in Alzheimer and Down pathologies, *Neuroreport* 4 (1), 69–72, 1993.

140. Flanders, K.C., Lippa, C.F., Smith, T.W., Pollen, D.A., and Sporn, M.B., Altered expression of transforming growth factor-beta in Alzheimer's disease, *Neurology* 45 (8), 1561–9, 1995.

141. Wyss-Coray, T., Masliah, E., Mallory, M., McConlogue, L., Johnson-Wood, K., Lin, C., and Mucke, L., Amyloidogenic role of cytokine TGF-beta1 in transgenic mice and in Alzheimer's disease, *Nature* 389 (6651), 603–6, 1997.

142. Flanders, K.C., Ren, R.F., and Lippa, C.F., Transforming growth factor-betas in neurodegenerative disease, *Prog Neurobiol* 54 (1), 71–85, 1998.

143. Brionne, T.C., Tesseur, I., Masliah, E., and Wyss-Coray, T., Loss of TGF-beta 1 leads to increased neuronal cell death and microgliosis in mouse brain, *Neuron* 40 (6), 1133–45, 2003.

144. Wyss-Coray, T., Lin, C., Yan, F., Yu, G.Q., Rohde, M., McConlogue, L., Masliah, E., and Mucke, L., TGF-beta1 promotes microglial amyloid-beta clearance and reduces plaque burden in transgenic mice, *Nat Med* 7 (5), 612–8, 2001.

145. Samii, A., Nutt, J.G., and Ransom, P.B.R., Parkinson's disease, *The Lancet* 363 (9423), 1783–1793, 2004.

146. Kitada, T., Asakawa, S., Hattori, N., Matsumine, H., Yamamura, Y., Minoshima, S., Yokochi, M., Mizuno, Y., and Shimizu, N., Mutations in the parkin gene cause autosomal recessive juvenile parkinsonism, *Nature* 392 (6676), 605–8, 1998.

147. Polymeropoulos, M.H., Lavedan, C., Leroy, E., Ide, S.E., Dehejia, A., Dutra, A., Pike, B., Root, H., Rubenstein, J., Boyer, R., Stenroos, E.S., Chandrasekharappa, S., Athanassiadou, A., Papapetropoulos, T., Johnson, W.G., Lazzarini, A.M., Duvoisin, R. C., Di Iorio, G., Golbe, L.I., and Nussbaum, R.L., Mutation in the alpha-synuclein gene identified in families with Parkinson's disease, *Science* 276 (5321), 2045–7, 1997.

148. Koutsilieri, E., Scheller, C., Grunblatt, E., Nara, K., Li, J., and Riederer, P., Free radicals in Parkinson's disease, *J Neurol* 249 Suppl 2, II1–5, 2002.

149. Kim, W.G., Mohney, R.P., Wilson, B., Jeohn, G.H., Liu, B., and Hong, J.S., Regional difference in susceptibility to lipopolysaccharide-induced neurotoxicity in the rat brain: role of microglia, *J Neurosci* 20 (16), 6309–16, 2000.

150. Liu, B., Jiang, J.W., Wilson, B.C., Du, L., Yang, S.N., Wang, J.Y., Wu, G.C., Cao, X.D., and Hong, J.S., Systemic infusion of naloxone reduces degeneration of rat substantia nigral dopaminergic neurons induced by intranigral injection of lipopolysaccharide, *J Pharmacol Exp Ther* 295 (1), 125–32, 2000.

151. Bronstein, D.M., Perez-Otano, I., Sun, V., Mullis Sawin, S.B., Chan, J., Wu, G.C., Hudson, P.M., Kong, L.Y., Hong, J.S., and McMillian, M.K., Glia-dependent neurotoxicity and neuroprotection in mesencephalic cultures, *Brain Res* 704 (1), 112–6, 1995.

152. Schapira, A.H., Cooper, J.M., Dexter, D., Clark, J.B., Jenner, P., and Marsden, C.D., Mitochondrial complex I deficiency in Parkinson's disease, *J Neurochem* 54 (3), 823–7, 1990.

153. Langston, J.W. and Irwin, I., MPTP: current concepts and controversies, *Clin Neuropharmacol* 9 (6), 485–507, 1986.

154. Langston, J.W., Forno, L.S., Tetrud, J., Reeves, A.G., Kaplan, J.A., and Karluk, D., Evidence of active nerve cell degeneration in the substantia nigra of humans years after 1-methyl-4-phenyl-1,2,3,6-tetrahydropyridine exposure, *Ann Neurol* 46 (4), 598–605, 1999.

155. Mirza, B., Hadberg, H., Thomsen, P., and Moos, T., The absence of reactive astrocytosis is indicative of a unique inflammatory process in Parkinson's disease, *Neuroscience* 95 (2), 425–32, 2000.

156. Liberatore, G.T., Jackson-Lewis, V., Vukosavic, S., Mandir, A.S., Vila, M., McAuliffe, W. G., Dawson, V.L., Dawson, T.M., and Przedborski, S., Inducible nitric oxide synthase stimulates dopaminergic neurodegeneration in the MPTP model of Parkinson disease, *Nat Med* 5 (12), 1403–9, 1999.

157. Kohutnicka, M., Lewandowska, E., Kurkowska-Jastrzebska, I., Czlonkowski, A., and Czlonkowska, A., Microglial and astrocytic involvement in a murine model of Parkinson's disease induced by 1-methyl-4-phenyl-1,2,3,6-tetrahydropyridine (MPTP), *Immunopharmacology* 39 (3), 167–80, 1998.

158. Sherer, T.B., Betarbet, R., Kim, J.H., and Greenamyre, J.T., Selective microglial activation in the rat rotenone model of Parkinson's disease, *Neurosci Lett* 341 (2), 87–90, 2003.

159. Vila, M., Jackson-Lewis, V., Guegan, C., Wu, D.C., Teismann, P., Choi, D.K., Tieu, K., and Przedborski, S., The role of glial cells in Parkinson's disease, *Curr Opin Neurol* 14 (4), 483–9, 2001.

160. Mogi, M., Harada, M., Kondo, T., Riederer, P., Inagaki, H., Minami, M., and Nagatsu, T., Interleukin-1 beta, interleukin-6, epidermal growth factor and transforming growth factor-alpha are elevated in the brain from Parkinsonian patients, *Neurosci Lett* 180 (2), 147–50, 1994.

161. Mogi, M., Harada, M., Riederer, P., Narabayashi, H., Fujita, K., and Nagatsu, T., Tumor necrosis factor-alpha (TNF-alpha) increases both in the brain and in the cerebrospinal fluid from Parkinsonian patients, *Neurosci Lett* 165 (1–2), 208–10, 1994.

162. Mogi, M., Harada, M., Kondo, T., Mizuno, Y., Narabayashi, H., Riederer, P., and Nagatsu, T., The soluble form of Fas molecule is elevated in Parkinsonian brain tissues, *Neurosci Lett* 220 (3), 195–8, 1996.

163. Mogi, M., Togari, A., Kondo, T., Mizuno, Y., Komure, O., Kuno, S., Ichinose, H., and Nagatsu, T., Caspase activities and tumor necrosis factor receptor R1 (p55) level are elevated in the substantia nigra from Parkinsonian brain, *J Neural Transm* 107 (3), 335–41, 2000.

164. Hunot, S., Dugas, N., Faucheux, B., Hartmann, A., Tardieu, M., Debre, P., Agid, Y., Dugas, B., and Hirsch, E.C., FcepsilonRII/CD23 is expressed in Parkinson's disease and induces, *in vitro*, production of nitric oxide and tumor necrosis factor-alpha in glial cells, *J Neurosci* 19 (9), 3440–7, 1999.

165. Grunblatt, E., Mandel, S., Maor, G., and Youdim, M.B., Gene expression analysis in N-methyl-4-phenyl-1,2,3,6-tetrahydropyridine mice model of Parkinson's disease using cDNA microarray: effect of R-apomorphine, *J Neurochem* 78 (1), 1–12, 2001.

166. Sriram, K., Matheson, J.M., Benkovic, S.A., Miller, D.B., Luster, M.I., and O'Callaghan, J.P., Mice deficient in TNF receptors are protected against dopaminergic neurotoxicity: implications for Parkinson's disease, *Faseb J* 16 (11), 1474–6, 2002.

167. Boka, G., Anglade, P., Wallach, D., Javoy-Agid, F., Agid, Y., and Hirsch, E.C., Immunocytochemical analysis of tumor necrosis factor and its receptors in Parkinson's disease, *Neurosci Lett* 172 (1–2), 151–4, 1994.

168. Hartmann, A., Hunot, S., Michel, P.P., Muriel, M.P., Vyas, S., Faucheux, B.A., Mouatt-Prigent, A., Turmel, H., Srinivasan, A., Ruberg, M., Evan, G.I., Agid, Y., and Hirsch, E. C., Caspase-3: a vulnerability factor and final effector in apoptotic death of dopaminergic neurons in Parkinson's disease, *Proc Natl Acad Sci U.S.A.* 97 (6), 2875–80, 2000.

169. Hunot, S., Brugg, B., Ricard, D., Michel, P.P., Muriel, M.P., Ruberg, M., Faucheux, B.A., Agid, Y., and Hirsch, E.C., Nuclear translocation of NF-kappaB is increased in dopaminergic neurons of patients with Parkinson disease, *Proc Natl Acad Sci U.S.A.* 94 (14), 7531–6, 1997.

170. Wu, D.C., Jackson-Lewis, V., Vila, M., Tieu, K., Teismann, P., Vadseth, C., Choi, D.K., Ischiropoulos, H., and Przedborski, S., Blockade of microglial activation is neuroprotective in the 1-methyl-4-phenyl-1,2,3,6-tetrahydropyridine mouse model of Parkinson disease, *J Neurosci* 22 (5), 1763–71, 2002.

171. Aubin, N., Curet, O., Deffois, A., and Carter, C., Aspirin and salicylate protect against MPTP-induced dopamine depletion in mice, *J Neurochem* 71 (4), 1635–42, 1998.

172. Teismann, P. and Ferger, B., Inhibition of the cyclooxygenase isoenzymes COX-1 and COX-2 provide neuroprotection in the MPTP-mouse model of Parkinson's disease, *Synapse* 39 (2), 167–74, 2001.

173. Vane, J.R., Bakhle, Y.S., and Botting, R.M., Cyclooxygenases 1 and 2, *Annu Rev Pharmacol Toxicol* 38, 97–120, 1998.

174. Mattila, K.M., Rinne, J.O., Lehtimaki, T., Roytta, M., Ahonen, J.P., and Hurme, M., Association of an interleukin 1B gene polymorphism (-511) with Parkinson's disease in Finnish patients, *J Med Genet* 39 (6), 400–2, 2002.

175. Nishimura, M., Mizuta, I., Mizuta, E., Yamasaki, S., Ohta, M., Kaji, R., and Kuno, S., Tumor necrosis factor gene polymorphisms in patients with sporadic Parkinson's disease, *Neurosci Lett* 311 (1), 1–4, 2001.

176. Togo, T., Iseki, E., Marui, W., Akiyama, H., Ueda, K., and Kosaka, K., Glial involvement in the degeneration process of Lewy body-bearing neurons and the degradation process of Lewy bodies in brains of dementia with Lewy bodies, *J Neurol Sci* 184 (1), 71–5, 2001.

177. Katsuse, O., Iseki, E., and Kosaka, K., Immunohistochemical study of the expression of cytokines and nitric oxide synthases in brains of patients with dementia with Lewy bodies, *Neuropathology* 23 (1), 9–15, 2003.

178. Gomez-Tortosa, E., Gonzalo, I., Fanjul, S., Sainz, M.J., Cantarero, S., Cemillan, C., Yebenes, J.G., and del Ser, T., Cerebrospinal fluid markers in dementia with lewy bodies compared with Alzheimer disease, *Arch Neurol* 60 (9), 1218–22, 2003.

179. Burke, R.E., Antonelli, M., and Sulzer, D., Glial cell line-derived neurotrophic growth factor inhibits apoptotic death of postnatal substantia nigra dopamine neurons in primary culture, *J Neurochem* 71 (2), 517–25, 1998.

180. Batchelor, P.E., Liberatore, G.T., Wong, J.Y., Porritt, M.J., Frerichs, F., Donnan, G. A., and Howells, D.W., Activated macrophages and microglia induce dopaminergic sprouting in the injured striatum and express brain-derived neurotrophic factor and glial cell line-derived neurotrophic factor, *J Neurosci* 19 (5), 1708–16, 1999.

181. Batchelor, P.E., Liberatore, G.T., Porritt, M.J., Donnan, G.A., and Howells, D.W., Inhibition of brain-derived neurotrophic factor and glial cell line-derived neurotrophic factor expression reduces dopaminergic sprouting in the injured striatum, *Eur J Neurosci* 12 (10), 3462–8, 2000.

182. Gash, D.M., Zhang, Z., Ovadia, A., Cass, W.A., Yi, A., Simmerman, L., Russell, D., Martin, D., Lapchak, P.A., Collins, F., Hoffer, B.J., and Gerhardt, G.A., Functional recovery in Parkinsonian monkeys treated with GDNF, *Nature* 380 (6571), 252–5, 1996.

183. Kordower, J.H., Emborg, M.E., Bloch, J., Ma, S.Y., Chu, Y., Leventhal, L., McBride, J., Chen, E.Y., Palfi, S., Roitberg, B.Z., Brown, W.D., Holden, J.E., Pyzalski, R., Taylor, M.D., Carvey, P., Ling, Z., Trono, D., Hantraye, P., Deglon, N., and Aebischer, P., Neurodegeneration prevented by lentiviral vector delivery of GDNF in primate models of Parkinson's disease, *Science* 290 (5492), 767–73, 2000.

184. Eberhardt, O., Coelln, R.V., Kugler, S., Lindenau, J., Rathke-Hartlieb, S., Gerhardt, E., Haid, S., Isenmann, S., Gravel, C., Srinivasan, A., Bahr, M., Weller, M., Dichgans, J., and Schulz, J.B., Protection by synergistic effects of adenovirus-mediated X-chromosome-linked inhibitor of apoptosis and glial cell line-derived neurotrophic factor gene transfer in the 1-methyl-4-phenyl-1,2,3,6-tetrahydropyridine model of Parkinson's disease, *J Neurosci* 20 (24), 9126–34, 2000.

185. Mazzio, E.A. and Soliman, K.F., Glioma cell antioxidant capacity relative to reactive oxygen species produced by dopamine, *J Appl Toxicol* 24 (2), 99–106, 2004.

186. Rosen, D.R., Mutations in Cu/Zn superoxide dismutase gene are associated with familial amyotrophic lateral sclerosis, *Nature* 364 (6435), 362, 1993.

187. Hadano, S., Hand, C.K., Osuga, H., Yanagisawa, Y., Otomo, A., Devon, R.S., Miyamoto, N., Showguchi-Miyata, J., Okada, Y., Singaraja, R., Figlewicz, D.A., Kwiatkowski, T., Hosler, B.A., Sagie, T., Skaug, J., Nasir, J., Brown, R.H., Jr., Scherer, S.W., Rouleau, G.A., Hayden, M.R., and Ikeda, J.E., A gene encoding a putative GTPase regulator is mutated in familial amyotrophic lateral sclerosis 2, *Nat Genet* 29 (2), 166–73, 2001.

188. Bruijn, L.I., Miller, T.M., and Cleveland, D.W., Unraveling the mechanisms involved in motor neuron degeneration in ALS, *Annu Rev Neurosci* 27, 723–49, 2004.

189. Engelhardt, J.I. and Appel, S.H., IgG reactivity in the spinal cord and motor cortex in amyotrophic lateral sclerosis, *Arch Neurol* 47 (11), 1210–6, 1990.

190. Hensley, K., Fedynyshyn, J., Ferrell, S., Floyd, R.A., Gordon, B., Grammas, P., Hamdheydari, L., Mhatre, M., Mou, S., Pye, Q.N., Stewart, C., West, M., West, S., and Williamson, K.S., Message and protein-level elevation of tumor necrosis factor alpha (TNF alpha) and TNF alpha-modulating cytokines in spinal cords of the G93A-SOD1 mouse model for amyotrophic lateral sclerosis, *Neurobiol Dis* 14 (1), 74–80, 2003.

191. Yoshihara, T., Ishigaki, S., Yamamoto, M., Liang, Y., Niwa, J., Takeuchi, H., Doyu, M., and Sobue, G., Differential expression of inflammation- and apoptosis-related genes in spinal cords of a mutant SOD1 transgenic mouse model of familial amyotrophic lateral sclerosis, *J Neurochem* 80 (1), 158–67, 2002.

192. Elliott, J.L., Cytokine upregulation in a murine model of familial amyotrophic lateral sclerosis, *Brain Res Mol Brain Res* 95 (1–2), 172–8, 2001.

193. Wilms, H., Sievers, J., Dengler, R., Bufler, J., Deuschl, G., and Lucius, R., Intrathecal synthesis of monocyte chemoattractant protein-1 (MCP-1) in amyotrophic lateral sclerosis: further evidence for microglial activation in neurodegeneration, *J Neuroimmunol* 144 (1–2), 139–42, 2003.

194. Wilms, H., Rosenstiel, P., Sievers, J., Deuschl, G., and Lucius, R., Cerebrospinal fluid from patients with neurodegenerative and neuroinflammatory diseases: no evidence for rat glial activation *in vitro*, *Neurosci Lett* 314 (3), 107–10, 2001.

195. Dangond, F., Hwang, D., Camelo, S., Pasinelli, P., Frosch, M.P., Stephanopoulos, G., Brown, R.H., Jr., and Gullans, S. R., Molecular signature of late-stage human ALS revealed by expression profiling of postmortem spinal cord gray matter, *Physiol Genomics* 16 (2), 229–39, 2004.

196. Clement, A.M., Nguyen, M.D., Roberts, E.A., Garcia, M.L., Boillee, S., Rule, M., McMahon, A.P., Doucette, W., Siwek, D., Ferrante, R.J., Brown, R.H., Jr., Julien, J.P., Goldstein, L.S., and Cleveland, D.W., Wild-type nonneuronal cells extend survival of SOD1 mutant motor neurons in ALS mice, *Science* 302 (5642), 113–7, 2003.

197. Sendtner, M., Kreutzberg, G.W., and Thoenen, H., Ciliary neurotrophic factor prevents the degeneration of motor neurons after axotomy, *Nature* 345 (6274), 440–1, 1990.

198. Sendtner, M., Schmalbruch, H., Stockli, K.A., Carroll, P., Kreutzberg, G.W., and Thoenen, H., Ciliary neurotrophic factor prevents degeneration of motor neurons in mouse mutant progressive motor neuronopathy, *Nature* 358 (6386), 502–4, 1992.

199. Oppenheim, R.W., Prevette, D., Yin, Q.W., Collins, F., and MacDonald, J., Control of embryonic motoneuron survival *in vivo* by ciliary neurotrophic factor, *Science* 251 (5001), 1616–8, 1991.

200. Anand, P., Parrett, A., Martin, J., Zeman, S., Foley, P., Swash, M., Leigh, P.N., Cedarbaum, J.M., Lindsay, R.M., Williams-Chestnut, R.E., and et al., Regional changes of ciliary neurotrophic factor and nerve growth factor levels in post mortem spinal cord and cerebral cortex from patients with motor disease, *Nat Med* 1 (2), 168–72, 1995.

201. DeChiara, T.M., Vejsada, R., Poueymirou, W.T., Acheson, A., Suri, C., Conover, J. C., Friedman, B., McClain, J., Pan, L., Stahl, N., and et al., Mice lacking the CNTF receptor, unlike mice lacking CNTF, exhibit profound motor neuron deficits at birth, *Cell* 83 (2), 313–22, 1995.

202. Giess, R., Holtmann, B., Braga, M., Grimm, T., Muller-Myhsok, B., Toyka, K.V., and Sendtner, M., Early onset of severe familial amyotrophic lateral sclerosis with a SOD-1 mutation: potential impact of CNTF as a candidate modifier gene, *Am J Hum Genet* 70 (5), 1277–86, 2002.

203. Al-Chalabi, A., Scheffler, M.D., Smith, B.N., Parton, M.J., Cudkowicz, M.E., Andersen, P.M., Hayden, D.L., Hansen, V.K., Turner, M.R., Shaw, C.E., Leigh, P.N., and Brown, R.H., Jr., Ciliary neurotrophic factor genotype does not influence clinical phenotype in amyotrophic lateral sclerosis, *Ann Neurol* 54 (1), 130–4, 2003.

204. Acarin, L., Gonzalez, B., Castellano, B., and Castro, A. J., Microglial response to N-methyl-D-aspartate-mediated excitotoxicity in the immature rat brain, *J Comp Neurol* 367 (3), 361–74, 1996.

205. Pennica, D., Arce, V., Swanson, T.A., Vejsada, R., Pollock, R.A., Armanini, M., Dudley, K., Phillips, H.S., Rosenthal, A., Kato, A.C., and Henderson, C.E., Cardiotrophin-1, a cytokine present in embryonic muscle, supports long-term survival of spinal motoneurons, *Neuron* 17 (1), 63–74, 1996.

206. Bordet, T., Lesbordes, J.C., Rouhani, S., Castelnau-Ptakhine, L., Schmalbruch, H., Haase, G., and Kahn, A., Protective effects of cardiotrophin-1 adenoviral gene transfer on neuromuscular degeneration in transgenic ALS mice, *Hum Mol Genet* 10 (18), 1925–33, 2001.

207. Metcalf, D., Leukemia inhibitory factor—a puzzling polyfunctional regulator, *Growth Factors* 7 (3), 169–73, 1992.

208. Murphy, G.M., Jr., Jia, X.C., Song, Y., Ong, E., Shrivastava, R., Bocchini, V., Lee, Y.L., and Eng, L.F., Macrophage inflammatory protein 1-alpha mRNA expression in an immortalized microglial cell line and cortical astrocyte cultures, *J Neurosci Res* 40 (6), 755–63, 1995.

209. Giess, R., Beck, M., Goetz, R., Nitsch, R.M., Toyka, K.V., and Sendtner, M., Potential role of LIF as a modifier gene in the pathogenesis of amyotrophic lateral sclerosis, *Neurology* 54 (4), 1003–5, 2000.

210. Feeney, S.J., Austin, L., Bennett, T.M., Kurek, J.B., Jean-Francois, M.J., Muldoon, C., and Byrne, E., The effect of leukaemia inhibitory factor on SOD1 G93A murine amyotrophic lateral sclerosis, *Cytokine* 23 (4–5), 108–18, 2003.

211. Aguzzi, A. and Polymenidou, M., Mammalian prion biology: one century of evolving concepts, *Cell* 116 (2), 313–27, 2004.

212. Budka, H., Neuropathology of prion diseases, *Br Med Bull* 66, 121–30, 2003.

213. Sasaki, A., Hirato, J., and Nakazato, Y., Immunohistochemical study of microglia in the Creutzfeldt-Jakob diseased brain, *Acta Neuropathol (Berl)* 86 (4), 337–44, 1993.

214. Prusiner, S.B., Scott, M.R., DeArmond, S.J., and Cohen, F.E., Prion protein biology, *Cell* 93 (3), 337–48, 1998.

215. Manuelidis, L., Transmissible encephalopathies: speculations and realities, *Viral Immunol* 16 (2), 123–39, 2003.

216. Bruce, M.E., TSE strain variation: an investigation into prion disease diversity, *Br Med Bull* 66 (1), 99–108, 2003.

217. Van Everbroeck, B., Dewulf, E., Pals, P., Lubke, U., Martin, J.J., and Cras, P., The role of cytokines, astrocytes, microglia and apoptosis in Creutzfeldt-Jakob disease, *Neurobiol Aging* 23 (1), 59–64, 2002.

218. Kordek, R., Nerurkar, V.R., Liberski, P.P., Isaacson, S., Yanagihara, R., and Gajdusek, D.C., Heightened expression of tumor necrosis factor alpha, interleukin 1 alpha, and glial fibrillary acidic protein in experimental Creutzfeldt-Jakob disease in mice, *Proc Natl Acad Sci U.S.A.* 93 (18), 9754–8, 1996.

219. Baker, C.A. and Manuelidis, L., Unique inflammatory RNA profiles of microglia in Creutzfeldt-Jakob disease, *Proc Natl Acad Sci U.S.A.* 100 (2), 675–9, 2003.

220. Williams, A., Van Dam, A.M., Ritchie, D., Eikelenboom, P., and Fraser, H., Immunocytochemical appearance of cytokines, prostaglandin E2 and lipocortin-1 in the CNS during the incubation period of murine scrapie correlates with progressive PrP accumulations, *Brain Res* 754 (1–2), 171–80, 1997.

221. Walsh, D.T., Betmouni, S., and Perry, V.H., Absence of detectable IL-1beta production in murine prion disease: a model of chronic neurodegeneration, *J Neuropathol Exp Neurol* 60 (2), 173–82, 2001.

222. Cunningham, C., Boche, D., and Perry, V.H., Transforming growth factor beta1, the dominant cytokine in murine prion disease: influence on inflammatory cytokine synthesis and alteration of vascular extracellular matrix, *Neuropathol Appl Neurobiol* 28 (2), 107–19, 2002.

223. Perry, V.H., Cunningham, C., and Boche, D., Atypical inflammation in the central nervous system in prion disease, *Curr Opin Neurol* 15 (3), 349–54, 2002.

224. Deininger, M., Meyermann, R., and Schluesener, H., Detection of two transforming growth factor-beta-related morphogens, bone morphogenetic proteins-4 and -5, in RNA of multiple sclerosis and Creutzfeldt-Jakob disease lesions, *Acta Neuropathol (Berl)* 90 (1), 76–9, 1995.

225. Baker, C.A., Lu, Z.Y., Zaitsev, I., and Manuelidis, L., Microglial activation varies in different models of Creutzfeldt-Jakob disease, *J Virol* 73 (6), 5089–97, 1999.

226. Tashiro, H., Dohura, K., and Iwaki, T., Differential expression of transforming growth factor-beta isoforms in human prion diseases, *Neuropathol Appl Neurobiol* 24 (4), 284–92, 1998.

227. Vonsattel, J.P. and DiFiglia, M., Huntington disease, *J Neuropathol Exp Neurol* 57 (5), 369–84, 1998.

228. A novel gene containing a trinucleotide repeat that is expanded and unstable on Huntington's disease chromosomes. The Huntington's Disease Collaborative Research Group, *Cell* 72 (6), 971–83, 1993.

229. MacDonald, M.E., Huntingtin: alive and well and working in middle management, *Sci STKE* 2003 (207), pe48, 2003.

230. Ona, V.O., Li, M., Vonsattel, J.P., Andrews, L.J., Khan, S.Q., Chung, W.M., Frey, A.S., Menon, A.S., Li, X. J., Stieg, P. E., Yuan, J., Penney, J.B., Young, A.B., Cha, J.H., and Friedlander, R.M., Inhibition of caspase-1 slows disease progression in a mouse model of Huntington's disease, *Nature* 399 (6733), 263–7, 1999.

231. Li, P., Allen, H., Banerjee, S., Franklin, S., Herzog, L., Johnston, C., McDowell, J., Paskind, M., Rodman, L., Salfeld, J., and et al., Mice deficient in IL-1 beta-converting enzyme are defective in production of mature IL-1 beta and resistant to endotoxic shock, *Cell* 80 (3), 401–11, 1995.

232. Mangiarini, L., Sathasivam, K., Seller, M., Cozens, B., Harper, A., Hetherington, C., Lawton, M., Trottier, Y., Lehrach, H., Davies, S.W., and Bates, G.P., Exon 1 of the HD gene with an expanded CAG repeat is sufficient to cause a progressive neurological phenotype in transgenic mice, *Cell* 87 (3), 493–506, 1996.

233. Chen, M., Ona, V.O., Li, M., Ferrante, R.J., Fink, K.B., Zhu, S., Bian, J., Guo, L., Farrell, L.A., Hersch, S.M., Hobbs, W., Vonsattel, J.P., Cha, J.H., and Friedlander, R.M., Minocycline inhibits caspase-1 and caspase-3 expression and delays mortality in a transgenic mouse model of Huntington disease, *Nat Med* 6 (7), 797–801, 2000.

234. Wang, X., Zhu, S., Drozda, M., Zhang, W., Stavrovskaya, I.G., Cattaneo, E., Ferrante, R.J., Kristal, B.S., and Friedlander, R.M., Minocycline inhibits caspase-independent and -dependent mitochondrial cell death pathways in models of Huntington's disease, *Proc Natl Acad Sci U.S.A.* 100 (18), 10483–7, 2003.

235. Keller, E.T., Wanagat, J., and Ershler, W.B., Molecular and cellular biology of interleukin-6 and its receptor, *Front Biosci* 1, d340–57, 1996.

236. Bensadoun, J.C., de Almeida, L.P., Dreano, M., Aebischer, P., and Deglon, N., Neuroprotective effect of interleukin-6 and IL6/IL6R chimera in the quinolinic acid rat model of Huntington's syndrome, *Eur J Neurosci* 14 (11), 1753–61, 2001.

237. Roman, G.C., Vascular dementia: distinguishing characteristics, treatment, and prevention, *J Am Geriatr Soc* 51 (5 Suppl Dementia), S296–304, 2003.

238. Reed, B.R., Vascular dementia, *Arch Neurol* 61 (3), 433–5, 2004.

239. Vinters, H.V., Ellis, W.G., Zarow, C., Zaias, B. W., Jagust, W.J., Mack, W.J., and Chui, H.C., Neuropathologic substrates of ischemic vascular dementia, *J Neuropathol Exp Neurol* 59 (11), 931–45, 2000.

240. Allan, S. and Stock, C., Cytokines in stroke, *Ernst Schering Res Found Workshop* (47), 39–66, 2004.

241. Tarkowski, E., Liljeroth, A.M., Minthon, L., Tarkowski, A., Wallin, A., and Blennow, K., Cerebral pattern of pro- and anti-inflammatory cytokines in dementias, *Brain Res Bull* 61 (3), 255–60, 2003.

242. Hallenbeck, J.M., The many faces of tumor necrosis factor in stroke, *Nat Med* 8 (12), 1363–8, 2002.

243. Cacabelos, R., Alvarez, X.A., Franco-Maside, A., Fernandez-Novoa, L., and Caamano, J., Serum tumor necrosis factor (TNF) in Alzheimer's disease and multi-infarct dementia, *Methods Find Exp Clin Pharmacol* 16 (1), 29–35, 1994.

244. Liva, S.M., Kahn, M.A., Dopp, J.M., and de Vellis, J., Signal transduction pathways induced by GM-CSF in microglia: significance in the control of proliferation, *Glia* 26 (4), 344–52, 1999.

245. Tarkowski, E., Wallin, A., Regland, B., Blennow, K., and Tarkowski, A., Local and systemic GM-CSF increase in Alzheimer's disease and vascular dementia, *Acta Neurol Scand* 103 (3), 166–74, 2001.

246. Skoog, T., van't Hooft, F.M., Kallin, B., Jovinge, S., Boquist, S., Nilsson, J., Eriksson, P., and Hamsten, A., A common functional polymorphism (C→A substitution at position-863) in the promoter region of the tumour necrosis factor-alpha (TNF-alpha) gene associated with reduced circulating levels of TNF-alpha, *Hum Mol Genet* 8 (8), 1443–9, 1999.

247. Pola, R., Gaetani, E., Flex, A., Aloi, F., Papaleo, P., Gerardino, L., De Martini, D., Flore, R., Pola, P., and Bernabei, R., -174 G/C interleukin-6 gene polymorphism and increased risk of multi-infarct dementia: a case-control study, *Exp Gerontol* 37 (7), 949–55, 2002.

248. Aisen, P.S., The potential of anti-inflammatory drugs for the treatment of Alzheimer's disease, *Lancet Neurol* 1 (5), 279–84, 2002.

249. Ferrer, I., Boada Rovira, M., Sanchez Guerra, M.L., Rey, M.J., and Costa-Jussa, F., Neuropathology and pathogenesis of encephalitis following amyloid-beta immunization in Alzheimer's disease, *Brain Pathol* 14 (1), 11–20, 2004.

10 Cytokine Expression and Signaling in Brain Tumors

Nandini Dey, Donald L. Durden,
and Erwin G. Van Meir

CONTENTS

I. INTRODUCTION

This chapter updates previous knowledge on cytokine expression in tumors of the central nervous system (CNS) and adds information relating to specific cytokine receptors and their cognate ligands as possible targets for brain tumor therapeutic modulation [1,2]. Due to space constraints, we focused our discussion and cited references mostly on work done in the last 10 years. Further readings on cytokine expression in relation to normal brain physiology and pathology can be found in other chapters of this book. Lastly, we discuss the potential of intracellular targeting of nonredundant "bottleneck" signaling events within the stromal and tumor microenvironment for more efficacious therapeutic activity *in vivo*.

A. EXPANDING COMPLEXITY OF THE CYTOKINE–CYTOKINE RECEPTOR SIGNALING PATHWAYS

The complexity of the cytokine signaling networks as manifested in the diverse array of biological effects has expanded dramatically in recent years with the discovery of additional cytokine–cytokine receptors and a greater understanding of the intracyto-plasmic events downstream of receptor engagement. Cytokines exert a diverse array of effects on multiple cell types and tissues (pleiotropy), and it is clear that dissimilar cytokine or chemokine molecules can have similar functions (functional redundancy). The functional redundancy is in part a result of similarities in the molecular structure and expression of the cytokine receptors present on the different target cells.

Three major families of cytokine receptors are defined by mechanistic aspects by which they transmit intracytoplasmic signals: (1) gp130 linked receptors, (2) common gamma subunit associated receptors, and (3) the interferon family. Considerable functional redundancy exists partly based on common structure function relations within the receptor subunits and downstream effector mechanisms involved in the transmission of intracellular signals. However, subtle differences in the quality of the signal transmitted exist, at least partly engendered by the intracellular components of signal transduction. These differences are translated into qualitatively different signaling events in different cell types, including glioma and neurectoder-mal tumor cells and the stromal elements of brain tumors. It will be important to define the molecular basis for cytokine–cytokine receptor effects in glioma and stromal compartments if we are to effectively utilize gene therapy or small molecule inhibitors for the treatment of brain tumors.

Most of the cytokine receptors are made up of several subunits, some of which are cytokine specific and others that are common to a group of cytokine receptors. For example, signal transducer gp130 is common in receptor complexes for IL-6, leukemia inhibitory factor (LIF), oncostatin M (OM), IL-11, cardiotrophin-1, and ciliary neurotrophic factor (CNTF). A second example is provided by the common β chain (βc, also called KH97), the common signal transducer for IL-3, IL-5, and GM-CSF. Finally, the IL-2 receptor γ-subunit is also part of IL-4, IL-7, IL-9, and IL-15 high affinity receptors and participates in signal transduction. Furthermore, recent research on postreceptor signal transducing molecules has shown that many cytokine receptors act via the janus family of tyrosine kinases (JAK 1,2,3, and Tyk2) and their transcription factor substrates (STAT 1,2,3,4,5a,5b, and 6; Figure 10.1 and Figure 10.2). The use of common signal transducers and similar intracellular signaling cascades helps explain why various cytokines have simi-lar/redundant effects. It can be appreciated that if the glioma cell would receive survival or proliferative signals from a large number of cytokines, growth factors, extracellular matrix proteins and other ligands *in vivo* that cell surface antagonism of one of these receptors will not be effective to halt such signals and will have limited efficacy. Moreover, as studies progress, there is an increasing appreciation that these different cytokine receptor family members also signal via the activation of other tyrosine kinases, serine–threonine kinases, small GTPases, and the acti-vation of a number of lipid kinases including the PI-3 kinase-AKT axis, thus expanding the complexity of cytokine receptor signaling to other areas of mam-malian biochemistry [3]. This concept has lead to the idea that a more effective way to control a large number of cytokine receptor signaling events is to target

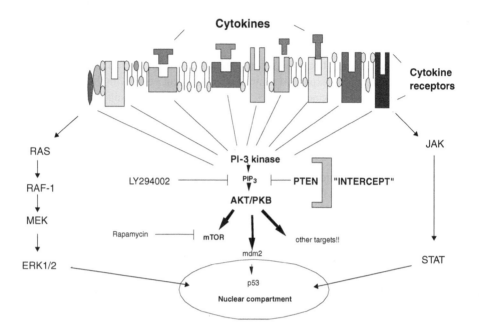

FIGURE 10.1 Intercept points for cytokine/chemokine signal antagonism. This schematic diagram is meant to highlight the major features of the intercept hypothesis for targeting multiple cytokine–cytokine receptor signaling events within the glioma and stromal cellular population to augment potential antitumor efficacy. For these therapeutic modalities to work, the targets must be sufficiently broad, e.g., PI-3 kinase inhibitors to affect a final common pathway downstream of multiple cell surface receptors, i.e., cytokines/chemokines, e.g., Achilles heel for multiple cytokine receptor readout. An example of a prominent final common pathway is the class IA PI-3 kinase node at a point targeted by the Pan-PI-3 kinase inhibitor, a derivative of LY294002, SF1126. Others include the RAS-RAF-ERK pathway and JAK-STAT signaling axis. PTEN is a natural brake for this PI-3 kinase-PIP$_3$ signaling axis and is a pivotal regulator of growth factor and cytokine signaling events mediated through the PI-3 kinase axis.

common components of the intracellular signaling cascade. This "intercept con-cept" is shown in Figure 10.1. Intercept points are relatively nonredundant focal points within the cell where signaling pathways for survival, migration, invasion and angiogenesis converge. One such point of convergence is the PI-3 kinase-AKT signaling axis controlled by the PTEN tumor suppressor gene. Others include the RAS-RAF-MEK-ERK pathway, the JAK-STAT pathway, and others that will likely serve a therapeutic target to control cytokine responses. Recent evidence suggests that the PTEN-PI-3 kinase-AKT pathway is frequently disrupted in human high grade glioma tumors (70–80%) as compared to the JAK-STAT signaling axis (<6% or tumors tested), thus supporting the importance of this signaling axis for con-sideration of molecular therapeutics. Recent evaluation of the RAS-RAF-ERK pathway activation in high grade gliomas has also revealed alterations in this pathway for growth and tumor progression in brain tumors [4].

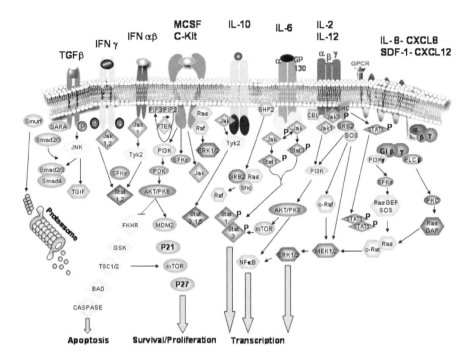

FIGURE 10.2 (Color insert follows page 208). Schematic representation of the cytokine–cytokine receptors grouped into functional categories The cytokine–cytokine receptors are related to the mechanisms by which each cytokine and receptor transmit intracytoplasmic signals in target cells. Shown is the complexity of cytokine signaling events beginning with receptor–ligand interaction at the plasma membrane, which leads to the upstream activation of tyrosine kinases; e.g., JAK kinases are activated leading to the phosphorylation of key substrates that assemble nucleotide exchange proteins, GEFs, for the purpose of activating small G proteins. Once these small G proteins (RAS, RAC, etc.) are activated, they bind specific downstream effectors, e.g., PI-3 kinase, RAF, and mediate the activation of serine threonine kinase cascades, which affect nuclear transcription factors (NF-kB) to control gene expression. Heterotrimeric G proteins linked cell surface receptors (GPCR) also signal through the exchange of GTP for GDP and are important for chemokine-related signal transduction. These heterotrimeric linked receptors can signal through the small G protein pathways and Src family kinases (SFK). Negative control elements include the degradation of signal-some components mediated by E3 ligases, e.g., SMURFs, CBL adapter proteins, which can carry signaling proteins to the proteasome for rapid degradation. Other negative feedback systems include lipid and protein phosphatases, e.g., PTEN, which can dephosphorylate key substrates and shut off cytokine and chemokine signaling cascades in the CNS. This coordinated and somewhat redundant network of signaling events is the essence of cytokine–cytokine receptor signaling in different cell types and contributes to the difficulty in formulating a clear understanding of cell type specific effects of cytokines in glioma systems.

1. Intercept Concept for Control of Multiple Cytokine Receptor Signaling Events in a Brain Tumor

Both tumor and stromal cells utilize important, in some cases redundant, groups of cytokine receptors to encode the signals required for (1) proliferation, (2) migration, (3) invasion and matrix degradation, (4) angiogenesis, (5) survival, and (6) chemo-radioinsensitivity *in vivo*. Below we provide a schematic that would pose the question: Can we identify potential nonredundant signaling "nodes" that may be better targets for therapeutic attack (1) PI-3 kinase-AKT axis, (2) The RAF-1-ERK pathway, and (3) JAK-STAT signaling axis, etc. For example, there might be 50–100 cytokine receptors providing survival, angiogenic and proliferative signals to the glioma cell as depicted in Figure 10.1. The antagonism of one such cytokine receptor would leave a majority of the signals involved in the activation of PI-3 kinase, AKT or ERK intact. In distinct contrast, a PAN-PI-3 kinase, RAF or mTOR inhibitor would potentially impact on 50–80% of these upstream cytokine signals. The downstream inhibition of the intercept point would be predicted to have a greater impact with less potential for cross-resistance within the tumor cell population. The capacity to target more than one nonredundant pathway, e.g., PI-3 kinase/AKT, RAF/MEK/ERK or JAK-STAT would potentially augment the efficacy of this intercept approach to therapeutics *in vivo*. It has been suggested that such broad spectrum antagonism would be associated with profound toxicity. In studies where PAN-PI-3 kinase inhibitors have been administered to mice the predicted toxicity of such PAN-PI-3 kinase inhibition has not been observed [5]. It is the basis for a renewed interest in the use of PAN-PI-3 kinase inhibitors, isoform specific PI-3 kinase inhibitors, RAF inhibitors or JAK-STAT pathway inhibitors in brain tumor therapy [5–9]. A more detailed diagram of the specific cytokine and chemokine signaling pathways is shown in Figure 10.2. Lastly, both RAF-ERK and JAK-STAT pathway inhibitors are in clinical or preclinical development and will potentially be active as antiglioma therapies [9,10].

Restricted expression of cytokine-specific receptor subunits, and transduction of independent signals by their cytoplasmic domains, allows each to retain unique functions in specific cell types under different physiologic stimulation conditions. The IL-2 receptor β subunit, for example, has two distinct cytoplasmic domains: one is serine-rich and responsible for c-myc induction and proliferating signals to Ba/F3 cells, whereas the second is acidic and mediates tyrosine activation and c-fos and c-jun induction. Moreover, in different cell types under differing conditions of receptor and coreceptor stimulation (e.g., cytokine–integrin or cytokine-growth factor receptor coactivation) the quality of the signal is distinctly altered to yield a more complex readout or phenotypic change in the cell.

Cytokines often display similar functions, likely because they share structural homology, found both at the receptor and cytokine levels. Receptors in the class 1 cytokine receptor family are characterized by two common structures within a 200 amino acid region of the extracellular domain: four conserved cysteine residues in the amino half of the homologous region, and a Trp-Ser-X-Trp-Ser motif (WS motif) at its carboxyl terminus. Crystallographic studies have shown that the WS motif is part of a hinge region connecting two barrel-like fibronectin type III modules, each consisting of seven antiparallel β strands. The receptor's ligand binding capacity lies in the hinge region; mutations here abolish binding capability. This family also includes receptors for erythropoietin and the more distantly related growth hormone (GH), prolactin and ciliary neurotrophic factor (CNTF). Structural homology is found at the ligands as well, and hematopoietic cytokines have two pairs of antiparallel α helices interconnected by peptide loops. In contrast, class 2 cytokine receptors do not contain this WSWxWS motif in their extracellular domain, e.g., IL-10 receptor family, interferon receptors. Many related cytokines derive from common ancestral genes that diverged after initial genomic duplication events. Examples of this are the clustered genes GM-CSF, IL-3, IL-4, IL-5, IL-9, IL-12, and IL-13 on chromosome 5 and the chemokine superfamily of cytokines, where human chromosome 4 and 17 each encode several homologous members of the CXC and CC families, respectively (see below). A common origin might explain some of the conserved functional characteristics between members of the same cytokine gene cluster. Two further levels of complexity are the capacity of cytokines to influence the production of other cytokines (cascade action) and modulate the expression of other cytokine receptors (receptor transmodulation). Any alterations to the expression of individual cytokines are expected to alter the globality of the system, the cytokine network, which will determine the biological outcome. Finally the cytokine receptor signaling pathways within the cytokine responsive cell will define the quality of the intracellular signals transmitted within the cell to affect different physiologic responses: (1) cytoskeletal changes, (2) migration, (3) proliferation, (4) angiogenesis, (5) differentiation, (6) apoptosis, and (7) cell to cell communication. These are likely related to the differential activation of distinct downstream intracytoplasmic effectors within the target cell (Figure 10.2).

II. "PRO-INFLAMMATORY" (TH1) CYTOKINES

These cytokines promote the development of a TH1-mediated immune response generally considered to activate the immune response and inflammatory cascades. TH1 lymphocytes produce IFNγ, IL-2, TNFα, and lymphotoxin and are involved in cell-mediated responses to intracellular pathogens. These pro-inflammatory cytokines, which include interleukin 1, TNFα, interferon, and the IL-2 family of cytokines, IL-2, IL-7, IL-9, IL-12, and other recently defined cytokines are involved in the activation of antitumor immunity. The presence or absence of this group of cytokines or their receptors within glioma tumors may be important in glioma biology. The expression pattern for inflammatory versus anti-inflammatory cytokines in glioma tumors, shows a trend toward a lack of pro-inflammatory cytokine expression in malignant glial tumors [11].

A. INTERLEUKIN-1α, β, AND IL-18

IL-1α, IL-1β, and IL-1Ra (see below) are three distinct genes forming the IL-1 gene cluster on human chromosome 2q13-21. Both IL-1α and IL-1β are synthesized as 31 KD precursor proteins lacking a signal peptide, yet they share only 26% amino acid identity. IL-1 is a multifunctional cytokine produced by numerous cell types, including cells of the CNS. Another highly related cytokine, IL-18 also signals through a homologous receptor and has been observed in glioma cells. The current literature supports a role for the IL-1 cytokine cascade in normal astrocyte physiology, tumor biology and induction of neurological manifestations. Immune and vascular host–tumor interactions play a pivotal role in the control of tumor development, and inflammatory mechanisms may participate in the host's defense against tumor cells. IL-1β is a pro-inflammatory cytokine remarkable for its wide range of influence on innate host defense as well as the host homeostasis characterized by incredible complexity of the physiology adjoining its activation, release, and signaling.

1. IL-1 in CNS Tumors

Expression of pro-inflammatory cytokines, like IL-1β has been noted in various types of CNS malignant tumors, raising the possibility that endogenous expression of cytokines and the resulting cytotoxic action may play a role in the control of tumor progression. Both high-grade astrocytomas and pilocytic astrocytomas express high levels of IL-1β localized mainly in astrocytic tumor cells and macrophages [12]. IL-1α, IL-1β, IL-1 receptor type I, and IL-1Ra mRNAs are found in pediatric astrocytomas, ependymomas, and primitive neuroectodermal tumors. IL-1 receptors are expressed by both infiltrating macrophages and neoplastic astrocytes [13]. Immunohistochemical analysis of reactive astrocytes around glioblastoma from postmortem cases showed that IL-1β was largely expressed in the glial cells at the periphery of the tumor [14]. IL-1β gene expression is closely correlated with IL-6 expression in biopsy tissue, xenografts, and cultures of human gliomas [15,16]. This suggests a potential synergistic action with other cytokines (IL-6, TNFα, IFNγ) for a wide range of effects via autocrine/paracrine mechanisms possibly through altering the balance between stimulatory and inhibitory cytokines. Cytokines that are produced in brain tumor can also generate neurological manifestations via paracrine action. Cytokine modulation in an *in vivo* brain tumor model (Rat C6 glioma cells) was examined using behavioral, morphological, and molecular approaches. Brain tumor formation was associated with increased amount of IL-1β, IL-1Ra, IL-1 receptor type I, TNFα, TGFβ1 mRNAs, and IL-1 receptor accessory proteins I and II mRNAs in different brain regions like cerebellum, hippocampus, and hypothalamus [17]. IL-1 along with tumor necrosis factor has been shown to mediate the regulation of C3 gene expression in human astroglioma cells [18]. These synergic/paracrine effects of IL-1 with such a wide range of other cytokines for different effects like cytokine expression, mRNA expression and gene induction strongly indicates that, as brain tumor evolves and progresses, the pattern of cytokine secretion also changes, which may play a major role in bringing tumor-related neurological and immunological manifestations. Another way IL-1 eventually shapes the evolution of tumors is largely contributing to its invasive and vascular characters. IL-β has been implicated in tumor progression by virtue of its direct effect on invasion via induction of MMPs. Intracerebral IL-1β impaired response to tumor invasion [19]. Recently induction of MMP-9 by IL-1 and TNFα in rat C6 glioma cells is shown to be mediated by PKC-zeta [20]. Interestingly, the effect of IL-1 on tumor progression by prompting invasion appears to intersect in the proangiogenic signaling. This effect is contributed by the induction of the two cardinal factors for angiogenesis, HIF-1α and VEGF by

IL-1. IL-1β (in combination with OSM, a hematopoietic cytokine), induced seven-fold higher VEGF expression in human astroglioma cell lines [21]. Recently, the action of IL-1β has been attributed to the upregulation of HIF-1α. HIF-1α is upregulated by IL-1β via NF-kB/COX-2 pathways [22]. It appears that IL-1 activates NF-kB through the PI-3 kinase–AKT pathway, and PI-3 kinase–AKT–mTOR is a signaling pathway leading to induction of HIF-1α in normoxia. IL-1β may be a cytokine link between the inflammatory and oncogenetic pathways via HIF-1α and VEGF. Hence tumor-derived IL-1 is critical for tumor growth, angiogenesis and invasiveness, and should be revisited in light of the IL-1β–PI-3 kinase–HIF-1α–VEGF connection for a potential point of therapeutic exploitation.

B. TUMOR NECROSIS FACTOR α

The TNFα gene is located on human chromosome 6 close to the HLA-B locus, 1.2 Kb from the lymphotoxin α (LTα) gene and 6 Kb from the LTβ gene. Polymorphisms exist in the TNFα gene and a decrease in TNFβ4 allele frequency was found in an Italian series of glioblastomas patients [23]. The TNF gene encodes a 233 amino acids precursor belonging to a family of type II membrane proteins comprising LTα, LTβ and the ligands for FAS, CD40, CD27 and CD30. Although bioactive as an anchored membrane protein, the majority of mature TNFα is located extracellularly and results from cleavage of the precursor by a proteolytic mechanism involving a zinc metalloprotease. TNFα signaling is mediated by two independent receptors: type I (TNFRI, CD120a) is encoded by chromosome 12, has a molecular weight of 55–60 KD and is ubiquitously expressed; type II (TNFRII, CD120b) is encoded by chromosome 1, has a molecular weight of 75–80 KD and is found mainly in monocytes and lymphocytes. TNF receptors intracellular sequences differ from each other and have no tyrosine or serine/threonine kinase domains, yet they activate several signal transducing pathways that activate transcription factors AP-1 and NF-kB. More recent results suggest additional complexity to the TNF receptor signaling cascade [24]. The TNF receptors interact with several proteins that either transmit death signals or the activation of inflammatory signals via the stimulation of MAP kinases. The engagement of TNF receptors in certain cell types activates apoptosis via the recruit-ment of death domain containing adapter proteins, TRADD. In other cases, the stimulation by TNFα will stimulate the p38 or JNK MAP kinase pathway and inflammation by activation of NF-kB through the recruitment of another adapter protein, TRAF1/2. This will result in the activation of the transcription factor AP-1 that affects cell proliferation and the inflammatory response. Moreover the TNF receptors also activate survival and proliferation signals in certain cell types by activating the PI-3 kinase-AKT kinase cascade that leads to the stimulation of NF-kB via the activation of IKKα kinase. This may provide a potential explanation of the differential effects of TNF receptor stimulation in different cell types. TNFRI signals most of TNF activities: cytotoxicity, antiviral activity, fibroblast proliferation and NF-kB induction. TNFRII induces thymocyte and CTL proliferation. Finally, both receptors can be proteolytically cleaved, yielding soluble peptides with antagonistic activity. These may attenuate the effects of TNFα release at sites of inflammation, or function as reservoirs for progressive release of bioactive TNFα.

TNFα has many biological functions. TNF was first named after its antitumor activity, the generation of haemorrhagic necrotic lesions. Subsequently, it became clear that TNFα is toxic to endothelial cells and is not cytotoxic to tumor cells. TNFα is believed essential to the host's inflammatory response to bacterial and parasitic challenge. In the circulation TNFα mediates cachexia, myalgia, anaemia and hypotension. Locally, TNFα increases immune functions such as cytotoxic T-lymphocyte development, IL-2 receptor expression, MHC class II antigen expression, IFNγ production, NK cell activity and B-cell and thymocyte proliferation. TNFα can induce cytokine synthesis (IL-1, IL-6, IL-8, IFNγ, GM-CSF, TGFβ, NGF, and PDGF) in many target cells; this is mediated in part by the induction of NF-kB-dependent transcriptional activation. TNFα synthesis is downregulated by TGFβ, IL-4, IL-10, IL-13, CNTF, and truncated TNFα receptor peptides. Finally, the secretion of TNFα by macrophages may play an indirect role in the induction of neovascularisation during wound repair, inflammation or tumor growth [25]. This may occur by stimulating the release of angiogenic factors such as IL-8 [26], PDGF by endothelial cells [27] and B61, the ligand for the Eck receptor protein tyrosine kinase [28]. TNF can induce the synthesis of proangiogenic factors VEGF, bFGF and IL-4 in different cell types [29,30]. Transfection of TNFα genes into tumor cells can lead to increased antitumor responses in different tumor models [31].

1. TNF in CNS Tumors

Glioma cell lines do not spontaneously produce TNFα, but synthesis can be induced by PMA, LPS or IL-1 albeit at low levels (10–40 pg/ml in 24 h/5×10^4 cells). TNF receptors are broadly expressed in most glioma cell lines as shown by TNF-induced gene expression (for example IL-6, IL-8, MCP-1, CD54/ICAM-1, VCAM-1, etc.) and by 125I-TNF binding experiments (32). Thus, there is no evidence for a TNFα autocrine growth loop in glioma. In the majority of glioma cell lines exogenous treatment of cells with TNFα either has no effect, or diminishes growth although in cases it was shown to upregulate EGFR [32] and could slightly augment thymidine incorporation (2–3-fold). The molecular basis for these variable responses to TNFα are unknown, but may somehow be linked to expression of various affinity TNF receptors, their antagonistic soluble forms and the elements of downstream signal transduction pathways.

Another very interesting property of TNF that warrants further investigation is its ability to modulate cell locomotion [33]. TNFα mRNA and protein are detected in primary brain tumors. Careful *in situ* analysis of the identity of TNFα-positive cells showed the strongest expression in stellate reactive astrocytes, infiltrating macrophages and a few perivascular microglial cells [34]. Protoplasmic tumor cells were also positive, but only in areas surrounding necrosis in glioblastoma. Very interestingly, a clear correlation was found between the local presence of TNF and the presence of infiltrated leukocytes, perhaps suggesting participation of TNF in diapedesis by increasing adhesion molecule expression (ICAM-1, LFA3, etc.) on endothelial cells. Staining was absent from normal peritumoral areas (34). TNFα is expected to have only local effects because it was not found in cyst fluids or the CSF of glioma patients (detection limit: 30 pg/ml; [35]. In contrast, systemic immunosuppressive effects could result from the presence of soluble forms of the TNF receptors in the CSF [36].

2. TNF in CNS Tumor Therapy

The majority of glioma primary cultures and cell lines are fairly resistant to cytotoxic effects of TNFα, showing at best cytostatic effects [37], whereas medulloblastoma cells can be more sensitive [38]. Exogenous TNFα treatment has the potential to promote T-cell immune responses in glioma and potentiate other anticancer effects by:

(i)   induction of MHC class I and II antigens [39]
(ii)   induction of CD54/ICAM-1 expression on glioma cells *in vitro* [39], which augments cytolytic activity of effector cells [40]
(iii)   downregulation of TGFβ1/2 [39]
(iv)   increasing susceptibility of glioma cells to Fas-mediated apoptosis [41]
(v)   potentiation of anticancer drugs by decreasing MDR-1 expression [42]
(vi)   decreasing MMP2 expression [43]
(vii)   decreasing CXCR4 expression [44] and
(viii) synergy with radiotherapy [45]

Toxicity of TNFα was evaluated in the RT-2 rat glioma model and showed alteration of BBB with associated edema, hemorrhagic necrosis in the tumors, and neutrophil adherence to vessels and intratumoral infiltration [46]. Increased survival was observed in a U87MG mouse glioma model with delivery of TNFα expressing gene therapy vectors [31,47].

TNFα has been used in several clinical trials. A Phase I trial where TNF was injected intracavitarily in 6 glioma patients showed 2 patient responses [48]. It is unclear whether TNF had a cytotoxic effect in the tumor cells or whether the responses were the indirect consequence of the immune reactions elicited by TNFα. A local timed increase in cytokine expression (IL-6, IL-1β, IL-8) initially accompanied by neutrophil and subsequently by CD4+ T cells and monocyte infiltrates was observed [48]. Other TNF clinical trails report safety with isolated cases of efficiency [49,50]. The disparity of effects seen clinically in glioma patients treated with this pleiotrophic cytokine may relate to amounts delivered, leukocytic infiltration, receptor stimulation and presence of soluble antagonistic forms of the TNF receptors.

C. INTERLEUKIN-6

Active IL-6 is a 21–28 KD protein that binds with low affinity to gp80, an 80-KD transmembrane protein (IL-6Rα). When associated with gp130, a 130-KD signal transducer glycoprotein (IL-6Rβ), this complex becomes a high affinity receptor. Gp130 is a shared signal transducer that can also bind to the receptors for IL-11, cardiotrophin-1 and the neuropoietic cytokines LIF (leukemia inhibitory factor), OSM (oncostatin M) and CNTF (ciliary neurotrophic factor). IL-6 receptor binding induces JAK/STAT as well as MAPK signal cascades and is silenced by inhibitors like SOCS (suppressor of cytokine signaling) and PIAS (protein inhibitor of activated STAT proteins) [51]. IL-6 production is induced by many factors including IL-1, TNFα, IL-17, IL-2, IFNβ, Fas L, and PDGF, and inhibited by IL-4 and IL-13 in various cell types. IL-6 participates in multiple defense mechanisms. It activates the immune response through B cell and cytotoxic T cell differentiation, and by activation of NK and cytotoxic T cells. It synergizes with IL-3 in haematopoiesis to stimulate stem cell growth and the maturation of megakaryocytes. Abnormal IL-6 synthesis is involved in inflammation and autoimmune diseases and IL-6 is a major mediator of acute phase protein synthesis in the liver.

1. IL-6 and Tumor Growth

IL-6 has been found expressed in numerous cancers and is directly involved in the autocrine or paracrine growth stimulation of myeloma, plasmacytoma, T cell lymphoma, EBV-transformed B cells, colon carcinoma, cervical cancer and renal cell carcinoma. However, depending on the dose, it can inhibit growth of certain cancers models *in vitro* or *in vivo* including certain myeloid leukemia, breast carcinoma, prostate cancer and pulmonary and hepatic micrometastases. These findings have prompted clinical studies with anti-IL-6 antibodies for multiple myeloma, renal cell carcinoma, and B-lymphoproliferative disorders that were well tolerated. The treatments led to a decrease in acute phase proteins, and cancer-related cachexia and anorexia [52]. IL-6 has been implicated in the modulation of the angiogenic process occurring in wound healing [7] and tumor growth [53], and transgenic mice overexpressing IL-6 in the brain showed neurological disease and angiogenesis [54]. IL-6 effects on angiogenesis are in part mediated by activation of VEGF, MMP-2 and MMP-9 expression, possibly through STAT3 activation [55]. Others report that IL-6 can inhibit angiogenesis in neuroblastoma [56]. There is also evidence for IL-6 mediating antiapoptotic functions [57].

2. IL-6 in CNS Tumors

Human glioma cell lines are known to secrete IL-6 under a variety of stimuli including inflammatory cytokines and Fas activation [16,58]. Transcriptional and posttranscriptional mechanisms are involved in IL-6 expression in gliomas [59]. Some genetic changes in tumor cells might be responsible: ploidy increase in chromosome 7 observed in astrocytomas augments the copy number of IL-6 genes [60,61], loss of RB function in some astrocytomas may reduce binding of the transcription factor NF-IL-6 to the promoters of many cytokines including IL-6, and the loss of wild-type p53 in these tumors may also alter IL-6 expression in gliomas because WTp53 can downregulate IL-6 promoter activity [62].

In vivo, brain tumor cells (especially glioblastoma) and infiltrating macrophages appear to be the main source of IL-6. The production of IL-6 is locally significant for it can be detected in cyst fluids and CSFs of patients with various brain tumors [16] and IL-6 levels were also found to increase with glioma grade [63].

What is the role of IL-6 in brain tumors? There is strong recent evidence that IL-6 can contribute to glioma growth. IL-6 deficient transgenic mice were bred to GFAP-viral src oncogene transgenic mice that are prone to glioma development. In the absence of IL-6, tumor formation was abolished [64]. The precise mechanisms by which IL-6 affects glioma growth are unknown but there is evidence for autocrine growth stimulation. Glioma cells have been reported to respond to IL-6 in culture [65,66], although expression of membrane-bound or soluble IL-6 receptors has not been consistently found in glioma. There is, however, good evidence that IL-6 can be mitogenic for human astrocytes in culture and lead to astrogliosis in transgenic mice [54,67]. An interesting model that tries to explain the start of an autocrine IL-6 loop in glioma cells has been proposed. It is postulated that development of hypoxia in the tumors activates Rac1 expression, which then activates NF-kB and IL-6 transcription. Secreted IL-6 then activates IL-6 receptors expressed on the tumor cells and induces STAT3 signaling [68]. Although the majority of studies suggest a protumorigenic action of IL-6 in glioma, there is also evidence to the contrary. Forced overexpression of IL-6 in a T9 rat glioma model stimulated antitumoral immune responses [69] and IL-6 stimulation of C6 rat glioma cells induced differentiation [70]. IL-6 has also been shown to be an autocrine growth factor for pituitary tumors [71]. Production of IL-6 by brain astrocytes appears to also favor metastasis of breast cancer cells to the brain [72].

3. IL-6 in CNS Tumor Therapy

The safety and anticachectic effects of IL-6 treatments observed in clinical trials combined with the recent results of knockout IL-6 mice being protective against glioma development warrant consideration of anti-IL-6 treatments for brain tumor patients.

D. INTERLEUKIN-8

IL-8 and other chemokines are released *in vivo* during infection and inflammation (see Section V).

E. INTERLEUKIN 2 AND 12

The interleukin-2 gene on human chromosome 4 encodes a 153 amino acid polypeptide; a mature IL-2 protein of 15.5 KD is produced after signal peptide cleavage, glycosylation and disulfide bond formation. The gene has *cis*-acting regulatory sequences for NFAT-1, NF-kB, AP-1, and octamer proteins, and the stability of transcribed IL-2 mRNA is decreased by AU rich 3' untranslated regions. Biological activity of IL-2 occurs via ligand receptor binding. Such a receptor signaling is mediated by the high-affinity IL-2 receptor (IL-2R), which consists of three subunits: the α-chain (IL-2Rα or CD25) of 55 KD, the β-chain (IL-2Rβ or CD122) of 70–75 KD and the common cytokine-receptor γ-chain (γc or CD132) of 64 KD. In T cells, the ligand binding to IL-2Rγ causes the association of this subunit with IL-2R and the γc to form a stable heterotrimer. Formation of a stable heterodimer initiates the signal through JAK3 (associated with the γc) and JAK1 (associated with IL-2Rβ). Following this association, MAPK and PI-3 kinase-AKT pathways were activated through adaptor molecule SHC. Tyrosine residues in STAT5 are also phosphorylated to form homodimers (Figure 10.2). STAT5 homodimer translocates to the nucleus and causes induction of STAT5-responsive genes [73].

IL-2 is a key element in the immune response: it mediates clonal expansion of antigen-stimulated resting mature T cells. Antigen recognition by a specific T-cell receptor induces IL-2 synthesis and secretion within the cell concerned, and concomitant expression of high affinity IL-2 receptors. IL-2 will then mediate clonal expansion of this specific T-cell population by autocrine growth stimulation. IL-2 further promotes IFNγ and IL-4 production by T-cells, and directly induces growth and differentiation of B cells, Natural Killer (NK) cells, monocytes, macrophages and lymphokine-activated killer (LAK) cells.

1. IL-2 in CNS Tumor Therapy

Effects of combined granulocyte–macrophage colony-stimulating factor (GM-CSF), interleukin-2, and interleukin-12 based immunotherapy using intracranial rat glioma model have been recently reported [74]. Intratumoral injection of IL-2-activated NK cells has shown beneficial effect [75]. IL-2 delivered from biodegradable polymer microspheres for local immunotherapy was attempted [76]. The result of the study favors a combination of local immunotherapy and chemotherapy for a better clinical outcome in patients with malignant brain tumors. Allogeneic IL-2 secreting fibroblasts can prevent the development of malignant brain tumors [77]. Safety and efficacy of high-dose IL-2 for the therapy of patients with brain metastases has been reported [78]. A more detailed discussion on IL-2 and other cytokine gene therapy is found in a recent review [79].

2. Interleukin-12 (IL-12)

Bioactive IL-12 is a 70 KD heterodimeric cytokine formed by covalently linked subunits of 35 and 40 KD. The p35 and p40 subunits are encoded by two separate genes on chromosomes 3 and 5, respectively. The sequence of p35 is homologous to that of IL-6 and G-CSF and indicates a four-α-helix bundle structure, typical of cytokines. The sequence of the p40 chain is homologous to the extracellular portion of members of the hemopoietin receptor family, particularly the IL-6 receptor α chain (IL-6Rα). This justifies the current proposition that IL-12 evolved from a cytokine of the IL-6 family covalently bound to the extracellular portion of its primordial α chain receptor. The membrane receptor complex of IL-12 is formed by the two chains IL-12Rβ1 and IL-12Rβ2, which are homologous to gp130. IL-12Rβ1 binds IL-12 p40, and it is associated with Tyk2, whereas IL-12Rβ2 recognizes either the heterodimer or the p35 chain and is associated with JAK2. Signaling through the IL-12 receptor complex induces phosphorylation, dimerization, and nuclear translocation of several STAT transcription factors (1,3,4,5), of which predominant biological responses to IL-12 are mediated by STAT4. IL-12 is mostly produced by activated hematopoietic phagocytic cells (monocytes, macrophages, neutrophils) and dendritic cells and in synergy with IL-2, it enhances proliferation and cytolytic activity of activated T, NK and LAK cells. IL-12 exhibits antitumoral and antimetastatic activity in several murine models and this activity appears dependent on the presence of CD8+ and CD4+ T cells, but not NK cells (reviewed in [80]).

3. IL-12 in CNS Tumor Therapy

IL-12 has been reported to be effective in glioma therapy by stimulating the cytotoxic activity of gammadelta T cells [81]. Recently cytotoxicity in gliomas due to IL-12 stimulation of macrophages has been shown to be mediated by IFNγ-regulated nitric oxide [82]. In this regard, adenoviral IL-12 gene transfer is reported to confer more potent, long-lasting cytotoxic immunity and antitumor immunological response in glioma [83] either alone or in combination with IL-18 [84]. Interestingly, higher intracavitary levels of VEGF and IL-8 and lower IL-12 levels have been correlated with shorter adjunctive survival times in glioma patients [85]. Different modalities of IL-12 delivery and gene therapy treatment have been studied in rodent glioma models using neural stem cells (NSCs) or engineered herpes simplex virus [86–88]. Neural stem cells engineered to secrete IL-12 had a better survival rate as compared to the controls. This antitumor effect is mediated by T-cell infiltration in tumor microsatellites and long-term antitumor immunity. IL-12 has also been used in combination with other cytokines (GM-CSF, IL-2) for immunotherapy against an intracranial glioma model in the rat [74,89,90]. The promise of these preclinical studies is now being evaluated in clinical trials. Recently immunogene therapy of recurrent glioblastoma multiforme with a liposomally encapsulated replication-incompetent Semliki forest virus vector containing the human interleukin-12 gene has been carried out in a phase I/II clinical trial [80]. It remains to be seen how the data from experimental results matches the outcome of the clinical trials. Clearly a better understanding of IL-12 action will be needed to close the gap between the preclinical results and clinical application in patients.

F. Interferon γ

IFNγ antagonizes IL-10 and inhibits proliferation of TH2 cells, favoring a TH1-mediated immune response (see Section VIB).

G. Other Cytokines and the CNS

In concert with the expanded genomic information there has been an explosion in the discovery of new cytokines. Hence it is anticipated that the increased complexity of cytokine action will evolve as it relates to glioma–stromal interaction in gliomagenesis. New members of the IL-12 family, IL-23, and IL-27, have been elucidated, however no information on expression levels in brain and glioma tissue are available. IL-13 was recently shown to be critical for autoimmune responses in the brain, suggesting that it may be exploited for anti CNS tumor immunity [79]. Several new members of the IL-10/class 2 α-helical cytokine family include IL-19, IL-20, IL-22, IL-24, IL-26, and IL-29 [91]. Interaction of these cytokines with their specific receptor molecules initiates a broad and varied signal in different cell types. There is evidence that IL-24 inhibits growth and enhances radiosensitivity of glioma cells via JNK signaling [92]. A new pro-inflammatory cytokine, IL-17, has been implicated in glioma biology in particular in the control of nitric oxide metabolism via the NF-kB pathway [93,94]. This cytokine appears to regulate glioma expression of IL-6 and IL-8. Moreover, new members of the IL-17 family recently identified include IL-25 and IL-27. There are a large number of new cytokines recently discovered; most have not been studied in great depth nor have they been investigated in the context of brain tumor pathobiology. As the potential for functional redundancy continues to expand, the concept of "intercept therapeutics" and the study of common signaling events following activation of multiple different cytokines as outlined above becomes more and more germaine to the development of efficacious therapies for malignant brain tumors.

III. ANTI-INFLAMMATORY (TH2/3) CYTOKINES

TH2 lymphocytes produce IL-4, IL-5, IL-6, IL-9, IL-10, and IL-13 and participate in immune responses against large extracellular pathogens such as helminthes. In contrast to immunostimulatory cytokines, a number of studies have examined the TH2/3 cytokine and cytokine receptor expression pattern in glioma tumors in an effort to determine if these immunomodulatory cytokines may play a role in glioma progression. In general, there appears to be a tendency in malignant glioma tumors for the expression of these immunosuppressive cytokines, a result that correlates with negative prognosis of these tumors [11].

A. INTERLEUKIN-1 RECEPTOR ANTAGONIST

IL-1Ra is a member of the IL-1 family and a naturally occurring inhibitor of IL-1. It is encoded by a gene on chromosome 2 and alternative splicing generates soluble and intracellular isoforms (sIL-1Ra and icIL-1Ra). IL-1Ra has the capacity to bind to IL-1 receptors without triggering an agonist response, thus functioning as a receptor antagonist. Hence IL-1Ra is a physiological IL-1 inhibitor that binds to IL-1RI and IL-1RII, preventing signal transduction. The mRNA of IL-1Ra is abundantly expressed by macrophages. Glioma cell lines and tumors produce both s and ic IL-1Ra. Immunohistochemistry identified the producer cells as small proliferating tumor cells surrounding vessels or localized in pseudopalisades [95]. As these cells express IL-1 receptors and can produce IL-1 under certain circumstances IL-1Ra may down regulate a potential autocrine/intracrine growth loop. Autoregulation of the interleukin-1 system and intercytokine interactions in primary human astrocytoma cells has been reported [96]. Human astrocytoma cells following neurosurgical resection respond to the direct application of human IL-1β with a significant upregulation of IL-1α, IL-1β, IL-1RI, and TNFα mRNAs. However IL-1Ra mRNA was not upregulated. Application of heat-inactivated IL-1β did not have any effect on any cytokine component examined, indicating specificity of action. The data suggest a positive autoregulatory IL-1β feedback system and synergistic IL-1β/TNFα interactions that can be involved in the development of pilocytic astrocytomas. These results together with their previous reports indicate that IL-1Ra or a compound with comparable cytokine inhibitory activities may be relevent for brain immunotherapy of patients with astrocytomas. A better understanding of the critical relevance of IL-1 in glioma, as well as the balance between IL-1 and IL-1Ra will be necessary before a rationale for therapy with IL-1 or IL-1Ra can be envisioned.

B. INTERLEUKIN 4 AND 14

IL-4 is secreted by activated T lymphocytes, basophils and mast cells. IL-4 promotes the differentiation of naive T cells towards the TH2 subset. It also antagonizes

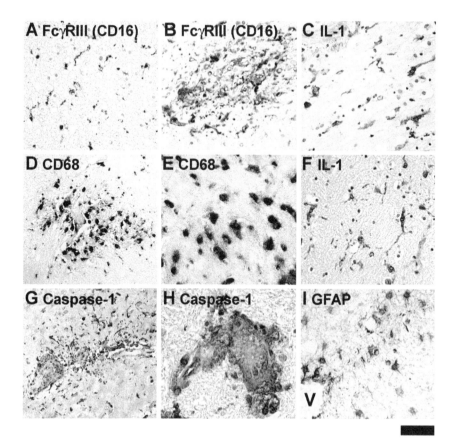

FIGURE 1.3 Microglia and astrocytes in human CNS. Immunohistochemistry on human postmortem brain tissues with diverse pathologies demonstrate microglia and astrocytes with distinct morphology and antigen expression. Expression of FcRIII α-chain (CD16) protein in ramified microglia in normal brain (A) and in activated microglia and macrophages in HIVE in a microglial nodule (B). IL-1 expression in ramified microglial cells in an acute infarct (C): the tissue is otherwise normal appearing on hematoxylin and eosin stain (not shown). CD68 is a macrophage-lineage antigen expressed in lysosomal membrane. A perivascular focus of CD68+ macrophages and microglia in HIVE is shown in (D). Lipid-laden macrophages in an active MS lesion are positive for CD68 (E). IL-1α expression in activated microglia in HIVE: cerebral white matter shows IL-1+ cells with a small bipolar cell body and delicate processes characteristic of microglia (F). Caspase-1, an enzyme that cleaves IL-1 to a biologically active form, is expressed in activated microglia and macrophages in human CNS: caspase-1 in HIVE in a microglial nodule (G) and in a multinucleated giant cell, a hallmark of HIV-1-infected cells (H). Reactive astrocytes in an active MS lesion display increased immunoreactivity for GFAP (I): a vessel (V) in the lower left corner is surrounded by astrocyte processes that define perivascular space (glia limitans). Immunocytochemistry on paraffin-embedded postmortem sections employing methods illustrated in references [8,57,107]. Chromogens were either diaminobenzidine (all except D and E) or nitroblue tetrazolium (D and E): all diaminobenzidine slides were also counterstained with hematoxylin. Scale bar represents 200 μm for B, D, and G; 50 μm for A, C, E, F and I; and 20 μm for H.

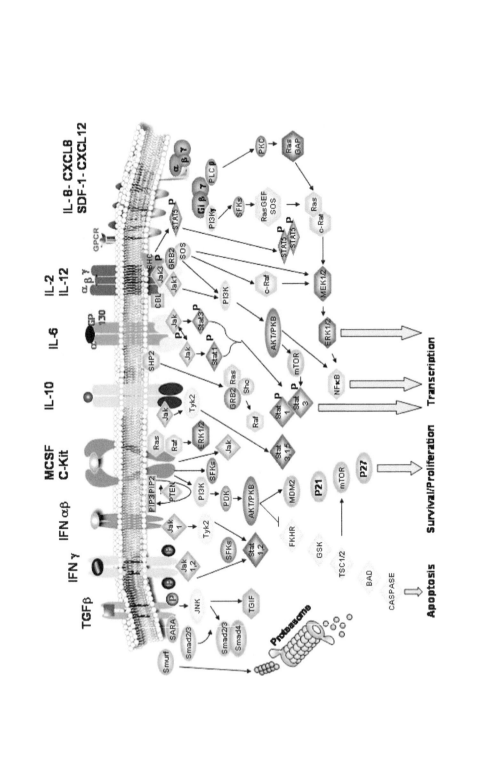

FIGURE 10.2 Schematic representation of the cytokine–cytokine receptors grouped into functional categories The cytokine–cytokine receptors are related to the mechanisms by which each cytokine and receptor transmit intracytoplasmic signals in target cells. Shown is the complexity of cytokine signaling events beginning with receptor–ligand interaction at the plasma membrane, which leads to the upstream activation of tyrosine kinases; e.g., JAK kinases are activated leading to the phosphorylation of key substrates that assemble nucleotide exchange proteins, GEFs, for the purpose of activating small G proteins. Once these small G proteins (RAS, RAC, etc.) are activated, they bind specific downstream effectors, e.g., PI-3 kinase, RAF, and mediate the activation of serine threonine kinase cascades, which affect nuclear transcription factors (NF-kB) to control gene expression. Heterotrimeric G proteins linked cell surface receptors (GPCR) also signal through the exchange of GTP for GDP and are important for chemokine-related signal transduction. These heterotrimeric linked receptors can signal through the small G protein pathways and Src family kinases (SFK). Negative control elements include the degradation of signalsome components mediated by E3 ligases, e.g., SMURFs, CBL adapter proteins, which can carry signaling proteins to the proteasome for rapid degradation. Other negative feedback systems include lipid and protein phosphatases, e.g., PTEN, which can dephosphorylate key substrates and shut off cytokine and chemokine signaling cascades in the CNS. This coordinated and somewhat redundant network of signaling events is the essence of cytokine–cytokine receptor signaling in different cell types and contributes to the difficulty in formulating a clear understanding of cell type specific effects of cytokines in glioma systems.

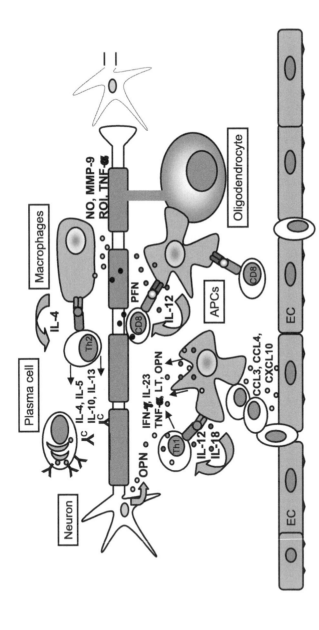

FIGURE 12.3 Cytokines control mechanisms of structural damage to myelin and axons. Immune cells (CD4, CD8, T cells, and B lymphocytes) are directed by chemokines to encounter APC (microglia or DC), which present their cognate autoantigen for reactivation. Pro-inflammatory cytokines (e.g., IFN-γ, IL–23, osteopontin OPN) are secreted by T cells and either directly affect myelin structures or activate macrophages to release nitric oxide (NO), reactive oxygen intermediates (ROI), matrix metalloproteinases (MMP), or TNF-α. CD8 cells may directly attack axons by release of the cytotoxic mediator perforin (PFN). B lymphocytes are induced to terminal plasma cell differentiation by Th2 cytokines, and upon activation release myelin-specific antibodies, which can induce complement (C) mediated demyelination.

IL-12 and IFNγ, thus stimulating a TH2-mediated immune response. IL-4 has also been reported to have mitogenic activity for endothelial cells [29]. The human IL-4 gene is located on a 500 kb cytokine gene cluster on chromosome 5q23-31, with the IL-3, IL-5, IL-9, IL-13, and GM-CSF genes. The gene encodes a main transcript and an alternatively spliced transcript lacking exon 2. The encoded precursor protein has 153 amino acids, gets glycosylated and secreted as a 129 amino acid polypeptide. IL-4 binds to two different heterodimeric receptor complexes made of two transmembrane proteins (Figure 10.3). The first is composed of the γ chain of the γ IL-2 receptor and a 140 KD IL-14Rβ protein. This receptor binds exclusively IL-4. The second is composed of a 140 KD IL-4Rβ subunit and an IL-13Rα' (IL-13Rα1) subunit. This receptor can also bind IL-13. IL-4 binding to the extracellular domain augments affinity of two conserved boxes in the cytoplasmic domain for the JAK1

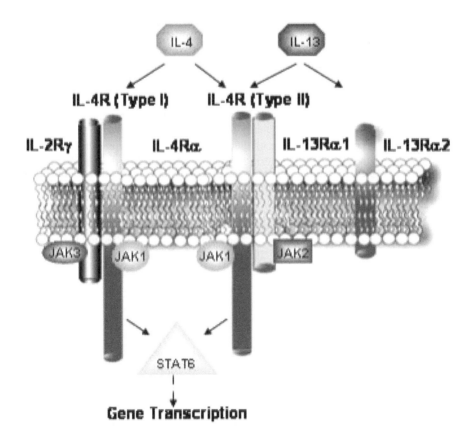

FIGURE 10.3 Structure and signaling components of IL-4 and IL-13 receptors. IL-4 binds exclusively to IL-4R type I, which is composed of an IL-4Rα subunit and the common IL-2R chain. This receptor uses JAK1 and 3 for STAT6 signal transduction. IL-4 and IL-13 share the type II IL-R4, which is composed of an IL-4Rα subunit and an IL-13Rα1 subunit. This receptor signals via JAK1 and 2 to activate STAT6. IL-13 also uniquely binds with high affinity to IL-13Rα2, a transmembrane protein that does not appear to induce intracellular signaling.

and JAK3 tyrosine kinases. This will induce phosphorylation of STAT6, a transcription factor essential for IL-4-mediated signaling as reviewed in [97].

1. IL-4 in CNS Tumor Therapy

Glioma cells do not express IL-4 [98], however IL-4 mRNA can be found in nearly 10% of primary brain tumors likely due to lymphoid infiltrates [99]. IL-4 receptors were found to be expressed in brain tumors [100]. Cytotoxins directed at IL-4 receptors were used as effective therapy for human brain tumors [101]. Safety, tolerability, and tumor response of IL-4-Pseudomonas exotoxin (NBI-3001) in patients with recurrent malignant glioma and medulloblastoma has been studied [101]. In patients with recurrent malignant glioma, a long term survival has been reported following intratumoral infusion of NBI-3001 [102]. Local expression of IL-4 can also prevent brain metastasis [103].

   IL-4 has also been used to stimulate antitumor immune responses. Immune gene therapy of brain tumor was performed with retrovirus and adenovirus-mediated gene transfer of murine IL-4 in a mouse model [104,105]. Autologous glioma cell vaccine admixed with IL-4 gene transfected fibroblasts is being tested for the treatment of patients with recurrent glioblastoma with preliminary encouraging results (reviewed in [79]). IL-4 delivered to brain tumors through IL-4 producing stem cells has shown potent antitumor effects in experimental rodent models [106]. Overall, these studies have demonstrated that IL-4 is one of the most potent antitumoral cytokines, with very significant induction of long term survival when used in subcutaneous vaccination protocols.

C. Interleukin 10 Family

IL-10 is an anti-inflammatory cytokine and immunosuppressive factor. It has been implicated in autoimmunity and tumorigenesis. New IL-10 related cytokines recently described include IL-19, IL-20, IL-22, IL-24, IL-26, and IL-29 (91). Much less is known about these cytokines and their role in CNS tumorigenesis. Elevated levels of IL-10 have been observed in the sera of patients with cancer. IL-10 produced by tumor or the stromal compartment might contribute to the observed tumor-associated T cell anergy, decreased MHC class I expression and general antagonism of T cell proliferation. In contrast, experiments performed in the IL-10$^{-/-}$ mouse model with the GL261 syngeneic mouse glioma suggests that IL-10 secretion in the CNS is required for glioma rejection following vaccination [107]. Other experiments have demonstrated that FasL–Fas interactions between glioma cells and T cells results in the induction of IL-10 production within the T cell, a process that is dependent upon caspase activation [108]. The tumor FasL effectively induces transcriptional activity of the IL-10 gene in T cells without causing cell death. The expression of IL-10 in certain tumors is positively correlated with the expression of FasL. Interestingly, gliomas express detectable levels of FasL that may contribute to the modulation of cytotoxic T cells in the CNS. It is possible that cross-talk between tumor and immune cells through the Fas/FasL system is mediated by the IL-10 signaling pathway. If so, this would provide glioblastomas tumor cells a mechanism by which Fas would induce via caspase activation the increased production of IL-10 in immune cells, and this would be a mechanism by which they might suppress an immune attack. Consistent with this idea is the observed correlation between clinical malignancy and cytokine gene expression within malignant gliomas especially regarding the relevancy of inhibitory cytokines, such as TGFβ and IL-10 [109]. Specific cytokine mRNA profiles in glioma patients might have prognostic significance for immunotherapy [110]. Other studies have shown that IL-10 is expressed in human gliomas *in vivo* and increases glioma cell proliferation and motility *in vitro*. These data indicates that IL-10 is involved in the progression of glial tumors, especially in the enhancement of tumor cell proliferation and migration, which promotes infiltration of the surrounding tissue [111,112].

However the molecular basis for this invasive effect of IL-10 is currently unclear. In summary, the data generated in human and mouse glioma systems suggest that IL-10 antagonism may be a useful therapeutic approach in the treatment of these tumors.

D. INTERLEUKIN 13

IL-13 is produced by various activated human T-cell subsets. It exerts profound effects on monocytes; these include changes in morphology, surface antigen expression, antigen presentation, antibody-dependent cellular cytotoxicity (ADCC) and cytokine synthesis. Like IL-4, IL-13 suppresses TH1 cell development by inhibiting transcription of IFNγ and both IL-12 subunits, thus favoring TH2 developmental pathways. Both IL-4 and IL-13 modulate surface phenotype, growth and induce both Ig production and class switch from IgG to IgE on B cells, suggesting participation in allergic responses. In contrast to IL-4, IL-13 does not activate human T cells. IL-13 can bind two receptors (Figure 10.3). IL-13Rα2 is a 42 KD transmembrane protein that binds IL-13 exclusively. The second receptor is a heterodimeric complex of two transmembrane receptors; a 45 KD IL-13Rα1 subunit that associates with a 140 KD IL-4Rα chain to form a complex binding both IL-4 and IL-13. IL-13 was also recently shown to negatively regulate CD8+ CTL-mediated tumors immunosurveillance [113].

Beside its immune modulatory activities, IL-13 has also direct effects on tumor growth. It is an autocrine growth factor for Hodgkin's lymphoma/Reed–Sternberg cells, can be antiapoptotic and promote metastasis (reviewed in [113]).

The vast majority of high grade gliomas bind IL-13 and expresses both IL-13 receptors. Signaling through the shared IL-4/IL-13 receptor induces STAT6 and VCAM expression in glioma cells. Gliomas are highly sensitive to IL-13 based cytotoxins [114]. This appears to be mediated to a large extent by the IL-13Rα2, which is specifically overexpressed on high grade glial tumor cells. The IL-13Rα2 receptor is known to bind IL-13 with high affinity, but does not transmit intracellular signals to activate STAT6 upon binding to its ligand (reviewed in [113]). The significance of this lack of JAK-STAT signaling for the IL-13Rα2 is not completely clear. It has been speculated that the IL-13Rα2 may antagonize IL-13Rα1 signaling in glioma cells.

E. INTERLEUKIN-15

IL-15 is a T cell growth factor that shares many of biological activities and receptor components with IL-2. In contrast to its pro-TH1 activity in mice, IL-15 supports a

TH2 response in humans. The IL-15 receptor is made of 3 components: a unique α subunit that is involved in ligand binding, and the β and γ subunits of IL-2R. These β and γ subunits are essential for signaling through the JAK-STAT pathway (Figure 10.2). The IL-2Rγ subunit also forms part of the receptors for other T-cell growth factors (IL-2, IL-4, IL-7, IL-9). The mature protein has a molecular weight of 14 to 15 KD and is likely folded in four helical bundles with two disulfide bonds and is closely related to IL-2.

IL-15 antagonizes apoptotic signals, protecting DCs and T cells from proapoptotic signals induced by tumor cells (reviewed in [79]). The IL-15 gene is constitutively expressed in the nerve tissue and during differentiation of neurons two distinct isoforms of IL-15 mRNA have been found. Results of the study provide evidence that in the normal brain, IL-15 is present in neurons only. Glial cells express the cytokine under activated and or transformed conditions following inflammatory insults [115]. The function of IL-15 and its role as growth factor has been reviewed [116]. Transcripts for IL-2 family cytokines (IL-2, IL-7, IL-9, IL-15) were reported to be essentially absent in 12 human glioblastomas tumors and 6 human glioblastoma cell lines [11]. In summary, insufficient information exists at this stage as to the role of IL-15 in the human CNS and its function in gliomagenesis.

F.   TGFβ1,2,3 (TH3)

TGFβ regulates the immune system, TH2/3 cells may produce TGFβ encouraging IgA class switching and anti-inflammatory type responses (see Section VII).

IV.   CSFS AND HEMATOPOIETIC GROWTH FACTORS

The growth and differentiation of hematopoietic cells are modulated by a growing number of colony stimulating factors (CSFs), e.g., GM-CSF, G-CSF, and M-CSF and growth factors, e.g., IL-3, IL-6, IL-9, IL-11, and SCF. There are conflicting reports as to the whether these factors are present in glial tumors and whether they play a role in the pathogenesis of glial malignancies. Some of these have become clinically important, such as in the treatment of chemotherapy-induced neutropenia or as a recovery aid after bone marrow transplantation. The factors that have been studied in relation to brain tumors are discussed in the following.

A.   GRANULOCYTE–MACROPHAGE COLONY STIMULATING
FACTOR (GM-CSF)

The GM-CSF gene lies on the long arm of human chromosome 5, about 10 Kbp 3' of the IL-3 gene. GM-CSF becomes an active secreted glycoprotein of 23 KD after cleavage of a 17 amino acid signal peptide. The receptor for GM-CSF is composed of a α subunit of 80 KD and a β-type subunit, βc (=KH97), that is also part of the IL-3 and IL-5 receptor complexes. GM-CSF is mainly produced by T-lymphocytes, macrophages, endothelial cells, fibroblasts, and stromal cells. GM-CSF synthesis is induced by cytokine stimulation (IL-1, TNF, etc.) and antigens or inflammatory agents (LPS). GM-CSF regulates the constitutive generation and functional activities of granulocyte and macrophage populations, but is not an essential growth factor for basal hematopoiesis. Activated T-cell-derived IL-3 has the same generative capacity, but is essentially released during immune and inflammatory responses. Tumor vaccination experiments have shown that GM-CSF is very potent in eliciting specific and long-lasting antitumor immunity. This requires activities of both CD4+ and CD8+ lymphocytes, suggesting that GM-CSF likely increases the function of host professional antigen-presenting cells. Indeed, it was found that GM-CSF, in conjunction with IL-4, increases the generation and maturation of DC.

Constitutive low level secretion of GM-CSF is found in astrocytoma and glioblastoma cell lines, which is heavily upregulated after TNFα or IL-1 stimulation [35]. There does not appear to be substantial GM-CSF production *in vivo* in glioma tumors. GM-CSF mRNA was not detectable by RNAse protection assays and GM-CSF bioactivity was not detected in tumor-derived cyst fluids [35]. The lack of GM-CSF production *in vivo* may be due to inhibition by tumor-derived inhibitory factors such as TGFβ2 and PGE2 as seen *in vitro* [35]. The absence of *in vivo* production of GM-CSF in astrocytoma despite the presence of inducing cytokines and successes obtained with vaccination strategies using GM-CSF genes in melanoma prompted similar investigations in glioma. Given the importance of GM-CSF in the generation of DC, vaccination attempts with tumor cells expressing GM-CSF were made. It was hoped that this would induce tumor infiltration by DCs, which could then cross present tumor antigens to antigen-specific T cells (reviewed in [79]). In the mouse GL261 glioma model, survival was tested following subcutaneous vaccination using GM-CSF-transduced glioma cells. The animals vaccinated with GM-CSF-secreting cells survived over 120 days as compared with the group of animals vaccinated with wild type cells (median survival time of 30 days; 117). In conclusion, GM-CSF is a very promising cytokine to deliver at the vaccination site in brain tumor immunotherapy strategies.

B.   Granulocyte Colony Stimulating Factor (G-CSF)

G-CSF is secreted by activated macrophages, endothelial cells and fibroblasts, and regulates neutrophil production from precursor cells and subsequent maturation. Therefore, it is an essential regulator of the organism's initial antibacterial defense; neutrophils comprise over 70% of white blood cells. Malignant tumors exhibiting constitutive G-CSF production display marked granulopoiesis even in the absence of bacterial infection [118]. Constitutive G-CSF production is seen in about 50% of glioblastoma cell lines, others require induction by IL-1β or TNFα for expression [119]. In these cells, G-CSF does not function as a direct growth factor because G-CSF receptor mRNA was undetectable by RT/PCR [119]. Immunostaining revealed G-CSF in astrocytoma and in reactive astrocytes associated with ischemic brain infarct or gliosis in the hippocampal formation. Glioblastoma, medulloblastoma and oligodendroglioma cells did not stain for G-CSF, but immunoreactivity was present on endothelial cells and reactive astrocytes in the peritumoral area [120]. Despite the presence of G-CSF, astrocytoma do not contain neutrophilic infiltrates. Treatment of malignant astrocytoma patients with high dose chemotherapy induces myelosuppression. To reduce neutropenia, glioma patients have been treated systemically with G-CSF. In a phase-2 study carried out in pediatric brain tumors, (POG 9237) idarubicin infusion was followed by G-CSF, and G-CSF was maintained until the blood count recovered [121]. Aternatively, G-CSF was used to mobilize (in conjunction with cyclophosphamide and etoposide) peripheral stem cells for cryopreservation before chemotherapy. This allowed us to test the safety and efficacy of high-dose chemotherapy with autologous stem cell transplantation in malignant astrocytomas, medulloblastoma and PNET. The study concluded that autologous stem cell transplantation can be safely used to treat patients and it can have a survival advantage [122,123].

C. MACROPHAGE-COLONY STIMULATING FACTOR (CSF-1)

The *M-CSF* (*CSF*-1) gene is located on human chromosome 1 and encodes a primary transcript associated with several alternative splicing events. Alternative splicing in exon 6 determines whether the mRNA translates into a membrane-bound (mM-CSF) or secreted glycoprotein (80–100 KD) or chondroitin sulfate-containing proteoglycan (130–160 KD). Alternative use of exons 9 or 10 modifies the 3' untranslated region of the mRNA, and probably controls mRNA stability; exon 10 encodes AU-rich regions signaling mRNA degradation [124]. The M-CSF receptor is a 165 KD transmembrane protein encoded by the *c-fms* gene on chromosome 5. Its extracellular domain has immunoglobulin-like regions, whereas the cytoplasmic domain contains a tyrosine kinase domain belonging to the class III family of tyrosine kinases. M-CSF is an essential regulator of the production of monocytes in the blood and of tissue macrophages in the liver, kidney, spleen and dermis. M-CSF further differentiates osteoclast progenitor cells and regulates implantation events in the female reproductive system.

Similar to normal astrocytes, human glioblastoma cell lines secrete M-CSF into culture medium; this capacity does not correlate with their ability to form tumors in *nu/nu* mice [125]. M-CSF mRNA is present in glioma and medulloblastoma specimens, but the presence of M-CSF in the cyst fluid or CSF of patients was not tested. Serum of these and other brain tumor patients was negative. M-CSF receptor transcripts were detected by Northern blot in glioma samples, but not in cell lines, probably reflecting expression by local microglial cells or infiltrating macrophages. Tumor secretion of M-CSF (and possibly MCP-1, TGFβ and other cytokines) could serve as the chemoattractants for the numerous macrophage infiltrates observed in these tumors. M-CSF might be of importance for tumor growth as cytokine cascades produced by recruited microglial or monocyte populations in the CNS may elicit proangiogenic signals. M-CSF was shown to greatly augment the antibody-dependent cellular cytotoxicity of peripheral blood mononuclear cells against a glioma cell line [126]. Such results might be exploited in cancer therapy approaches to increase effectiveness of unconjugated antibodies. M-CSF was found to be tumoricidal in two subcutaneous glioma models (9L and T9) when used at high dose (16 million units/kg; 4 days a week for 3 weeks) [127]. More recent studies demonstrated that glioma cells expressing mM-CSF were rejected when implanted subcutaneously, an effect mediated by neutrophils and macrophages. Furthermore, these cells brought out a vaccination effect and protective immunity against intracranial challenge [128,129].

D. Interleukin-3 (Multi-CSF)

The *IL-3* gene is located within a cluster of cytokine genes on human chromosome 5q23–31. It encodes a 17 KD precursor protein from which a 19 amino acid signal peptide is cleaved to give mature IL-3. The human interleukin-3 receptor is a heterodimer of a specific IL-3R α chain and a β chain common to the IL-3, IL-5, and GM-CSF receptors (Figure 10.2). Signal transduction is mediated by a ligand binding-dependent conformational change resulting in the activation of JAK1 and JAK2 tyrosine kinases, leading to RAS/MAP kinase cascade activation and subsequent c-fos induction.

The growth promoting activity of IL-3 on brain microglial cells has been confirmed both *in vitro* and *in vivo*. Its presence was identified in the brain and even in cultured brain cells. Isolated rat microglia express IL-3 mRNA and synthesize the IL-3 polypeptide. RT/PCR studies also found IL-3 mRNA in C6 glioblastoma, in rat brain aggregate cultures, and in newborn and adult rat brain [130]. IL-3 has been evaluated in cytokine-assisted tumor vaccines to examine what cytokines would be most efficacious against tumors within the CNS. Irradiated B16 murine melanoma cells that produces murine IL-2, IL-3, IL-4, IL-6, IFNγ, or GM-CSF were used as subcutaneous vaccines against melanoma implanted in the brain. Results showed that cells producing IL-3, IL-6, or GM-CSF prolonged the survival of mice when challenged with viable B16 cells in the brain as compared to the controls [131].

E. Stem Cell Factor (SCF)

SCF, a hematopoietic growth factor, is the ligand of the tyrosine kinase receptor encoded by the c-kit proto-oncogene. SCF is an essential haemopoietic progenitor cell growth factor with both proliferative and antiapoptotic functions. Molecular biologists have now dissected some of the various pathways through which this cytokine signals to the nucleus. The capacity of SCF to synergize with other cytokines has been exploited for therapeutic promise. SCF is also produced by neurons and its receptor (c-kit receptor), encoded by the proto-oncogene c-kit, is expressed in microglia, suggesting a possibility that SCF/c-kitR signaling is involved in neuron–microglia interactions. SCF supports the microglial cells, in their culture, process-bearing morphology, and inhibits proliferation induced by CSF-1. SCF enhances microglial expression of the mRNAs of different growth factors like, nerve growth factor, brain-derived neurotrophic factor and ciliary neurotrophic factor, and down-regulates inflammation-associated cytokines, TNFα and IL-1β. The control of c-Kit expression, its isoforms, functions and the nature of the biological effects elicited by this receptor–ligand pair have been reviewed [132]. The local production of SCF may be an important mechanism for regulating proliferation, differentiation, and migration of various cells bearing c-kit receptors, and might be susceptible to the cytokines in inflammation and tissue repair. Different aspects of cellular transformation and progression are regulated by the autocrine stimulation of growth factor receptors by autonomously generated ligands.

In different types of tumors, including gliomas, multiple autocrine systems are activated and may exert several functions in the process of malignant transformation. SCF and its receptor have been shown to be expressed in brain tumors. The expression of SCF and its receptor c-Kit were found in 20 human malignant glioma cell lines examined. These results suggest that SCF and c-Kit can mediate nonproliferative signals or may contribute to the downstream signaling mechanisms for autocrine growth regulation of glioma cells [95,133]. Although c-kit and SCF may have limited contribution in the transformation and progression of glial tumors they may have secondary role in glioma biology. It can be hypothesized that the simultaneous activation of different autocrine systems including those that have found less attention in gliomas, like c-kit/SCF, could compromise the therapeutic outcome targeting possibly other autocrine loops [133].

V. CHEMOKINES

Chemokines are a superfamily of over 50 small cytokines (8–10 KD in size) that are central to the mediation of inflammatory responses as they directionally attract and activate leukocytes through chemokine gradients. Moreover, chemokines function in tumor cells to drive migration, invasion and survival responses. The members of the chemokine superfamily share structural homology and conserved cysteines in their amino terminus. The latter feature allows subdivision of the superfamily into four families: CXC, CC, C and CX3C, depending on the location and number of conserved cysteines. These chemokines bind to over 18 seven-transmembrane chemokine receptors, most of which belong to the G-protein coupled receptor family. Signaling through the chemokine receptor is not confined to the activation of the α, β and γ subunits of the heterotrimeric G protein complex, and involves other downstream effectors including but not restricted to PI-3 kinase/AKT, ERK and PKC (Figure 10.2). The complexity of chemokine signaling is further increased by the ability of these receptors to bind multiple chemokines. The majority of chemokines are part of the CXC family in which the first two cysteines are separated by one amino acid or the CC family where the first two cysteines are adjacent. The CXC chemokines attract mainly neutrophils, and are encoded by human chromosome 4. They include among others CXCL8 (IL-8), CXCL1-3 (Gro alpha, beta, gamma), neutrophil activating protein (ENA78), granulocyte chemotactic protein (GCP-2), stromal cell-derived factor-1 (CXCL12 or SDF-1), and platelet factor 4 (CXCL4 or PF4). CXC chemokines transduce their biological signals by binding to five different specific G protein-coupled CXC chemokine receptors (CXCR1-5). The CC chemokines attract mainly monocytes, but also eosinophils, basophils and some T-cell subsets, and are encoded by human chromosome 17. The prototype member of the CC family is monocyte chemoattractant protein 1 (MCP-1, now renamed CCL2). CC chemokines bind 10 specific receptors abbreviated CCR1-10. CCR6 and CCR9 are the only receptors having a single ligand, all others are shared.

The function of chemokines in cancer is emerging and early studies indicate that the different functions identified to date can both promote and inhibit cancer malignancy depending on tumor type [134–136]. Chemokines can have cancer cell intrinsic effects through autocrine growth and survival pathways activation. They also play major roles in host–tumor cell interactions by attracting leukocyte infiltrates, influencing angiogenesis, mediating immune evasion, enabling tumor cell motility and defining the site-specific spread of metastatic cancer cells. We will focus our discussion on four chemokines that have best been studied in brain tumors: CXCL8, CXCL4, CXCL12, and CCL2.

A. INTERLEUKIN-8 (IL-8, CXCL8)

The IL-8 gene encodes a primary peptide of 99 amino acids that matures through signal sequence cleavage and N-terminal proteolysis into several active IL-8 molecules; the 72 and 77 amino acid forms are the most biologically relevant. IL-8 is the major chemoattractant for neutrophils; it mediates changes in neutrophil shape, transendothelial migration, induces exocytosis of granule enzymes such as elastase and myeloperoxidase and a respiratory burst where hydrogen peroxide and superoxide radical are produced.

Two homologous seven transmembrane domain G protein-coupled receptors bind IL-8 with high affinity and mediate IL-8 biological responses. CXCR1 (IL-8RA, 67 KD) binds only IL-8 with high affinity, whereas CXCR2 (IL-8RB, 59 KD) binds IL-8, hGROα/MGSA, hGROβhGROγ, ENA78, GCP2, and NAP2 [137]. A third receptor, Duffy Antigen receptor for cytokines (DARC), binds a gamut of CXC and CC cytokines, including IL-8. This promiscuous receptor does not appear to elicit any intracellular signaling. It is currently debated whether it functions on endothelial cells as a molecular sink for binding and degrading of chemokines through receptor internalization, acts as a dampening chemokine reservoir for the plasma or has a chemokine transport function [138]. IL-8 shows further low affinity binding to cell surface glycosaminoglycans, although it is unclear at this stage whether this has any functional significance.

IL-8 expression is strongly upregulated by pro-inflammatory signals such as IL-1, TNFα, and IL-17 [93]. Other inducers include LPS, phytohemagglutinin (PHA), and phorbol 12-myristate 13-acetate (PMA). Inhibitory signals include anti-inflammatory cytokines such as IFNγ, IL-4, and IL-10 and glucocorticoids (dexamethasone). IL-8 upregulation is mediated at both transcriptional and posttranscriptional levels. The IL-8 gene promoter is believed to span the −1 to −2500 5' region. Within the first 500 bp, binding sites for transcription factors including AP-1, NF-kB, C-EBP/NF-IL-6, glucocorticoid receptor, hepatocyte nuclear factor-1, and interferon regulatory factor-1 have been found [139]. In many tumor cell lines, IL-8 expression is constitutively activated, although the mechanisms have only been partially elucidated [140]. A number of signaling cascades can participate in IL-8 gene activation. These include Ras, MAP kinase kinase 7 (MKK7) and NF-kB-inducing kinase (NIK) that activate the stress-activated protein kinase/c-jun N-terminal kinase (SAPK/JNK). Another interesting mechanism of IL-8 upregulation is through activation of the Fas or DR5 death receptors in apoptosis-resistant glioma cells, a process that involves ERK 1/2 and p38 MAPK pathways [141]. IL-8 is also regulated at the RNA level. The IL-8 mRNA contains an AUUUA instability element in the 3' untranslated region that is responsible for IL-8's short half-life. MAP kinase kinase kinase (MEKK1) and p38 MAP kinase can regulate IL-8 mRNA stability [142].

1. IL-8 Expression in CNS Tumors

It has been known since the early 1990s that human glioma cell lines and primary astrocyte cultures produce large amounts of IL-8 after IL-1β and TNFα stimulation. IL-8 mRNA and protein are also found in brain tumors and derived cyst fluid, as well as patient CSF. IL-8 production is not restricted to glial tumors as it was also detected in the CSF of metastatic CNS lymphomas, perhaps in relation to the associated gliosis [1,143].

The global mechanism by which IL-8 secretion is induced in human gliomas has been partly elucidated. In glioma, IL-8 appears to derive from two main sources: tumor cells and stromal cells such as macrophages/microglial cells [98]. The increase in IL-8 expression could occur during tumor progression by both direct mechanisms as the result of a change in the genetic program of the cells, and indirect mechanisms that are secondary to changes in tissue physiology in the tumor (hypoxia) or infiltration of stromal elements (leukocytes). Whether IL-8 expression is directly linked to any genetic defects occurring during glioma formation has not been examined. It will be important to examine whether *PTEN* gene loss affects IL-8 expression in glioma as Ras activation was recently shown to activate IL-8 [140]. Physiological changes occurring in the tumor appear to also play a significant role in the induction of IL-8 in glioblastoma. We found elevated levels of IL-8 in pseudopalisading cells surrounding necrosis, suggesting that this particular tumor environment with low oxygen tension induces IL-8 production. *In vitro* results confirmed that IL-8 mRNA was induced by anoxic culture conditions in glioblastoma cell lines [143]. Finally, an antitumoral inflammatory response similar to that observed in CNS infections can induce the secretion of cytokines such as IL-1 or TNF that then can trigger IL-8 secretion. In gliomas, this appears to be mainly mediated by tumor infiltrating macrophages.

2. IL-8 Receptor Expression in CNS Tumors

High affinity IL-8 receptors have not been detected on glioma cells in culture using ^{125}I-IL-8 binding studies [98], suggesting that IL-8 is not an autocrine growth factor. *Ex vivo* glioblastoma samples show CXCR1 and CXCR2 transcripts by RT/PCR (144) and *in situ* hybridization studies [145]. Immunohistochemistry with antibodies specific for CXCR1 and CXCR2 demonstrated receptor expression on infiltrating leukocytes around the microvasculature [145], supporting paracrine effects for IL-8 in glioma. These cells appeared to be infiltrating T-lymphocytes. Although IL-8 is primarily a chemoattractant for neutrophils (at 10–100 ng/ml), these cells are rare in glial tumors [98]. A possible reason for this is the absence of the homing factor E-selectin [146]. Lymphocyte infiltrates are frequent in gliomas, but the observed *in vitro* T-lymphocyte chemoattractant property of IL-8 (at 0.1-1 ng/ml) [147] was not clearly confirmed *in vivo* [137].

3. Function of IL-8 in Brain Tumors

Much of the ongoing research efforts on IL-8 in CNS tumors have tried to establish its precise role in tumor formation or progression. Beside its initial roles in regulating inflammatory processes, IL-8 was also shown to be a proangiogenic factor in 1992 [26], when it was shown to be expressed in glial tumors [35]. This prompted investigation into IL-8's role as a proangiogenic factor in glioma., and it was shown that supernatant derived from glioma cells induced tubular morphogenesis by human microvascular endothelial cells, an effect involving either IL-8 or VEGF, depending on the cell line [148]. Further work showed that IL-8 is upregulated by anoxia [145] and that this is mediated at the transcriptional level, principally through AP-1 transcription complex activation [143]. Most recently, new studies using the growth of human glioblastoma cell lines in cranial open chamber windows allowed for a direct analysis of the growing tumor neovasculature upon IL-8 modulation by siRNA or inhibition of NF-kB signaling using ING4, a negative regulator of the p65 subunit of NF-kB [149]. This study clearly demonstrated that IL-8 modulation affects glioma angiogenesis. Further studies are necessary to establish the precise mechanisms through which IL-8 influences angiogenesis in gliomas as CXCR1 and 2 were not found to be expressed on tumor vasculature by immunohistochemistry. The third chemokine receptor, DARC, was found on glioma endothelium, but its role in glioma is currently not well defined.

4. IL-8 in CNS Tumor Therapy

With the recent functional demonstration that IL-8 can contribute to angiogenesis in mouse models of vascular development in human glioblastoma [149], combined with the findings of significant IL-8 expression in human specimens [145], the design of IL-8 targeted therapeutic interventions is warranted. This is compounded by the fact that irradiation can augment IL-8 production *in vitro* [150,151]. Anti-IL-8 treatments should be best used in combination with other antiangiogenic strategies (anti-VEGF, MMP) so that the tumor cannot easily switch to alternative proangiogenic means. It is important to mention in this regard that other CXC chemokines can also have angiogenic properties. It was found that all CXC chemokines containing the sequence Glu-Leu-Arg (ELF motif) before their CXC motif are proangiogenic [152]. These include CXCL8/IL-8, ENA-78, GCP2, and CXCL1-3 (GRO-α,β,γ), whereas CXC chemokines lacking the ELF motif (CXCL4/PF4, Mig and γ-IP10) can be antiangiogenic. These findings suggest a complex regulation of angiogenesis by CXC chemokines and require further study in CNS tumors. A more in-depth discussion of IL-8 functions in gliomas is found in a recent review [153].

B. Platelet Factor 4 (CXCL4, PF-4)

CXCL4 is a unique cytokine for which no G-protein coupled transmembrane receptor has been identified to date. It was the first chemokine shown to have antiangiogenic activity. The mechanism through which CXCL4 can inhibit VEGF, IL-8 and bFGF induced proangiogenic responses of endothelial cells is unclear. It has been proposed to occur through binding to cell surface glycosaminoglycans although studies with mutant CXCL4 abrogating this binding could still inhibit angiogenesis. It was also advanced that CXCL4 may directly bind IL-8 and bFGF [154]. More recent evidence points to CXCR3-B, an isoform of CXCR3 generated by alternative CXCR3 gene splicing. This receptor was found to be expressed on microvascular cells and can mediate the antiangiogenic effects of CXCL4 as well as other antiangiogenic chemokines (CXCL9, CXCL10, CXCL11) [155].

CXCL4's ability to inhibit angiogenesis has been exploited for therapeutic purposes in brain tumors. A secreted form of CXCL4 was engineered in viral expression vectors and used for gene therapy. Prolonged survival of intracerebral gliomas was obtained using this strategy [156]. Similar antiglioma effects were obtained using small peptides derived from the minimal antiangiogenic COOH-terminal segment of CXCL4 [157].

C. STROMAL CELL-DERIVED FACTOR 1 (CXCL12, SDF1)

Perhaps one of the most interesting chemokine ligand–receptor pair for cancer involvement is CXCL12/CXCR4. The salient feature of CXCR4 is that it is found expressed in over 23 different types of tumors of different histological origin. The study of this receptor–ligand pair is greatly facilitated by the fact that CXCR4 is the only receptor CXCL12 binds to, and CXCL12 is the only ligand for CXCR4. Furthermore, the role of CXCR4 in AIDS has stimulated the development of small molecule inhibitors such as AMD3100 that can be applied to elucidating CXCR4's role in cancer. Increased CXCR4 expression in cancer can occur by different mechanisms in different cancer types [136]. In gliomas it is likely to be the result of hypoxia, through HIF-1 activation and inflammatory cytokines activating NF-kB [36,158]. The CXCR4-CXCL12 interaction in glioma cells induces activation of the ERK2 and AKT signaling pathways, and an elevation in intracellular calcium levels. Interestingly, this interaction led to the release of other chemokines (CXCL8/IL-8, CCL2/MCP-1, and CXCL10/γIP-10) [159]. CXCR4 activation has been implicated in glioma cell survival, migration and invasion [158] and is an important mechanism contributing to the survival and multiplication of metastastic tumor cells in organs releasing CXCL12 [136]. The first report clearly implicating CXCL12-CXCR4 in brain tumor growth used the AMD3100 CXCR4 antagonist to inhibit the growth of intracranial models of both glioma and medulloblastoma [160]. These results warrant further exploration in the clinic. Another area of interest that needs further investigation is the ability of CXCL12 to recruit dendritic cell precursors to the tumor and reduce dendritic cell maturation, preventing antigen presentation [161]. Furthermore, CXCR4 has recently been identified as a molecule responsible for glioma cell tracking by tumortropic Neural Stem Cells. The presence of CXCR4 on these cells is responsible for their migratory behavior. Use of these tumortropic NSCs to deliver therapeutic gene products is an attractive novel strategy for targeting residual neoplastic foci that are normally refractory to conventional treatments in glioma [162].

D. MACROPHAGE CHEMOATTRACTANT PROTEIN 1 (CCL2, MCP-1)

CCL2 is the prototype member of the CC chemokine family and is chemotactic for monocytes, basophils and memory T-lymphocytes (163). CCL2 is a potent monocyte activating and chemotactic factor, and a primary regulator of the macrophage response in solid tumors [136]. CCL2 can bind to two signaling receptors, CCR2 and CCR4, as well as two nonsignaling receptors, DARC and D6 [138].

What about CCL2 in CNS tumors? One of the histological characteristics of gliomas is the presence of leukocytic infiltrates, mostly macrophages and T lympho-cytes, although they have no clear prognostic value. The precise biologic mechanism at the origin of these infiltrates is unknown, but it likely involves the release of chemoattractants by the tumor. Supernatant from cultured human glioma cell lines contains monocyte chemotactic activity that is at least in part due to CCL2 [144]. More importantly, CCL2 has been qualitatively identified in gliomas and meningiomas and expression was found on tumor cells, tumor-associated macrophages, as well as microglia and reactive astrocytes at the tumor periphery [164]. Tumoral production of CCL2 is significant and can be detected in CSF and cyst fluid derived from patients with malignant glioma (144, 165). In an experimental intracranial glioma model (rat CNS1 cells), forced CCL2 expression increased microglial infiltration and aggressiveness of the tumor [166], although suppression of endogenous levels of CCL2 in this model did not alter macrophage infiltration, and macrophage depletion did not alter tumor growth [167]. It is arguable that if CCL2 regulates tumor leukocyte infiltration, then expression of the specific CCL2 receptors will be found expressed on tumor infiltrating macrophages and lymphocytes. In the single study that addressed this issue in glioma, CCR2 expression was found markedly upregulated on peritumoral microglia in the 9L intracranial rat glioma model [168]. Unexpectedly, this study also demonstrated that the predominant CCR2 expression was on the tumor cells. The mechanisms inducing CCL2 upregulation on the tumor have not been elucidated, but likely involve inflammatory signals such as IL-1 and TNF as well as Fas-mediated signals [141]. In neuroblastoma, CCL2 expression was found highest in tumors lacking N-myc amplification. The production of CCL2 in these tumors was responsible for infiltration of natural killer T cells expressing CCR2 [169].

Overall, the study of chemokine receptor signaling pathways in CNS tumors has already identified CXCR4 as an important target for glioma and medulloblastoma and further studies will likely yield new downstream targets for glioma therapeutics. Importantly, little is known about the expression of chemokines and their receptors on tumor cell infiltrates. Such studies will be essential to define the tumor–stroma interactions that may contribute to brain tumor growth.

VI. INTERFERONS

Interferons (IFNs) are a family of cytokines that have a broad range of activities, yet it is their unique role as antiviral defense agents that distinguish them from other cytokines. IFNs have also become very popular as therapeutic anticancer agents for patients with carcinoma, melanoma, myeloma, ovarian cancer and hairy cell leukemia. Direct IFN action occurs through the inhibition of cell replication; IFNs also act indirectly by inducing macrophage activation, stimulation of T and NK cells, or increasing MHC antigen expression. IFNs downregulate c-myc and H-RAS-1 oncogenes in certain cells, and can stimulate the differentiation of malignant cells.

A. INTERFERONS α AND β

IFNα and β belong to the type I interferons, as does the less well studied IFNω. All type I *IFN* genes are devoid of introns and probably derive from an ancestral gene by duplication. The human genome contains single copies of *IFNβ* and *IFNα(II)* (=*IFNω*) genes; and at least 14 genes and 9 pseudogenes for *IFNα(I)*, all located in a cluster on chromosome 9p22. All three type I IFNs have retained sufficient structural homology to bind to a single receptor, the IFNα/βR. IFNα binding to its receptor induces phosphorylation and activation of JAK1 and Tyk2 tyrosine kinases that phosphorylate transcription factor subunits STAT1 and STAT2, respectively (Figure 10.2). Oligomerization of STAT1 and 2 with IRF9, a DNA binding adaptor protein, forms a trimeric transcription factor called ISGF3 (IFN-stimulated gene factor 3), which migrates to the nucleus. ISGF3 binds to the DNA of IFNα/β-responsive genes, at binding sites called ISRE for IFN-stimulated response elements [170]. IFN signal transduction activates transcription of genes encoding proteins involved in the antiviral state (for example 2'–5' oligo-adenylate synthetase). Importantly, more recent results confirm a more complex network of signaling events downstream of IFN receptor activation including: (1) recruitment of adapter proteins like CBL, (2) activation of the PI-3 kinase-AKT pathway, and (3) activation of RAS/RAF/ERK signaling, suggesting additional complexity to this cytokine network.

IFNα and β secretion is the first response of the host against viral infections, probably induced by the formation of double stranded RNA in the infected cells. Growth factors and cytokines can equally induce these IFNs. Type I interferons also have immunomodulatory activities: they regulate delayed-type hypersensitivity, modulate MHC class I expression and boost NK cell activity. They prevent normal and tumor cell replication, and exhibit antitumoral activity mediated by T cells, NK cells and macrophages.

1. IFNα/β in CNS Tumors

Some glioblastoma cell lines produce an IFNβ-like activity, whereas others do not. The frequent absence of IFN expression originates from genomic deletions at the chromosomal 9p21 region in glioblastoma involving the IFNα/β cluster. The main targets of these deletions are tumor suppressor genes *p16CDKN2A* and *p15CDKN2B*, two inhibitors of cyclin-dependent kinases involved in the negative regulation of cell cycle progression. It is unclear whether the deletion of the *IFN type I* genes has any influence on tumor growth or whether it is simply a consequence of being localized close to the p16 gene [171].

The biological consequence of IFNα/β treatment of brain tumor cells in culture has been evaluated. IFNβ administration alters the growth and antigenic phenotype of glioblastoma and meningioma cells in culture. All these effects are likely mediated through IFN receptors, and IFNα/βR transcripts are found in most glioblastomas [95]. Loss of IFN α/β genes in glioma cells may also benefit the tumor in relieving the antiangiogenic constraints mediated by IFNα/β [172]. Under certain circumstances, IFNα can induce tumor formation. IFNα was found to induce medulloblastoma in mice via IFNgamma-induced production of sonic hedgehog in cerebellar granule neurons of mice [173].

2. IFNα/β in CNS Tumor Therapy

The direct growth inhibitory effects of IFN *in vitro* encouraged human applications. Clinical trials with recombinant IFNα or IFNβ failed to demonstrate any efficacy.

Moreover, a significant number of severe side effects was noted in these clinical trials [174]. Due to the toxicity of IFN when delivered systematically, novel approaches using local delivery with gene or cell therapy have been attempted. These studies did not try to directly inhibit tumor cell growth but rather aimed at improving an antitumor immune response. It was demonstrated that intratumoral cationic liposome delivery of IFNβ genes elicited cellular immunity activation in preclinical models. These findings were translated into a phase I clinical trial that demonstrated safety and a temporary response in a few patients [175]. It was found that intracranial delivery of dendritic cells genetically engineered to produce IFNα augmented the antitumor efficacy of peripheral peptide-based vaccines [79,176]. It is generally agreed upon that a more detailed understanding of the signaling pathways downstream of IFN receptor stimulation in different cell types, e.g., tumor and stroma, will be needed to more optimally exploit IFNs for brain tumor therapy.

B. INTERFERON γ

IFNγ is the only class II interferon representative. It is a single gene with 4 exons and 3 introns on human chromosome 12q24.1, and encodes a protein that has no obvious evolutionary relationship with the class I IFNs. IFNγ binds to a specific receptor (IFNγR) composed of two functional subunits: an IFNγ-binding subunit (IFNγRα) and a signal transducing subunit (IFNγRβ). IFNγ-receptor binding signals through JAK1 and JAK2 tyrosine kinases, which phosphorylate the signal transducer and activator of transcription STAT1. This leads to the formation of GAF (gamma-activated transcription factor), a complex that binds directly to gamma-activated sequences (GAS) in the promoter of IFNγ-responsive genes [170]. IFNγ production is restricted to cytotoxic T cells, TH1 helper cells and NK cells. During the course of viral infection, it is T-cell sensitization to viral antigens that triggers IFNγ production. In contrast to IFNα/β, the primary role of IFNγ is to amplify the antiviral immune response, and secondarily mediate direct antiviral effects. IFNγ has unique immunomodulatory activities: (i) IFNγ inhibits the growth of TH2 cells and thus favors a TH1-mediated immune response; (ii) IFNγ augments the phagocytic activity of macrophages; (iii) IFNγ regulates MHC class II antigen expression on macrophages, T cells, B cells and tumor cells, promoting antibody formation and cytotoxic T cell development; and (iv) stimulates production of other cytokines including IL-6 and MCP-1.

1. IFNγ in CNS Tumors

IFNγ has growth inhibitory as well as proapoptotic properties on glioma cells [177]. IFNγ transcripts are detected in glioblastoma [178], likely produced by intratumoral T cells and NK cells. IFNγ may promote antigen presentation in the CNS: IFNγ stimulates glioma cell expression of MHC class I and II antigens, and of CD54/ICAM-1 and VCAM-1, the respective binding partners for T-cell homing factors LFA-1 and VLA-4. Under particular circumstances IFNγ can also be protumorogenic by inducing expression of the sonic hedgehog gene as recently demonstrated by induction of medulloblastoma in a mouse model [173].

2. IFNγ in CNS Tumor Therapy

Both the potential for direct growth interference and immunostimulation prompted the use of IFNγ in brain tumor immunotherapy. Injection of IFN in brain tumors leads to recruitment of antigen-specific T cells and increased expression of MHC class II antigens on tumor-infiltrating APCs [179,180] and class I on tumor cells [181]. However clinical trials with intravenous or intrathecal IFNγ were conducted in the 1980s and failed therapeutically, in part due to IFNγ's short half-life and poor blood brain barrier permeability. In addition, serious systemic side effects were observed [182]. Current approaches with local delivery through IFN gene therapy will likely overcome these toxicity issues so that IFN can be integrated in an immune therapy strategy that will bolster optimal antigen presentation.

VII. TRANSFORMING GROWTH FACTOR β1, β2, β3, AND TGFα

Members of the transforming growth factor β (TGFβ) are homodimeric molecules that bind to and sequentially activate a receptor serine threonine kinase, TGFR2, which phosphorylates TGFR1 (Figure 10.2). The activated TGFR1 then phosphorylates receptor associated SMADs. Once activated, SMADs accumulate in the nucleus and form transcriptional complexes with SMAD4 and different DNA binding partners, coactivators, and corepressors to activate or repress transcriptional activity of various promoters of specific target genes (Id proteins, p21/WAF1/Cip1 promoter, c-*Jun* promoter). Recently, FoxO proteins, members of the Forkhead box (Fox) family, have been identified as key cofactors of SMADs (3 and 4) in TGFβ-mediated growth arrest. Smad 3/4 and FoxO form a transcription complex that activates the *CDKN1A/p21Cip1* gene [183]. This p21 cyclin-dependent kinase inhibitor-mediated growth arrest pathway is negatively regulated by PI3 kinase/Akt and FoxG1 pathways. Other proteins involved in TGFβ signaling include an adapter protein, SARA and the SMURF proteins that are E3 ubiquitin ligases and lead to degradation of the TGFβ signaling complex via the proteosome (Figure 10.2). The TGFβ family includes an array of factors including TGFβ1-3 and bone morphogenic proteins (BMPs), which through interaction with different receptors mediate a diverse array of biological events. TGFβ1 is a prototype of a group of 35 related cytokines including TGFβ1, TGFβ2, TGFβ3, activins, nodals, and BMPs.

The role of the TGFβ family in CNS tumorigenesis is just now being addressed in the literature. Most of this analysis is focused on the prototypical members of this family, TGFβ1 and TGFβ2, with relatively minimal detail on the large number of other TGFβ related proteins or receptors. Both of the TGFs have been implicated in glioma biology. TGFβ is a pleiotropic cytokine that delivers cytostatic signals to neuronal and immune cells. TGFβ stimulation has also been linked to multiple downstream effects including: proliferation, differentiation, migration, and apoptosis, and is intimately involved in the regulation of all these processes. The role of TGFβ in human cancer has been reviewed [184].

1. TGFβ in CNS Tumors

Expression of TGFβ isoforms, and TGFβ receptors at different stages of human glioma pathogenesis has been reviewed [2]. The release of TGFβ has been shown to be isoform and glioma-specific [185]. In fact TGFβ1 has been shown to regulate the growth of astrocytes and some glioma cell lines. One study suggested that exposure of C6 glioma cells to TGFβ inhibits gap junction intercellular communications, due to an altered phosphorylation of connexin 43. This was correlated with an increased proliferative response of this cell line [186]. TGFβ mediates effects on adhesion molecules, glioma cell migration as well as invasion [187,188]. The TGFβ-inducible secretory protein beta (ig)-h3 promotes cell adhesion of human astrocytes and astrocytoma cells *in vitro* and blocking experiments suggested involvement of the α6β4 integrin. This effect appears to involve transduction of intracellular signals through the nonreceptor kinase, SRC and focal adhesion proteins, which may regulate some aspects of astrocyte response following brain injury [189]. Glioma cell migration and invasion are facilitated by TGFβ mediated upregulation of αvβ3 integrin and MMP2 and 9 expression, as well as downregulation of tissue inhibitor of metalloproteinases (TIMP)-2 expression [190,191]. TGFβ has also been correlated with the metastatic potential of the tumor cells. The expression of cytokine receptors (IL-6 receptor, TGFβ receptor and IGF receptor) were significantly increased in Jyg MC/B (brain metastasizing subline of breast cancer cell) cells. These results suggest that cytokines that are produced *in vivo* from glial cells can contribute to the development of brain metastases in a paracrine manner [192].

TGFβ may also participate in macrophage recruitment into brain tumors *in vivo*. Study of cytokine expression by the rat astrocytoma cell line, CNS-1, revealed a constitutive expression of TGFβ but not other proinflammatory cytokines [167]. The importance of antiangiogenic strategies in glioma therapy has become a new arena for cancer research. The role of angiogenic factors in glioma angiogenesis was recently reviewed and new strategies to inhibit glioma growth by application of antiangiogenic substances with a focus on TGFβ were discussed [193]. Endothelin-1 and TGFβ1 expression have been correlated with the malignancy of the tumor and tumor vascularity in gliomas. Endothelial cells within all tumors were positive for ET-1. ET-1 and TGFβ1 are present in human astrocytomas and their expression correlates with tumor vascularity and malignancy. This work suggests that both ET-1 and TGFβ1 plays crucial roles in the tumor growth and tumor induced angiogenesis in human glioma [194]. Moreover, the participation of thrombospondin-1 in the activation of TGFβ in malignant glioma cells has been reported. The results show a predominant role of TSP-1 in the activation of latent TGFβ in malignant glioma cell lines [195]. In a murine high-grade glioma model, endogenous expression of TGFβ1 inhibits both growth and tumorigenicity and increased Fas-mediated programmed cell death [196]. Neuroepithelial tumors might overcome TGFβ induced SMAD 3/4/FoxO activation of p21-mediated growth arrest. This can occur through two mechanisms: (1) PI-3 kinase/AKT signaling leads to FoxO phosphorylation and (2) overexpression of FoxG1 as a neural progenitor proliferation program. Therefore, three signaling pathways, namely, SMAD, PI-3 kinase, and FoxG1 merge on FoxO factors to contribute to glial oncogenesis and establish a signaling link between the TGFβ/SMAD and PI-3 kinase/AKT pathways [183]. In summary, TGFβ exerts a complex set of effects in cancers with an early tumor suppressive effect through growth inhibition, but later effects in cancer development that are tumorigenic, including increased tumor cell motility and invasion, induction of angiogenesis and immune suppression.

2. TGFβ in CNS Tumor Therapy

The prognostic value of plasma TGFβ in glioblastoma multiforme patients was tested but no correlation was found between TGFβ and survival, or between plasma TGFβ and the diameter of the postoperative contrast-enhancing lesion. Plasma TGFβ revealed no prognostic value, partly due to a wide range of variation in plasma levels of TGFβ in between patients [197]. Nevertheless, both early preclinical as well as clinical studies reveal a promise of anti-TGFβ strategies to treat malignant gliomas, indicating TGFβ may still be a potential therapeutic target in the field of neuro-oncology [198]. Better survival of rats with intracranial C6 gliomas by treatment with TGFβ antisense [199] support this argument. An increased understanding of the role of TGFβ2 has been helpful to develop an antisense immunotherapy targeting this cytokine [200]. These promising results may at least in part be related to antagonizing the immunosuppressive effects of TGFβ. These encompass inhibition of maturation and antigen presentation by dendritic and other APCs, impairment of T cell activation and differentiation towards effector cells and the effector function of T cells themselves (reviewed in [79]). In the field of malignant glioma biology, a role for TGFβ in growth, motility, angiogenesis, and immune escape remains to be explored for therapeutic potentials.

3. Transforming Growth Factor Alpha

The potential role of TGFα and other growth factors/their receptors (EGFR, PDGF-A, PDGF-B, PDGFRα, PDGFRβ, and bFGF) in gliomas has been reported. The role of growth factors like EGF, TGFβ, PDGF and VEGF ligands as well as receptors in the vascularization of several brain tumor types, including gliomas and meningiomas in tumor growth and progression and their potential influence on future therapeutic strategies have been reviewed extensively [201]. TGFα, a cytokine that binds to the EGF tyrosine kinase receptor (EGFR) plays an essential role in controlling human glioma cell proliferation, and may serve as a potential target for treatment of malignant brain tumors [202]. Correlation of TGFα and EGFR expression with proliferative activity in human astrocytic gliomas suggested the presence of an autocrine loop of the mitogenic pathway through the TGFα/EGFR system [203,204]. Gliomas are characterized by a deregulation of both production of growth factors as well as expression of growth factor receptors, e.g., excess production of TGFα and overexpression or constitutive activation of EGFR [204]. Glioma cell migration is brought out by a complex interaction between soluble mitogens and extracellular matrix components, and TGFs play cardinal roles in glioma migration [204,205]. TGFα and its receptor, EGFR, have been known to correlate with malignancy grades in gliomas. TGFα was found to have the second most potent chemotactic effect for glioblastoma cell lines (up to 17-fold stimulation), analyzed in Boyden chamber assays [206]. The effects of TGFα on cell growth were studied in human glioma U251 cells transfected with antisense TGFα vectors and growth inhibition was observed [207]. Downregulation of cadherin-11, a calcium dependent cell adhesion protein, has been identified as a common response of astrocytoma cells to TGFα treatment. This downregulation of cadherin-11 is a frequent molecular event observed in the neoplastic transformation of astrocytes and TGFα autocrine/paracrine loops are reported to be responsible for such events during tumor progression [208]. Coexpression of TGFα and EGFR has been reported in capillary hemangioblastomas of the central nervous system [209]. A Phase I study of the intracerebral microinfusion of a recombinant chimeric protein made up of TGFα and a mutated form of the Pseudomonas exotoxin, PE-38 (TP-38) has been published for the treatment of malignant brain tumors [210]. Following TP-38, the survival for all patients was 23 weeks, however those without radiographic evidence of residual disease (when the therapy started) survived for 31.9 weeks.

In summary, it appears that TGFβ and TGFα interactions in glioma cells are relevant to both the tumor and its microenvironment. Further analysis of TGFβ family members including BMPs, and their respective receptors will provide additional information regarding the role of this expansive signaling network in glioma progression and identify new targets for drug intervention. The development of TGFα/EGFR or TGFRβ signaling antagonists is underway and will likely be tested soon in glioma therapeutics.

VIII. SUMMARY AND CONCLUSIONS

The study of cytokine–cytokine signaling networks within CNS tumors, currently in its early stages of development, will likely provide evidence to support their role in the various biological aspects of tumor initiation or progression. Knowledge of how these different cytokines transmit signals within CNS tumors will lead to the development of novel antitumor therapeutic strategies (Figure 10.1 and Figure 10.2). Cytokines affect brain tumorigenesis via a variety of means, including a complex set of cellular interactions between tumor cells, endothelial cells, vascular smooth muscle cells, lymphocytes, reactive astrocytes, microglial cells, etc. Each cell type is a potential producer and target for the action of cytokines or chemokines. The important biological parameters associated with tumor initiation, growth, and treatment are proliferation, resistance to apoptosis, immortalization, transformation, neovascularization, invasion, resistance to immune destruction, host genetic susceptibility, and therapeutic resistance. Cytokines can modulate all of these phenomena. At the present time there is clear evidence for a general mitogenic effect of IL-6 in brain tumors, whereas the potential for other autocrine growth loops involving IL-1, TNFα, IFNβ, TGFβ, M-CSF, and IL-8 needs further investigation.

Clearly, the effects of cytokines on all the biological processes associated with CNS tumor development need to be further studied. The initial phenomenological descriptions have been made, and now the second generation of experiments will face the much more difficult task to establish the precise biological function of these players in the brain tumor system. Moreover the molecular basis for how the cytokine–cytokine receptor interaction affects the tumor and stromal signaling response is of critical importance if we are to devise effective therapies.

Recently, several laboratories have used transgenic or knockout technologies where spontaneous and predictable types of CNS tumors appear at a reasonable frequency [211–213]. Genetic crosses with transgenic mice overexpressing cytokines or cytokine knockout mice would permit one to evaluate the impact of altering cytokine expression on tumor development. The use of such animal models would of course reach far beyond CNS tumor cytokinology and be invaluable to study other aspects of brain tumor biology, including epidemiology, genetic susceptibility, tumor biology, and efficiency of all currently available treatment modalities. Ideally, the genetic changes used should be those occurring in human disease (p53 mutation, p16 gene deletion, PTEN mutations, alterations in RAS, and EGFR rearrangement and amplification for astrocytoma)—exemplified by the wide ranging transcriptional effects of PTEN and p53 affecting cytokine genes such as IL-6 or TGFβ [62,214]. Because many of these genes cause other lethal cancers, it might be necessary to direct expression only to the CNS (transgenics or tissue-specific knockout). As we go forward, it is clear that the greatest advances in brain tumor therapeutics will derive from a more complete under-standing of the fundamental signaling events within the tumor or stromal cells that generate the malignant brain tumor phenotype and block antitumor immune responses. Once sufficient information is collected on how the intracellular environment controls the response to cytokine- or chemokine-driven extracellular milieu, we will be in a position to rationally devise better glioma therapy. These will include the large number of cell signaling targeted therapeutic agents that have been developed in recent years as well as cytokine gene therapies to bolster antitumor immune responses [79].

ACKNOWLEDGMENTS

Support for the preparation of this chapter include NIH grants CA94233 (to DLD), CA 86335, and CA87830 (to EGVM) and support from the Georgia Cancer Coalition and AFLAC Cancer Center and Blood Disorders Services (to DLD) and the Musella Foundation (to EGVM).

REFERENCES

1. Van Meir, E.G., Cytokines and tumors of the central nervous system. *Glia, 15:* 264–288, 1995.
2. Van Meir, E., Cytokine Expression in Brain Tumors: its Role in Tumor Biology and Tumor-associated Immune Responses. *In:* R.R.J. Ruffolo, G.Z., Feuerstein, A.J., Hunter, G., Poste, and B.W., Metcalf (eds.), Inflammatory Cells and Mediators in CNS Diseases, pp. 169–243: Harwood Academic Publishers, 1999.
3. Stroud, R.M. and Wells, J.A., Mechanistic diversity of cytokine receptor signaling across cell membranes. *Sci STKE, 2004:* re7, 2004.
4. Ding, H., Shannon, P., Lau, N., Wu, X., Roncari, L., Baldwin, R.L., Takebayashi, H., Nagy, A., Gutmann, D.H., and Guha, A., Oligodendrogliomas result from the expression of an activated mutant epidermal growth factor receptor in a RAS transgenic mouse astrocytoma model. *Cancer Res, 63:* 1106–1113, 2003.
5. Su, J.D., Mayo, L.D., Donner, D.B., and Durden, D.L., PTEN and phosphatidylinositol 3'-kinase inhibitors up-regulate p53 and block tumor-induced angiogenesis: evidence for an effect on the tumor and endothelial compartment. *Cancer Res, 63:* 3585–3592, 2003.

6. Hood, J.D. and Cheresh, D.A., Role of integrins in cell invasion and migration. *Nat Rev Cancer*, *2:* 91–100, 2002.

7. Adjei, A.A. and Rowinsky, E.K., Novel anticancer agents in clinical development. *Cancer Biol Ther*, *2:* S5–15, 2003.

8. Bjornsti, M.A. and Houghton, P.J., Lost in translation: dysregulation of cap-dependent translation and cancer. *Cancer Cell*, *5:* 519–523, 2004.

9. Sudbeck, E.A., Liu, X.P., Narla, R.K., Mahajan, S., Ghosh, S., Mao, C., and Uckun, F.M., Structure-based design of specific inhibitors of Janus kinase 3 as apoptosis-inducing antileukemic agents. *Clin Cancer Res*, *5:* 1569–1582, 1999.

10. Wilhelm, S.M., Carter, C., Tang, L., Wilkie, D., McNabola, A., Rong, H., Chen, C., Zhang, X., Vincent, P., McHugh, M., Cao, Y., Shujath, J., Gawlak, S., Eveleigh, D., Rowley, B., Liu, L., Adnane, L., Lynch, M., Auclair, D., Taylor, I., Gedrich, R., Voznesensky, A., Riedl, B., Post, L.E., Bollag, G., and Trail, P.A., BAY 43-9006 exhibits broad spectrum oral antitumor activity and targets the RAF/MEK/ERK pathway and receptor tyrosine kinases involved in tumor progression and angiogenesis. *Cancer Res*, *64:* 7099–7109, 2004.

11. Hao, C., Parney, I.F., Roa, W.H., Turner, J., Petruk, K.C., and Ramsay, D.A., Cytokine and cytokine receptor mRNA expression in human glioblastomas: evidence of Th1, Th2 and Th3 cytokine dysregulation. *Acta Neuropathol (Berl)*, *103:* 171–178, 2002.

12. Sasaki, A., Tamura, M., Hasegawa, M., Ishiuchi, S., Hirato, J., and Nakazato, Y., Expression of interleukin-1beta mRNA and protein in human gliomas assessed by RT-PCR and immunohistochemistry. *J Neuropathol Exp Neurol*, *57:* 653–663, 1998.

13. Giometto, B., Bozza, F., Faresin, F., Alessio, L., Mingrino, S., and Tavolato, B., Immune infiltrates and cytokines in gliomas. *Acta Neurochir (Wien)*, *138:* 50–56, 1996.

14. Nagashima, G., Suzuki, R., Asai, J., and Fujimoto, T., Immunohistochemical analysis of reactive astrocytes around glioblastoma: an immunohistochemical study of post-mortem glioblastoma cases. *Clin Neurol Neurosurg*, *104:* 125–131, 2002.

15. Lichtor, T. and Libermann, T.A., Coexpression of interleukin-1 beta and interleukin-6 in human brain tumors. *Neurosurgery*, *34:* 669–672; discussion 672–663, 1994.

16. Van Meir, E., Sawamura, Y., Diserens, A.C., Hamou, M.F., and de Tribolet, N., Human glioblastoma cells release interleukin 6 *in vivo* and *in vitro*. *Cancer Res*, *50:* 6683–6688, 1990.

17. Ilyin, S.E., Gayle, D., Gonzalez-Gomez, I., Miele, M.E., and Plata-Salaman, C.R., Brain tumor development in rats is associated with changes in central nervous system cytokine and neuropeptide systems. *Brain Res Bull*, *48:* 363–373, 1999.

18. Barnum, S.R., Jones, J.L., and Benveniste, E.N., Interleukin-1 and tumor necrosis factor-mediated regulation of C3 gene expression in human astroglioma cells. *Glia*, *7:* 225–236, 1993.

19. Hodgson, D.M., Yirmiya, R., Chiappelli, F., and Taylor, A.N., Intracerebral interleukin-1beta impairs response to tumor invasion: involvement of adrenal catecholamines. *Brain Res*, *816:* 200–208, 1999.

20. Esteve, P.O., Chicoine, E., Robledo, O., Aoudjit, F., Descoteaux, A., Potworowski, E.F., and St-Pierre, Y., Protein kinase C-zeta regulates transcription of the matrix metalloproteinase-9 gene induced by IL-1 and TNF-alpha in glioma cells via NF-kappa B. *J Biol Chem*, *277:* 35150–35155, 2002.

21. Repovic, P., Fears, C.Y., Gladson, C.L., and Benveniste, E.N., Oncostatin-M induction of vascular endothelial growth factor expression in astroglioma cells. *Oncogene*, *22:* 8117–8124, 2003.

22. Jung, Y.J., Isaacs, J.S., Lee, S., Trepel, J., and Neckers, L., IL-1beta-mediated up-regulation of HIF-1alpha via an NFkappaB/COX-2 pathway identifies HIF-1 as a critical link between inflammation and oncogenesis. *Faseb J, 17:* 2115–2117, 2003.

23. Frigerio, S., Ciusani, E., Pozzi, A., Silvani, A., Salmaggi, A., and Boiardi, A., Tumor necrosis factor microsatellite polymorphisms in Italian glioblastoma patients. *Cancer Genet Cytogenet, 109:* 172–174, 1999.

24. Baud, V. and Karin, M., Signal transduction by tumor necrosis factor and its relatives. *Trends Cell Biol, 11:* 372–377, 2001.

25. Leibovich, S.J., Polverini, P.J., Shepard, H.M., Wiseman, D.M., Shively, V., and Nuseir, N., Macrophage-induced angiogenesis is mediated by tumour necrosis factor-alpha. *Nature, 329:* 630–632, 1987.

26. Koch, A.E., Polverini, P.J., Kunkel, S.L., Harlow, L.A., DiPietro, L.A., Elner, V.M., Elner, S.G., and Strieter, R.M., Interleukin-8 as a macrophage-derived mediator of angiogenesis. *Science, 258:* 1798–1801, 1992.

27. Hajjar, K.A., Hajjar, D.P., Silverstein, R.L., and Nachman, R.L., Tumor necrosis factor-mediated release of platelet-derived growth factor from cultured endothelial cells. *J Exp Med, 166:* 235–245, 1987.

28. Pandey, A., Shao, H., Marks, R.M., Polverini, P.J., and Dixit, V.M., Role of B61, the ligand for the Eck receptor tyrosine kinase, in TNF-alpha-induced angiogenesis. *Science, 268:* 567–569, 1995.

29. Klein, N.J., Rigley, K.P., and Callard, R.E., IL-4 regulates the morphology, cytoskeleton, and proliferation of human umbilical vein endothelial cells: relationship between vimentin and CD23. *Int Immunol, 5:* 293–301, 1993.

30. Ryuto, M., Ono, M., Izumi, H., Yoshida, S., Weich, H.A., Kohno, K., and Kuwano, M., Induction of vascular endothelial growth factor by tumor necrosis factor alpha in human glioma cells. Possible roles of SP-1. *J Biol Chem, 271:* 28220–28228, 1996.

31. Yamini, B., Yu, X., Gillespie, G.Y., Kufe, D.W., and Weichselbaum, R.R., Transcriptional targeting of adenovirally delivered tumor necrosis factor alpha by temozolomide in experimental glioblastoma. *Cancer Res, 64:* 6381–6384, 2004.

32. Adachi, K., Belser, P., Bender, H., Li, D., Rodeck, U., Benveniste, E.N., Woo, D., Schmiegel, W.H., and Herlyn, D., Enhancement of epidermal growth factor receptor expression on glioma cells by recombinant tumor necrosis factor alpha. *Cancer Immunol Immunother, 34:* 370–376, 1992.

33. Chicoine, M.R., Madsen, C.L., and Silbergeld, D.L., Modification of human glioma locomotion in vitro by cytokines EGF, bFGF, PDGFbb, NGF, and TNF alpha. *Neurosurgery, 36:* 1165-1170; discussion 1170–1161, 1995.

34. Roessler, K., Suchanek, G., Breitschopf, H., Kitz, K., Matula, C., Lassmann, H., and Koos, W.T., Detection of tumor necrosis factor-alpha protein and messenger RNA in human glial brain tumors: comparison of immunohistochemistry with *in situ* hybridization using molecular probes. *J Neurosurg, 83:* 291–297, 1995.

35. Frei, K., Piani, D., Malipiero, U.V., Van Meir, E., de Tribolet, N., and Fontana, A., Granulocyte-macrophage colony-stimulating factor (GM-CSF) production by glioblastoma cells. Despite the presence of inducing signals GM-CSF is not expressed *in vivo. J Immunol, 148:* 3140–3146, 1992.

36. Ammirato, M., Rao, S., and Granger, G., Detection of TNF inhibitors (soluble receptors) in the sera and tumor cyst fluid of patients with malignant astrocytomas of the brain. *Front Biosci, 6:* B17–24, 2001.

37. Sakuma, S., Sawamura, Y., Tada, M., Aida, T., Abe, H., Suzuki, K., and Taniguchi, N., Responses of human glioblastoma cells to human natural tumor necrosis factor-alpha: susceptibility, mechanism of resistance and cytokine production studies. *J Neurooncol, 15:* 197–208, 1993.

38. Dufay, N., Reboul, A., Touraine-Moulin, F., Belin, M.F., and Giraudon, P., Soluble factors, including TNF alpha, secreted by human T cells are both cytotoxic and cytostatic for medulloblastoma cells. *J Neurooncol, 43:* 115–126, 1999.

39. Chen, T.C., Hinton, D.R., Apuzzo, M.L., and Hofman, F.M., Differential effects of tumor necrosis factor-alpha on proliferation, cell surface antigen expression, and cytokine interactions in malignant gliomas. *Neurosurgery, 32:* 85–94, 1993.

40. Yoshida, J., Takaoka, T., Mizuno, M., Momota, H., and Okada, H., Cytolysis of malignant glioma cells by lymphokine-activated killer cells combined with anti-CD3/antiglioma bifunctional antibody and tumor necrosis factor-alpha. *J Surg Oncol, 62:* 177–182, 1996.

41. Weller, M., Malipiero, U., Rensing-Ehl, A., Barr, P.J., and Fontana, A., Fas/APO-1 gene transfer for human malignant glioma. *Cancer Res, 55:* 2936–2944, 1995.

42. Walther, W., Stein, U., and Pfeil, D., Gene transfer of human TNF alpha into glioblastoma cells permits modulation of mdr1 expression and potentiation of chemosensitivity. *Int J Cancer, 61:* 832–839, 1995.

43. Qin, H., Moellinger, J.D., Wells, A., Windsor, L.J., Sun, Y., and Benveniste, E.N., Transcriptional suppression of matrix metalloproteinase-2 gene expression in human astroglioma cells by TNF-alpha and IFN-gamma. *J Immunol, 161:* 6664–6673, 1998.

44. Han, Y., Wang, J., He, T., and Ransohoff, R.M., TNF-alpha downregulates CXCR4 expression in primary murine astrocytes. *Brain Res, 888:* 1–10, 2001.

45. Gridley, D.S., Li, J., Kajioka, E.H., Andres, M.L., Moyers, M.F., and Slater, J.M., Combination of pGL1-TNF-alpha gene and radiation (proton and gamma-ray) therapy against brain tumor. *Anticancer Res, 20:* 4195–4203, 2000.

46. Kido, G., Wright, J.L., and Merchant, R.E., Acute effects of human recombinant tumor necrosis factor-alpha on the cerebral vasculature of the rat in both normal brain and in an experimental glioma model. *J Neurooncol, 10:* 95–109, 1991.

47. Niranjan, A., Moriuchi, S., Lunsford, L.D., Kondziolka, D., Flickinger, J.C., Fellows, W., Rajendiran, S., Tamura, M., Cohen, J.B., and Glorioso, J.C., Effective treatment of experimental glioblastoma by HSV vector-mediated TNF alpha and HSV-tk gene transfer in combination with radiosurgery and ganciclovir administration. *Mol Ther, 2:* 114–120, 2000.

48. Tada, M., Sawamura, Y., Sakuma, S., Suzuki, K., Ohta, H., Aida, T., and Abe, H., Cellular and cytokine responses of the human central nervous system to intracranial administration of tumor necrosis factor alpha for the treatment of malignant gliomas. *Cancer Immunol Immunother, 36:* 251–259, 1993.

49. Maruno, M., Yoshimine, T., Nakata, H., Nishioka, K., Kato, A., and Hayakawa, T., Complete regression of anaplastic astrocytoma by intravenous tumor necrosis factor-alpha (TNF alpha) after recurrence: a case report. *Surg Neurol, 41:* 482–485, 1994.

50. Fukushima, T., Yamamoto, M., Oshiro, S., Tsugu, H., Hirakawa, K., and Soma, G., Recombinant mutant human tumor necrosis factor-alpha (TNF-SAM2) immunotherapy with ranimustine chemotherapy and concurrent radiation therapy for malignant astrocytomas. *Anticancer Res, 23:* 4473–4481, 2003.

51. Heinrich, P.C., Behrmann, I., Haan, S., Hermanns, H.M., Muller-Newen, G., and Schaper, F., Principles of interleukin (IL)-6-type cytokine signalling and its regulation. *Biochem J, 374:* 1–20, 2003.

52. Trikha, M., Corringham, R., Klein, B., and Rossi, J.F., Targeted anti-interleukin-6 monoclonal antibody therapy for cancer: a review of the rationale and clinical evidence. *Clin Cancer Res, 9:* 4653–4665, 2003.

53. Liu, T.F., Cohen, K.A., Willingham, M.C., Tatter, S.B., Puri, R.K., and Frankel, A.E., Combination fusion protein therapy of refractory brain tumors: demonstration of efficacy in cell culture. *J Neurooncol, 65:* 77–85, 2003.

54. Campbell, I.L., Abraham, C.R., Masliah, E., Kemper, P., Inglis, J.D., Oldstone, M.B., and Mucke, L., Neurologic disease induced in transgenic mice by cerebral overexpression of interleukin 6. *Proc Natl Acad Sci U.S.A., 90:* 10061-10065, 1993.

55. Cohen, T., Nahari, D., Cerem, L.W., Neufeld, G., and Levi, B.Z., Interleukin 6 induces the expression of vascular endothelial growth factor. *J Biol Chem, 271:* 736–741, 1996.

56. Hatzi, E., Murphy, C., Zoephel, A., Rasmussen, H., Morbidelli, L., Ahorn, H., Kunisada, K., Tontsch, U., Klenk, M., Yamauchi-Takihara, K., Ziche, M., Rofstad, E.K., Schweigerer, L., and Fotsis, T.N, myc oncogene overexpression downregulates IL-6; evidence that IL-6 inhibits angiogenesis and suppresses neuroblastoma tumor growth. *Oncogene, 21:* 3552–3561, 2002.

57. Yamagiwa, Y., Marienfeld, C., Meng, F., Holcik, M., and Patel, T., Translational regulation of x-linked inhibitor of apoptosis protein by interleukin-6: a novel mechanism of tumor cell survival. *Cancer Res, 64:* 1293–1298, 2004.

58. Choi, C., Kutsch, O., Park, J., Zhou, T., Seol, D.W., and Benveniste, E.N., Tumor necrosis factor-related apoptosis-inducing ligand induces caspase-dependent interleukin-8 expression and apoptosis in human astroglioma cells. *Mol Cell Biol, 22:* 724–736, 2002.

59. Nabors, L.B., Gillespie, G.Y., Harkins, L., and King, P.H., HuR, a RNA stability factor, is expressed in malignant brain tumors and binds to adenine- and uridine-rich elements within the 3′ untranslated regions of cytokine and angiogenic factor mRNAs. *Cancer Res, 61:* 2154–2161, 2001.

60. Bigner, S.H., Mark, J., Burger, P.C., Mahaley, M.S., Jr., Bullard, D.E., Muhlbaier, L.H., and Bigner, D.D., Specific chromosomal abnormalities in malignant human gliomas. *Cancer Res, 48:* 405–411, 1988.

61. Tchirkov, A., Rolhion, C., Bertrand, S., Dore, J.F., Dubost, J.J., and Verrelle, P., IL-6 gene amplification and expression in human glioblastomas. *Br J Cancer, 85:* 518–522, 2001.

62. Santhanam, U., Ray, A., and Sehgal, P.B., Repression of the interleukin 6 gene promoter by p53 and the retinoblastoma susceptibility gene product. *Proc Natl Acad Sci U.S.A., 88:* 7605–7609, 1991.

63. Rolhion, C., Penault-Llorca, F., Kemeny, J.L., Lemaire, J.J., Jullien, C., Labit-Bouvier, C., Finat-Duclos, F., and Verrelle, P., Interleukin-6 overexpression as a marker of malignancy in human gliomas. *J Neurosurg, 94:* 97–101, 2001.

64. Weissenberger, J., Loeffler, S., Kappeler, A., Kopf, M., Lukes, A., Afanasieva, T.A., Aguzzi, A., and Weis, J., IL-6 is required for glioma development in a mouse model. *Oncogene, 23:* 3308–3316, 2004.

65. Goswami, S., Gupta, A., and Sharma, S.K., Interleukin-6-mediated autocrine growth promotion in human glioblastoma multiforme cell line U87MG. *J Neurochem, 71:* 1837–1845, 1998.

66. Schaefer, L.K., Menter, D.G., and Schaefer, T.S., Activation of stat3 and stat1 DNA binding and transcriptional activity in human brain tumour cell lines by gp130 cytokines. *Cell Signal, 12:* 143–151, 2000.

67. Brunello, A.G., Weissenberger, J., Kappeler, A., Vallan, C., Peters, M., Rose-John, S., and Weis, J., Astrocytic alterations in interleukin-6/Soluble interleukin-6 receptor alpha double-transgenic mice. *Am J Pathol, 157:* 1485–1493, 2000.

68. Faruqi, T.R., Gomez, D., Bustelo, X.R., Bar-Sagi, D., and Reich, N.C., Rac1 mediates STAT3 activation by autocrine IL-6. *Proc Natl Acad Sci U.S.A., 98:* 9014–9019, 2001.

69. Graf, M.R. and Merchant, R.E., Interleukin-6 transduction of a rat T9 glioma clone results in attenuated tumorigenicity and induces glioma immunity in Fischer F344 rats. *J Neurooncol, 45:* 209–218, 1999.

70. Takanaga, H., Yoshitake, T., Hara, S., Yamasaki, C., and Kunimoto, M., cAMP-induced astrocytic differentiation of C6 glioma cells is mediated by autocrine interleukin-6. *J Biol Chem, 279:* 15441–15447, 2004.

71. Renner, U., Gloddek, J., Arzt, E., Inoue, K., and Stalla, G.K., Interleukin-6 is an autocrine growth factor for folliculostellate-like TtT/GF mouse pituitary tumor cells. *Exp Clin Endocrinol Diabetes, 105:* 345–352, 1997.

72. Sierra, A., Price, J.E., Garcia-Ramirez, M., Mendez, O., Lopez, L., and Fabra, A., Astrocyte-derived cytokines contribute to the metastatic brain specificity of breast cancer cells. *Lab Invest, 77:* 357–368, 1997.

73. Malek, T.R. and Bayer, A.L., Tolerance, not immunity, crucially depends on IL-2. *Nat Rev Immunol, 4:* 665–674, 2004.

74. Jean, W.C., Spellman, S.R., Wallenfriedman, M.A., Flores, C.T., Kurtz, B.P., Hall, W.A., and Low, W.C., Effects of combined granulocyte-macrophage colony-stimulating factor (GM-CSF), interleukin-2, and interleukin-12 based immunotherapy against intracranial glioma in the rat., *J Neurooncol, 66:* 39–49, 2004.

75. Ishikawa, E., Tsuboi, K., Takano, S., Uchimura, E., Nose, T., and Ohno, T., Intratumoral injection of IL-2-activated NK cells enhances the antitumor effect of intradermally injected paraformaldehyde-fixed tumor vaccine in a rat intracranial brain tumor model. *Cancer Sci, 95:* 98–103, 2004.

76. Rhines, L.D., Sampath, P., DiMeco, F., Lawson, H.C., Tyler, B.M., Hanes, J., Olivi, A., and Brem, H., Local immunotherapy with interleukin-2 delivered from biodegradable polymer microspheres combined with interstitial chemotherapy: a novel treatment for experimental malignant glioma. *Neurosurgery, 52:* 872–879; discussion 879–880, 2003.

77. Glick, R.P., Lichtor, T., Panchal, R., Mahendra, A., and Cohen, E.P., Treatment with allogeneic interleukin-2 secreting fibroblasts protects against the development of malignant brain tumors. *J Neurooncol, 64:* 139–146, 2003.

78. Guirguis, L.M., Yang, J.C., White, D.E., Steinberg, S.M., Liewehr, D.J., Rosenberg, S.A., and Schwartzentruber, D.J., Safety and efficacy of high-dose interleukin-2 therapy in patients with brain metastases. *J Immunother, 25:* 82–87, 2002.

79. Okada, H. and Pollack, I.F., Cytokine gene therapy for malignant glioma. *Expert Opin Biol Ther, 4:* 1609–1620, 2004.

80. Ren, H., Boulikas, T., Lundstrom, K., Soling, A., Warnke, P.C., and Rainov, N.G., Immunogene therapy of recurrent glioblastoma multiforme with a liposomally encapsulated replication-incompetent Semliki forest virus vector carrying the human interleukin-12 gene—a phase I/II clinical protocol. *J Neurooncol, 64:* 147–154, 2003.

81. Fujimiya, Y., Suzuki, Y., Katakura, R., Miyagi, T., Yamaguchi, T., Yoshimoto, T., and Ebina, T., *In vitro* interleukin 12 activation of peripheral blood CD3(+)CD56(+) and CD3(+)CD56(-) gammadelta T cells from glioblastoma patients. *Clin Cancer Res, 3:* 633–643, 1997.

82. Kito, T., Kuroda, E., Yokota, A., and Yamashita, U., Cytotoxicity in glioma cells due to interleukin-12 and interleukin-18-stimulated macrophages mediated by interferon-gamma-regulated nitric oxide. *J Neurosurg*, 98: 385–392, 2003.

83. Liu, Y., Ehtesham, M., Samoto, K., Wheeler, C.J., Thompson, R.C., Villarreal, L.P., Black, K.L., and Yu, J.S., *In situ* adenoviral interleukin 12 gene transfer confers potent and long-lasting cytotoxic immunity in glioma. *Cancer Gene Ther*, 9: 9–15, 2002.

84. Yamanaka, R., Tsuchiya, N., Yajima, N., Honma, J., Hasegawa, H., Tanaka, R., Ramsey, J., Blaese, R.M., and Xanthopoulos, K.G., Induction of an antitumor immunological response by an intratumoral injection of dendritic cells pulsed with genetically engineered Semliki Forest virus to produce interleukin-18 combined with the systemic administration of interleukin-12. *J Neurosurg*, 99: 746–753, 2003.

85. Salmaggi, A., Eoli, M., Frigerio, S., Silvani, A., Gelati, M., Corsini, E., Broggi, G., and Boiardi, A., Intracavitary VEGF, bFGF, IL-8, IL-12 levels in primary and recurrent malignant glioma. *J Neurooncol*, 62: 297–303, 2003.

86. Roy, E.J., Gawlick, U., Orr, B.A., Rund, L.A., Webb, A.G., and Kranz, D.M., IL-12 treatment of endogenously arising murine brain tumors. *J Immunol*, 165: 7293–7299, 2000.

87. Ehtesham, M., Kabos, P., Kabosova, A., Neuman, T., Black, K.L., and Yu, J.S., The use of interleukin 12-secreting neural stem cells for the treatment of intracranial glioma. *Cancer Res*, 62: 5657–5663, 2002.

88. DiMeco, F., Rhines, L.D., Hanes, J., Tyler, B.M., Brat, D., Torchiana, E., Guarnieri, M., Colombo, M.P., Pardoll, D.M., Finocchiaro, G., Brem, H., and Olivi, A., Paracrine delivery of IL-12 against intracranial 9L gliosarcoma in rats. *J Neurosurg*, 92: 419–427, 2000.

89. Chen, B., Timiryasova, T.M., Haghighat, P., Andres, M.L., Kajioka, E.H., Dutta-Roy, R., Gridley, D.S., and Fodor, I., Low-dose vaccinia virus-mediated cytokine gene therapy of glioma. *J Immunother*, 24: 46–57, 2001.

90. Desaknai, S., Lumniczky, K., Hidvegi, E.J., Hamada, H., and Safrany, G., Brain tumor treatment with IL-2 and IL-12 producing autologous cancer cell vaccines. *Adv Exp Med Biol*, 495: 369–372, 2001.

91. Pestka, S., Krause, C.D., Sarkar, D., Walter, M.R., Shi, Y., and Fisher, P.B., Interleukin-10 and related cytokines and receptors. *Annu Rev Immunol*, 22: 929–979, 2004.

92. Yacoub, A., Mitchell, C., Hong, Y., Gopalkrishnan, R.V., Su, Z.Z., Gupta, P., Sauane, M., Lebedeva, I.V., Curiel, D.T., Mahasreshti, P.J., Rosenfeld, M.R., Broaddus, W.C., James, C.D., Grant, S., Fisher, P.B., and Dent, P., MDA-7 Regulates Cell Growth and Radiosensitivity *In Vitro* of Primary (Non-Established) Human Glioma Cells. *Cancer Biol Ther*, 3, 2004.

93. Kehlen, A., Thiele, K., Riemann, D., Rainov, N., and Langner, J., Interleukin-17 stimulates the expression of IkappaB alpha mRNA and the secretion of IL-6 and IL-8 in glioblastoma cell lines. *J Neuroimmunol*, 101: 1–6, 1999.

94. Miljkovic, D. and Trajkovic, V., Inducible nitric oxide synthase activation by interleukin-17. *Cytokine Growth Factor Rev*, 15: 21–32, 2004.

95. Tada, M., Diserens, A.C., Desbaillets, I., and de Tribolet, N., Analysis of cytokine receptor messenger RNA expression in human glioblastoma cells and normal astrocytes by reverse-transcription polymerase chain reaction. *J Neurosurg*, 80: 1063–1073, 1994.

96. Ilyin, S.E., Gonzalez-Gomez, I., Romanovicht, A., Gayle, D., Gilles, F.H., and Plata-Salaman, C.R., Autoregulation of the interleukin-1 system and cytokine-cytokine interactions in primary human astrocytoma cells. *Brain Res Bull*, 51: 29–34, 2000.

97. Leonard, W.J., STATs and cytokine specificity. *Nat Med*, 2: 968–969, 1996.

98. Van Meir, E., Ceska, M., Effenberger, F., Walz, A., Grouzmann, E., Desbaillets, I., Frei, K., Fontana, A., and de Tribolet, N., Interleukin-8 is produced in neoplastic and infectious diseases of the human central nervous system. *Cancer Res, 52:* 4297–4305, 1992.

99. Merlo, A., Juretic, A., Zuber, M., Filgueira, L., Luscher, U., Caetano, V., Ulrich, J., Gratzl, O., Heberer, M., and Spagnoli, G.C., Cytokine gene expression in primary brain tumours, metastases and meningiomas suggests specific transcription patterns. *Eur J Cancer, 29A:* 2118–2125, 1993.

100. Joshi, B.H., Leland, P., Asher, A., Prayson, R.A., Varricchio, F., and Puri, R.K., *In situ* expression of interleukin-4 (IL-4) receptors in human brain tumors and cytotoxicity of a recombinant IL-4 cytotoxin in primary glioblastoma cell cultures. *Cancer Res, 61:* 8058–8061, 2001.

101. Kawakami, M., Kawakami, K., and Puri, R.K., Interleukin-4-Pseudomonas exotoxin chimeric fusion protein for malignant glioma therapy. *J Neurooncol, 65:* 15–25, 2003.

102. Rainov, N.G. and Heidecke, V., Long term survival in a patient with recurrent malignant glioma treated with intratumoral infusion of an IL4-targeted toxin (NBI-3001). *J Neurooncol, 66:* 197–201, 2004.

103. Weilemann, F., Steinmetz, A., Kirsch, M., Buttler, A., Kunze, S., Kuhlisch, E., Schackert, H.K., and Schackert, G., Prevention of brain metastasis formation by local expression of interleukin-4 or hemagglutinin antigen. *Zentralbl Neurochir, 64:* 65–70, 2003.

104. Benedetti, S., Bruzzone, M.G., Pollo, B., DiMeco, F., Magrassi, L., Pirola, B., Cirenei, N., Colombo, M.P., and Finocchiaro, G., Eradication of rat malignant gliomas by retroviral-mediated, *in vivo* delivery of the interleukin 4 gene. *Cancer Res, 59:* 645–652, 1999.

105. Yoshikawa, K., Kajiwara, K., Ideguchi, M., Uchida, T., and Ito, H., Immune gene therapy of experimental mouse brain tumor with adenovirus-mediated gene transfer of murine interleukin-4. *Cancer Immunol Immunother, 49:* 23–33, 2000.

106. Benedetti, S., Pirola, B., Pollo, B., Magrassi, L., Bruzzone, M.G., Rigamonti, D., Galli, R., Selleri, S., Di Meco, F., De Fraja, C., Vescovi, A., Cattaneo, E., and Finocchiaro, G., Gene therapy of experimental brain tumors using neural progenitor cells. *Nat Med, 6:* 447–450, 2000.

107. Segal, B.M., Glass, D.D., and Shevach, E.M., Cutting Edge: IL-10-producing CD4+ T cells mediate tumor rejection. *J Immunol, 168:* 1–4, 2002.

108. Yang, B.C., Lin, H.K., Hor, W.S., Hwang, J.Y., Lin, Y.P., Liu, M.Y., and Wang, Y.J., Mediation of enhanced transcription of the IL-10 gene in T cells, upon contact with human glioma cells, by Fas signaling through a protein kinase A-independent pathway. *J Immunol, 171:* 3947–3954, 2003.

109. Nitta, T., Cytokine gene expression within the central nervous system. *Cell Mol Neurobiol, 18:* 703–708, 1998.

110. Nitta, T., Hishii, M., Sato, K., and Okumura, K., Selective expression of interleukin-10 gene within glioblastoma multiforme. *Brain Res, 649:* 122–128, 1994.

111. Huettner, C., Czub, S., Kerkau, S., Roggendorf, W., and Tonn, J.C., Interleukin 10 is expressed in human gliomas *in vivo* and increases glioma cell proliferation and motility *in vitro*. *Anticancer Res, 17:* 3217–3224, 1997.

112. Huettner, C., Paulus, W., and Roggendorf, W., Messenger RNA expression of the immunosuppressive cytokine IL-10 in human gliomas. *Am J Pathol, 146:* 317–322, 1995.

113. Terabe, M., Park, J.M., and Berzofsky, J.A., Role of IL-13 in regulation of anti-tumor immunity and tumor growth. *Cancer Immunol Immunother, 53:* 79–85, 2004.

114. Mintz, A., Gibo, D.M., Slagle-Webb, B., Christensen, N.D., and Debinski, W., IL-13Ralpha2 is a glioma-restricted receptor for interleukin-13. *Neoplasia, 4:* 388–399, 2002.

115. Maslinska, D., The cytokine network and interleukin-15 (IL-15) in brain development. *Folia Neuropathol, 39:* 43–47, 2001.

116. Weng, N.P., Liu, K., Catalfamo, M., Li, Y., and Henkart, P.A., IL-15 is a growth factor and an activator of CD8 memory T cells. *Ann N Y Acad Sci, 975:* 46–56, 2002.

117. Herrlinger, U., Aulwurm, S., Strik, H., Weit, S., Naumann, U., and Weller, M., MIP-1alpha antagonizes the effect of a GM-CSF-enhanced subcutaneous vaccine in a mouse glioma model. *J Neurooncol, 66:* 147–154, 2004.

118. Hintzen, R.Q., Voormolen, J., Sonneveld, P., and van Duinen, S.G., Glioblastoma causing granulocytosis by secretion of granulocyte-colony-stimulating factor. *Neurology, 54:* 259–261, 2000.

119. Kikuchi, T., Nakahara, S., and Abe, T., Granulocyte colony-stimulating factor (G-CSF) production by astrocytoma cells and its effect on tumor growth. *J Neurooncol, 27:* 31–38, 1996.

120. Stan, A.C., Walter, G.F., Welte, K., and Pietsch, T., Immunolocalization of granulocyte-colony-stimulating factor in human glial and primitive neuroectodermal tumors. *Int J Cancer, 57:* 306–312, 1994.

121. Dreyer, Z.E., Kadota, R.P., Stewart, C.F., Friedman, H.S., Mahoney, D.H., Kun, L.E., McCluggage, C.W., Burger, P.C., Kepner, J., and Heideman, R.L., Phase 2 study of idarubicin in pediatric brain tumors: Pediatric Oncology Group study POG 9237. *Neuro-oncol, 5:* 261–267, 2003.

122. Chen, B., Ahmed, T., Mannancheril, A., Gruber, M., and Benzil, D.L., Safety and efficacy of high-dose chemotherapy with autologous stem cell transplantation for patients with malignant astrocytomas. *Cancer, 100:* 2201–2207, 2004.

123. Perez-Martinez, A., Quintero, V., Vicent, M.G., Sevilla, J., Diaz, M.A., and Madero, L., High-dose chemotherapy with autologous stem cell rescue as first line of treatment in young children with medulloblastoma and supratentorial primitive neuroectodermal tumors. *J Neurooncol, 67:* 101–106, 2004.

124. Shaw, G. and Kamen, R.A., conserved AU sequence from the 3' untranslated region of GM-CSF mRNA mediates selective mRNA degradation. *Cell, 46:* 659–667, 1986.

125. Alterman, R.L. and Stanley, E.R., Colony stimulating factor-1 expression in human glioma. *Mol Chem Neuropathol, 21:* 177–188, 1994.

126. Ragnhammar, P., Frodin, J.E., Trotta, P.P., and Mellstedt, H., Cytotoxicity of white blood cells activated by granulocyte-colony-stimulating factor, granulocyte/macrophage-colony-stimulating factor and macrophage-colony-stimulating factor against tumor cells in the presence of various monoclonal antibodies. *Cancer Immunol Immunother, 39:* 254–262, 1994.

127. Matsuoka, T., Uozumi, T., Kurisu, K., Maeda, H., Kawamoto, K., and Monden, S., Antitumor effects of human recombinant macrophage colony-stimulating factor against rat brain tumors. *Biotherapy, 8:* 51–62, 1994.

128. Jadus, M.R., Chen, Y., Boldaji, M.T., Delgado, C., Sanchez, R., Douglass, T., Al-Atar, U., Schulz, W., Lloyd, C., and Wepsic, H.T., Human U251MG glioma cells expressing the membrane form of macrophage colony-stimulating factor (mM-CSF) are killed by human monocytes in vitro and are rejected within immunodeficient mice via paraptosis that is associated with increased expression of three different heat shock proteins. *Cancer Gene Ther, 10:* 411–420, 2003.

129. Sanchez, R., Williams, C., Daza, J.L., Dan, Q., Xu, Q., Chen, Y., Delgado, C., Arpajirakul, N., Jeffes, E.W., Kim, R.C., Douglass, T., Al Atar, U., Terry Wepsic, H., and Jadus, M.R., T9 glioma cells expressing membrane-macrophage colony stimulating factor produce CD4+ T cell-associated protective immunity against T9 intracranial gliomas and systemic immunity against different syngeneic gliomas. *Cell Immunol, 215:* 1–11, 2002.

130. Appel, K., Honegger, P., and Gebicke-Haerter, P.J., Expression of interleukin-3 and tumor necrosis factor-beta mRNAs in cultured microglia. *J Neuroimmunol, 60:* 83–91, 1995.

131. Sampson, J.H., Archer, G.E., Ashley, D.M., Fuchs, H.E., Hale, L.P., Dranoff, G., and Bigner, D.D., Subcutaneous vaccination with irradiated, cytokine-producing tumor cells stimulates CD8+ cell-mediated immunity against tumors located in the "immunologically privileged" central nervous system. *Proc Natl Acad Sci U.S.A., 93:* 10399–10404, 1996.

132. Ashman, L.K., The biology of stem cell factor and its receptor C-kit. *Int J Biochem Cell Biol, 31:* 1037–1051, 1999.

133. Hamel, W. and Westphal, M., The road less travelled: c-kit and stem cell factor. *J Neurooncol, 35:* 327–333, 1997.

134. Opdenakker, G. and Van Damme, J., The countercurrent principle in invasion and metastasis of cancer cells. Recent insights on the roles of chemokines. *Int J Dev Biol, 48:* 519–527, 2004.

135. Strieter, R.M., Belperio, J.A., Phillips, R.J., and Keane, M.P., CXC chemokines in angiogenesis of cancer. *Semin Cancer Biol, 14:* 195–200, 2004.

136. Balkwill, F., Cancer and the chemokine network. *Nat Rev Cancer, 4:* 540–550, 2004.

137. Baggiolini, M., Dewald, B., and Moser, B., Interleukin-8 and related chemotactic cytokines—CXC and CC chemokines. *Adv Immunol, 55:* 97–179, 1994.

138. Nibbs, R., Graham, G., and Rot, A., Chemokines on the move: control by the chemokine "interceptors" Duffy blood group antigen and D6. *Semin Immunol, 15:* 287–294, 2003.

139. Hoffmann, E., Dittrich-Breiholz, O., Holtmann, H., and Kracht, M., Multiple control of interleukin-8 gene expression. *J Leukoc Biol, 72:* 847–855, 2002.

140. Sparmann, A. and Bar-Sagi, D., Ras-induced interleukin-8 expression plays a critical role in tumor growth and angiogenesis. *Cancer Cell, 6:* 447–458, 2004.

141. Choi, C., Gillespie, G.Y., Van Wagoner, N.J., and Benveniste, E.N., Fas engagement increases expression of interleukin-6 in human glioma cells. *J Neurooncol, 56:* 13–19, 2002.

142. Winzen, R., Gowrishankar, G., Bollig, F., Redich, N., Resch, K., and Holtmann, H., Distinct domains of AU-rich elements exert different functions in mRNA destabilization and stabilization by p38 mitogen-activated protein kinase or HuR. *Mol Cell Biol, 24:* 4835–4847, 2004.

143. Desbaillets, I., Diserens, A.C., de Tribolet, N., Hamou, M.F., and Van Meir, E.G., Regulation of interleukin-8 expression by reduced oxygen pressure in human glioblastoma. *Oncogene, 18:* 1447–1456, 1999.

144. Desbaillets, I., Tada, M., de Tribolet, N., Diserens, A.C., Hamou, M.F., and Van Meir, E.G., Human astrocytomas and glioblastomas express monocyte chemoattractant protein-1 (MCP-1) *in vivo* and *in vitro. Int J Cancer, 58:* 240–247, 1994.

145. Desbaillets, I., Diserens, A.C., Tribolet, N., Hamou, M.F., and Van Meir, E.G., Upregulation of interleukin 8 by oxygen-deprived cells in glioblastoma suggests a role in leukocyte activation, chemotaxis, and angiogenesis. *J Exp Med, 186:* 1201–1212, 1997.

146. Moynagh, P.N., Williams, D.C., and O'Neill, L.A., Activation of NF-kappa B and induction of vascular cell adhesion molecule-1 and intracellular adhesion molecule-1 expression in human glial cells by IL-1. Modulation by antioxidants. *J Immunol, 153:* 2681–2690, 1994.

147. Larsen, C.G., Anderson, A.O., Appella, E., Oppenheim, J.J., and Matsushima, K., The neutrophil-activating protein (NAP-1) is also chemotactic for T lymphocytes. *Science, 243:* 1464–1466, 1989.

148. Wakabayashi, Y., Shono, T., Isono, M., Hori, S., Matsushima, K., Ono, M., and Kuwano, M., Dual pathways of tubular morphogenesis of vascular endothelial cells by human glioma cells: vascular endothelial growth factor/basic fibroblast growth factor and interleukin-8. *Jpn J Cancer Res, 86:* 1189–1197, 1995.

149. Garkavtsev, I., Kozin, S.V., Chernova, O., Xu, L., Winkler, F., Brown, E., Barnett, G.H., and Jain, R.K., The candidate tumour suppressor protein ING4 regulates brain tumour growth and angiogenesis. *Nature, 428:* 328–332, 2004.

150. Yamanaka, R., Tanaka, R., and Yoshida, S., Effects of irradiation on cytokine production in glioma cell lines. *Neurol Med Chir (Tokyo), 33:* 744–748, 1993.

151. Yamanaka, R., Tanaka, R., and Yoshida, S., Effects of irradiation on the expression of the adhesion molecules (NCAM, ICAM-1) by glioma cell lines. *Neurol Med Chir (Tokyo), 33:* 749–752, 1993.

152. Strieter, R.M., Chemokines: not just leukocyte chemoattractants in the promotion of cancer. *Nat Immunol, 2:* 285–286, 2001.

153. Brat, D.J., Bellail, A.C., and Van Meir, E., The Role of Interleukin-8 (IL-8, CXCL8) and its Receptors in Gliomagenesis and Tumoral Angiogenesis. *Neurooncology, In Press,* 2004.

154. Struyf, S., Burdick, M.D., Proost, P., Van Damme, J., and Strieter, R.M., Platelets release CXCL4L1, a nonallelic variant of the chemokine platelet factor-4/CXCL4 and potent inhibitor of angiogenesis. *Circ Res, 95:* 855–857, 2004.

155. Lasagni, L., Francalanci, M., Annunziato, F., Lazzeri, E., Giannini, S., Cosmi, L., Sagrinati, C., Mazzinghi, B., Orlando, C., Maggi, E., Marra, F., Romagnani, S., Serio, M., and Romagnani, P., An alternatively spliced variant of CXCR3 mediates the inhibition of endothelial cell growth induced by IP-10, Mig, and I-TAC, and acts as functional receptor for platelet factor 4. *J Exp Med, 197:* 1537–1549, 2003.

156. Tanaka, C., Kamata, H., Takeshita, H., Yagisawa, H., and Hirata, H., Redox regulation of lipopolysaccharide (LPS)-induced interleukin-8 (IL-8) gene expression mediated by NF kappa B and AP-1 in human astrocytoma U373 cells. *Biochem Biophys Res Commun, 232:* 568–573, 1997.

157. Hagedorn, M., Zilberberg, L., Wilting, J., Canron, X., Carrabba, G., Giussani, C., Pluderi, M., Bello, L., and Bikfalvi, A., Domain swapping in a COOH-terminal fragment of platelet factor 4 generates potent angiogenesis inhibitors. *Cancer Res, 62:* 6884–6890, 2002.

158. Rempel, S.A., Dudas, S., Ge, S., and Gutierrez, J.A., Identification and localization of the cytokine SDF1 and its receptor, CXC chemokine receptor 4, to regions of necrosis and angiogenesis in human glioblastoma. *Clin Cancer Res, 6:* 102–111, 2000.

159. Oh, J.W., Drabik, K., Kutsch, O., Choi, C., Tousson, A., and Benveniste, E.N., CXC chemokine receptor 4 expression and function in human astroglioma cells. *J Immunol, 166:* 2695–2704, 2001.

160. Rubin, J.B., Kung, A.L., Klein, R.S., Chan, J.A., Sun, Y., Schmidt, K., Kieran, M.W., Luster, A.D., and Segal, R.A., A small-molecule antagonist of CXCR4 inhibits intracranial growth of primary brain tumors. *Proc Natl Acad Sci U.S.A., 100:* 13513–13518, 2003.

161. Zou, W., Machelon, V., Coulomb-L'Hermin, A., Borvak, J., Nome, F., Isaeva, T., Wei, S., Krzysiek, R., Durand-Gasselin, I., Gordon, A., Pustilnik, T., Curiel, D.T., Galanaud, P., Capron, F., Emilie, D., and Curiel, T.J., Stromal-derived factor-1 in human tumors recruits and alters the function of plasmacytoid precursor dendritic cells. *Nat Med, 7:* 1339–1346, 2001.

162. Ehtesham, M., Yuan, X., Kabos, P., Chung, N.H., Liu, G., Akasaki, Y., Black, K.L., and Yu, J.S., Glioma tropic neural stem cells consist of astrocytic precursors and their migratory capacity is mediated by CXCR4. *Neoplasia, 6:* 287–293, 2004.

163. Carr, M.W., Roth, S.J., Luther, E., Rose, S.S., and Springer, T.A., Monocyte chemoattractant protein 1 acts as a T-lymphocyte chemoattractant. *Proc Natl Acad Sci U.S.A., 91:* 3652–3656, 1994.

164. Leung, S.Y., Wong, M.P., Chung, L.P., Chan, A.S., and Yuen, S.T., Monocyte chemoattractant protein-1 expression and macrophage infiltration in gliomas. *Acta Neuropathol (Berl), 93:* 518–527, 1997.

165. Kuratsu, J., Yoshizato, K., Yoshimura, T., Leonard, E.J., Takeshima, H., and Ushio, Y., Quantitative study of monocyte chemoattractant protein-1 (MCP-1) in cerebrospinal fluid and cyst fluid from patients with malignant glioma. *J Natl Cancer Inst, 85:* 1836–1839, 1993.

166. Platten, M., Kretz, A., Naumann, U., Aulwurm, S., Egashira, K., Isenmann, S., and Weller, M., Monocyte chemoattractant protein-1 increases microglial infiltration and aggressiveness of gliomas. *Ann Neurol, 54:* 388–392, 2003.

167. Kielian, T., van Rooijen, N., and Hickey, W.F., MCP-1 expression in CNS-1 astrocytoma cells: implications for macrophage infiltration into tumors *in vivo. J Neurooncol, 56:* 1–12, 2002.

168. Galasso, J.M., Stegman, L.D., Blaivas, M., Harrison, J.K., Ross, B.D., and Silverstein, F.S., Experimental gliosarcoma induces chemokine receptor expression in rat brain. *Exp Neurol, 161:* 85–95, 2000.

169. Metelitsa, L.S., Wu, H.W., Wang, H., Yang, Y., Warsi, Z., Asgharzadeh, S., Groshen, S., Wilson, S.B., and Seeger, R.C., Natural killer T cells infiltrate neuroblastomas expressing the chemokine CCL2. *J Exp Med, 199:* 1213–1221, 2004.

170. Lau, J.F. and Horvath, C.M., Mechanisms of Type I interferon cell signaling and STAT-mediated transcriptional responses. *Mt Sinai J Med, 69:* 156–168, 2002.

171. Nobori, T., Miura, K., Wu, D.J., Lois, A., Takabayashi, K., and Carson, D.A., Deletions of the cyclin-dependent kinase-4 inhibitor gene in multiple human cancers. *Nature, 368:* 753–756, 1994.

172. Sidky, Y.A. and Borden, E.C., Inhibition of angiogenesis by interferons: effects on tumor- and lymphocyte-induced vascular responses. *Cancer Res, 47:* 5155–5161, 1987.

173. Wang, J., Pham-Mitchell, N., Schindler, C., and Campbell, I.L., Dysregulated Sonic hedgehog signaling and medulloblastoma consequent to IFN-alpha-stimulated STAT2-independent production of IFN-gamma in the brain. *J Clin Invest, 112:* 535–543, 2003.

174. Mahaley, M.S., Jr., Urso, M.B., Whaley, R.A., Blue, M., Williams, T.E., Guaspari, A., and Selker, R.G., Immunobiology of primary intracranial tumors. Part 10: Therapeutic efficacy of interferon in the treatment of recurrent gliomas. *J Neurosurg, 63:* 719–725, 1985.

175. Yoshida, J., Mizuno, M., Fujii, M., Kajita, Y., Nakahara, N., Hatano, M., Saito, R., Nobayashi, M., and Wakabayashi, T., Human gene therapy for malignant gliomas (glioblastoma multiforme and anaplastic astrocytoma) by *in vivo* transduction with human interferon beta gene using cationic liposomes. *Hum Gene Ther, 15:* 77–86, 2004.

176. Okada, H., Tsugawa, T., Sato, H., Kuwashima, N., Gambotto, A., Okada, K., Dusak, J.E., Fellows-Mayle, W.K., Papworth, G.D., Watkins, S.C., Chambers, W.H., Potter, D.M., Storkus, W.J., and Pollack, I.F., Delivery of interferon-alpha transfected dendritic cells into central nervous system tumors enhances the anti-tumor efficacy of peripheral peptide-based vaccines. *Cancer Res, 64:* 5830–5838, 2004.

177. Choi, C., Jeong, E., and Benveniste, E.N., Caspase-1 mediates Fas-induced apoptosis and is up-regulated by interferon-gamma in human astrocytoma cells. *J Neurooncol, 67:* 167–176, 2004.

178. Nitta, T., Ebato, M., Sato, K., and Okumura, K., Expression of tumour necrosis factor-alpha, -beta and interferon-gamma genes within human neuroglial tumour cells and brain specimens. *Cytokine, 6:* 171–180, 1994.

179. Phillips, L.M. and Lampson, L.A., Site-specific control of T cell traffic in the brain: T cell entry to brainstem vs. hippocampus after local injection of IFN-gamma. *J Neuroimmunol, 96:* 218–227, 1999.

180. Takamura, Y., Ikeda, H., Kanaseki, T., Toyota, M., Tokino, T., Imai, K., Houkin, K., and Sato, N., Regulation of MHC class II expression in glioma cells by class II transactivator (CIITA). *Glia, 45:* 392–405, 2004.

181. Yang, I., Kremen, T.J., Giovannone, A.J., Paik, E., Odesa, S.K., Prins, R.M., and Liau, L.M., Modulation of major histocompatibility complex Class I molecules and major histocompatibility complex-bound immunogenic peptides induced by inter-feron-alpha and interferon-gamma treatment of human glioblastoma multiforme. *J Neurosurg, 100:* 310–319, 2004.

182. Mahaley, M.S., Jr., Bertsch, L., Cush, S., and Gillespie, G.Y., Systemic gamma-interferon therapy for recurrent gliomas. *J Neurosurg, 69:* 826–829, 1988.

183. Seoane, J., Le, H.V., Shen, L., Anderson, S.A., and Massague, J., Integration of Smad and forkhead pathways in the control of neuroepithelial and glioblastoma cell pro-liferation. *Cell, 117:* 211–223, 2004.

184. Gold, L.I., The role for transforming growth factor-beta (TGF-beta) in human cancer. *Crit Rev Oncog, 10:* 303–360, 1999.

185. Dhandapani, K.M., Wade, M.F., Mahesh, V.B., and Brann, D.W., Basic fibroblast growth factor induces TGF-beta release in an isoform and glioma-specific manner. *Neuroreport, 13:* 239–241, 2002.

186. Robe, P.A., Rogister, B., Merville, M.P., and Bours, V., Growth regulation of astro-cytes and C6 cells by TGFbeta1: correlation with gap junctions. *Neuroreport, 11:* 2837–2841, 2000.

187. Chen, T.C., Hinton, D.R., Yong, V.W., and Hofman, F.M., TGF-B2 and soluble p55 TNFR modulate VCAM-1 expression in glioma cells and brain derived endothelial cells. *J Neuroimmunol, 73:* 155–161, 1997.

188. Mori, T., Abe, T., Wakabayashi, Y., Hikawa, T., Matsuo, K., Yamada, Y., Kuwano, M., and Hori, S., Up-regulation of urokinase-type plasminogen activator and its receptor correlates with enhanced invasion activity of human glioma cells mediated by transforming growth factor-alpha or basic fibroblast growth factor. *J Neurooncol, 46:* 115–123, 2000.

189. Kim, M.O., Yun, S.J., Kim, I.S., Sohn, S., and Lee, E.H., Transforming growth factor-beta-inducible gene-h3 (beta(ig)-h3) promotes cell adhesion of human astro-cytoma cells *in vitro*: implication of alpha6beta4 integrin. *Neurosci Lett, 336:* 93–96, 2003.

190. Wick, W., Platten, M., and Weller, M., Glioma cell invasion: regulation of metallo-proteinase activity by TGF-beta. *J Neurooncol, 53:* 177–185, 2001.

191. Rooprai, H.K., Rucklidge, G.J., Panou, C., and Pilkington, G.J., The effects of exogenous growth factors on matrix metalloproteinase secretion by human brain tumour cells. *Br J Cancer, 82:* 52–55, 2000.

192. Nishizuka, I., Ishikawa, T., Hamaguchi, Y., Kamiyama, M., Ichikawa, Y., Kadota, K., Miki, R., Tomaru, Y., Mizuno, Y., Tominaga, N., Yano, R., Goto, H., Nitanda, H., Togo, S., Okazaki, Y., Hayashizaki, Y., and Shimada, H., Analysis of gene expression involved in brain metastasis from breast cancer using cDNA microarray. *Breast Cancer, 9:* 26–32, 2002.

193. Mentlein, R. and Held-Feindt, J., Angiogenesis factors in gliomas: a new key to tumour therapy? *Naturwissenschaften, 90:* 385–394, 2003.

194. Stiles, J.D., Ostrow, P.T., Balos, L.L., Greenberg, S.J., Plunkett, R., Grand, W., and Heffner, R.R., Jr., Correlation of endothelin-1 and transforming growth factor beta 1 with malignancy and vascularity in human gliomas. *J Neuropathol Exp Neurol, 56:* 435–439, 1997.

195. Sasaki, A., Naganuma, H., Satoh, E., Kawataki, T., Amagasaki, K., and Nukui, H., Participation of thrombospondin-1 in the activation of latent transforming growth factor-beta in malignant glioma cells. *Neurol Med Chir (Tokyo), 41:* 253–258; discussion 258–259, 2001.

196. Ashley, D.M., Kong, F.M., Bigner, D.D., and Hale, L.P., Endogenous expression of transforming growth factor beta1 inhibits growth and tumorigenicity and enhances Fas-mediated apoptosis in a murine high-grade glioma model. *Cancer Res, 58:* 302–309, 1998.

197. Hulshof, M.C., Sminia, P., Barten-Van Rijbroek, A.D., and Gonzalez Gonzalez, D., Prognostic value of plasma transforming growth factor-beta in patients with glioblastoma multiforme. *Oncol Rep, 8:* 1107–1110, 2001.

198. Rich, J.N., The role of transforming growth factor-beta in primary brain tumors. *Front Biosci, 8:* e245–260, 2003.

199. Liau, L.M., Fakhrai, H., and Black, K.L., Prolonged survival of rats with intracranial C6 gliomas by treatment with TGF-beta antisense gene. *Neurol Res, 20:* 742–747, 1998.

200. Lou, E., Oncolytic viral therapy and immunotherapy of malignant brain tumors: two potential new approaches of translational research. *Ann Med, 36:* 2–8, 2004.

201. Nieder, C., Schlegel, J., Andratschke, N., Thamm, R., Grosu, A.L., and Molls, M., The role of growth factors in central nervous system tumours. *Anticancer Res, 23:* 1681–1686, 2003.

202. Jennings, M.T., Kaariainen, I.T., Gold, L., Maciunas, R.J., and Commers, P.A., TGF beta 1 and TGF beta 2 are potential growth regulators for medulloblastomas, primitive neuroectodermal tumors, and ependymomas: evidence in support of an autocrine hypothesis. *Hum Pathol, 25:* 464–475, 1994.

203. von Bossanyi, P., Sallaba, J., Dietzmann, K., Warich-Kirches, M., and Kirches, E., Correlation of TGF-alpha and EGF-receptor expression with proliferative activity in human astrocytic gliomas. *Pathol Res Pract, 194:* 141–147, 1998.

204. Tang, P., Steck, P.A., and Yung, W.K., The autocrine loop of TGF-alpha/EGFR and brain tumors. *J Neurooncol, 35:* 303–314, 1997.

205. Chicoine, M.R. and Silbergeld, D.L., Mitogens as motogens. *J Neurooncol, 35:* 249–257, 1997.

206. Brockmann, M.A., Ulbricht, U., Gruner, K., Fillbrandt, R., Westphal, M., and Lamszus, K., Glioblastoma and cerebral microvascular endothelial cell migration in response to tumor-associated growth factors. *Neurosurgery, 52:* 1391–1399; discussion 1399, 2003.

207. Tang, P., Jasser, S.A., Sung, J.C., Shi, Y., Steck, P.A., and Yung, W.K., Transforming growth factor-alpha antisense vectors can inhibit glioma cell growth. *J Neurooncol*, *43:* 127–135, 1999.

208. Zhou, R. and Skalli, O., Identification of cadherin-11 downregulation as a common response of astrocytoma cells to transforming growth factor-alpha. *Differentiation*, *66:* 165–172, 2000.

209. Reifenberger, G., Reifenberger, J., Bilzer, T., Wechsler, W., and Collins, V.P., Coexpression of transforming growth factor-alpha and epidermal growth factor receptor in capillary hemangioblastomas of the central nervous system. *Am J Pathol*, *147:* 245–250, 1995.

210. Sampson, J.H., Akabani, G., Archer, G.E., Bigner, D.D., Berger, M.S., Friedman, A.H., Friedman, H.S., Herndon, J.E., 2nd, Kunwar, S., Marcus, S., McLendon, R.E., Paolino, A., Penne, K., Provenzale, J., Quinn, J., Reardon, D.A., Rich, J., Stenzel, T., Tourt-Uhlig, S., Wikstrand, C., Wong, T., Williams, R., Yuan, F., Zalutsky, M.R., and Pastan, I., Progress report of a Phase I study of the intracerebral microinfusion of a recombinant chimeric protein composed of transforming growth factor (TGF)-alpha and a mutated form of the Pseudomonas exotoxin termed PE-38 (TP-38) for the treatment of malignant brain tumors. *J Neurooncol*, *65:* 27–35, 2003.

211. Holland, E.C., Gliomagenesis: genetic alterations and mouse models. *Nat Rev Genet*, *2:* 120–129, 2001.

212. Weissenberger, J., Steinbach, J.P., Malin, G., Spada, S., Rulicke, T., and Aguzzi, A., Development and malignant progression of astrocytomas in GFAP-v-src transgenic mice. *Oncogene*, *14:* 2005–2013, 1997.

213. Reilly, K.M. and Jacks, T., Genetically engineered mouse models of astrocytoma: GEMs in the rough? *Semin Cancer Biol*, *11:* 177–191, 2001.

214. Mack, D.H., Vartikar, J., Pipas, J.M., and Laimins, L.A., Specific repression of TATA-mediated but not initiator-mediated transcription by wild-type p53. *Nature*, *363:* 281–283, 1993.

11 Cytokines and Defense and Pathology of the CNS

Valéry Combes and Georges E. Grau

CONTENTS

I. INTRODUCTION

Homeostasis implies constantly operational defense mechanisms, against both outside and inside aggressions. Among these, infectious agents represent a major functional compartment. During any infection, the release of a vast array of cytokines is elicited in the host. These pleiomorphic mediators are of prime importance in the defense of the host, both in the systemic circulation and at sites of tissue injury, particularly the blood–brain barrier. Here we will focus on mechanisms by which cytokines can achieve this defense of the CNS against pathogens, but we also discuss the conditions under which cytokines can trigger tissue lesions, i.e., mediate immunopathological reactions. We will address malaria parasites vs. other pathogens and the involvement of cytokines in the pathogenesis of other insults to the CNS, such as ischemia, inflammation, and oxidative stress.

II.   CYTOKINES AND INFECTIOUS AGENTS AT THE BBB LEVEL: THE CASE OF MALARIA PARASITES VS. OTHER PATHOGENS

A.   MALARIA PARASITES

Extensive studies focusing on CNS pathology have been undertaken in HIV/AIDS. Neuroscience has provided a substantial contribution to these studies. In contrast, few studies have been carried out on the neurobiology of malaria, even though this disease constitutes a major neurological problem as well as a major problem of public health. In CM, the involvement of the nervous system is dramatic, but the treatment of the neurological complications is inefficient or hampered by severe side effects [1]. Indeed, the CNS is able to modulate the cytokine production and, conversely, cytokines are produced by cells from the immune system and thus can modulate neurological responses.

In CM, CNS signs and symptoms such as seizures, raised intracranial pressure, or coma predominate in African children, whereas in adults, mutiorgan system failure is more common. Immunohistochemistry of autopsy brain tissues showed macrophage and endothelial activation and disruption of endothelial intercellular junctions in vessel sequestered parasitized red blood cells (PRBC) but no gross leakage of plasma protein. Impaired BBB function was evidenced by the detection of albumin both in plasma and cerebrospinal fluid, suggesting that BBB breakdown occurs in areas of parasite sequestration in African children [2]. Conversely, in Vietnamese adults, no evidence of BBB breakdown was found [3], suggesting that pediatric CM significantly differs from the equivalent syndrome in adults.

Several theories exist concerning the pathogenesis of CM [4]. One of them relies on the sequestration of PRBC in the brain vasculature. The different parasite molecules at the PRBC surface, as well as and their role in sequestration have been amply reviewed elsewhere [5,6]. Similarly, host molecules mediating PRBC sequestration have also been characterized (reviewed in [7]). More recently, a membrane bound form of fractalkine (FKN/CX3CL1), which is expressed on the surface of endothelial cells, has been proposed to act, via its chemokine domain, as a new potential receptor for PRBC cytoadherence [8]. These sequestered PRBCs are thought to obstruct microvessels, induce anoxia and metabolic disturbances, and lead to coma and then death. However, this succession of events by itself does not appear sufficient to explain all the features observed in CM. Another theory, which is not exclusive from the previous one, suggests that following infection, an exaggerated immune response, first protective then deleterious, could be implicated [4,9]. Moreover, in addition to vascular alterations that are observed during development of CM, morphological and functional changes have also been observed in resident cells of the CNS, notably glial cells. The degree of these changes is associated to the neurological complications observed in murine CM. It therefore seems important to take into account in the pathogenesis of CM not only the vascular side but also the implication of the CNS itself [10].

The potent anti-malarial properties of cytokines have been extensively studied, as reviewed in [11–13], but such properties are not necessarily occurring at the blood–brain barrier (BBB) level. In contrast, in malarial infections, the potential risk that cytokines produced participate in BBB damage is high, in view of intrinsic property of these mediators, as recently discussed [4,9,14,15]. Cytokines, but also several cytokine-induced molecules, can be protective against malaria. For example, in a murine model of cerebral malaria (CM), the inhibition of COX-2 (cycloxygenase 2 enzyme), which is normally up-regulated in the brain, resulted in an earlier onset of CM. Expression of heme oxygenase-1 (HO-1), an enzyme degrading heme into CO and antioxidant compounds, is considered to be a protective reaction against inflammation in the brain. Its expression was found in the brain lesion during advanced CM in Dürck's granuloma in association with monocytic cells. Although the expression of HO-1 occurs too late to have a protective effect on the development of cerebral lesions, understanding the generation of local tissue protective cellular reactions might help in the development of drugs limiting the formation of these CM lesions [16].

Among all the cytokines involved in the functioning of the CNS, tumour necrosis factor (TNF) is a major mediator involved in these communications between the CNS and the immune system, but the pathways involved are not completely understood [13,17]. Its role has been extensively studied and appears to be major in the pathogenesis of CM both in patients and in murine models [4,9]. TNF is produced by host cells in response to various malaria antigens and is responsible for the overexpression of adhesion molecules. Immunohistochemical and RT-PCR studies performed on human postmortem samples revealed TNF mRNA only in CM patients although the protein expression was not restricted to this disease and was also found in meningitis [18]. In neuropathologies including CM, multiple sclerosis (MS), and HIV-encephalitis, elevated TNF expression can be observed, albeit with different histopathological consequences, as recently discussed [15]. Using MRI, it has been shown that focal intrastriatal injection of TNF is able to reduce the cerebral volume, damage the BBB, and disrupt tissue homeostasis via an endothelin- and TNFR2-dependent pathway [19]. The requirement of the TNFR2 (and not TNFR1) in the TNF-induced neurovascular lesion of CM was also evidenced using an experimental CM model, infection by *Plasmodium berghei* ANKA (PbA). This was shown in TNFR2 knockout mice [20] and, more recently, in bone marrow chimaeras, indicating that the endothelial expression of the receptor is predominant in pathogenesis [21]. The BBB integrity is altered in CM and is associated with an increased binding of cells, including parasitized erythrocytes and monocytes, which usually do not interact with the brain microvascular endothelium. This adhesion is, at least in part, mediated by the proinflammatory-cytokine-induced ICAM-1 overexpression [22]. In humans, an important role of ICAM-1 was demonstrated through the discovery of a mutation in the *P. falciparum* binding domain of ICAM-1 (named ICAM-1 *Kilifi* mutation), which increased the susceptibility of the African population to CM [23]. In brain microvascular endothelial cells (MVEC) isolated from either CM-susceptible (CBA/J) or CM-resistant mice (BALB/c) mice, distinct differences in cytokine responsiveness was observed between the two cell lines. Indeed, CM-resistant-derived brain MVEC displayed a hypoinduction of ICAM-1 and VCAM-1 expression, as well as a reduced IL-6 production, in response to TNF stimulation, compared to CM-susceptible-derived cells [24]. This was related to the presence of a kinase pathway specific to the BALB/c strain, implying that differential reactivity to TNF reflects the genetic susceptibility to CM [24]. The relationship between altered BBB integrity and ICAM-1 up-regulation in response to TNF stimulation has also been demonstrated on various cell lines, including HUVEC and ECV304 [25]. However, although ICAM-1 appears to be important for the development of experimental CM (as evidenced by the protection against the PbA-induced mortality of ICAM-1 deficient mice and the coincidence between the increased ICAM-1 expression and the increased microvascular permeability in the brain and the lung in the ICAM-1 intact strain), the involvement of ICAM-1 in leucocyte rolling was not demonstrated [26]. Indeed, an increase in leucocyte rolling and adhesion, as assessed by brain intravital microscopy, was demonstrated in PbA-infected ICAM-1 deficient mice compared to noninfected animals, indicating that ICAM-1 is not required for leucocyte rolling and adhesion during PbA malaria and that other ligands are used [26]. Infection of mice with another lethal strain plasmodium, *Plasmodium yoelii* 17XL, also induced an up-regulation of ICAM-1 in small brain venules and capillaries and, to a lesser extent, of VCAM-1 in larger venules. In parallel, TNF-deficient mice were infected with this plasmodium strain and with its nonlethal counterpart (*P. yoelii* 17XNL). Both strains displayed higher parasitaemia than controls, but the three groups of mice were equally susceptible to death, suggesting that in this model TNF is not essential in mediating CM [27].

It is clear that high plasma TNF levels cannot be the only element responsible for the signs and symptoms unique to cerebral and lethal *P. falciparum* malaria. Among proinflammatory cytokines, the importance of lymphotoxin (LT, previously called TNF-beta) may have been underestimated, even if elevated plasma levels have been reported in CM patients [28]. Indeed, in the PbA model, it appeared that whereas TNF-deficient mice were susceptible to PbA-induced CM, LTalpha- deficient mice were resistant, suggesting that LT-alpha but not TNF, which both interact with a common receptor, TNFR2, is the principal mediator of murine CM [29]. A recent study compared, on biochemical, bioenergetical, and metabolic levels, the cerebral (PbA) and the noncerebral (PbK) strains of Plasmodium *berghei*. An increased production of hypoxia markers such as lactate and alanine accompanied by compromized brain bioenergetics, mRNA from LT-alpha, TNF and IFN-gamma upregulation was observed in the brain of PbA-infected mice. These changes were consistent with a state of cytopathic hypoxia. In contrast, these modifications were not observed in PbK-infected mice, the brain of which showed an increased metabolic rate and an up-regulation of metabolic enzymes with no deleterious effect on bioenergetics [30].

However, mediators acting downstream of TNF/LT production, such as migration inhibitory factor (MIF) and iNOS, are also of importance. A study on postmortem samples from malarial and nonmalarial patients showed, among malaria cases, either no histological changes with or without sequestered PRBC or microhaemorrhages associated with intravascular mononuclear accumulations and sequestered PRBC.

A strong labelling of iNOS was observed in the cerebral vascular wall of the latter group, whereas it was almost absent from the other groups. This iNOS overexpression was not restricted to the brain and was also found in the lung but with no differences between the groups. Conversely, MIF, often associated with iNOS, was common in the lung but was absent from brain vessels. These results emphasize the fact that in CM in African children, death occurs from overlapping syndromes acting through different organ systems, with several mechanisms, not necessarily associated with cerebral vascular inflammation and damage [31,32]. This overexpression of iNOS in the brain was also observed in the simian model of infection of *Macaca mulatta* monkeys with *Plasmodium coatneyi*. The expression of iNOS together with TNF, IFN-gamma, IL-1 beta, and ICAM-1 were highest in the cerebellum of infected animals and were correlated with the sequestration of PRBC [33]. The expression of iNOs was also observed, *in vitro*, after co-culture of brain EC with mature PRBC [34].

It is also proposed that the role of iNOS induction in CM is protective, its primary purpose being to inhibit the side effects of brain indoleamine 2,3-dioxygenase (IDO) induction and quinolinic acid accumulation during hypoxia [35]. The IDO is the first and limiting-rate enzyme in the kinurenin pathway of tryptophan metabolism. It may be induced in some infectious diseases. Its expression is primarily induced by IFN-gamma and may act as part of an initial antimicrobial defense mechanism. In the murine model of CM, IDO was studied by immunohistochemistry and also through its gene expression by laser capture microdissection. IDO expression was found in vascular endothelium and was systemic, and it was also possible to detect changes in gene expression due to malaria infection. Although high levels of IDO production may not cause pathology, it is possible that it combines with other features of CM to cause the pathology observed [36,37]. When nitrate levels, and not its inducible synthase, were determined in CSF and sera from comatose and noncomatose falciparum malaria patients, no difference was observed between the two groups of patients and no correlation could be observed with either coma depth, parasitaemia, or time of recovery, implying that these parameters do not allow differentiation between severe from nonsevere malaria[11].

In the mouse model, in addition to the expression of cytokines, the early expression chemokine expression was explored comparing CM-susceptible and CM-resistant mice. In both strains, the gene expression of IFN-inducible protein (IP)-10/CXCL10 and monocytic chemotactic protein (MCP-1)/CCL2 was induced early after infection (24 h), whereas RANTES/CCL5 was enhanced only in CM-susceptible mice. In addition, *in vitro* studies using an astrocyte cell line showed that IP-10/CXCL10 and MCP-1/CCL2 expressions could be induced by malarial antigens, suggesting a direct involvement of brain parenchymal cells in the response to the malarial infection [38].

Cytokine-inducible molecules have been studied. P-selectin-deficient mice are protected against PbA-induced CM [39,40] and P-selectin levels are increased in various organs of PbA-infected P-selectin-intact mice. However, P-selectin levels were also increased in PbA-infected CM-resistant BALB/c mice, a feature that was not evidenced in another study [40]. Altogether these data demonstrate that P-selectin is important in the development of the neurological syndrome but is not sufficient to mediate malarial pathogenesis [39]. Moreover, development by bone marrow grafting of chimeric mice deficient in either endothelial or platelet P-selectin demonstrated that only the endothelial P-selectin was important in the pathogenesis of experimental CM [40].

Finally, an additional piece of evidence for a massive endothelial activation in human CM has been provided recently. Upon activation, EC produces microparticles, a process called vesiculation. Dramatically elevated levels of CD51-positive endothelial microparticles have recently been reported in Malawian children with neurological involvement. These microparticles are not specific to cerebral endothelium, as they are likely to originate from the vesiculation throughout the circulation, but, in view of their proinflammatory and procoagulant properties, they could play a major role in CM pathogenesis [41].

B. OTHER PARASITES

Trypanosomes are the pathogens responsible for sleeping sickness, the pathogenetic mechanisms of which remain incompletely understood. An experimental rodent model of this human disease is infection by *Trypanosoma brucei brucei* (Tbb). Tbb can, directly or indirectly via the cytokine production that it induces, activate brain EC, as evidenced by ICAM-1 and VCAM-1 upregulation upon infection. Interestingly, this EC activation and passage of the pathogen into the CNS can take place, but without obvious disruption of the BBB, as assessed by tight junction preservation [42]. At various times after Tbb infection in rats, the location of the parasite in the central nervous system was examined in relation to the brain vascular endothelium, visualized with an antiglucose transporter-1 antibody. At 12 and 22 days postinfection, the large majority of parasites were confined within blood vessels. At this stage, however, some parasites were also clearly observed in the brain parenchyma. This was accompanied by an upregulation of ICAM-1/VCAM-1. At later stages, 42, 45, and 55 days postinfection, parasites could still be detected within or in association with blood vessels. In addition, the parasite was frequently found, at this time, in the brain parenchyma and the extravasation of parasites was more prominent in the white matter than the cerebral cortex. A marked penetration of parasites was seen in the septal nuclei. In spite of this, occludin and zonula occludens 1 staining of the vessels was preserved. The Tbb parasite is able to cross the blood–brain barrier *in vivo,* without a generalized loss of tight junction proteins.

In the case of infection by *Toxoplama gondii* also, numerous cytokines are induced. Among these, IFN-gamma and TNF have been extensively studied. The respective role of their cell surface receptors was analyzed in a series of experiments involving gene KO mice [43]. Brain endothelial cells of wild-type mice reacted in response to *Toxoplasma* infection with a strong upregulation of VCAM-1, ICAM-1, and major histocompatibility complex (MHC) class I and II antigens. A similar response was seen in mice genetically deficient for either TNFR1, TNFR2, or both TNFRs, whereas IFN-gammaR-deficient (IFN-gammaR0/0) mice were found to be defective in the up-regulation of these molecules. However, recruitment of leukocytes to the brain and their intracerebral movement were not impaired in IFN-gammaR0/0 mice.

Thus, IFN-gammaR, but not TNFR signalling, is the major pathway for the activation of endothelial cells and microglia in murine toxoplasmic encephalitis.

The production and release of cytokines in *Toxoplasma* infection is likely to be under the control of some Toll-like receptor (TLR). Therefore, TLR2-, TLR4-, and MyD88-deficient mice were infected with the avirulent cyst-forming Fukaya strain of *T. gondii* [44]. It was found that peritoneal macrophages from T. gondii-infected TLR2- and MyD88-deficient mice did not produce any detectable levels of NO. T. gondii loads in the brain tissues of TLR2- and MyD88-deficient mice were higher than in those of TLR4-deficient and wild-type mice. Furthermore, high levels of IFN-gamma and IL-12 were produced in peritoneal exudate cells (PEC) of TLR4-deficient and wild-type mice after infection, but low levels of cytokines were produced in PEC of TLR2- and MyD88-deficient mice. On the other hand, high levels of IL-4 and IL-10 were produced in PEC of TLR2- and MyD88-deficient mice after infection, but low levels of cytokines were produced in PEC of TLR4-deficient and wild-type mice. The most remarkable histological changes with infiltration of inflammatory cells were observed in the lungs of TLR2-deficient mice infected with *T. gondii*, where severe interstitial pneumonia occurred and abundant *T. gondii* were found.

Interestingly, a dissociation between TNF and NO can occur, as evidenced by the results of experiments conducted in TNFR1 and TNFR2 KO mice infected by T. gondii [45]. Upon i.p. inoculation with the avirulent ME49 strain, knockout mice were capable of limiting acute infection, but succumbed within 3 to 4 weeks to a fulminant necrotizing encephalitis. Receptor-deficient mice harbored higher cyst burdens and exhibited uncontrolled tachyzoite replication in the brain. The lack of TNF receptors did not adversely affect the development of a type 1 IFN-gamma response. *In vitro* studies with peritoneal macrophages stimulated with IFN-gamma and tachyzoites indicated that under limiting concentrations of IFN-gamma, nitric oxide-mediated toxoplasmastatic activity is TNF-alpha dependent. However, this requirement is overcome by increasing the dose of IFN-gamma. Furthermore, both *ex vivo* and *in vivo* studies demonstrated that inducible nitric oxide synthase induction in the peritoneal cavity and brain is unimpaired in receptor-deficient mice. Thus, TNF-dependent immune control of *T. gondii* expansion in the brain involves an effector function distinct from inducible nitric oxide synthase activation. Several fine mechanisms by which *Toxoplasma spp.* can induce cytokines have been deciphered: for instance, the IL-12 production and the brain immunopathology are mediated by the parasite-induced lipoxin A4 [46].

C. BACTERIA

1. Extracellular Bacteria

Escherichia coli, the most common Gram-negative bacterium that causes meningitis in neonates, invades human brain microvascular endothelial cells (HB-MVEC). An important invasion mechanism involves the disassembly of vascular-endothelial cadherins at tight junctions [47]. More precisely, a dissociation of beta-catenins from vascular-endothelial cadherins is induced by the pathogen, resulting in an increase in permeability of HB-MVEC and a decrease in transendothelial electrical resistance. The intracellular signalling pathways have been defined: *E. coli* rearranges host cell actin via the activation of phosphatidylinositol 3-kinase (PI3K) and PKC-alpha. The pathogen can invade brain EC via PLC-gamma1 activation in a PI3K-dependent manner, resulting in increased Ca2+ levels in HB-MVEC [48], but the possible modulation of this phenomenon by cytokines remains unclear. However, because *E. coli* internalization involves caveolae [49] and because cytokines such as TNF are able to qualitatively, if not quantitatively, alter caveolae [50], it may be hypothesized that invasion of brain EC by *E. coli* could be modulated by cytokines.

The molecular mechanisms of the *E. coli* translocation at the blood–brain barrier have been extensively reviewed [51,52]. Bacterial factors that play essential roles in brain EC invasion include K1 capsule, *ompA*, *Ibe* proteins, and *TraJ* genes. In addition, some bacteria implicate toxin production [53]. In such a setting, TNF facilitates the intracellular compartmentalization of fluorescent bacteria and leads to higher levels of apoptosis. In terms of host factors, several molecules have been ascribed a major role in bacterial invasion, such as C1qRp, CNF1, and CD46, a cell-surface protein involved in regulation of complement activation that interacts with the pili of *N. meningitidis* [54].

Even if the effects of cytokines are not yet well known, some peculiar effects of TNF have been observed in the context of bacterial invasion of EC. TNF is capable of upregulating a receptor for a toxin on HB-MVEC [55]: TNF enhances the surface expression of globotriaosylceramide (Gb3), the functional receptor for Shiga toxin, produced by some *E. coli* strains. Functional consequences of this were clear, as TNF treatment of HB-MVEC enhanced Shiga toxin binding and sensitivity to toxin. This represents an important mechanism of CNS pathology occurring in the haemolytic-uremic syndrome caused by enterohaemorragic *E. coli*.

Other toxins, such as the beta-hemolysin/cytolysin toxin (beta-h/c), produced by human meningeal pathogen group B *Streptococcus* (GBS), can profoundly affect HB-MVEC. Oligonucleotide microarray studies indicate that GBS infection induces a highly specific and coordinate set of genes including IL-8, Groalpha/CXCL1, Grobeta/CXCL2, IL-6, GM-CSF, myeloid cell leukemia sequence-1 (Mcl-1), and ICAM-1, which act to orchestrate neutrophil recruitment, activation, and enhanced survival. Cell-free bacterial supernatants containing beta-h/c activity induced IL-8 release, identifying this toxin as a principal provocative factor for BBB activation. These findings were further substantiated *in vitro* and *in vivo*. Neutrophil migration across polar HB-MVEC monolayers was stimulated by GBS and its beta-h/c through a process involving IL-8 and ICAM-1. In a murine model of hematogenous meningitis, mice infected with beta-h/c mutants exhibited lower mortality and decreased brain bacterial counts compared with mice infected with the corresponding WT GBS strains [56].

Further *in vivo* appraisal of a direct involvement of TNF in meningitis was provided by experiments conducted in the *Streptococcus pneumoniae* model [57]. After intraperitoneal infection of *S. pneumoniae* type 6, the BBB opening was increased continuously from 6 h and the mice died of septic shock within 36 h due to bacterial overgrowth. The bacteria crossed the BBB and began to deposit in brain at 6 h postinfection. There was strong staining of TNF-alpha on blood vessels of brain from 6 h to 24 h postinfection. Anti-TNF antibody blocked both the BBB opening and the entrance of circulatory *S. pneumoniae* type 6 into brain, indicating that TNF played an important role in controlling the opening of BBB. Furthermore, an adult murine model of hematogenous pneumococcal meningitis was developed that is based on opening of the BBB by TNF and controlling the degree of bacteraemia by cefazolin antibiotic. Thus, hematogenous meningitis developed as TNF initiated BBB opening, peripheral bacteria entered into the brain and formed bacterial emboli, and then progressed to meningitis.

In addition to direct invasion of brain EC by bacteria, mediated by known and as-yet-unidentified factors [58], an important mechanism of penetration of these pathogens into the CNS is the "utilization" of host leucocytes. Taking advantage of leucocyte trafficking to and through the blood–brain or blood–cerebrospinal fluid barrier is another strategy employed by some bacteria, as reviewed in [59].

Apart from potential effects on invasion, cytokines can modulate the proliferation of some bacteria. This has been shown by using HB-MVEC cultured with *S. aureus*, as an experimental model of cerebral abscesses [60]. The originality of this work was to show that the functional link between the cytokine, IFN-gamma, and bacterial proliferation is an enzyme induction in brain MVEC. IFN-gamma inhibited *S. aureus* replication by the induction of indoleamine 2,3-dioxygenase (IDO) in HB-MVEC. This activation of IDO in HB-MVEC could be shown by RT-PCR and by detection of kynurenine in culture supernatants of activated cells. Resupplementation of L-tryptophan abrogated the inhibitory effect of IFN-gamma on the growth of staphylococci, hence confirming the activation of IDO as being responsible for the induced bacteriostasis. Addition of TNF enhanced the IFN-gamma mediated antibacterial effects, whereas TNF alone had no influence on staphylococcal growth. Stimulation of HB-MVEC with IFN-gamma failed to activate inducible nitric oxide synthase (iNOS) and subsequent production of nitric oxide (NO). Thus, intra- and extracellular depletion of L-tryptophan seems to be an important process in the defense against staphylococcal brain abscesses by means of creating an unfavorable microenvironment.

In view of the wide range of cytokines and chemokines abundantly produced during the acute phase of brain abcess [61], it is likely that more complex modulations of the interactions between bacteria and HB-MVEC take place and are crucial in CNS infection. Among the molecules released upon *S. aureus* infection, macrophage-inflammatory protein (MIP)-1alpha/CCL3, MIP-1beta/CCL4, MIP-2/CXCL1, MCP-1/CCL2, and TCA-3/CCL1, were detected in the brain within 6 h [62].

Cytokine-induced CAMs are essential in the host defense against infection, as demonstrated by studies with CAM-deficient mice infected with *S. pneumoniae* [63]. All *S. pneumoniae* infected, E-, P-, or E-/P-selectin-deficient mice showed a more pronounced morbidity, a significantly higher mortality associated with persistent bacteremia, and a higher bacterial load when compared with wild-type mice. These differences were most remarkable in the E-selectin-deficient mice, which showed the highest rate of mortality and bacteraemia. No significant differences were observed among the groups in the inflammatory response present in the peritoneal cavity, brain, liver, spleen, or kidney at 48 h after inoculation. Extensive hepatic and splenic necrosis and thrombosis were noted in E- and P-selectin-deficient mice.

Although the absence of endothelial selectins did not substantially impair leucocyte emigration to sites of infection 48 h after pneumococcal sepsis, it resulted in increased mortality and a higher bacterial load in the bloodstream of selectin-deficient mice. These results demonstrate a definitive phenotypic abnormality in E-selectin-deficient mice and suggest that E- and P-selectin are important in the host defense against S. pneumoniae infection.

The functional importance of CAM is further illustrated by the demonstration of alternative bacterial strategies to invade the CNS: even when streptococci do not invade HB-MVEC, adherence is crucial. Ensuing cytotoxicity of the pathogen for BMEC can occur, even more than with invading bacteria. This appears to be mediated by a specific haemolysin from S. suis, which is important in pathogenesis [64]. However, there clearly is a hierarchy of importance among cytokines, in terms of efficacy in host defense. In pneumococcal meningitis, IL-1 but not IL-10 is crucial in host defense, as shown by in vivo experiments in IL-1R type I gene-deficient mice [65], whereas the absence of IL-10 was associated with higher cytokine and chemokine concentrations and a more pronounced infiltrate, but antibacterial defense or survival was not influenced [66].

2. Intracellular Bacteria

The facultative intracellular bacterium *Listeria monocytogenes* is an invasive pathogen that crosses the vascular endothelium and disseminates to the placenta and the central nervous system [67]. Its interaction with endothelial cells is crucial for the pathogenesis of listeriosis. By infecting *in vitro* human umbilical vein endothelial cells (HUVEC) with *L. monocytogenes*, it was found that wild-type bacteria induced the expression of the adhesion molecules (ICAM-1 and E-selectin), chemokine secretion (IL-8 and MCP-1), and NF-kappa B nuclear translocation. The activation of HUVEC required viable bacteria and was abolished in *prfA*-deficient mutants of *L. monocytogenes*, suggesting that virulence genes are associated with endothelial cell activation. Direct evidence that listeriolysin O (LLO) is involved in NF-kappa B activation in transgenic mice was provided by injecting intravenously purified LLO, thus inducing stimulation of NF-kappa B in endothelial cells of blood capillaries. Our results demonstrate that functional LLO secreted by bacteria contributes as a potent inflammatory stimulus to inducing endothelial cell activation during the infectious process.

L. monocytogenes invasion of HB-MVEC and its role as a stimulus for endothelial cell activation were studied [68]. Binding and invasion of intact HB-MVEC monolayers were independent of the *L. monocytogenes* inlAB invasion locus. Cytochalasin D abrogated invasion of BMEC, whereas genistein effected only a 53% decrease in invasion, indicating a requirement for rearrangement of actin microfilaments but less dependence on tyrosine kinase activity. *L. monocytogenes* stimulated surface expression of E-selectin, ICAM-1, and to a lesser extent, VCAM-1, whereas *L. monocytogenes prfA*- and *Deltahly* mutants were severely compromised in this respect. Other experiments showed that HB-MVEC infection stimulated monocyte and neutrophil adhesion and that CD18-mediated binding was the predominant mechanism for neutrophil adhesion to infected HB-MVEC under static conditions. These data suggest that invasion of HB-MVEC is a mechanism for triggering inflammation and leukocyte recruitment into the central nervous system during bacterial meningitis.

In the field of mycobacteria, the pathogenesis of tuberculous meningitis remains unclear, and there are few data describing the kinetics of the immune response during the course of its treatment. Blood and CSF concentrations of pro- and anti-inflammatory cytokines, TNF receptors, and MMP-9 and its tissue inhibitor were measured, and blood–brain barrier permeability was assessed by the albumin and IgG partition indices. CSF concentrations of lactate, IL-8, and IFN-gamma were high before treatment and then decreased rapidly with antituberculosis chemotherapy. However, significant immune activation and BBB dysfunction were still apparent after 60 days of treatment. Death was associated with high initial CSF concentrations of lactate, low numbers of white blood cells, in particular neutrophils, and low CSF glucose levels [69].

Aside from varied and potent effects of cytokines and chemokines, the CAMs induced also have a central role in CNS infections. Elegant *in vivo* intervention experiments have pointed toward differences between inflammatory and infectious insults to the BBB: whereas anti-JAM mAbs prevent leucocyte efflux in the brain in models using TNF or IL-1 as inducing agents, these mAbs appear to be deleterious when *L. monocytegenes* or LCMV are used as inducers [70].

D. Viruses

Viruses also use specific host molecules as entry receptors. For instance, measles virus (MV) utilizes membrane cofactor protein (CD46) and the signaling lymphocyte-activation molecule (SLAM) as receptors. [71].There is a marked expression of these molecules on EC lining blood vessels, on epithelial cells and tissue macrophages in a wide range of peripheral tissues, as well as in Langerhans' and squamous cells in the skin. Strong CD46 staining was observed on cerebral endothelium throughout the brain and also on ependymal cells lining the ventricles and choroid plexus. Comparatively weaker CD46 staining was observed on subsets of neurons and oligodendrocytes. In SSPE brain sections, the areas distant from lesion sites and negative for MV by immunocytochemistry showed the same distribution for CD46 as in normal brain. However, cells in lesions, positive for MV, were negative for CD46. Other mechanisms, clearly independent of CD46 and CD150, can also be involved in the endothelial cell infection with wild-type measles viruses [72].

In the case of HIV, a number of chemokine receptors including but not limited to APJ, CCR3, CXCR4, and CCR5 may act, in addition to CD4, as coreceptors for HIV-1/SIV, not only in peripheral blood and lymphoid tissues but also in the CNS. Isolated HB-MVEC appear to express significant amounts of the chemokine receptors, APJ, CCR3, CXCR4, and CCR5, plus C-type lectins DC-SIGN and L-SIGN [73]. As these MVECs do not express CD4, this suggests a CD4-independent HIV/SIV entry/infection of these cells, which are the major cells constituting the human BBB. In addition, since chemokines for cognate chemokine receptors individually were unable to block binding of HIV-1 to brain MVECs, viral attachment is mediated by a possible previously unknown receptor or by cooperative activity of various receptors. Overall, molecular mechanisms of HIV/BBB inter-actions, with particular attention to cytokines and chemokines, have been reviewed recently [74]. Among the mechanisms of entry, macropinocytosis dependent on lipid rafts and the mitogen-activated protein kinase signaling pathway seem to be important [75]. In this case, the chemokine AOP-RANTES inhibited virus entry, whereas anti-CD4 mAbs had no effect [75]. Numerous cytokines and chemokines [76] are elicited by HIV infection, but the respective roles of these molecules and their receptors [77] remain to be defined [78]. A peculiar role of fractalkine (FKN) has been proposed. FKN is markedly up-regulated in neurons and neuropil in brain tissue from pediatric patients with HIV-1 encephalitis (HIVE) compared with those without HIVE or that were HIV-1 seronegative. In patients with HIVE, FKN receptors (CX3CR1) are expressed on neurons and perhaps on microglial cells. FKN is able to potently induce the migration of primary human monocytes across an endothelial cell/primary human fetal astrocyte trans-well bilayer and is neuro-protective to cultured neurons when coadministered with either the HIV-1 neuro-toxin, platelet activating factor (PAF) or the regulatory HIV-1 gene product Tat. Thus focal inflammation in brain tissue with HIVE may up-regulate neuronal FKN levels, which in turn may be a neuroimmune modulator recruiting peripheral macrophages into the brain, and in a paracrine fashion protecting glutamatergic neurons [79].

Infections with human T-cell leukemia virus type 1 implicate adhesion and migra-tion of the pathogen to brain EC, as well as infected lymphocyte mediated enhancement of paracellular permeability of the endothelial monolayer. Such HTLV-1 infected lymphocytes produce large amounts of TNF in the presence of brain endothelial cells [80].

As the Toll-like receptors (TLRs) may play an important role in the induction of inflammatory cytokines in response to viruses, studies in KO mice were conducted in herpes simplex virus 1 (HSV-1) infection models [81]. It was demonstrated that TLR2 mediates the inflammatory cytokine response to HSV-1 by using both trans-fected cell lines and knockout mice. Studies of infected mice revealed that HSV-1 induced a blunted cytokine response in TLR2(−/−) mice. Brain levels of the chemok-ine, MCP-1 were significantly lower in TLR2(−/−) mice than in either wild-type or TLR4(−/−) mice. TLR2(−/−) mice had reduced mortality compared with wild-type mice. The differences between TLR2(−/−) mice and both wild-type and TLR4(−/−) mice in the induction of monocyte chemoattractant protein 1, brain inflammation, or mortality could not be accounted for on the basis of virus levels. Thus, these studies suggest the TLR2-mediated cytokine response to HSV-1 is detrimental to the host.

Cytokines can be directly responsible for recovery after viral infection of the CNS, as shown in the vesicular stomatitis virus model [82]. Decreased VSV titers in brain homogenate of IL-12-injected mice were associated with increased expres-sion of iNOS in the CNS, enhanced expression of both MHC class I and class II Ags in the CNS, increased T cell infiltration in the CNS, especially in the olfactory bulb, and diminished VSV-induced apoptosis in olfactory bulb. Exogenously added IL-12, even when injected peripherally, may thus be useful in recovery from VSV infection of the CNS.

E. OTHER INFECTIOUS AGENTS

The binding of *Candida albicans* to HB-MVEC is mediated by an interaction between the pathogen enolase plasminogen/plasmin, and this may contribute to invasion of the tissue barrier [83]. In the case of *Chlamydia pneumoniae* infection, it has been shown that transmigration of monocytes through human brain MVEC is increased. This represents a mechanism for breaching the BBB, and may be of pathogenic importance in Alzheimer's disease (AD) [84]. The entry of the pathogen appears to occur in the context of a substantial activation of both EC and leucocyte, as evidenced by the up-regulation of VCAM-1 and ICAM-1 on HB-MVEC and a corresponding increase of LFA-1, VLA-4, and Mac-1 on monocytes, respectively [84]. *C. pneumoniae* infection of HB-MVECs resulted in increased expression of the zonula adherens proteins beta-catenin, N-cadherin, and VE-cadherin, and decreased expression of the tight junctional protein occludin [85]. These events may underlie a mechanism for the regulation of paracellular permeability while maintaining barrier integrity. Thus, infection by *C. pneumoniae* at the level of the vasculature may be a key initiating factor in the pathogenesis of neurodegenerative diseases such as sporadic AD.

The fate of pathogens such as *Cryptococcus neoformans* can be potently modulated by IL-4, responsible for weaker resistance, or IFN-gamma, IL-12, and IL-18, capable of protecting the host, but this has been shown for the lung pathology, not the CNS involvement [86,87]. The yeast are detected in macrophages in the leptomeningeal space, in monocytes circulating in leptomeningeal capillaries, or in the endothelium itself, strengthening the hypothesis that monocytes and brain EC play key roles in the pathogenesis of cryptococcal meningitis [88]. In fact, cryptococcal capsular polysaccharides induce not only circulating cells, but also cells inside the CNS to produce chemokines. This can lead to complex settings in which the functional consequence of chemokines, namely transmigration, can be blocked. For instance, microglial cells are brought to secrete IL-8 which could mediate leucocyte entry into the CNS, but this is inhibited by the effect of the same cryptococcal polysaccharides on circulating cells which, in turn, block transmigration toward IL-8 [89].

The pathogenesis of infections by *Rickettsia spp.* and related organisms remains captivating, albeit incompletely understood. The need to elucidate such mechanisms in typhus and spotted fever rickettsioses is even more crucial in view of the specter of bioterrorism employing genetically engineered Rickettsia resistant to all antibiotics [90]. It is clear that pathogenic processes include rickettsial entry into the dermis, hematogenous dissemination to vascular endothelial cells (most critically in brain and lungs), increased vascular permeability, edema, and immunity mediated by NK cells, IFN-gamma, TNF, RANTES, antibodies, and cytotoxic T lymphocytes. Also, reactive oxygen species (ROS) produced by *R. rickettsii*-infected endothelial cells lead to peroxidative damage to cell membranes *in vitro*, and complex interactions between inflammation and coagulation are major features of rickettsiosis.

Cytokines produced by T cells during typhus group rickettsiosis have been analyzed, but not yet those produced by the target cell itself, namely the endothelial cell.

In vivo intervention studies in mice have demonstrated that IFN-gamma and CD8 T lymphocytes are crucial to clearance of the rickettsiae and recovery from infection [91]. In a model using *R. typhi*, an effective antirickettsial immune response was associated with elevated serum concentrations of IL-12 on day 5 and increased secretion of IL-12 by concanavalin-A-stimulated spleen cells on day 5. Evidence for transient suppression of the immune response consisted of marked reduction in the secretion of IL-2 and IL-12 by concanavalin-A-stimulated spleen cells on days 10 and 15.

Studies carried out in experimental infection by *Ehrlichia chaffeensis*, the agent of human monocytic ehrlichiosis, led to the discovery of original mechanisms: brain EC can be altered in a subtle way by pathogens, i.e., without involving direct alteration of the BBB resistivity [92]. In an *in vitro* model based on brain MVEC, it was found that uninfected monocyte/macrophages crossed endothelial cell barriers six times more efficiently than neutrophils and that more *E. chaffeensis*-infected monocytes transmigrated than uninfected monocytes. Differences were not due to barrier dysfunction, as transendothelial cell resistivities were the same for uninfected cell controls, but were related to the ability of the pathogen to modify cell migratory capacity. These are unique pathogen-specific host cell functional alterations that are likely important for pathogen survival, pathogenesis, and disease induction.

III. CYTOKINES AS PLAYERS IN THE MODULATION OF INFECTION-INDUCED CNS DAMAGE

A. Defense vs. Pathology: Immunopathology

Besides their potent effects in protection, cytokines can also be responsible for pathology, notably in the CNS. TNF can be viewed as a prototype of the double-edged sword, particularly in the context of infection, as previously discussed [93]. The protective pathogenic role of TNF depends on numerous parameters, including amounts that are produced in a given setting and site and rate of production. IL-1 also is a central player in brain inflammatory pathology, including psychoneuroimmunological consequences [94,17]. The CNS regulates the immune system by elaborating anti-inflammatory hormone cascades in response to pathogens and immune mediators. Further levels of complexity are added to the system by the discoveries that the CNS also responds via acetylcholine-mediated efferent signals to immune cells and that nicotinic cholinergic receptors expressed on monocyte/macrophages detect these signals and respond with a dampened cytokine response [95,96]. Neuroinflammation also involves vast numbers of other cytokines, including GM-CSF [97]. Cytokines indeed do exert major effects on the endocrine system, as recently evidenced by the demonstration of anti-TNF mAb therapy-induced changes on the hypothalamic–pituitary axis [98].

The net effects of given cytokines will depend on these modulatory properties, and paradoxical effects also have to be taken into account. For example the proinflammatory cytokines IL-1 and IL-6 can be neuroprotective *in vitro* probably via prevention of oxidative damage [99]. Equally paradoxical and unexpected phenomena are observed at the level of cytokine signalling itself, as illustrated by the exacerbation of cytokine-induced pathology in STAT-1 deficient mice [100].

In addition, it recently has been suggested that the determinant cytokines may not be the ones previously identified: IL-23 p40 subunit expression by CNS resident cells is important in the Th1 bias [101]. In fact, IL-23, rather than IL-12, appears to be the critical cytokine for autoimmune inflammation in the brain [102].

B. Effector Mechanisms of Cytokines

The effector mechanisms of TNF and other cytokines include direct toxicity, induction of CAM upregulation, MMP secretion, oxidative stress, and metabolic changes, notably neuromediator production. However, TNF alone, even if it induces various chemokine and chemokine receptors, does not suffice to cause leucocyte accumulation in the CNS parenchyma, as shown in a model of cytokine microinjection in mouse brain [103].

Via the upregulation of CAMs on brain EC, cytokines induce modified platelet–endothelial interactions [15,104]. For instance, in human pediatric CM, platelets have been found to accumulate in large numbers in brain microvessels [105], as previously shown in the mouse PbA model [106]. The interactions between endothelial cells, platelets, and leucocytes also play determinant roles in cerebrovascular reactivity [107]. Endothelial reorganization with the expression of different pro-thrombotic factors and activation of platelets and leucocytes, combined, lead to blood cell adhesion to the endothelial monolayer, aggregation as thrombi, and the formation of numerous spasmogenic substances. Activation of the blood cells in the vicinity of the endothelium may induce endothelial dysfunction/injury, resulting in impairment of normal endothelial antispasmodic control. Within the microcirculatory bed, intravascular activation of the blood cells leads to scattered microvessel plugging, increased vascular permeability, edema formation, and cytotoxic actions of blood cell-released agents on the underlying CNS tissue.

Cytokine-driven metabolic alterations of neuromediators may actively participate to the pathogenesis of immunopathological complications of infectious diseases, such as CM [4], and conversely cytokines are potent modulators of behavioral and neuropathological consequences of neuromediator administration, as shown in the kainic acid injection model of epilepsy [108].

C. Cerebral Ischaemia

A body of evidence supports the view that TNF is a mediator of brain ischaemia. This cytokine, which rapidly upregulates in the brain after injury, increases the percent hemispheric infarct induced by permanent MCAO in a dose-related manner, and anti-TNF mAbs significantly reduce focal ischaemic brain injury [109]. TNF toxicity does not appear to be due to a direct effect on neurons or modulation of neuronal sensitivity to glutamate or oxygen radicals and apparently is mediated through nonneuronal cells. More generally, stroke is followed by both acute and prolonged inflammatory responses characterized by the production of inflammatory cytokines and leukocyte infiltration into the brain. A debate on whether inflammation after stroke is neurotoxic or participates in brain repair remains unresolved. However, the need to pharmacologically control inflammatory amplification has been commonly acknowledged [110].

Beyond ischemic brain injury, TNF clearly is important in neuropathology, consisting of EC activation, meningeal inflammation or glial cell damage, as shown by Akassoglou and colleagues [111], who demonstrated, using transgenic mice, that the differential expression of the two TNF receptors is crucial in the shaping of TNF signalling in neuroimmune interactions.

However, once again, unexpectedly and paradoxically, TNF can serve a neuroprotective function. Damage to neurons caused by focal cerebral ischaemia and epileptic seizures was exacerbated in TNFR-KO mice. Oxidative stress was increased and levels of an antioxidant enzyme reduced in brain cells of TNFR-KO mice, indicating that TNF protects neurons by stimulating antioxidant pathways [112]. Thus, TNF also seems to be important in limiting the ischaemia-induced brain damage [113].

D. Oxidative Stress

The role of cytokine-induced processes, such as oxidative stress and protease release, particularly MMP-9, both play important roles in the post-hypoxic disturbances of the cerebral microcirculation [114]. As for cytokines, the pleiotropic effects of nitric oxide render this mediator extremely difficult to appraise: NO can be neuroprotective, as indicated by *in vitro* and *in vivo* results, by (i) suppressing iron-induced generation of hydroxyl radicals (.OH) via the Fenton reaction, (ii) interrupting the chain reaction of lipid peroxidation, (iii) augmenting the antioxidative potency of reduced glutathione (GSH) and (iv) inhibiting cysteine proteases. To exert these effects, the short-lived NO induces the more stable S-nitrosoglutathione (GSNO), which is 100-fold more potent than GSH; it completely inhibits the weak peroxidative effect of ONOO-. Neuroprotective effects of GSNO and .NO have been demonstrated in brain preparations *in vivo*. This putative GSNO pathway (GSH→GS.→GSNO→.NO + GSSG→GSH) may be an important part of endogenous antioxidative defense system, which could protect neurons and other brain cells against oxidative stress caused by oxidants, iron complexes, proteases and cytokines. NO thus is a potent antioxidant against oxidative damage caused by reactive oxygen species.

However, NO also appears to be capable of altering the functional integrity of the BBB, and thereby of being pathogenic. It is via an inhibition of glyceraldehyde-3-phosphate dehydrogenase that NO causes a reduction in endothelial ATP content, with ensuing energy deficiency in brain EC and a delayed barrier dysfunction [115]. Also, the cytokine-modulated enzyme iNOS is implicated in brain EC damage and in neurodegeneration, implying that NO can act as a neurotoxin, via the tyrosine nitration of a 32-kDa protein in murine leukemia virus-infected brain MVEC [116].

The brain EC itself suffers from oxidative stress, and there is evidence accumulating that brain microvasculature is involved critically in oxidative stress-mediated

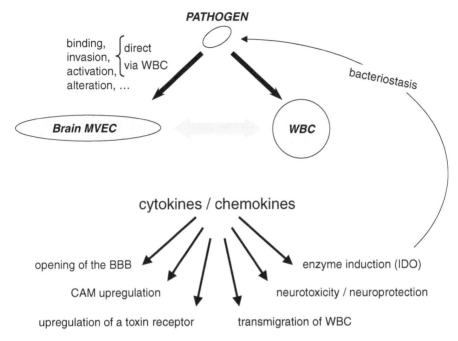

FIGURE 11.1 Some mechanisms by which cytokines may mediate defense and protection at the BBB level. WBC: white blood cells. MVEC: microvascular endothelial cell.

brain damage. In a model of glucose deprivation *in vitro*, HB-MVEC were shown to undergo chromosomal aberrations, induction of micronuclei and apoptosis. Oxidative stress thus exerts considerable genotoxic and cytotoxic effects on brain endothelium, which might contribute to the progression of tissue damage in the CNS [117].

Because of such EC changes, reducing the adverse effects due to oxidative stress is thus desirable in several pathologies, and could be achieved by numerous ways, including methylene blue, aspirin, or alpha-lipoic acid. These compounds are at least capable of reducing fever, without decreasing circulating TNF levels [118], but it remains to be demonstrated that they can limit the extent of cerebral tissue injury due to oxidative stress.

IV. CONCLUSIONS

Cytokines, produced notably upon infection, as developed here, are at the interface between blood and CNS. They do so by acting in particular on brain ECs, which occupy a strategically important location between circulating blood and underlying parenchyma. Not only do cytokines exert their effects via vast numbers of mechanisms of action, but also complex interactions exist among them, as well as among the molecules they induce. Some of the relevant mechanisms are schematically presented in Figure 11.1.

Thus, the pleiomorphic nature of cytokines and of their effector molecules makes it difficult, if not impossible, to predict the final effect in a given situation, as illustrated by some examples discussed here. Finally, complex intrications between inflammation and coagulation are essential to pathogenesis, and there is a need for experimental models that reproduce as efficiently as possible the alterations occurring in pathology. There also is a need to establish a hierarchy of importance in the molecules involved in pathogenesis. Only an integrated understanding of immunopathology, taking into accounts the various parameters evoked here can allow the development of better therapies in thrombotic and inflammatory diseases.

REFERENCES

1. Mhlanga JD, Bentivoglio M, Kristensson K: Neurobiology of cerebral malaria and African sleeping sickness. Brain Res Bull 1997, 44:579–589.
2. Brown H, Rogerson S, Taylor T, Tembo M, Mwenechanya J, Molyneux M, Turner G: Blood–brain barrier function in cerebral malaria in Malawian children. Am J Trop Med Hyg 2001, 64:207–213.
3. Brown HC, Chau TT, Mai NT, Day NP, Sinh DX, White NJ, Hien TT, Farrar J, Turner GD: Blood–brain barrier function in cerebral malaria and CNS infections in Vietnam. Neurology 2000, 55:104–111.
4. Hunt NH, Grau GE: Cytokines: accelerators and brakes in the pathogenesis of cerebral malaria. Trends Immunol 2003, 24:491–499.
5. Craig A: Malaria: a new gene family (clag) involved in adhesion. Parasitol Today 2000, 16:366–367; discussion 405.
6. Craig A, Scherf A: Molecules on the surface of the Plasmodium falciparum infected erythrocyte and their role in malaria pathogenesis and immune evasion. Mol Biochem Parasitol 2001, 115:129–143.
7. Rasti N, Wahlgren M, Chen Q: Molecular aspects of malaria pathogenesis. FEMS Immunol Med Microbiol 2004, 41:9–26.
8. Hatabu T, Kawazu S, Aikawa M, Kano S: Binding of Plasmodium falciparum-infected erythrocytes to the membrane-bound form of Fractalkine/CX3CL1. Proc Natl Acad Sci U.S.A 2003, 100:15942–15946.
9. de Souza JB, Riley EM: Cerebral malaria: the contribution of studies in animal models to our understanding of immunopathogenesis. Microbes Infect 2002, 4:291–300.
10. Medana IM, Chaudhri G, Chan-Ling T, Hunt NH: Central nervous system in cerebral malaria: 'innocent bystander' or active participant in the induction of immunopathology? Immunol Cell Biol 2001, 79:101–120.
11. Langhorne J, Quin SJ, Sanni LA: Mouse models of blood-stage malaria infections: immune responses and cytokines involved in protection and pathology. Chem Immunol 2002, 80:204–228.
12. Artavanis-Tsakonas K, Tongren JE, Riley EM: The war between the malaria parasite and the immune system: immunity, immunoregulation and immunopathology. Clin Exp Immunol 2003, 133:145–152.
13. Furlan R, Villa P, Senaldi G, Martino G: TNFalpha in experimental diseases of the CNS. Methods Mol Med 2004, 98:171–190.
14. Clark IA, Cowden WB: The pathophysiology of falciparum malaria. Pharmacol Ther 2003, 99:221–260.
15. Coltel N, Combes V, Hunt NH, Grau GE: Cerebral malaria—a neurovascular pathology with many riddles still to be solved. Curr Neurovasc Res 2004, 1:91–110.

16. Schluesener HJ, Kremsner PG, Meyermann R: Heme oxygenase-1 in lesions of human cerebral malaria. Acta Neuropathol (Berl) 2001, 101:65–68.

17. Tracey KJ, Czura CJ, Ivanova S: Mind over immunity. Faseb J 2001, 15:1575–1576.

18. Brown H, Turner G, Rogerson S, Tembo M, Mwenechanya J, Molyneux M, Taylor T: Cytokine expression in the brain in human cerebral malaria. J Infect Dis 1999, 180:1742–1746.

19. Sibson NR, Blamire AM, Perry VH, Gauldie J, Styles P, Anthony DC: TNF-alpha reduces cerebral blood volume and disrupts tissue homeostasis via an endothelin- and TNFR2-dependent pathway. Brain 2002, 125:2446–2459.

20. Lucas R, Juillard P, Decoster E, Redard M, Burger D, Donati Y, Giroud C, Monso-Hinard C, De Kesel T, Buurman WA, Moore MW, Dayer JM, Fiers W, Bluethmann H, Grau GE: Crucial role of tumor necrosis factor (TNF) receptor 2 and membrane-bound TNF in experimental cerebral malaria. Eur J Immunol 1997, 27:1719–1725.

21. Stoelcker B, Hehlgans T, Weigl K, Bluethmann H, Grau GE, Mannel DN: Require-ment for tumor necrosis factor receptor 2 expression on vascular cells to induce experimental cerebral malaria. Infect Immun 2002, 70:5857–5859.

22. Dietrich JB: The adhesion molecule ICAM-1 and its regulation in relation with the blood–brain barrier. J Neuroimmunol 2002, 128:58–68.

23. Craig A, Fernandez-Reyes D, Mesri M, McDowall A, Altieri DC, Hogg N, Newbold C: A functional analysis of a natural variant of intercellular adhesion molecule-1 (ICAM-1Kilifi). Hum Mol Genet 2000, 9:525–530.

24. Lou J, Gasche Y, Zheng L, Critico B, Monso-Hinard C, Juillard P, Morel P, Buurman WA, Grau GE: Differential reactivity of brain microvascular endothelial cells to TNF reflects the genetic susceptibility to cerebral malaria. Eur J Immunol 1998, 28:3989–4000.

25. Dobbie MS, Hurst RD, Klein NJ, Surtees RA: Upregulation of intercellular adhesion molecule-1 expression on human endothelial cells by tumour necrosis factor-alpha in an *In vitro* model of the blood–brain barrier. Brain Res 1999, 830:330–336.

26. Li J, Chang WL, Sun G, Chen HL, Specian RD, Berney SM, Kimpel D, Granger DN, van der Heyde HC: Intercellular adhesion molecule 1 is important for the development of severe experimental malaria but is not required for leukocyte adhesion in the brain. J Investig Med 2003, 51:128–140.

27. Shear HL, Marino MW, Wanidworanun C, Berman JW, Nagel RL: Correlation of increased expression of intercellular adhesion molecule-1, but not high levels of tumor necrosis factor-alpha, with lethality of Plasmodium yoelii 17XL, a rodent model of cerebral malaria. Am J Trop Med Hyg 1998, 59:852–858.

28. Clark IA, Gray KM, Rockett EJ, Cowden WB, Rockett KA, Ferrante A, Aggarwal BB: Increased lymphotoxin in human malarial serum, and the ability of this cytokine to increase plasma interleukin-6 and cause hypoglycaemia in mice: implications for malarial pathology. Trans R Soc Trop Med Hyg 1992, 86:602–607.

29. Engwerda CR, Mynott TL, Sawhney S, De Souza JB, Bickle QD, Kaye PM: Locally up-regulated lymphotoxin alpha, not systemic tumor necrosis factor alpha, is the principle mediator of murine cerebral malaria. J Exp Med 2002, 195:1371–1377.

30. Rae C, McQuillan JA, Parekh SB, Bubb WA, Weiser S, Balcar VJ, Hansen AM, Ball HJ, Hunt NH: Brain gene expression, metabolism, and bioenergetics: interrelationships in murine models of cerebral and noncerebral malaria. Faseb J 2004, 18:499–510.

31. Clark I, Awburn M: Migration inhibitory factor in the cerebral and systemic endot-helium in sepsis and malaria. Crit Care Med 2002, 30:S263–267.

32. Clark IA, Awburn MM, Whitten RO, Harper CG, Liomba NG, Molyneux ME, Taylor TE: Tissue distribution of migration inhibitory factor and inducible nitric oxide synthase in falciparum malaria and sepsis in African children. Malar J 2003, 2:6.

33. Tongren JE, Yang C, Collins WE, Sullivan JS, Lal AA, Xiao L: Expression of proinflammatory cytokines in four regions of the brain in Macaque mulatta (rhesus) monkeys infected with Plasmodium coatneyi. Am J Trop Med Hyg 2000, 62:530–534.

34. Maneerat Y, Wilairatana P, Udomsangpetch R, Looareesuwan S: Lack of association between CSF nitrate and sera nitrate in falciparum malaria infection. Southeast Asian J Trop Med Public Health 2001, 32:268–274.

35. Sanni LA: The role of cerebral oedema in the pathogenesis of cerebral malaria. Redox Rep 2001, 6:137–142.

36. Hansen AM, Driussi C, Turner V, Takikawa O, Hunt NH: Tissue distribution of indoleamine 2,3-dioxygenase in normal and malaria-infected tissue. Redox Rep 2000, 5:112–115.

37. Ball HJ, McParland B, Driussi C, Hunt NH: Isolating vessels from the mouse brain for gene expression analysis using laser capture microdissection. Brain Res Brain Res Protoc 2002, 9:206–213.

38. Hanum PS, Hayano M, Kojima S: Cytokine and chemokine responses in a cerebral malaria-susceptible or -resistant strain of mice to Plasmodium berghei ANKA infection: early chemokine expression in the brain. Int Immunol 2003, 15:633–640.

39. Chang WL, Li J, Sun G, Chen HL, Specian RD, Berney SM, Granger DN, van der Heyde HC: P-selectin contributes to severe experimental malaria but is not required for leukocyte adhesion to brain microvasculature. Infect Immun 2003, 71:1911–1918.

40. Combes V, Rosenkranz AR, Redard M, Pizzolato G, Lepidi H, Vestweber D, Mayadas TN, Grau GE: Pathogenic role of P-selectin in experimental cerebral malaria: importance of the endothelial compartment. Am J Pathol 2004, 164:781–786.

41. Combes V, Taylor TE, Juhan-Vague I, Mege JL, Mwenechanya J, Tembo M, Grau GE, Molyneux ME: Circulating endothelial microparticles in malawian children with severe falciparum malaria complicated with coma. Jama 2004, 291:2542–2544.

42. Mulenga C, Mhlanga JD, Kristensson K, Robertson B: Trypanosoma brucei brucei crosses the blood–brain barrier while tight junction proteins are preserved in a rat chronic disease model. Neuropathol Appl Neurobiol 2001, 27:77–85.

43. Deckert-Schluter M, Bluethmann H, Kaefer N, Rang A, Schluter D: Interferon-gamma receptor-mediated but not tumor necrosis factor receptor type 1- or type 2-mediated signaling is crucial for the activation of cerebral blood vessel endothelial cells and microglia in murine Toxoplasma encephalitis. Am J Pathol 1999, 154:1549–1561.

44. Mun HS, Aosai F, Norose K, Chen M, Piao LX, Takeuchi O, Akira S, Ishikura H, Yano A: TLR2 as an essential molecule for protective immunity against Toxoplasma gondii infection. Int Immunol 2003, 15:1081–1087.

45. Yap GS, Scharton-Kersten T, Charest H, Sher A: Decreased resistance of TNF receptor p55- and p75-deficient mice to chronic toxoplasmosis despite normal activation of inducible nitric oxide synthase in vivo. J Immunol 1998, 160:1340–1345.

46. Aliberti J, Serhan C, Sher A: Parasite-induced lipoxin A4 is an endogenous regulator of IL-12 production and immunopathology in Toxoplasma gondii infection. J Exp Med 2002, 196:1253–1262.

47. Sukumaran SK, Prasadarao NV: Escherichia coli K1 invasion increases human brain microvascular endothelial cell monolayer permeability by disassembling vascular-endothelial cadherins at tight junctions. J Infect Dis 2003, 188:1295–1309.

48. Sukumaran SK, McNamara G, Prasadarao NV: Escherichia coli K-1 interaction with human brain micro-vascular endothelial cells triggers phospholipase C-gamma1 activation downstream of phosphatidylinositol 3-kinase. J Biol Chem 2003, 278:45753–45762.

49. Sukumaran SK, Quon MJ, Prasadarao NV: Escherichia coli K1 internalization via caveolae requires caveolin-1 and protein kinase Calpha interaction in human brain microvascular endothelial cells. J Biol Chem 2002, 277:50716–50724.

50. Grigsby RJ, Dobrowsky RT: Inhibition of ceramide production reverses TNF-induced insulin resistance. Biochem Biophys Res Commun 2001, 287:1121–1124.

51. Kim KS: Escherichia coli translocation at the blood–brain barrier. Infect Immun 2001, 69:5217–5222.

52. Kim KS: Strategy of Escherichia coli for crossing the blood–brain barrier. J Infect Dis 2002, 186 Suppl 2:S220–224.

53. Kita E, Yunou Y, Kurioka T, Harada H, Yoshikawa S, Mikasa K, Higashi N: Pathogenic mechanism of mouse brain damage caused by oral infection with Shiga toxin-producing Escherichia coli O157:H7. Infect Immun 2000, 68:1207–1214.

54. Johansson L, Rytkonen A, Bergman P, Albiger B, Kallstrom H, Hokfelt T, Agerberth B, Cattaneo R, Jonsson AB: CD46 in meningococcal disease. Science 2003, 301:373–375.

55. Eisenhauer PB, Chaturvedi P, Fine RE, Ritchie AJ, Pober JS, Cleary TG, Newburg DS: Tumor necrosis factor alpha increases human cerebral endothelial cell Gb3 and sensitivity to Shiga toxin. Infect Immun 2001, 69:1889–1894.

56. Doran KS, Liu GY, Nizet V: Group B streptococcal beta-hemolysin/cytolysin activates neutrophil signaling pathways in brain endothelium and contributes to development of meningitis. J Clin Invest 2003, 112:736–744.

57. Tsao N, Chang WW, Liu CC, Lei HY: Development of hematogenous pneumococcal meningitis in adult mice: the role of TNF-alpha. FEMS Immunol Med Microbiol 2002, 32:133–140.

58. Nassif X, Bourdoulous S, Eugene E, Couraud PO: How do extracellular pathogens cross the blood–brain barrier? Trends Microbiol 2002, 10:227–232.

59. Drevets DA, Leenen PJ: Leukocyte-facilitated entry of intracellular pathogens into the central nervous system. Microbes Infect 2000, 2:1609–1618.

60. Schroten H, Spors B, Hucke C, Stins M, Kim KS, Adam R, Daubener W: Potential role of human brain microvascular endothelial cells in the pathogenesis of brain abscess: inhibition of Staphylococcus aureus by activation of indoleamine 2,3-dioxygenase. Neuropediatrics 2001, 32:206–210.

61. Kielian T, Hickey WF: Proinflammatory cytokine, chemokine, and cellular adhesion molecule expression during the acute phase of experimental brain abscess development. Am J Pathol 2000, 157:647–658.

62. Kielian T, Barry B, Hickey WF: CXC chemokine receptor-2 ligands are required for neutrophil-mediated host defense in experimental brain abscesses. J Immunol 2001, 166:4634–4643.

63. Munoz FM, Hawkins EP, Bullard DC, Beaudet AL, Kaplan SL: Host defense against systemic infection with Streptococcus pneumoniae is impaired in E-, P-, and E-/P-selectin-deficient mice. J Clin Invest 1997, 100:2099–2106.

64. Charland N, Nizet V, Rubens CE, Kim KS, Lacouture S, Gottschalk M: Streptococcus suis serotype 2 interactions with human brain microvascular endothelial cells. Infect Immun 2000, 68:637–643.

65. Zwijnenburg PJ, van der Poll T, Florquin S, Roord JJ, Van Furth AM: IL-1 receptor type 1 gene-deficient mice demonstrate an impaired host defense against pneumococcal meningitis. J Immunol 2003, 170:4724–4730.

66. Zwijnenburg PJ, van der Poll T, Florquin S, Roord JJ, van Furth AM: Interleukin-10 negatively regulates local cytokine and chemokine production but does not influence antibacterial host defense during murine pneumococcal meningitis. Infect Immun 2003, 71:2276–2279.

67. Kayal S, Lilienbaum A, Poyart C, Memet S, Israel A, Berche P: Listeriolysin O-dependent activation of endothelial cells during infection with Listeria monocytogenes: activation of NF-kappa B and upregulation of adhesion molecules and chemokines. Mol Microbiol 1999, 31:1709–1722.

68. Wilson SL, Drevets DA: Listeria monocytogenes infection and activation of human brain microvascular endothelial cells. J Infect Dis 1998, 178:1658–1666.

69. Thwaites GE, Simmons CP, Than Ha Quyen N, Thi Hong Chau T, Phuong Mai P, Thi Dung N, Hoan Phu N, White NP, Tinh Hien T, Farrar JJ: Pathophysiology and prognosis in Vietnamese adults with tuberculous meningitis. J Infect Dis 2003, 188:1105–1115.

70. Lechner F, Sahrbacher U, Suter T, Frei K, Brockhaus M, Koedel U, Fontana A: Antibodies to the junctional adhesion molecule cause disruption of endothelial cells and do not prevent leukocyte influx into the meninges after viral or bacterial infection. J Infect Dis 2000, 182:978–982.

71. McQuaid S, Cosby SL: An immunohistochemical study of the distribution of the measles virus receptors, CD46 and SLAM, in normal human tissues and subacute sclerosing panencephalitis. Lab Invest 2002, 82:403–409.

72. Andres O, Obojes K, Kim KS, ter Meulen V, Schneider-Schaulies J: CD46- and CD150-independent endothelial cell infection with wild-type measles viruses. J Gen Virol 2003, 84:1189–1197.

73. Mukhtar M, Harley S, Chen P, BouHamdan M, Patel C, Acheampong E, Pomerantz RJ: Primary isolated human brain microvascular endothelial cells express diverse HIV/SIV-associated chemokine coreceptors and DC-SIGN and L-SIGN. Virology 2002, 297:78–88.

74. Annunziata P: Blood–brain barrier changes during invasion of the central nervous system by HIV-1. Old and new insights into the mechanism. J Neurol 2003, 250:901–906.

75. Liu NQ, Lossinsky AS, Popik W, Li X, Gujuluva C, Kriederman B, Roberts J, Pushkarsky T, Bukrinsky M, Witte M, Weinand M, Fiala M: Human immunodeficiency virus type 1 enters brain microvascular endothelia by macropinocytosis dependent on lipid rafts and the mitogen-activated protein kinase signaling pathway. J Virol 2002, 76:6689–6700.

76. Boven LA, Middel J, Breij EC, Schotte D, Verhoef J, Soderland C, Nottet HS: Interactions between HIV-infected monocyte-derived macrophages and human brain microvascular endothelial cells result in increased expression of CC chemokines. J Neurovirol 2000, 6:382–389.

77. Berger O, Gan X, Gujuluva C, Burns AR, Sulur G, Stins M, Way D, Witte M, Weinand M, Said J, Kim KS, Taub D, Graves MC, Fiala M: CXC and CC chemokine receptors on coronary and brain endothelia. Mol Med 1999, 5:795–805.

78. Becher B, Prat A, Antel JP: Brain-immune connection: immuno-regulatory properties of CNS-resident cells. Glia 2000, 29:293–304.

79. Tong N, Perry SW, Zhang Q, James HJ, Guo H, Brooks A, Bal H, Kinnear SA, Fine S, Epstein LG, Dairaghi D, Schall TJ, Gendelman HE, Dewhurst S, Sharer LR, Gelbard HA: Neuronal fractalkine expression in HIV-1 encephalitis: roles for macrophage recruitment and neuroprotection in the central nervous system. J Immunol 2000, 164:1333–1339.

80. Romero IA, Prevost MC, Perret E, Adamson P, Greenwood J, Couraud PO, Ozden S: Interactions between brain endothelial cells and human T-cell leukemia virus type 1-infected lymphocytes: mechanisms of viral entry into the central nervous system. J Virol 2000, 74:6021–6030.

81. Kurt-Jones EA, Chan M, Zhou S, Wang J, Reed G, Bronson R, Arnold MM, Knipe DM, Finberg RW: Herpes simplex virus 1 interaction with Toll-like receptor 2 contributes to lethal encephalitis. Proc Natl Acad Sci U.S.A 2004, 101:1315–1320.

82. Bi Z, Quandt P, Komatsu T, Barna M, Reiss CS: IL-12 promotes enhanced recovery from vesicular stomatitis virus infection of the central nervous system. J Immunol 1995, 155:5684–5689.

83. Jong AY, Chen SH, Stins MF, Kim KS, Tuan TL, Huang SH: Binding of Candida albicans enolase to plasmin(ogen) results in enhanced invasion of human brain microvascular endothelial cells. J Med Microbiol 2003, 52:615–622.

84. MacIntyre A, Abramov R, Hammond CJ, Hudson AP, Arking EJ, Little CS, Appelt DM, Balin BJ: Chlamydia pneumoniae infection promotes the transmigration of monocytes through human brain endothelial cells. J Neurosci Res 2003, 71:740–750.

85. MacIntyre A, Hammond CJ, Little CS, Appelt DM, Balin BJ: Chlamydia pneumoniae infection alters the junctional complex proteins of human brain microvascular endothelial cells. FEMS Microbiol Lett 2002, 217:167–172.

86. Kawakami K, Qureshi MH, Zhang T, Okamura H, Kurimoto M, Saito A: IL-18 protects mice against pulmonary and disseminated infection with Cryptococcus neoformans by inducing IFN-gamma production. J Immunol 1997, 159:5528–5534.

87. Kawakami K, Hossain Qureshi M, Zhang T, Koguchi Y, Xie Q, Kurimoto M, Saito A: Interleukin-4 weakens host resistance to pulmonary and disseminated cryptococcal infection caused by combined treatment with interferon-gamma-inducing cytokines. Cell Immunol 1999, 197:55–61.

88. Chretien F, Lortholary O, Kansau I, Neuville S, Gray F, Dromer F: Pathogenesis of cerebral Cryptococcus neoformans infection after fungemia. J Infect Dis 2002, 186:522–530.

89. Lipovsky MM, Gekker G, Hu S, Ehrlich LC, Hoepelman AI, Peterson PK: Cryptococcal glucuronoxylomannan induces interleukin (IL)-8 production by human microglia but inhibits neutrophil migration toward IL-8. J Infect Dis 1998, 177:260–263.

90. Walker DH, Valbuena GA, Olano JP: Pathogenic mechanisms of diseases caused by Rickettsia. Ann NY Acad Sci 2003, 990:1–11.

91. Walker DH, Popov VL, Feng HM: Establishment of a novel endothelial target mouse model of a typhus group rickettsiosis: evidence for critical roles for gamma interferon and CD8 T lymphocytes. Lab Invest 2000, 80:1361–1372.

92. Park J, Choi KS, Grab DJ, Dumler JS: Divergent interactions of Ehrlichia chaffeensis- and Anaplasma phagocytophilum-infected leukocytes with endothelial cell barriers. Infect Immun 2003, 71:6728–6733.

93. Beutler B, Grau GE: The role of TNF in the pathogenesis of infectious disease. CritCareMed 1993, 21:S423–S435.

94. Dantzer R: Innate immunity at the forefront of psychoneuroimmunology. Brain Behav Immun 2004, 18:1–6.

95. Borovikova LV, Ivanova S, Zhang M, Yang H, Botchkina GI, Watkins LR, Wang H, Abumrad N, Eaton JW, Tracey KJ: Vagus nerve stimulation attenuates the systemic inflammatory response to endotoxin. Nature 2000, 405:458–462.

96. Pavlov VA, Wang H, Czura CJ, Friedman SG, Tracey KJ: The cholinergic anti-inflammatory pathway: a missing link in neuroimmunomodulation. Mol Med 2003, 9:125–134.

97. Franzen R, Bouhy D, Schoenen J: Nervous system injury: focus on the inflammatory cytokine 'granulocyte-macrophage colony stimulating factor.' Neurosci Lett 2004, 361:76–78.

98. Straub RH, Pongratz G, Scholmerich J, Kees F, Schaible TF, Antoni C, Kalden JR, Lorenz HM: Long-term anti-tumor necrosis factor antibody therapy in rheumatoid arthritis patients sensitizes the pituitary gland and favors adrenal androgen secretion. Arthritis Rheum 2003, 48:1504–1512.

99. Bissonnette CJ, Klegeris A, McGeer PL, McGeer EG: Interleukin 1alpha and interleukin 6 protect human neuronal SH-SY5Y cells from oxidative damage. Neurosci Lett 2004, 361:40–43.

100. Wang J, Schreiber RD, Campbell IL: STAT1 deficiency unexpectedly and markedly exacerbates the pathophysiological actions of IFN-alpha in the central nervous system. Proc Natl Acad Sci U.S.A 2002, 99:16209–16214.

101. Becher B, Durell BG, Noelle RJ: IL-23 produced by CNS-resident cells controls T cell encephalitogenicity during the effector phase of experimental autoimmune encephalomyelitis. J Clin Invest 2003, 112:1186–1191.

102. Cua DJ, Sherlock J, Chen Y, Murphy CA, Joyce B, Seymour B, Lucian L, To W, Kwan S, Churakova T, Zurawski S, Wiekowski M, Lira SA, Gorman D, Kastelein RA, Sedgwick JD: Interleukin-23 rather than interleukin-12 is the critical cytokine for autoimmune inflammation of the brain. Nature 2003, 421:744–748.

103. Glabinski AR, Bielecki B, Kolodziejski P, Han Y, Selmaj K, Ransohoff RM: TNF-alpha microinjection upregulates chemokines and chemokine receptors in the central nervous system without inducing leukocyte infiltration. J Interferon Cytokine Res 2003, 23:457–466.

104. Mannel DN, Grau GE: Role of platelet adhesion in homeostasis and immunopathology. Mol Pathol 1997, 50:175–185.

105. Grau GE, Mackenzie CD, Carr RA, Redard M, Pizzolato G, Allasia C, Cataldo C, Taylor TE, Molyneux ME: Platelet accumulation in brain microvessels in fatal pediatric cerebral malaria. J Infect Dis 2003, 187:461–466.

106. Piguet PF, Da Laperrousaz C, Vesin C, Tacchini-Cottier F, Senaldi G, Grau GE: Delayed mortality and attenuated thrombocytopenia associated with severe malaria in urokinase- and urokinase receptor-deficient mice. Infect Immun 2000, 68:3822–3829.

107. Akopov S, Sercombe R, Seylaz J: Cerebrovascular reactivity: role of endothelium/platelet/leukocyte interactions. Cerebrovasc Brain Metab Rev 1996, 8:11–94.

108. Oprica M, Eriksson C, Schultzberg M: Inflammatory mechanisms associated with brain damage induced by kainic acid with special reference to the interleukin-1 system. J Cell Mol Med 2003, 7:127–140.

109. Barone FC, Arvin B, White RF, Miller A, Webb CL, Willette RN, Lysko PG, Feuerstein GZ: Tumor necrosis factor-alpha. A mediator of focal ischemic brain injury. Stroke 1997, 28:1233–1244.

110. Zhang W, Stanimirovic D: Current and future therapeutic strategies to target inflammation in stroke. Curr Drug Targets Inflamm Allergy 2002, 1:151–166.

111. Akassoglou K, Douni E, Bauer J, Lassmann H, Kollias G, Probert L: Exclusive tumor necrosis factor (TNF) signaling by the p75TNF receptor triggers inflammatory ischemia in the CNS of transgenic mice. Proc Natl Acad Sci U.S.A 2003, 100:709–714.

112. Bruce AJ, Boling W, Kindy MS, Peschon J, Kraemer PJ, Carpenter MK, Holtsberg FW, Mattson MP: Altered neuronal and microglial responses to excitotoxic and ischemic brain injury in mice lacking TNF receptors. Nat Med 1996, 2:788–794.

113. Rothwell NJ, Luheshi GN: Brain TNF: damage limitation or damaged reputation? Nature Med 1996, 2:746–747

114. Kolev K, Skopal J, Simon L, Csonka E, Machovich R, Nagy Z: Matrix metallopro-teinase-9 expression in post-hypoxic human brain capillary endothelial cells: H2O2 as a trigger and NF-kappaB as a signal transducer. Thromb Haemost 2003, 90:528–537

115. Hurst RD, Azam S, Hurst A, Clark JB: Nitric-oxide-induced inhibition of glyceral-dehyde-3-phosphate dehydrogenase may mediate reduced endothelial cell monolayer integrity in an *in vitro* model blood–brain barrier. Brain Res 2001, 894:181–188

116. Jinno-Oue A, Wilt SG, Hanson C, Dugger NV, Hoffman PM, Masuda M, Ruscetti SK: Expression of inducible nitric oxide synthase and elevation of tyrosine nitration of a 32-kilodalton cellular protein in brain capillary endothelial cells from rats infected with a neuropathogenic murine leukemia virus. J Virol 2003, 77:5145–5151

117. Bresgen N, Karlhuber G, Krizbai I, Bauer H, Bauer HC, Eckl PM: Oxidative stress in cultured cerebral endothelial cells induces chromosomal aberrations, micronuclei, and apoptosis. J Neurosci Res 2003, 72:327–333

118. Riedel W, Lang U, Oetjen U, Schlapp U, Shibata M: Inhibition of oxygen radical formation by methylene blue, aspirin, or alpha-lipoic acid, prevents bacterial-lipopolysaccharide-induced fever. Mol Cell Biochem 2003, 247:83–94

12 Cytokines and Multiple Sclerosis

P. Rieckmann

CONTENTS

I. INTRODUCTION

Multiple sclerosis (MS) has long been regarded as a classic organ-specific Th1-mediated autoimmune disease, like rheumatoid arthritis or type-1 diabetes mellitus. Recently the situation has become more complicated. First, there are at least four histopathological subtypes that constitute MS lesions, suggesting different pathophysiological pathways leading to inflammation, demyelination with oligodendrocyte death, astrogliosis, and axonal loss—the classic hallmarks of microscopically evaluated plaques [1]. Second, the concept of "protective autoimmunity" emerged, which puts inflammation into a new perspective. Beyond destruction, the invading immune cells possess the capacity to induce clearance of debris by microglia and convey trophic support (factors) to the target tissue. Both mechanisms can promote functional reorganization within the CNS [2]. Third, therapeutic strategies to neutralize pro-inflammatory cytokines, like tumor necrosis factor-alpha (TNF-a) thought to be a major culprit of MS, failed in clinical studies [3].

The role of cytokines as mediators of immunoregulation in the disease process is therefore difficult to assess. They are present at various stages in different cell types and may exert a plethora of functions depending on the amount of cytokines released, influence of the local environment, and timely regulated (in) activation. Here, an overview will be presented on the accumulating knowledge about cytokines in MS, their role during initiation and perpetuation of the disease process as well as mediators of tissue destruction, and cytokines as potential activity markers of disease and the target for therapeutic interventions.

II. MULTIPLE SCLEROSIS—THE DISEASE

MS is still the major disabling neurological disease of young adults, affecting 0.05–0.15% of the Caucasian population. It usually starts in young adulthood with a clinical bout encompassing symptoms like visual perturbation, numbness or tingling, difficulties in walking or coordination, and sometimes unspecific complaints, like fatigue, attention, or endurance problems [4]. The disease affects women more frequently than men. In 80–90% of cases, MS starts with a relapsing-remitting course (RR-MS).

Over time, the number of relapses decreases, and most patients develop progressive neurological deficits independent of relapses (secondary progressive phase [SPMS]). The transition to SPMS occurs in 40% of patients during the first 10 years of disease. Only in a minority of patients (10–20%), MS begins with a primary progressive course (PP-MS) without acute relapses. In general, the progression rate in RR-MS is comparable to that of PP-MS as soon as patients acquire a certain stage of disability or enter the secondary progressive phase [5].

The diagnostic work-up starts with a detailed history because quite frequently patients will reveal preceding ailments, which might have occurred years ago, suggesting that the clinical presenting symptom is not the onset of disease. These findings are supported by magnetic resonance imaging (MRI) studies, which usually demonstrate multiple signal intensities in the white matter of the central nervous system (CNS) in more than two-thirds of patients with clinical isolated syndromes (CIS) [6].

Imaging studies revealed differences between RR-MS and PP-MS. During the relapsing phase of the disease (RR-MS), acute CNS lesions are frequently detected on MRI, even in the absence of clinical attacks [7]. These lesions are usually located in areas of the white matter, orientated perpendicular to the ventricles, and are often characterized by a disturbance of the blood–brain barrier, local edema, and demyelination—features that are compatible with an inflammatory process [7]. Lesions with gadolinium enhancement on MRI have also been identified at sites of perivascular mononuclear cell infiltration [8]. During the last 10 years, intensive research efforts have been devoted to the mechanisms that control immune cell entry into the CNS. Several factors, including adhesion molecules, proteases, cytokines, and chemokines, with distinct functions in the sequential process of transmigration across the blood–brain barrier have been identified. Proinflammatory cytokines, like TNF-α and IFN-γ are regarded as key elements in this process and have therefore attracted much interest as potential targets for MS therapy [9]. During the progressive phase, such inflammatory activity is much less conspicuous.

Although global brain atrophy can be detected during the early phase of the disease [10], it is much more prominent in the progressive stage and seems to correlate with disability. These findings indicate that early in the disease, ongoing inflammatory activity is most likely responsible for the relapsing-remitting course, whereas the atrophic process indicative for ongoing tissue destruction [11] might be predominantly operative in the progressive phase of the disease when inflammatory activity diminishes [12]. Therefore, immunomodulatory treatment strategies are more likely to be effective if ongoing inflammatory disease activity is still present.

The inflammatory process of MS is also revealed by a lympho-monocytic pleocytosis and the continuous presence of oligoclonal IgG bands in the cerebrospinal fluid (CSF) of most patients, suggesting an ongoing immune process confined to the CNS. CSF alteration can be detected in over 90% of all MS patients, and the cellular pattern in the lumbar spinal fluid may be a reflection of chemokine-mediated attraction of these cells to the CNS [13].

Modern histopathological studies defined at least four patterns of oligodendrocyte damage and demyelination with various extents of immune cell infiltrates (mainly activated T lymphocytes, macrophages, and some plasma cells) as well as antibody-mediated demyelination [1]. There is still no satisfactory explanation for how the disease develops and what determines the different clinical courses and histopathological findings. Large cross-sectional and longitudinal epidemiological studies suggest a combined genetic and environmental influence, but individual risk factors have not yet been established.

The prevalence of MS varies significantly depending on the genetic background of the patient. MS is highly prevalent in Caucasians but only rarely observed in Asians or Africans. Even in areas with high MS prevalence, these ethnic groups are at much lower risk than Caucasians. Moreover, the risk of developing the disease is significantly higher in family members of patients with MS. The strongest genetic association in Caucasians has been described for human leukocyte antigen (HLA) class II alleles *DR15/DQw6* (*HLA-DRB1*1501; *HLA-DQB1*0601*) [14]. The HLA-locus is located on the short arm of chromosome 6 and codes for molecules that participate in antigen recognition by T lymphocytes. This genomic region contains other relevant immune mediators, like complement and the pro-inflammatory cytokines TNF-α and lymphotoxin (LT). Chromosome 6q is still regarded as a hot spot with high lod score in genome screens for MS susceptibility genes [15]. Epidemiological studies have provided evidence for an additional, environmental component in the disease process. MS relapses are often associated with upper respiratory or other viral infections, and in areas with homogeneous ethnic populations, MS prevalence varies, being higher in areas with moderate climates.

There is an ongoing debate on the role of infections as the driving force for the autoimmune reactions observed in MS [16], but no single agent has so far stood the test of time. Pregnancies seem to have an ameliorating effect on the clinical course; however, an increased relapse frequency was detected within the first 6 months postpartum [17]. These results may be explained by the roles of cytokines during the course of pregnancies [18,19].

Overall, both genetic and environmental factors seem to contribute synergistically to the manifestation and progression of the disease. Most findings from clinical studies point to a central role of a distorted immunoregulation as a central process in the pathogenesis of MS. Mediators of immunoregulation, like cytokines, appear as major players in the fugal themes of inflammation, demyelination, and axonal loss and repair in MS. Understanding the interdependence of these themes and intelligent conduction of this distorted orchestra will likely result in the development of advanced treatment strategies for MS patients.

III. WHAT DID WE LEARN FROM EARLY STUDIES OF CYTOKINES IN MS?

In the early 1980s, immunohistological studies demonstrated the presence of cytokines in brain lesion from MS patients. Hofman et al. described IL-1, IL-2, and IL-2 receptor positive cells in frozen brain sections [20]. A few years later, interferon- and TNF-α positive cells were also detected in active inflammatory plaques [21,22]. The authors concluded that "their presence in MS lesions suggests a significant role for cytokines and the immune response in disease progression." These observations ignited a plethora of studies on cytokine expression in brain lesions, cerebrospinal fluid, or blood of MS patients, which resulted in a better understanding of the pathophysiological events in MS and provided targets for innovative treatment strategies.

Another important observation has largely influenced our thinking about the role of cytokines in MS: the discovery of polarized T cell subsets in mice, which were characterized by their cytokine secretion pattern. This phenomenon was first demonstrated by Mosmann and colleagues [23] and since then has become a guiding principle for T cell activation. Although polarization is relatively easy to observe in mice, the paradigm has never been as clear cut in the human system. Th1 and Th2 cells can certainly be found in human disease [24]. However, there is a growing recognition that in many diseases, clear distinctions cannot be made and that T cells of both persuasions can often be generated simultaneously [25]. For example, in mice IL-10 is considered to be a Th2 cytokine, whereas in humans both Th1 and Th2 cells can make IL-10 [26]. A third group of cytokine secreting cells has been proposed that mainly produces transforming growth factor beta1 (TGF-β1) [27]. These so-called Th3 cells can be induced by oral tolerization with myelin antigens and were regarded as potent therapeutic vehicles to reduce inflammation and demyelination in MS, but, unfortunately, the clinical study using this approach failed to demonstrate a significant benefit for patients with relapsing–remitting MS.

The other principle of dissecting cytokine functions into pro- and anti-inflammatory activities is not as stringent as initially anticipated. This is particularly clear for the divergent role of TNF-α in MS pathogenesis [3]. Therefore, the multifunctional role of cytokines depending on the local environment, receptor subsets expressed on target cells, and their activation state [28] should always be considered in the interpretation of studies linking cytokine levels in body fluids or target tissue to disease activity.

A third observation, which has greatly influenced cytokine research in MS, was the different effects of interferons on disease activity. Although intrathecal application of type-1 interferon (fibroblast interferon) clearly reduced the relapse rate [29] and large clinical studies of subcutaneous or intramuscular recombinant IFN-β reduced inflammatory disease activity in MS, which finally resulted in the approval of recombinant IFN-β for relapsing MS [30], an open, randomized trial with intravenous application of recombinant IFN-γ had a worse outcome [31]. In this study 18 patients with clinically definite, relapsing-remitting MS received 1 μg, 30 μg, or 1000 μg of recombinant IFN-γ twice a week for 4 weeks. Seven patients had exacerbations during treatment. This exacerbation rate, compared both with the pretreatment and posttreatment rate, was significantly greater than expected [31]. It was therefore concluded that IFN-γ mediates pro-inflammatory activities resulting in increased relapse rate, whereas IFN-β most likely acts as a modulator of inflammation. These opposing properties of class 1 and class 2 interferons much better explained the different results of the clinical trials than their antiviral potential, which was initially the rationale to use interferons in MS therapy [32].

These three observations described above were published during the 1980s and resulted in an intense research into the role of cytokines as potential mediators of MS pathogenesis and promising candidates for therapeutic interventions. This chapter will give an overview on the accumulating current knowledge about cytokines in MS: their role during initiation and perpetuation of the disease process, cytokines as mediators of structural damage, potential activity markers of disease, and targets for therapeutic intervention.

IV. MEASURING CYTOKINES IN MULTIPLE SCLEROSIS

Several studies have addressed the question of cytokine levels and their relation to disease course or activity in MS. These studies have to a large extent produced conflicting results. Elevated, normal, or decreased levels of most cytokines were reported and various conclusions were drawn from these studies. A note of caution should therefore be put to the interpretation of cytokine studies in MS in general [33]. Contradictory results with regard to cytokine levels most likely reflect methodological issues as well as the complex biology of cytokines [34]. One major problem is that the target compartment of MS—the CNS—can only be assessed indirectly. Lumbar puncture may be performed for diagnostic purposes but is not a source for repeated evaluation. Studies of cytokine levels in the blood (serum) only reflect the regulatory pathway of immune cell activation and are not very sensitive to changes. Another possible source of cytokines are the circulating immune cells, which can be isolated; and cell bound cytokines may be measured using different assays, like RT PCR, ELISPOT, or intracellular cytokine staining (Table 12.1). Different methods focus on different stages of cytokine production and secretion. Protein levels in extracellular fluids can be directly measured by enzyme-linked immunoassay (ELISA). However, most cytokines are seldom present in an unbound state but rather bound to soluble cytokine receptors, antibodies, or other binding proteins. Because cytokines typically have a short half-life and are rapidly degraded, negative results from ELISA studies are difficult to interpret. Unlike hormones, cytokines mainly exert

TABLE 12.1
Methods to Detect Cytokines in Blood Cells, Target Tissue, or Body Fluids

Level	Method	Advantage	Disadvantage
mRNA	Quant. RT PCR	High sensitivity Multiple cytokines can be measured	Does not account for posttranscriptional or posttranslational modifications
	In situ hybridization	Combination with immunocytochem Localization in tissue samples	Time consuming No exact quantification
	Expression profiling using cDNA arrays	Multiple transcripts can be analyzed	Needs verification Complex data handling
Intracellular cytokines	Flow cytometry Immunocytoche mistry	Individual cell can be detected	Insensitive, needs stimulation *in vitro*
Secreted cytokines	ELISPOT	Quantification of individual cytokine producing cells, up to 2 cytokines	Does not discriminate between membrane bound and secreted cytokines
	ELISA	Standardized procedure, Quantification at protein level	Reflects concentrations in large compartment (serum, CSF), not very sensitive (short half life of cytokines, binding to soluble receptor or other binding proteins often not detected

their effects at short distances from their site of production. Therefore, the overall concentration of a cytokine measured in a highly diluted compartment like the blood and even the CSF may not reflect its rate of production or secretion. Another round of complexity has been recently added to our understanding about the effects of cytokines. The response of immune cells to individual cytokines is not only regulated by the expression of different cytokine receptors on the surface of the target cell but also by intracellular mediators, a family of recently discovered suppressors of cytokine signaling (SOCS) [35]. Therefore, this complexity of cytokine effector regulation as well as technical constraints of individual assays used to measure cytokines should be considered when interpreting results from different studies.

V.   THE ROLE OF CYTOKINES DURING INITIATION AND PERPETUATION OF THE DISEASE

The initial event in the pathogenesis of MS is still unknown, but perivascular inflammatory infiltrates in the white matter of the CNS with focal demyelination and axonal damage is an early histopathological finding, which argues for an autoaggressive process. The development of autoimmune reactions against components of the central myelin may rather be secondary to an unknown infection or degenerative process of the myelin-forming oligodendrocyte, but it is considered the driving force for the perpetuation of an inflammatory process, which ultimately leads to tissue destruction and progressive impairment. Most of our initial knowledge about the proposed CD4-mediated autoimmune reaction against central myelin was derived from the artificial model of experimental autoimmune encephalomyelitis (EAE). Both in rodents and nonhuman primates, EAE can be induced by s.c. application of myelin antigens together with unspecific inflammatory adjuvants or infusion of syngeneic, autoreactive CD4+ T cell clones [36]. These T cell clones were regarded as the main immunological vehicle to transfer the disease to susceptible hosts. Cytokines came into place when it was demonstrated that mainly T cell clones, which produce Th1-type cytokines, induce EAE upon transfer into syngeneic animals (reviewed in [37]).

Sophisticated culture techniques were developed to allow *in vitro* propagation of human antigen-specific T cells. Using this method, myelin reactive T cells were cultured both from MS patients and healthy controls [38]. Subsequently, it was shown that their precursor frequency, clonal restriction, and functional state may be more important for the disease process than their mere presence. Several studies provided evidence that myelin reactive cells from MS patients are functionally different from healthy donors. These differences include higher precursor frequency [39], clonal restriction [40], memory phenotype [41], and a cytokine secretion profile with more IFN-γ + cells on ELISPOT analysis, which is already present during the very early stage of clinical isolated syndrome [42]. These data clearly indicate that the presence of autoreactive immune cells alone is not sufficient for an autoimmune disease, but rather their functional state is important and defective control mechanisms of tolerance or anergy may be associated with an increased risk to develop MS.

The preponderance of myelin-specific CD4+ T cells secreting IFN-γ and TNF-γ in early MS suggests an early differentiation bias toward the pro-inflammatory, Th-1 type reaction profile, which can be best explained by the micromillieu provided by antigen-presenting cells (APC) at the time of T cell activation. These signals can either be mediated by microbial products via Toll-like receptors (TLR) present on the surface of professional APCs [43] or viral transformation of these cells, as has been demonstrated for Epstein–Barr virus transformed B cells [16]. Several cytokines have been described that are important triggers of T cell priming upon contact with APCs. Most of them belong to the IL-12 cytokine family (IL-12, IL-23, IL-27) [44] and are inducible by bacterial products. IL-18, another potential inductor of pro-inflammatory T cell differentiation, is upregulated and functionally active in MS [45, 46], but its role as a pure Th1-inducer has recently been challenged [47].

The IL-12 related pathway of T cell differentiation including IL-23 and IL-27 constitutes a crucial link between innate defense mechanisms and adaptive immunity [48]. The redundancy of the IL-12 system in the induction of Th1-mediated autoimmunity of the CNS has recently been demonstrated [49]. Stimulation via TLRs and CD40-CD40 ligand interaction both induce IL-12 secretion from activated APC [50], which will then induce differentiation of T cells into an IFN-γ secreting phenotype. Recently, it was demonstrated that treatment of highly purified activated CD4(+)

T cells with the dsRNA synthetic analog poly(I:C) and CpG oligodeoxy–nucleotides (CpG DNA), respective ligands for TLR-3 and TLR-9, could directly enhance survival of activated CD4+ T cells without augmenting proliferation [51]. This scenario of autoreactive T cell activation via mechanisms of innate immunity within a pro-inflammatory microenvironment likely occurs in MS and is supported by further experimental evidence:

- The cytoarchitectural requirements for the above mentioned immunological reactions are provided in the draining lymph nodes (LN) of the target
  tissue. Dendritic cells, as the major APC, transfer central antigens and
  peripheral signals to the draining cervical LN. In a recent study, redistribution of myelin autoantigens from brain lesions to cervical LN has been
  demonstrated both in EAE of nonhuman primates and in MS [52]. Immu-

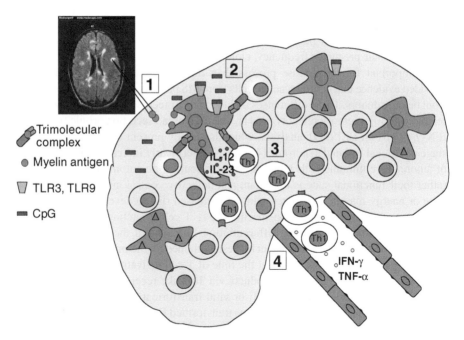

FIGURE 12.1 Cytokines modulate priming of autoreactive immune cells in lymphoid tissue. Proteins from the CNS cross-reactive with foreign antigens (autoimmune hypothesis) or generated after primary degeneration (degeneration hypothesis) are released into the periphery. They reach cervical lymph nodes (and tonsils?) of the brain draining area to set off an acquired immune response. T cell antigens are probably processed and presented by antigen-presenting cells, such as dendritic cells (DCs). DCs can load endocytotic and cytosolic proteins onto (MHC) class II molecules, allowing the priming of CD4+ T cells (similar also for MHC class I molecules and CD8 priming). Within a pro-inflammatory cytokine environment generated by microbial product, stimulation of Toll-like receptors (TLRs) on APCs specific recognition of the antigens by single T cells will result in clonal expansion and acquisition of Th1 effector functions. Primed cells will migrate through the body and accumulate in the CNS at sites where they encounter their target antigens.

TABLE 12.2
Cytokines and Disease Activity in MS

Increased Values During Disease Activity*	Decreased Values During Disease Activity
TNF-α, sTNF-RI, LT, IFN-γ, IL-2, sIL-2R, IL-6, IL-12p40, IL-13, IL-15, IL-18, IL-17, IL-23, Osteopontin, Oncostatin M	IL-4, IL-10, IL-12, TGF-β,

*Disease activity defined by clinical relapse or Gd+ lesion on MRI

---

nohistochemical analysis revealed significantly more myelin autoantigen-containing cells in LN of individuals with MS compared with those of control individuals. These cells expressed dendritic cell/macrophage-specific markers, MHC class II, and costimulatory molecules. They were directly juxtaposed to T cells, suggesting that cognate interactions between myelin-containing APC and T cells are taking place in brain-draining LN. These findings suggest that during MS, modulation of T cell reactivity against brain-derived antigens also takes place in cervical LN and not necessarily only inside the brain.

• Upper respiratory infections are a major risk factor for newly occurring relapses in MS [53]. These infections may provide an external stimulus to tonsillar or cervical lymph node cells via activation of TLRs on APCs [54] and thereby initiate a new round of immune cell differentiation and liberation of these cells into the blood circulation. As an example, myelin oligodendrocyte glycoprotein (MOG)-specific T cells, which secrete large amounts of IFN-γ, are detectable in the blood of MS patients during febrile upper respiratory infections [55]. In addition, several studies have demonstrated activated immune cells in the peripheral circulation that produce inflammatory cytokines, like IFN-γ [56], TNF-α [57], lymphotoxin [58], or IL-12 [59] prior to clinical relapses or flair ups of disease on serial magnetic resonance images (MRI; [60-62]; Table 12.2).

Many other studies on cross sectional cytokine analysis have been performed in MS patients. These data are difficult to interpret because association of a given factor with disease activity at only one point in time may obscure the whole picture and misleading conclusions may be drawn solely on the "guilt by association" hypothesis [63]. Other mechanisms, which are involved in the active suppression of autoreactive T cells, could be impaired in MS and thereby could escape the induction of peripheral tolerance *in vivo*. The deletion of one important cell type, the CD4+CD25+ suppressor cells resulted in the onset of organ-specific autoimmune disease [64]. CD4+CD25hi regulatory T cells are present in human peripheral blood and thymus. They synthesize IL-10 and TGF-β and are active suppressors of IL-2-mediated T cell proliferation [65]. Dysfunction or reduced numbers of these cells are increasingly accused for the development of autoimmune reactions. A loss of functional suppression by CD4CD25hi regulatory T cells has recently been described in patients with MS and may contribute to the breakdown of immunologic self-tolerance [66].

Although these data clearly provide evidence that activation of autoreactive immune cells in association with disease activity is present in the blood of MS patients and can be detected by measuring their cytokine profile *ex vivo*, it is still not known whether this event is causative for the incident myelin damage and axonal degeneration within the CNS or merely represents an adaptive immune reaction against myelin debris liberated into the draining LN after an unknown insult to oligodendrocytes. A recent histopathological study demonstrated oligodendrocyte apoptosis in actively demyelinating lesions without immune cell infiltration within 17 hours after onset of clinical symptoms [67]. The early detection of activated CD4+T cells reactive against βB-crystallin [68], a heat shock protein of oligodendrocytes, in patients with MS is supportive for this hypothesis. Further longitudinal studies on the early immunological events during clinical isolated syndromes—suggestive of MS—are important to define the major culprits for the development of a perpetuating autodestructive process. It is hoped that these studies will also provide us with new targets for early immunointerventions that may help to ameliorate the disease course or even control pathogenetic autoaggressive T cell clones.

VI. CYTOKINES AND CHEMOKINES IN IMMUNE CELL TRAFFIC ACROSS THE BLOOD–BRAIN BARRIER

Activated immune cells both of T- and B-lymphocyte origin may escape blood circulation and enter their target tissue once they are closely tethered to adhesion molecules expressed on the endothelial cells of postcapillary venules and receive chemotactic signals. With the development of advanced magnetic resonance technology, these areas of newly forming lesions can now be visualized. Periventricular, inflammatory lesions as seen on T2 weighted MRI are a hallmark of MS. Serial MRI studies revealed that these lesions occur much more frequently than clinical relapses. Newly formed lesions can be detected by enhancement with gadolinium-diethylenetriamine pentaacetic acid (Gd-DTPA), a paramagnetic contrast agent, which is an indicator for acute blood–brain barrier disturbance. In histopathological studies it was demonstrated that these lesions are associated with immune cell infiltration [8]. Therefore, it is suggested that during lesion formation transmigration of activated lymphocytes and monocytes across the blood–brain barrier must be regulated by appropriate molecular signals between invading lymphocytes, endothelial cells, and the surrounding brain tissue.

Transmigration of activated autoreactive T cells is an active process and requires upregulation of adhesion molecules on postcapillary brain endothelial cells within the white matter of the CNS (Figure 12.2). Major players on the endothelial cell surface are the integrin ligands intercellular adhesion molecule-1 (ICAM-1), vascular cell adhesion molecule (VCAM), and P-selectin. Increased expression of these molecules is mediated by luminal or abluminal TNF-α [69,70]. The process of immune cell transmigration can therefore be initiated by activated immune cells circulating in the blood via secretion of pro-inflammatory cytokines at the blood–brain barrier. Another likely event leading to a proadhesive microendothelium is focal secretion of cytokines by perivascular microglia [71]. In a recent *in vitro* study, adhesion of CD8+ T cells from MS patients to brain venules critically depended on P-selectin,

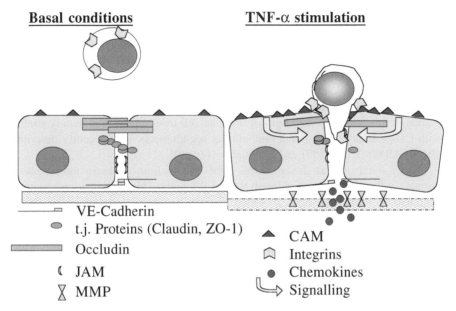

FIGURE 12.2 Immune cell adhesion and migration through the blood–brain barrier The process of immune cell transmigration is initiated by pro-inflammatory cytokines that induce upregulation of adhesion molecules on endothelial cells of the blood–brain barrier. Firm adhesion to endothelial cells requires functional interaction between integrins on the surface of immune cells and adhesion molecules, with LFA-1/ICAM-1 interaction playing a main role and being critically dependent on G-protein-mediated signals This signaling cascade affects cytoskeleton formation and may lead to alteration of tight junctions. TNF-α and IFN-γ downregulate the expression of occludin, a sealing molecule that is mandatory for intact barrier function. Both factors also cause redistribution of junctional adhesion molecules (JAM) in endothelial cells, resulting in an effective interaction of the lymphocyte integrin LFA-1 with JAM. Breaking through the basal lamina requires degradation of the extracellular matrix by matrix metalloproteinases, which are upregulated in blood mononuclear blood cells from MS patients and can be directly induced by pro-inflammatory cytokines.

Once beyond the blood–brain barrier, chemokines play a critical role for further immune cell traffic. Expression of chemokines CXCR3 and CCR5 on perivascular lymphocytes and the appropriate ligand CXCL10 (IP-10) CCL3 and CCL4 guide the route of invaded cells.

whereas this was not observed for CD4+ T cells [72]. Firm adhesion to endothelial cells requires functional interaction between integrins on the surface of immune cells and adhesion molecules, with LFA-1/ICAM-1 interaction playing a main role and being critically dependent on G-protein-mediated signals [73]. This signaling cascade affects cytoskeleton formation and may lead to alteration of tight junctions. These apical components of intercellular junctional complexes constitute the major anatomical structure of the blood–brain barrier. Cytokines may serve as extracellular stimuli on tight junctions: TNF-α and IFN-γ downregulate the expression of occludin [74], a sealing molecule that is mandatory for intact barrier function. Both factors also cause redistribution of junctional adhesion molecules (JAM) in endothelial cells [75], which may lead to effective interactions of the lymphocyte integrin LFA-1 with

JAM, the next step of successful immune cell transmigration [76]. Interestingly, TGF-β, IL-10, and recombinant IFN-β prevent cytokine-induced decrease in barrier function [77,78]. The cytokine regulated sequence of immune cell adhesion, disturbance of tight junction formation, and transmigration may be important for lesion formation in MS, as tight junction abnormality associated with blood–brain barrier leakage and active demyelination has recently been demonstrated in frozen sections from plaques and normal-appearing white matter (NAWM) of MS patients [79]. Although this multistep paradigm of immune cell transmigration is mainly studied with CD4+ T cells and monocytes, similar requirements were recently described also for CD8+ T cells [72] and B lymphocytes [80].

Having transmigrated across the tight endothelium of brain microvessels, immune cells need to break through the basal lamina. This active process requires degradation of the extracellular matrix by matrix metalloproteinases, which are upregulated in blood mononuclear blood cells from MS patients [81] and can be directly induced by pro-inflammatory cytokines.

Once beyond the blood–brain barrier, chemokines play a critical role for further immune cell traffic. Again, TNF-α seems to be a major cytokine for the induction of chemokines, which are essential for normal secondary lymphoid tissue development. Interestingly, ectopic lymphoid aggregation may also be present as a feature of CNS inflammation during MS. These small molecule chemoattractants are mainly produced by perivascular glial cells, direct leukocyte migration along their concentration gradient, and thereby define the cellular composition of inflammatory infiltrates. Immunohistopathological studies have described expression of chemokines CXCR3 and CCR5 on perivascular lymphocytes in MS autopsy material. IFN-γ as well as TNF-α secreting T cells (both cytokines increase chemokine production) are present in the blood of MS patients with active disease [82]. The appropriate ligand CXCL10 (IP-10) is expressed by astrocytes, whereas CCL3 and CCL4 are found on parenchymal inflammatory cells (macrophages and microglia) [83]. Chemokine receptor expression was also detected on mononuclear phagocytes in MS lesions. Their differential distribution, particularly CCR1 and CCR5, seems to be associated with histopathological subtypes II and III of MS lesions [84].

Investigations of chemokine receptor expression on mononuclear cells in CSF and peripheral blood revealed interesting results, which are instrumental for our understanding of immune cell trafficking into the CNS. In the CSF of MS patients, CXCR3 expression was found on >90% of T cells, whereas this receptor was only present on 40% of peripheral T cells. Increased levels of the ligand CXCL10 were measured in the CSF of patients with MS compared to control CSF [85]. This led to the conclusion that T cells entering the CSF compartment express CXCR3, and the differential distribution of this receptor on T cells in the two compartments are best explained by a chemokine-mediated process. CXCL10 and CXCR3 are therefore regarded as mediators of retention of activated T and B cells in the inflamed CNS [86].

Another important regulator of the immune cell infiltrate is oncostatin M, a cytokine of the IL-6 family, which is expressed on perivascular microglia and infiltrating lymphocytes in active lesions. Oncostatin M (OSM) significantly suppressed TNF-α induced secretion of CXCL8 (IL-8) by cerebral endothelial cells but synergizes with TNF-α in the induction of CCL2 (MCP-1) and ICAM-1, which favors monocyte rather than neutrophil recruitment to the inflamed CNS [87]. This cytokine is also produced spontaneously to high levels by blood lymphomononuclear cells in patients with active relapsing disease type [88]. We may encounter other factors in the future that are important in directing immune cells to brain tissue compartments. Altogether, they constitute promising targets for therapeutic interventions aimed at the redistribution of circulating immune cells and distracting them from entry into the CNS.

VII.   CYTOKINES AS EFFECTOR MECHANISMS IN MS

Permanent disability in progressive forms of MS has been linked to tissue destruction detectable as persistent hypointense T1 lesions on MRI [89]. It is still one of the major questions in MS research: what mediates tissue destruction? The concept of a purely immune-mediated process that will ultimately lead to axonal damage has recently been challenged because (1) widespread tissue damage may occur in some lesions without any immune cell infiltration [67], (2) neuroprotective factors are produced by infiltrating immune cells [90], and (3) reduced defense mechanisms or lack of trophic support contribute to the extent of destruction [91]. Therefore, any immune cell infiltrate and the detection of inflammatory mediators in CNS lesions should be interpreted within this context (Figure 12.3). The mere presence of a given

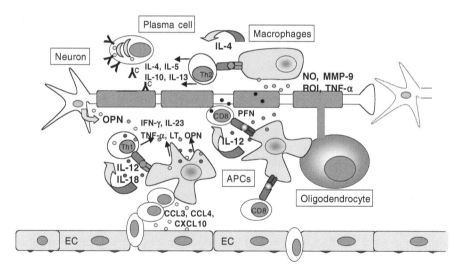

FIGURE 12.3 (Color insert follows page 208). Cytokines control mechanisms of structural damage to myelin and axons. Immune cells (CD4, CD8, T cells, and B lymphocytes) are directed by chemokines to encounter APC (microglia or DC), which present their cognate autoantigen for reactivation. Pro-inflammatory cytokines (e.g., IFN-γ, IL-23, osteopontin OPN) are secreted by T cells and either directly affect myelin structures or activate macrophages to release nitric oxide (NO), reactive oxygen intermediates (ROI), matrix metalloproteinases (MMP), or TNF-α. CD8 cells may directly attack axons by release of the cytotoxic mediator perforin (PFN). B lymphocytes are induced to terminal plasma cell differentiation by Th2 cytokines, and upon activation release myelin-specific antibodies, which can induce complement (C) mediated demyelination.

factor in the blood, CSF, or destructive lesion of MS patients could be a hint for a causal relationship, but it could also represent an endogenous regulatory process to reduce tissue damage or indicate an innocent epiphenomenon. Considering these three options, any therapeutic approach counteracting a specific factor may result in different outcomes depending on its temporal and spatial role during inflammation or tissue destruction. These problems are best exemplified by the story of TNF-α in MS.

This cytokine has been detected early in active MS lesions [22]; its presence in the CSF correlates with disability [92], and expression of TNF-α in blood mononuclear cells precedes clinical relapses by a few weeks [57]. Supporting evidence for an important role of TNF-α in immune-mediated tissue damage could be obtained from *in vitro* studies. TNF-α upregulates adhesion molecules on cerebral endothelial cells, induces chemokine production, mediates blood–brain barrier damage, and contributes to oligodendrocyte destruction [93] as well as axonal damage [94]. Another mechanism for TNF-α- mediated neurodegeneration has been suggested by interference with trophic factor signaling [95]. In addition, animal studies provided supporting evidence that neutralization of this factor ameliorates EAE [96]. As a conclusion of these experimental studies, neutralization of TNF-α appeared as a promising therapeutic strategy for MS. Unfortunately, all trials using TNF-α neutralizing strategies so far failed in MS [3]. To an even greater surprise, MS-like diseases manifested in a number of patients successfully treated with anti-TNF antibodies for Crohn's disease or rheumatoid arthritis [97]. Looking at other properties of TNF-α that have been described in the literature, the outcome of these trials is not as surprising as initially claimed. TNF-α induces apoptosis of invading T cells [98], it mediates release of soluble adhesion molecules from the surface of endothelial cells [99], and it induces production of neurotrophic factors by different cell types [2]. These pleiotropic effects of a single factor are not unusual and have been described for other cytokines as well. The more important determinants of cytokine action are rather the presence of other cytokines, distribution of the particular cytokine receptors on tissue cells, and the cross-talk of their signal transduction pathways within the cell than the specific properties of the cytokine itself [28].

There are only a few studies that directly correlated cytokine production of blood mononuclear cells with evidence for tissue destruction on cranial MRI. Prat et al. demonstrated increased levels of IFN-γ producing cells and a higher *in vitro* migration capacity of these cells in PPMS patients who were preselected for high T2 lesion load on cranial MRI [100]. Killestein and coworkers observed a negative correlation between IL-2, IL-4, IL-10, and IL-13 producing CD8+ T cells with accumulation of T1 lesion load over a 3-year period of repeated MRI measurements [101].

In another longitudinal exploratory trial, the expression of IL-12 (p40 and p35), TNF-α, IFN-γ, and IL-10 mRNA was quantified in blood cells from RRMS and SPMS in relation to monthly clinical and magnetic resonance imaging monitoring. In this study, MS patients had increased levels of IL-12p40 and decreased levels of IL-10 mRNA compared with controls; this difference was most pronounced in SPMS patients. Increased levels of IL-12p40 mRNA were also detected during the development of active, Gd+ MRI lesions. Moreover, in RRMS an increase was found prior to relapses. In RRMS, IL-10 mRNA was low 4 weeks before magnetic resonance imaging activity and 6 weeks before relapse; a significant increase to normal levels was noted when active lesions became apparent. In contrast, SPMS patients showed low IL-10 mRNA levels constitutively, suggesting that IL-10 plays an important role in the control of disease progression. [62].

Summarizing the results of studies correlating peripheral cytokine expression with disease progression on MRI, the negative correlations of anti-inflammatory cytokine-producing cells with increased T1 lesion load suggest a dysfunction in immunoregulation. Failure in controlling elements, which contribute to demyelination and axonal loss, could be the result. Not a direct proof but an important association, a study was conducted by careful examination of the immune factors present in lesions with recent or ongoing axonal damage. This study by Bitsch et al. suggests that macrophages and CD8+ T cells, but not TNF-α or the inducible nitric oxide synthase (iNOS), are significantly more frequent in these lesions than in less destructive areas [102]. It can therefore be hypothesized that a subgroup of CD8+ T cells accumulate in destructive MS lesions [103] and release of cytotoxic factors, e.g., granzyme-B or perforin are likely mediators of the acute axonal damage [104]. Therefore, pro-inflammatory cytokines probably have an indirect effect on tissue destruction, perhaps by shaping the appropriate inflammatory infiltrate.

Comparing the cytokine profile across the clinical spectrum of MS, in some studies, unique mRNA expression patterns for the different MS forms as compared to controls were described in blood cells. In one study, both RRMS and SPMS patients displayed increased levels of IL-12p40, IL-18, and TGF-β mRNA compared to controls, whereas PPMS patients showed only increased IL-18 mRNA levels. Both in PPMS and SPMS patients, IFN-γ and IL-10 mRNA were decreased compared to RRMS patients and controls. PPMS patients were unique in that they showed decreased IL-12Rβ1 mRNA [105].

In another study focusing on cell-associated cytokine production, it was shown that PPMS patients had a significant decrease in CD4+ T cells producing IL-2, IL-13, and TNF-α and a significant increase of CD8+ T cells producing IL-4 and IL-10 [106]. Such findings suggest that different cytokines may be involved in the pathogenesis of different forms of MS. However, the number of patients investigated until now is still very small, and further longitudinal studies are necessary to accurately define cytokine profiles and their specific role in different disease types.

VIII. NEUROPROTECTIVE ASPECTS OF INFLAMMATION MEDIATED BY CYTOKINES

Inflammatory reactions in the CNS usually are considered detrimental, but recent evidence suggests that they can also be beneficial and even have neuroprotective effects [2]. *In vitro*, immune cells and also cerebral endothelial cells can be induced by cytokines to produce neurotrophic factors, like brain derived neurotrophic factor (BDNF) and nerve growth factor (NGF) [107,108]. In response, these factors are able to directly suppress cytokine production of antigen-specific T cells [109]. Variable responses of immune cells to neurotrophic factors may depend on receptor expression. BDNF and NGF positive immune cells and their respective signaling transducing receptors (tyrosine kinase receptors trkB and trkA) were detected in inflammatory MS lesions [90]. Interestingly, full-length trkB (gp145trkB) is prominently expressed on neurons in the immediate vicinity of MS plaques, whereas trkA (the receptor for NGF) is expressed only on inflammatory cells in MS lesions [110]. Moreover, recent studies provided evidence that the p75 neurotrophin receptor (p75NTR) in MS lesions is also upregulated on oligodendrocytes and oligodendrocyte precursors as well as macrophages and microglia. It was suggested earlier that this receptor may play a role in the modulation of the inflammatory response, because NGF reduced HLA-antigen and costimulatory molecule expression on microglia via p75NTR [111]. In chronic active MS lesions, the low molecular weight receptor p75NTR and the tyrosine kinase receptor A (trkA) were found on immune cells, suggesting that these cells are mainly responsive to NGF. Thus, NGF signaling via p75NTR or trkA is likely to play a role in neurotrophin-mediated immunomodulation.

Ciliary neurotrophic factor (CNTF), a member of the neuropoietic, IL-6-like cytokine family, was shown to be protective against TNF-α- induced oligodendrocyte damage [112]. Furthermore, defective CNTF production both in EAE and MS is associated with an earlier onset and more severe course of the disease [91,113]. Oncostatin M (OSM) is another member of the IL-6 type cytokines that is expressed by immune cells within active MS lesions, and high amounts of this cytokine are spontaneously produced by PBMCs from MS patients compared to healthy donors. It is not only involved in the orchestration of the inflammatory infiltrate but has a considerable potential to induce myelination *in vitro* [114]. Owing to this intimate cross-talk between the immune and the nervous system, which confers beneficial outcomes on the target tissue, the concept of "neuroprotective immunity" has been proposed. This new view on the differential effects of inflammation in the CNS will likely have consequences for the pathogenesis and treatment of neuroinflammatory diseases such as MS. As an example, the immunomodulatory agent, glatiramer acetate, with strong homology to myelin basic protein (MBP) induces specific T helper cells, which regardless of their cytokine profile produces significant amounts of BDNF [115].

Whether modulation of neurotrophic factors production by immune cells can be used as a Trojan horse strategy for neuroprotection in MS is still under debate, and future trials using innovative endpoints will address that issue (see also Chapter 16).

IX. LESSONS FROM GENE TARGETING EXPERIMENTS IN THE ANIMAL MODEL OF EAE

The expression or upregulation of inflammatory mediators in the target tissue or within the circulating blood compartment in correlation with relapse, remission, or progression led to the guilt by association hypothesis. These mediators were then tested in different EAE models by strategies designed to block production or effect of the factor of interest using biological or pharmacological inhibitors. These studies were subject to concerns about specificity, side effects, and—of course—access of the inhibitory agent to the CNS. In order to better define the role of individual cytokines for immune-mediated demyelination, studies on cytokine knock-out animals were performed, and numerous experiments involving transgenic animals have identified causative roles for some cytokines but have also raised questions about the tissue specificity and cytokine expression kinetics [63]. Problems with the transgene approach have been appraised and include interference with normal developmental kinetics and spontaneous phenotypes with alterations in the immune or nervous system even in target-specific transgenes. These changes occur independent of autoimmune reactions induced by myelin antigen challenge or adoptive T cell transfer and may result in upregulation of other cytokines or signaling pathways, substituting at least in part for functions of the knock-out gene.

Again, TNF-α serves as a good example for the pleomorphic effects also observed in transgenic animal models. Most TNF-α transgenic animals showed spontaneous pathology in the CNS with progressive demyelination and macrophage infiltration, whereas TNF-receptor knock-out mice behave differently upon induction of EAE: TNFRI-deficient mice show a delayed disease onset but TNFRII-deficient animals develop more severe forms, arguing for a protective role of TNF-signaling via TNFRII in brain inflammation (summarized in [116]). These data suggest that the dualistic, pro-inflammatory and immunomodulatory role of TNF-α is best explained by a heterogeneity of TNF receptor usage in immune-mediated demyelination. In the case of TNF-α, it was therefore suggested to put forward strategies to block TNFRI-signaling rather than anti-TNF treatment in general. Blocking of the TNFRI in inflammatory diseases of the brain could inhibit the deleterious pro-inflammatory and tissue-damaging activities of TNF-α while allowing TNF to exert its beneficial immunomodulatory functions [28].

A more recent example about the role of cytokines, which gene targeting experiments taught us, is redundancy, which exists within the pro-inflammatory cytokine network. Experiments with IL-12 receptor knock-out animals revealed surprisingly a more severe EAE. These findings indicate that IL-12 responsiveness is not required for the pathogenesis of inflammatory demyelination in the CNS. In the absence of IL-12Rβ2, increased IL-23 levels and other inflammatory molecules could step in and may be responsible for increased severity of EAE [117] observed in this model. The obvious drawback of this approach is that expression of a given cytokine is constitutively "on" in transgenic animals or "off" in knock-out animals throughout development. Control of transgene expression by inducible and cell-specific promoters is a feasible strategy, particularly for studying diseases like MS where the initiating stimulus is unknown. Lessons from the transgenic animal models, which were performed in recent years, will guide the creation of more sophisticated genetically engineered animal models to better reflect the pathophysiological cascade associated with inflammatory-mediated demyelination [63].

X. POLYMORPHIC MARKERS IN CYTOKINE GENES AND DISEASE COURSE

Genetic factors have been involved in the pathogenesis of MS, but beyond the association of disease susceptibility with human leukocyte antigen (HLA) class II alleles DR15/DQw6 (HLA-DRB1*1501; HLA-DQB1*0601) [14], which code for molecules that participate in antigen recognition by T lymphocytes, no definite cluster of additional genes has been defined. Large multicenter studies have performed whole genome screening for the detection of susceptibility loci, but no conclusive answer about individual genes could be given until now. Several regions on individual chromosomes with high lod scores were identified (e.g., Chr. 6, 14, and 19), but sequencing of cytokine genes within these regions (e.g., TNF-α, LT-α, IFN-γ, IL-22, IL-26, and TGF-β2) did not reveal any significant association with disease susceptibility across different studies (Table 12.3).

Cytokine production may be affected by the HLA-type as has been demonstrated for T cell lines (TCL) from HLA-matched donors [118]. Increased secretion of lymphotoxin (LT) and TNFα was found in T cell lines from HLA-DR2-positive patients compared to other HLA-types. As both cytokines together with the HLA-genes are encoded on the short arm of chromosome 6, they may be linked to other polymorphic genes in this region. The results of that study suggest that the association of MS with HLA-DR2 implies a genetically determined propensity of T cells to produce increased amounts of LT and TNF-α [119].

Another way to study the effect of individual genes in relation to disease progression is the candidate gene approach. Gene products of factors known or suspected to be involved in the pathophysiology of immune-mediated structural damage to the CNS are regulated by specific sequences located in untranslated regions of that gene. These regions control expression levels or stability of the transcripts. Allelic polymorphisms that occur in these regions may alter the inducibility of a cytokine and could even result in tissue-specific alterations of expression, depending on the availability of transcription factors. Therefore, these polymorphisms can be regarded as biologically relevant alterations, which regulate gene expression levels and could result in different disease courses, progression indices, or recovery status, thereby fulfilling the criteria for disease modifying genes. Several polymorphic alleles were described in cytokine genes and have been investigated in blood cells from MS patients (see Table 12.3). Some of these studies provided additional information on gene expression levels or used appropriate functional assays to detect biological differences induced by the gene alterations.

The major goal of this candidate gene approach is to select for functional alterations, which are linked to disease phenotypes and may then define the gene of interest as a relevant functional modifier. Although associations of certain alleles with different expression levels were found (e.g., TNF-α, IL-10, and IL-6), none of these polymorphisms were unanimously associated with susceptibility to or course of the disease. Most of these studies were performed on relatively small populations, and only in a very few trials confirmatory analysis of independent populations were performed. Furthermore, the determination of disease phenotypes differed considerably among the various studies. For future studies it will be important to compare genotypes also to results from sophisticated MRI analysis or quantitative neuropsychological assessment in order to better define the correlation with relevant parameters of disease activity. Other limitations of the candidate gene approach are evident as the rationale to investigate functional polymorphisms in selected genes is highly dependent on the actual role the chosen factors play in the pathophysiology of disease. Less biased approaches that allow for large-scale screening procedures of multiple gene polymorphisms in comparison to expression analysis in blood cells or target tissue will hopefully increase our knowledge about the contribution of genetic variations to disease susceptibility or course.

TABLE 12.3
Cytokine Gene Polymorphisms and MS

Cytokine	Position of Polymorphism	Altered function	MS course or Susceptibility	Reference
IL-1	IL-1β C/T (Exon-5) + 85 bp VNTR intron-4 (IL-1ra)		Better outcome	[120]
IL-1	IL-1RN allele 2 IL-1β+3959allele2	Higher IL-1RA production *in vitro*	Worse disease course	[121]
IL-1α/β and IL-1RA	IL-1α-889,IL-1β-511 Promoter IL-1RA 85 bp VNTR		No effect on susceptibility, clinical course or MRI	[122]
IL-1α/β	IL-1α -889 C/T Promoter IL-1β -511 C/T Promoter		No effect on susceptibility or disease course	[123]
IL-2, IL-4, IL-10	Polymorphic microsatellite marker		No effect on susceptibility	[124]
IL-1α, IL-2, IL-2Rβ, IL-4, IL-9, IL-10	polymorphic microsatellite markers		No effect on susceptibility	[125]
IL-2	-475 A/T Promoter -631 G/A Promoter		No effect on susceptibility	[126]
IL-4	VNTR (IL-4B1)		Late onset	[127]
IL-4	I3(709)*VNTR		Susceptibility	[128]
IL-4R	R551		Associated with PPMS	[129]
IL-4R, TGF-β2, IFN-γ	Polymorphic microsatellite marker		Possible linkage on pedigree analysis	[124]
IL-6	-174/-597 Promoter	Less active	No effect	[130]
IL-6	VNTR (different alleles)		Association with disease course in Sardinians	[131]
IL-6	VNTR (C-allele)	Less active *in vivo*	No effect	[132]
IL-10	-2849 G/A Promoter -1082 A/G Promoter	Low IL-10 production	Less in PPMS	[133]

(*continued*)

TABLE 12.3 (Continued)
Cytokine Gene Polymorphisms and MS

Cytokine	Position of Polymorphism	Altered function	MS course or Susceptibility	Reference
IL-10	-1082 A/G Promoter		Less progressive disease	[134]
IL-10	-1082 A/G Promoter	Low IL-10 production	No effect	[135]
IL-11	Polymorphic microsatellite marker		No effect on susceptibility	[136]
IL-12p40	1188 (3′UTR)		No effect on susceptibility	[137]
IL-12p40	3′UTR (B-allele)	Less IL-12 production	Less in MS patients	[138]
IL-13	-1055 C/T Promoter -1024 C/T Promoter		No effect	[139]
IL-18	-607 C/A Promoter -137 G/C Promoter	Less IL-18 and IFN-γ production	No effect	[140]
IL-22, IL-26, IFN-γ	Polymorphic microsatellite marker on Chr.12q14		Gender differences	[141]
IFN-γ	IFN-γ alleles 11/12		Susceptibility to disease in Sardinian pop. not in other areas	[142]
IFN-γ	-333 C/T Promoter		No effect	[143]
IFN receptors	8 polymorphisms in IFNAR1, IFNAR2		No effect on treatment response to IFN-β	[144]
TNF-α	-376 G/A Promoter		A allele: increased susceptibility	[145]
TNF-α	Sequence of the TNF-gene revealed 4 sequence change none in protein coding regions		No association with disease severity and course	[146]

(continued)

TABLE 12.3 (Continued)
Cytokine Gene Polymorphisms and MS

Cytokine	Position of Polymorphism	Altered function	MS course or Susceptibility	Reference
TNF-α	-308 G/A Promoter	Affects TNF-α production	No effect	[147]
TNF-α LT-α	TNFα -308 G/A Promoter LTα Exon 3 C/A		Combined polymorphisms associated with susceptibility	[148]
TGF-β1	Polymorphic microsatellite marker		Milder disease course	[136]
TGFβ1	4 polymorphic marker in TGFb1 gene		No effect	[149]
OPN	4 SNP in exon 4+5		No effect	[150]
OPN	+1284 A/C		Effect on disease course	[151]
OPN	+8090 C/T		Associated with Susceptibility in Japanese	[152]

XI. NEW DEVELOPMENTS

A plethora of studies on measurement of individual cytokines or small series of Th1 vs. Th2 cytokines in the blood, CSF, or tissue of MS patients has been published, but no definite answer can be given as to what extent the results represent the overall inflammatory activity of the compartment of interest and how this actually relates to pathophysiology of the disease. With the advent of new powerful molecular techniques, it is now feasible to look at the gene expression levels of any given factor in comparison with over a thousand different parameters at the same time in compartments of interest. These gene arrays, either on filter or glass chip, consist of hundreds or thousands of DNA sequences representing defined genes that can be hybridized to RNA or cDNA from cells of interest. The cDNA probes labeled either radioactively or with a fluorescent dye bind to the matching gene spots on the microarray and identify all expressed genes of a given sample. Using this approach, systematic analyses of the gene expression patterns of EAE or MS brain tissue as well as blood cells have been performed, and molecules involved in specific biological processes (such as cellular activation, cell adhesion, intracellular signaling transmigration, or cell–cell interaction) were characterized. The few studies that looked at transcriptional profiles in MS lesions compared either acute lesions to normal brain [153] or different regions from the same MS brain [154]. Among several overlapping genes that were detected, transcripts encoding the cytokine osteopontin (OPN) were frequently detected in and exclusive to the mRNA population of MS plaques [155]. OPN is a pro-inflammatory cytokine and enhances IFN-γ and IL-12 production. The cellular expression pattern revealed a prominent staining with antibodies against OPN on microvascular endothelial cells and macrophages. Reactive astrocytes, microglial, and some neurons within MS brains were also positive for OPN [153], suggesting that this cytokine may play an important role in modulating the inflammatory response and mediating demyelination, thereby influencing the clinical severity of the disease. In a recent study, significantly increased OPN protein levels were detected in the plasma of relapsing–remitting MS patients. In contrast, OPN protein levels in primary progressive and secondary progressive MS patients were similar to healthy control levels. Interestingly, active relapsing–remitting patients had higher OPN protein levels than patients without relapses [156], further arguing for an important role of this cytokine in the inflammatory cascade during MS.

Several other genes related to structural components of the CNS, the neuroendocrine system, and allergy-associated genes were detected by this array approach [155]. It will be of special interest to evaluate these new genes in functional assays to gain further insights into the complex pathogenetic process of MS.

Expression profiling as a tool to monitor immunomodulatory treatment in MS has been applied to predict better outcome in individual patients and challenged our view on the concept of dichotomous effects of pro- and anti-inflammatory cytokines on the disease process [157]. In this study, it was demonstrated that high expression levels of the pro-inflammatory chemokine MCP-1 by IFNβ-1b was associated with a better outcome and less MRI activity. Increased levels of circulating MCP-1 and IP-10 were detected within 24 hours after subcutaneous or intramuscular injection of recombinant IFNβ [158], and anti-inflammatory properties of chemokines have already been described in the past [159]. Cytokines and chemokines have probably not yet uncovered their full spectrum of functions, which are relevant to understand their role in chronic neuroinflammation. Therefore, carefully designed longitudinal studies using the new approach of expression profiling together with careful phenotyping of clinical and MRI disease course will be important to provide new targets for therapeutic interventions in order to better control disease in the future.

REFERENCES

1. Lucchinetti, C. et al., Distinct patterns of multiple sclerosis pathology indicates heterogeneity in pathogenesis. *Brain Pathol.*, 6, 259–274, 1996.
2. Kerschensteiner, M. et al., Neurotrophic cross-talk between the nervous and immune systems: implications for neurological diseases. *Ann. Neurol.*, 53(3), 292–304, 2003.
3. Schwid, S.R. and Noseworthy, J.H., Targeting immunotherapy in multiple sclerosis— a near hit and a clear miss. *Neurology*, 53, 444–445, 1999.
4. Compston, A. et al., McAlpine's multiple sclerosis, 3rd ed. Churchill Livingstone, London, 1998.
5. Confavreux, C. et al., Relapses and progression of disability in multiple sclerosis. *N. Engl. J. Med.*, 343, 1430–38, 2000.

6. Dalton, C. et al., Application of the new McDonald criteria to patients with clinically isolated syndromes suggestive of multiple sclerosis. *Ann. Neurol.*, 52, 47–53, 2002.

7. McFarland, H.F., Frank, J.A., and Albert, P.S., Using gadolinium-enhanced magnetic resonance imaging lesions to monitor disease activity in multiple sclerosis. *Ann. Neurol.*, 32, 758–766, 1992.

8. Katz, D. et al., Correlation between magnetic resonance imaging findings and lesion development in chronic, active multiple sclerosis. *Ann. Neurol.*, 34(5), 661–9, 1993.

9. Hohlfeld, R., Biotechnological agents for the immunotherapy of multiple sclerosis. *Brain*, 120, 865–916, 1997.

10. Brex, P.A. et al., Detection of ventricular enlargement in patients at the earliest clinical stage of MS. *Neurology*, 54, 1689–1691, 2000.

11. Trapp, B. et al., Axonal transection in multiple sclerosis lesions. *N. Engl. J. Med.*, 338, 278–285, 1998.

12. Rieckmann, P. and Smith, K., Multiple Sclerosis: more than inflammation and demyelination. *Trends Neurosci.*, 24, 435–437, 2001.

13. Sellebjerg, F. and Sorensen, T.L., Chemokines and matrix metalloproteinase-9 in leukocyte recruitment to the central nervous system. *Brain Res. Bull.*, 61(3), 347–55, 2003.

14. Olerup, O. and Hillert, J., HLA class II-associated genetic susceptibility in multiple sclerosis: a critical evaluation. *Tissue Antigens*, 38, 1–15, 1991.

15. Sawcer, S. et al., A genome screen in multiple sclerosis reveals susceptibility loci on chromosome 6p21 and 17q22. *Nat. Genet.*, 13, 464–468, 1996.

16. Pender, M.P., Infection of autoreactive B lymphocytes with EBV, causing chronic autoimmune diseases. *Trends Immunol.*, 24(11), 584–8, 2003.

17. Confavreux, C. et al., Rate of pregnancy-related relapse in multiple sclerosis. *N. Engl. J. Med.*, 339, 285–291, 1998.

18. Raghupathy, R., Th1-type immunity is incompatible with successful pregnancy. *Immunol. Tod.*, 18, 478–482, 1997.

19. Kruse, N. et al., Variations in cytokine mRNA expression during normal human pregnancy. *Clin. Exp. Immunol.*, 119, 317–322, 2000.

20. Hofman, F.M. et al., Immunoregulatory molecules and IL 2 receptors identified in multiple sclerosis brain. *J. Immunol.*, 136(9), 3239–45, 1986.

21. Traugott, U. and Lebon, P., Demonstration of alpha, beta, and gamma interferon in active chronic multiple sclerosis lesions. *Ann. NY Acad. Sci*,. 540, 309–11, 1988.

22. Hofman, F.M. et al., Tumor necrosis factor identified in multiple sclerosis brain. *J. Exp. Med.*, 170, 607–612, 1989.

23. Mosmann, T. et al., Two types of murine helper T cell clone. I. Definition according to profiles of lymphokine activities and secreted proteins. *J. Immunol.*, 136, 2348–2355, 1986.

24. Romagnani, S., Human TH1 and TH2 subsets: doubt no more. *Immunol. Today*, 12, 256–262, 1991.

25. Gor, D., Rose, N. and Greenspan, N., TH1-TH2: a procrustean paradigm. *Nat. Immunol.*, 4, 503–507, 2003.

26. Del Prete, G. et al., Human IL-10 is produced by both type 1 helper (Th1) and type 2 helper (Th2) T cell clones and inhibits their antigen-specific proliferation and cytokine production. *J. Immunol.*, 150, 353–359, 1993.

27. Fukaura, H. et al., Induction of circulating myelin basic protein and proteolipid protein-specific transforming growth factor-beta1-secreting Th3 T cells by oral administration of myelin in multiple sclerosis patients. *J. Clin. Invest.*, 98(1), 70–7, 1996.

28. Aggarwal, B., Signaling pathways of the TNF superfamily: a double-edged sword. *Nat. Rev. Immunol.*, 3, 745–756, 2004.
29. Jacobs, L. et al., Intrathecally administered natural human fibroblast interferon reduces exacerbations of multiple sclerosis. Results of a multicenter, double-blind study. *Arch. Neurol.*, 44(6), 589–95, 1987.
30. Noseworthy, J., Gold, R., and Hartung, H.P., Treatment of multiple sclerosis: recent trials and future perspectives. *Curr. Opin. Neurol*, 12, 279–293, 1999.
31. Panitch, H.S. et al., Exacerbations of multiple sclerosis in patients treated with gamma interferon. *The Lancet*, 893–894, 1987.
32. Knobler, R.L., Systemic interferon therapy of multiple sclerosis: the pros. *Neurology*, 38(7 Suppl 2), 58–61, 1988.
33. Link, H., The cytokine storm in multiple sclerosis. *Mult. Scler.*, 4(1), 12–5, 1998.
34. Ozenci, V., Kouwenhoven, M., and Link, H., Cytokines in multiple sclerosis: methodological aspects and pathogenic implications. *Mult. Scler.*, 8(5), 396–404, 2002.
35. Kubo, M., Hanada, T., and Yoshimura, A., Suppressors of cytokine signaling and immunity. *Nat. Immunol.*, 4, 1169–1176, 2003.
36. Gold, R., Hartung, H., and Toyka, K., Animal models for autoimmune demyelinating disorders of the nervous system. *Mol Med Today*, 6, 88–91, 2000.
37. Olsson, T., Critical influences of the cytokine orchestration on the outcome of myelin antigen-specific T-cell autoimmunity in experimental autoimmune encephalomyelitis and multiple sclerosis. *Immunol Rev*, 144, 245–68, 1995.
38. Pette, M. et al., Myelin basic protein-specific T lymphocyte lines from MS patients and healthy individuals. *Neurology*, 43, 1770–1776, 1990.
39. Hong, J. et al., Ex vivo detection of myelin basic protein-reactive T cells in multiple sclerosis and controls using specific TCR oligonucleotide probes. *Eur. J. Immunol.*, 34, 870–881, 2004.
40. Vandevyver, C. et al., Clonal expansion of myelin basic protein-reactive T cells in patients with multiple sclerosis: restricted T cell receptor V gene rearrangements and CDR3 sequence. *Eur. J. Immunol.*, 25, 958–968, 1995.
41. Muraro, P. et al., Human autoreactive CD4+ T cells from naive CD45RA+ and memory CD45RO+ subsets differ with respect to epitope specificity and functional antigen avidity. *J. Immunol.*, 164, 5474–5481, 2000.
42. Correale, J., de los Milagros Bassani, A., and Molinas, M., Time course of T-cell responses to MOG and MBP in patients with clinically isolated syndromes. *J. Neuroimmunol.*, 136, 162–171, 2003.
43. Schjetne, K.W. et al., Cutting edge: link between innate and adaptive immunity: Toll-like receptor 2 internalizes antigen for presentation to CD4+ T cells and could be an efficient vaccine target. *J. Immunol.*, 171(1), 32–6, 2003.
44. Vandenbroeck, K. et al., Inhibiting cytokines of the interleukin-12 family: recent advances and novel challenges. *J. Pharm. Pharmacol.*, 56(2), 145–60, 2004.
45. Nicoletti, F. et al., Increased serum levels of interleukin-18 in patients with multiple sclerosis. *Neurology*, 57(2), 342–4, 2001.
46. Karni, A. et al., IL-18 is linked to raised IFN-gamma in multiple sclerosis and is induced by activated CD4(+) T cells via CD40-CD40 ligand interactions. *J. Neuroimmunol.*, 125(1-2), 134–40, 2002.
47. Nakanishi, K. et al., Interleukin-18 is a unique cytokine that stimulates both Th1 and Th2 responses depending on its cytokine milieu. *Cytokine Growth Factor Rev*, 12(1), 53–72, 2001.
48. Trinchieri, G., Interleukin-12 and the regulation of innate resistance and adaptive immunity. *Nat Rev Immunol*, 3(2), 133-46, 2003.

49. Becher, B., Durell, B.G., and Noelle, R.J., IL-23 produced by CNS-resident cells controls T cell encephalitogenicity during the effector phase of experimental autoimmune encephalomyelitis. *J Clin Invest*, 112(8), 1186–91, 2003.

50. Krug, A. et al., Toll-like receptor expression reveals CpG DNA as a unique microbial stimulus for plasmacytoid dendritic cells which synergizes with CD40 ligand to induce high amounts of IL-12. *Eur J Immunol*, 31(10), 3026–37, 2001.

51. Gelman, A.E. et al., Toll-like receptor ligands directly promote activated CD4+ T cell survival. *J Immunol*, 172(10), 6065–73, 2004.

52. de Vos, A. et al., Transfer of central nervous system autoantigens and presentation in secondary lymphoid organs. *J. Immunol.*, 169, 5415–23, 2002.

53. Sibley, A., Bamford, C.R., and Clark, K., Clinical viral infections and multiple sclerosis. *The Lancet*, 1313–1315, 1985.

54. Claeys, S. et al., Human beta-defensins and Toll-like receptors in the upper airway. *Allergy*, 58(8), 748–53, 2003.

55. Moriabadi, N.F. et al., Influenza vaccination in MS: absence of T-cell response against white matter proteins. *Neurology*, 56(7), 938–43, 2001.

56. Beck, J. et al., Increased production of interferon gamma and tumor necrosis factor precedes clinical manifestation in multiple sclerosis: do cytokines trigger off exacerbations? *Acta Neurol Scand*, 78(4), 318–23, 1988.

57. Rieckmann, P. et al., Tumor necrosis factor-a messenger RNA expression in patients with relapsing-remitting multiple sclerosis is asociated with disease activity. *Ann. Neurol.*, 37, 82–88, 1995.

58. Calabresi, P.A. et al., ELI-spot of Th-1 cytokine secreting PBMC's in multiple sclerosis: correlation with MRI lesions. *J Neuroimmunol*, 85(2), 212–9, 1998.

59. Matusevicius, D. et al., Interleukin-12 and perforin mRNA expression is augmented in blood mononuclear cells in multiple sclerosis. *Scand J Immunol*, 47(6), 582–90, 1998.

60. Martino, G. et al., Tumor necrosis factor alpha and its receptors in relapsing-remitting multiple sclerosis. *J Neurol Sci*, 152(1), 51–61, 1997.

61. Khoury, S.J. et al., Changes in serum levels of ICAM and TNF-R correlate with disease activity in multiple sclerosis. *Neurology*, 53(4), 758–64, 1999.

62. van Boxel-Dezaire, A.H. et al., Decreased interleukin-10 and increased interleukin-12p40 mRNA are associated with disease activity and characterize different disease stages in multiple sclerosis. *Ann Neurol*, 45(6), 695–703, 1999.

63. Owens, T., Wekerle, H., and Antel, J., Genetic models for CNS inflammation. *Nat Med*, 7(2), 161–6, 2001.

64. Sakaguchi, S. et al., Organ-specific autoimmune diseases induced in mice by elimination of T cell subset. I. Evidence for the active participation of T cells in natural self-tolerance; deficit of a T cell subset as a possible cause of autoimmune disease. *J Exp Med*, 161(1), 72–87, 1985.

65. Shevach, E.M., Regulatory, T, cells in autoimmmunity*. *Annu Rev Immunol*, 18, 423–49, 2000.

66. Viglietta, V. et al., Loss of functional suppression by CD4+CD25+ regulatory T cells in patients with multiple sclerosis. *J.Exp. Med.*, 199, 971–979, 2004.

67. Barnett, M.H. and Prineas, J.W., Relapsing and remitting multiple sclerosis: pathology of the newly forming lesion. *Ann Neurol*, 55(4), 458–68, 2004.

68. Chou, Y.K. et al., CD4 T-cell epitopes of human alpha B-crystallin. *J Neurosci Res*, 75(4), 516–23, 2004.

69. Stins, M.F., Gilles, F., and Kim, K.S., Selective expression of adhesion molecules on human brain microvascular endothelial cells. *J Neuroimmunol*, 76(1–2), 81–90, 1997.

70. Piccio, L. et al., Molecular mechanisms involved in lymphocyte recruitment in inflamed brain microvessels: critical roles for P-selectin glycoprotein ligand-1 and heterotrimeric G(i)-linked receptors. *J Immunol*, 168(4), 1940–9, 2002.

71. Woodroofe, M.N. and Cuzner, M.L., Cytokine mRNA expression in inflammatory multiple sclerosis lesions: detection by non-radioactive in situ hybridization. *Cytokine*, 5(6), 583–8, 1993.

72. Battistini, L. et al., CD8+ T cells from patients with acute multiple sclerosis display selective increase of adhesiveness in brain venules: a critical role for P-selectin glycoprotein ligand-1. *Blood*, 101(12), 4775–82, 2003.

73. Greenwood, J. et al., Intracellular domain of brain endothelial intercellular adhesion molecule-1 is essential for T lymphocyte-mediated signaling and migration. *J Immunol*, 171(4), 2099–108, 2003.

74. Mankertz, J. et al., Expression from the human occludin promoter is affected by tumor necrosis factor alpha and interferon gamma. *J Cell Sci*, 113 (Pt 11), 2085–90, 2000.

75. Ozaki, H. et al., Cutting edge: combined treatment of TNF-alpha and IFN-gamma causes redistribution of junctional adhesion molecule in human endothelial cells. *J Immunol*, 163(2), 553–7, 1999.

76. Ostermann, G. et al., JAM-1 is a ligand of the beta(2) integrin LFA-1 involved in transendothelial migration of leukocytes. *Nat Immunol*, 3(2), 151–8, 2002.

77. Sawada, N. et al., Tight junctions and human diseases. *Med Electron Microsc*, 36(3), 147–56, 2003.

78. Minagar, A. et al., Interferon (IFN)-beta 1a and IFN-beta 1b block IFN-gamma-induced disintegration of endothelial junction integrity and barrier. *Endothelium*, 10(6), 299–307, 2003.

79. Kirk, J. et al., Tight junctional abnormality in multiple sclerosis white matter affects all calibres of vessel and is associated with blood-brain barrier leakage and active demyelination. *J Pathol*, 201(2), 319–27, 2003.

80. Alter, A. et al., Determinants of human B cell migration across brain endothelial cells. *J Immunol*, 170(9), 4497–505, 2003.

81. Lichtinghagen, R. et al., Expression of matrix metalloproteinase-9 and its inhibitors in mononuclear blood cells of patients with multiple sclerosis. *J Neuroimmunol*, 99(1), 19–26, 1999.

82. Strunk, T. et al., Increased numbers of CCR5+ interferon gamma. and tumor necrosis factor alpha-secreting T lymphocytes in multiple sclerosis patients. *Ann. Neurol.*, 47, 269–273, 2000.

83. Trebst, C. and Ransohoff, R.M., Investigating chemokines and chemokine receptors in patients with multiple sclerosis: opportunities and challenges. *Arch Neurol*, 58(12), 1975–80, 2001.

84. Mahad, D.J. et al., Expression of chemokine receptors CCR1 and CCR5 reflects differential activation of mononuclear phagocytes in pattern II and pattern III multiple sclerosis lesions. *J Neuropathol Exp Neurol*, 63(3), 262–73, 2004.

85. Sörensen, T. et al., Expression of specific chemokines and chemokine receptors in the central nervous system of multiple sclerosis patients. *J Clin Invest*, 103, 807–815, 1999.

86. Sorensen, T.L., Roed, H., and Sellebjerg, F., Chemokine receptor expression on B cells and effect of interferon-beta in multiple sclerosis. *J Neuroimmunol*, 122(1-2), 125–31, 2002.

87. Ruprecht, K. et al., Effects of Oncostatin M on Human Cerebral Endothelial Cells and Expression in Inflammatory CNS Lesions. *J. Neuropath. Exp. Neurol.*, 60, 1087–1098, 2001.

88. Ensoli, F. et al., Lymphomononuclear cells from multiple sclerosis patients spontaneously produce high levels of oncostatin M, tumor necrosis factors alpha and beta, and interferon gamma. *Mult Scler*, 8(4), 284–8, 2002.

89. van Walderveen, M. et al., Neuronal damage in T1-hypointense multiple sclerosis lesions demonstrated *in vivo* using proton magnetic resonance spectroscopy. *Ann. Neurol.*, 46, 79–87, 1999.

90. Kerschensteiner, M. et al., Activated human T cells, B cells, and monocytes produce brain-derived neurotrophic factor *in vitro* and in inflammatory brain lesions: a neuroprotective role of inflammation? *J. Exp. Med.*, 189(5), 865–70, 1999.

91. Linker, R. et al., CNTF is a major protective factor in demyelinating CNS disease: a neurotrophic cytokine as modulator in neuroinflammation. *Nature Med.*, 8, 620–24, 2002.

92. Sharief, M.K. and Hentges, R., Association between tumor necrosis factor-alpha and disease progression in patients with multiple sclerosis. *N. Engl. J. Med.*, 325, 467–472, 1991.

93. Selmaj, K.W. and Raine, C.S., Tumor necrosis factor mediates myelin and oligodendrocyte damage *in vitro*. *Ann. Neurol*, 23, 339–346, 1988.

94. Gimsa, U. et al., Axonal damage induced by invading T cells in organotypic central nervous system tissue *in vitro*: involvement of microglial cells. *Brain Pathol*, 10(3), 365–77, 2000.

95. Venters, H.D., Dantzer, R., and Kelley, K.W., A new concept in neurodegeneration: TNF-alpha is a silencer of survival signals. *Trends Neurosci.*, 23, 175–180, 2000.

96. Selmaj, K., Raine, C.S., and Cross, A.H., Anti-tumor necrosis factor therapy abrogates autoimmune demyelination. *Ann. Neurol.*, 30, 694–700, 1991.

97. Sicotte, N.L. and Voskuhl, R.R., Onset of multiple sclerosis associated with anti-TNF therapy. *Neurology*, 57(10), 1885–8, 2001.

98. Pender, M.P. and Rist, M.J., Apoptosis of inflammatory cells in immune control of the nervous system: role of glia. *Glia*, 36(2), 137–44, 2001.

99. Kallmann, B. et al., Cytokine-induced modulation of cellular adhesion to human cerebral endothelial cells is mediated by soluble vascular cell adhesion molecule-1. *Brain*, 123, 687–697, 2000.

100. Prat, A. et al., Heterogeneity of T-lymphocyte function in primary progressive multiple sclerosis: relation to magnetic resonance imaging lesion volume. *Ann Neurol*, 47(2), 234–7, 2000.

101. Killestein, J. et al., Cytokine producing CD8+ T cells are correlated to MRI features of tissue destruction in MS. *J Neuroimmunol*, 142(1-2), 141–8, 2003.

102. Bitsch, A. et al., Acute axonal injury in multiple sclerosis. Correlation with demyelination and inflammation. *Brain*, 123, 1174–1183, 2000.

103. Babbe, H. et al., Clonal expansions of CD8(+) T cells dominate the T cell infiltrate in active multiple sclerosis lesions as shown by micromanipulation and single cell polymerase chain reaction. *J Exp Med*, 192(3), 393–404, 2000.

104. Neumann, H., Molecular mechanisms of axonal damage in inflammatory central nervous system diseases. *Curr Opin Neurol*, 16(3), 267–73, 2003.

105. van Boxel-Dezaire, A.H. et al., Cytokine and IL-12 receptor mRNA discriminate between different clinical subtypes in multiple sclerosis. *J Neuroimmunol*, 120(1-2), 152–60, 2001.

106. Killestein, J. et al., Intracellular cytokine profile in T-cell subsets of multiple sclerosis patients: different features in primary progressive disease. *Mult Scler*, 7(3), 145–50, 2001.

107. Moalem, G. et al., Production of neurotrophins by activated T cells: implications for neuroprotective autoimmunity. *J Autoimmun*, 15, 331–345, 2000.

108. Bayas, A. et al., Human cerebral endothelial cells are a potential source for bioactive BDNF. *Cytokine* 19, 55–58, 2002.

109. Bayas, A. et al., Modulation of cytokine mRNA expression by BDNF and NGF in human immune cells. *Neurosci. Lett.* 335, 755–758, 2003.

110. Stadelmann, C., Kerschensteiner, M., and Misgeld, T., BDNF and gp145trkB in multiple sclerosis brain lesions: neuroprotective interactions between immune and neuronal cells? *Brain*, 125, 74–85, 2002.

111. Neumann, H. et al., Neurotrophins inhibit major histocompatibility class II inducibility of microglia: involvement of the p75 neurotrophin receptor. *PNAS*, 95, 5779-84, 1998.

112. D'Souza, S., Alinauskas, K., and Antel, J., Ciliary neurotrophic factor selectively protects human oligodendrocytes from tumor necrosis factor-mediated injury. *J Neurosci Res 43*, 43, 289–298, 1996.

113. Gieß, R. et al., A null mutation in the CNTF gene is associated with early onset of multiple sclerosis. *Arch. Neurol.*, 59, 407–409, 2002.

114. Stankoff, B. et al., Ciliary neurotrophic factor (CNTF) enhances myelin formation: a novel role for CNTF and CNTF-related molecules. *J Neurosci*, 22(21), 9221–7, 2002.

115. Ziemssen, T. et al., Glatiramer acetate-specific T-helper 1- and 2-type cell lines produce BDNF: implications for multiple sclerosis therapy. Brain-derived neurotrophic factor. *Brain*, 125(Pt 11), 2381–91, 2002.

116. Kollias, G. and Kontoyiannis, D., Role of TNF/TNFR in autoimmunity: specific TNF receptor blockade may be advantageous to anti-TNF treatments. *Cytokine Growth Factor Rev*, 13, 315–21, 2002.

117. Zhang, G.X. et al., Induction of experimental autoimmune encephalomyelitis in IL-12 receptor-beta 2-deficient mice: IL-12 responsiveness is not required in the pathogenesis of inflammatory demyelination in the central nervous system. *J Immunol*, 170(4), 2153–60, 2003.

118. Zipp, F. et al., Genetic control of multiple sclerosis: increased production of lymphotoxin and tumor necrosis factor-alpha by HLA-DR2+ T cells. *Ann Neurol*, 38(5), 723–30, 1995.

119. Epplen, C. et al., Genetic predisposition to multiple sclerosis as revealed by immunoprinting. *Ann Neurol*, 41(3), 341–52, 1997.

120. Kantarci, O.H. et al., Association of two variants in IL-1beta and IL-1 receptor antagonist genes with multiple sclerosis. *J Neuroimmunol*, 106(1–2), 220–7, 2000.

121. Schrijver, H.M. et al., Association of interleukin-1beta and interleukin-1 receptor antagonist genes with disease severity in MS. *Neurology*, 52(3), 595–9, 1999.

122. Hooper-van Veen, T. et al., The interleukin-1 gene family in multiple sclerosis susceptibility and disease course. *Mult Scler*, 9(6), 535–9, 2003.

123. Ferri, C. et al., Lack of association between IL-1A and IL-1B promoter polymorphisms and multiple sclerosis. *J Neurol Neurosurg Psychiatry*, 69(4), 564–5, 2000.

124. He, B. et al., Linkage and association analysis of genes encoding cytokines and myelin proteins in multiple sclerosis. *J Neuroimmunol*, 86(1), 13–9, 1998.

125. McDonnell, G.V. et al., An evaluation of interleukin genes fails to identify clear susceptibility loci for multiple sclerosis. *J Neurol Sci*, 176(1), 4–12, 2000.

126. Fedetz, M. et al., Analysis of -631 and -475 interleukin-2 promoter single nucleotide polymorphisms in multiple sclerosis. *Eur J Immunogenet*, 29(5), 389–90, 2002.

127. Vandenbroeck, K. et al., Occurrence and clinical relevance of an interleukin-4 gene polymorphism in patients with multiple sclerosis. *J Neuroimmunol*, 76(1–2), 189–92, 1997.

128. Kantarci, O.H. et al., A population-based study of IL4 polymorphisms in multiple sclerosis. *J Neuroimmunol*, 137(1–2), 134–9, 2003.

129. Hackstein, H. et al., The interleukin-4 receptor variant R551 influences susceptibility to major clinical forms of multiple sclerosis. *J. Neuroimmunol.*, in press, 2001.

130. Fedetz, M. et al., The -174/-597 promoter polymorphisms in the interleukin-6 gene are not associated with susceptibility to multiple sclerosis. *J Neurol Sci*, 190(1–2), 69–72, 2001.

131. Vandenbroeck, K. et al., High-resolution analysis of IL-6 minisatellite polymorphism in Sardinian multiple sclerosis: effect on course and onset of disease. *Genes Immun*, 1(7), 460–3, 2000.

132. Schmidt, S. et al., Investigation of a genetic variation of a variable number tandem repeat polymorphism of interleukin-6 gene in patients with multiple sclerosis. *J Neurol*, 250(5), 607–11, 2003.

133. de Jong, B.A. et al., Frequency of functional interleukin-10 promoter polymorphism is different between relapse-onset and primary progressive multiple sclerosis. *Hum Immunol*, 63(4), 281–5, 2002.

134. Luomala, M. et al., Promoter polymorphism of IL-10 and severity of multiple sclerosis. *Acta Neurol Scand*, 108(6), 396–400, 2003.

135. Mäurer, M. et al., Genetic variation at position −1082 of the interleukin 10 (IL-10) Promoter and the outcome if multiple sclerosis. *J. Neuroimmunol.*, 104, 98–100, 2000.

136. Green, A.J. et al., Sequence variation in the transforming growth factor-beta1 (TGFB1) gene and multiple sclerosis susceptibility. *J Neuroimmunol*, 116(1), 116–24, 2001.

137. Hall, M.A. et al., Genetic polymorphism of IL-12 p40 gene in immune-mediated disease. *Genes Immun*, 1(3), 219–24, 2000.

138. van Veen, T. et al., Interleukin-12p40 genotype plays a role in the susceptibility to multiple sclerosis. *Ann Neurol*, 50(2), 275, 2001.

139. Hummelshoj, T. et al., Association between an interleukin-13 promoter polymorphism and atopy. *Eur J Immunogenet*, 30(5), 355–9, 2003.

140. Giedraitis, V. et al., Cloning and mutation analysis of the human IL-18 promoter: a possible role of polymorphisms in expression regulation. *J Neuroimmunol*, 112(1-2), 146–52, 2001.

141. Goris, A. et al., Linkage disequilibrium analysis of chromosome 12q14-15 in multiple sclerosis: delineation of a 118-kb interval around interferon-gamma (IFNG) that is involved in male versus female differential susceptibility. *Genes Immun*, 3(8), 470–6, 2002.

142. Goris, A. et al., Analysis of an IFN-gamma gene (IFNG) polymorphism in multiple sclerosis in Europe: effect of population structure on association with disease. *J Interferon Cytokine Res*, 19(9), 1037–46, 1999.

143. Giedraitis, V., He, B., and Hillert, J., Mutation screening of the interferon-gamma gene as a candidate gene for multiple sclerosis. *Eur J Immunogenet*, 26(4), 257–9, 1999.

144. Sriram, U. et al., Pharmacogenomic analysis of interferon receptor polymorphisms in multiple sclerosis. *Genes Immun*, 4(2), 147–52, 2003.

145. Fernandez-Arquero, M. et al., Primary association of a TNF gene polymorphism with susceptibility to multiple sclerosis. *Neurology*, 53(6), 1361–3, 1999.

146. Weinshenker, B.G. et al., Genetic variation in the tumor necrosis factor alpha gene and the outcome of multiple sclerosis. *Neurology*, 49(2), 378–85, 1997.

147. Maurer, M. et al., Gene polymorphism at position -308 of the tumor necrosis factor alpha Promoter is not associated with disease progression in multiple sclerosis patients. *J Neurol*, 246(10), 949–54, 1999.

148. Mycko, M. et al., Multiple sclerosis: the frequency of allelic forms of tumor necrosis factor and lymphotoxin-alpha. *J Neuroimmunol*, 84(2), 198–206, 1998.

149. Weinshenker, B.G. et al., Genetic variation in the transforming growth factor beta1 gene in multiple sclerosis. *J Neuroimmunol*, 120(1–2), 138–45, 2001.

150. Hensiek, A.E. et al., Osteopontin gene and clinical severity of multiple sclerosis. *J Neurol*, 250(8), 943–7, 2003.

151. Caillier, S. et al., Osteopontin polymorphisms and disease course in multiple sclerosis. *Genes Immun*, 4(4), 312–5, 2003.

152. Niino, M. et al., Genetic polymorphisms of osteopontin in association with multiple sclerosis in Japanese patients. *J Neuroimmunol*, 136(1–2), 125–9, 2003.

153. Chabas, D. et al., The influence of the proinflammatory cytokine, osteopontin, on autoimmune demyelinating disease. *Science*, 294(5547), 1731–5, 2001.

154. Mycko, M.P. et al., cDNA microarray analysis in multiple sclerosis lesions: detection of genes associated with disease activity. *Brain*, 126(Pt 5), 1048–57, 2003.

155. Steinman, L. and Zamvil, S., Transcriptional analysis of targets in multiple sclerosis. *Nat. Rev. Immunol.*, 3, 483–492, 2003.

156. Vogt, M.H. et al., Elevated osteopontin levels in active relapsing-remitting multiple sclerosis. *Ann Neurol*, 53(6), 819–22, 2003.

157. Sturzebecher, S. et al., Expression profiling identifies responder and non-responder phenotypes to interferon-beta in multiple sclerosis. *Brain*, 126(Pt 6), 1419–29, 2003.

158. Buttmann, M., Merzyn, C., and Rieckmann, P., IFN-beta induces systemic chemokine release in patients with multiple sclerosis: an important mode of action? *J. Neuroimmunol.* 156, 195–203, 2003.

159. Rutledge, B. et al., High level monocyte chemoattractant protein-1 expression in transgenic mice increases their susceptibility to intracellular pathogens. *J. Immunol.*, 155, 4838–4843, 1995.

13 Cytokines in Brain Trauma and Spinal Cord Injury

Roberta Brambilla, Valerie Bracchi-Ricard, and John R. Bethea

CONTENTS

Following traumatic injury to the central nervous system (CNS), cytokine levels rapidly increase at the site of lesion. Cytokines are produced at high concentrations by resident cells of the brain and spinal cord parenchyma (astrocytes, neurons, microglial cells) and by immune cells (macrophages, leukocytes), infiltrating the injured tissue through a compromised blood–brain barrier (BBB). Cytokines are commonly categorized into pro- and anti-inflammatory, based on the concept that they promote either positive or negative effects on the recovery of the injured CNS. However, an increasing body of evidence, provided mainly by recent studies in knock-out and transgenic models (see a summary of these studies in Table 13.1 and Table 13.2), reveals the existence of a more complicated scenario, as cytokines promoting negative inflammatory events at early stages following CNS trauma are found to participate in beneficial regenerative events at later phases.

TABLE 13.1
Cytokine Knock-Out and Transgenic Studies in Traumatic Brain Injury

Targeted Cytokine	Origin	Type of Trauma	Outcome	Reference
TNF KO	138	Controlled cortical impact	Better neurological/motor score acutely (2 days); worse neurological/motor score, higher tissue loss chronically (3 weeks)	17
TNFR KO	139	Controlled cortical impact	Increased lesion volume and BBB disruption	18
IL-1β KO	140	Corticectomy	Lack of CNTF elevation	52
GFAP-IL-1ra TG	138	Closed head injury	Reduced cytokine expression; Better neurological recovery	50
IL-6 KO	Jackson Labs, 141	ACI right parietal cortex	Slower rate of recovery and healing; leaky blood vessels	70
	141	Focal cryo-injury to fronto–parietal cortex	Decreased astrogliosis and macrophage recruitment; reduced expression of MTI+II and GM-CSF; increased MTIII; increased oxidative stress; increased apoptosis	142, 72
GFAP-IL-6 TG	67	Focal cryo-injury to fronto–parietal cortex	Decreased oxidative stress; decreased apoptosis; increased expression of MTI+II; faster tissue repair	69

(*continued*)

TABLE 13.1 (Continued)
Cytokine Knock-Out and Transgenic Studies in Traumatic Brain Injury

Targeted Cytokine	Origin	Type of Trauma	Outcome	Reference
LIF KO	143	Sagittal cortical laceration	Slower microglial/ astroglial responses	74
LIF/IL-6 KO	141, 143	Sagittal cortical laceration	Slower microglial/ astroglial responses; similar to single KO	74

After briefly introducing the characteristics of CNS trauma in humans and how they have been replicated in animal models, in this chapter, in the attempt to shed light on the complex role of cytokines in CNS trauma, we will present an overview of the data available in the literature on cytokines in various *in vivo* models of traumatic spinal cord and brain injury. Particularly, we will focus our attention on the TNF, IL-1, IL-6, IL-10, and TGFβ families of cytokines, which represent the classes of molecules mostly implicated in CNS injury. Finally, we will review the most recent clinical studies in humans as a basis of discussion on whether targeting the cytokine system might be of potential therapeutic interest for the treatment of CNS trauma.

TABLE 13.2
Cytokine Knock-Out and Transgenic Studies in Traumatic Spinal Cord Injury

Targeted Cytokine	Origin	Type of Trauma	Outcome	Reference
TNF KO	144	Extradural compression	No improvement of motor score; Trend to decreased white matter preservation	41
TNFR1 KO	145	Contusion injury	Increased apoptosis; Increased lesion volume; Worse functional recovery	42
TNFR2 KO	146	Contusion injury	Increased apoptosis; Increased lesion volume; Worse functional recoveryb (less pronounced than in TNFR1 KO)	42
IL-6 KO	141	Axotomy of sciatic nerve and crushed dorsal column 7d later	Failure in regeneration of dorsal column axons; Lack of GAP43 upregulation	83
IL-10 KO	Jackson Labs	Excitotoxic injury (quisqualic acid)	Greater damage and increased peripheral neutrophils acutely (1–7 days); Equal to WT chronically (14 days)	95

I. TRAUMATIC BRAIN INJURY (TBI) AND SPINAL CORD INJURY (SCI): FROM HUMAN PATHOLOGY TO ANIMAL MODELS

TBI is one of the leading causes of death and severe disability among young individuals (below 45 years of age) in western countries. As an immediate result of the initial mechanical impact, which produces tissue deformation and, frequently, laceration, the brain undergoes a phase of primary damage consisting in focal or diffuse vascular and axonal injury. At this early stage, the vascular damage often develops into foci of intracerebral, subdural, or extradural hemorrhage compromising the blood supply to the traumatized area. As a consequence of these initial events, a phase of secondary damage is initiated during which the ischemic–hypoxic tissue undergoes extensive swelling and the intracranial pressure is elevated. At the cellular level, inflammatory, neurochemical, and metabolic alterations take place, ultimately leading to diffuse cell death and, often, tissue cavitation.

In order to study the pathophysiology of human TBI, a number of animal models have been developed. Even though most of them apply controlled mechanical inputs, which result in highly reproducible, and hence quantifiable, injuries, none of them is sufficient to fully replicate the various aspects of the human pathology. Animal models of TBI can be generally classified into acceleration models (with or without impact) and direct impact models, extensively reviewed by Gennarelli [1] and Finnie and Blumbergs [2]. Acceleration models have been primarily used in higher mammals, such as primates, sheep, pigs, and cats. The ones with impact highly resemble the features and consequences of the human injury but have a lower degree of reproducibility. The ones without impact mainly replicate the consequences of the whiplash motion. They are highly reproducible, but are not representative of the effects of the head impact, which accompanies most motor vehicle accidents in humans. Direct impact models induce brain deformation by various means, such as fluid pulse (fluid percussion models) or pneumatic impactors (e.g., controlled cortical impact models). In these models, a rapid compression of the brain occurs by direct hit of the tissue on the dura surface, exposed following craniotomy. Impact models produce highly reproducible and localized injuries and have been widely used in rats. More recently, since the introduction of transgenic and knock-out mice, several of these models have been modified to be applied to the mouse, particularly direct impact models (Table 13.1), allowing for the study of the contribution of specific proteins to the pathophysiology of brain injury.

It must be noted that human head injury is never a clean and simple event as it is reproduced in any experimental model. In the majority of cases, trauma in humans is produced by multiple lesions occurring simultaneously at different sites and is often accompanied by lacerations. With this in mind, scientists need to select the specific animal model that best represents the feature or aspect of human brain trauma they are interested in investigating. Furthermore, in interpreting the results of such studies, species differences have to be taken into account because the anatomical and neurobiological differences (e.g., neurotransmitter receptor distribution, signal transduction pathways) between humans and other species are rather profound.

SCI is a devastating condition resulting in permanent motor dysfunction accompanied by a wide array of equally disabling secondary complications. Similarly to

TBI, SCI in humans is mainly caused by motor vehicle accidents (about 50% of cases) and involves, to a larger extent, young individuals. During the acute phase of injury, the initial mechanical impact to the spinal cord induces an immediate tissue disruption causing focal hemorrhage, local edema, release of excitatory amino acids and oxygen radicals, leading to massive cell death. Within 1 day after injury, the spinal cord is invaded by neutrophils, macrophages, and lymphocytes, infiltrating the parenchyma through a damaged BBB. The inflammatory response is activated, and the initial lesion expands as the injury progresses into a chronic phase. At this stage, both gray matter damage and white matter demyelination are evident, often associated with the appearance of small cystis or large syrinx cavities.

Most of the experimental studies on SCI have been conducted in rat models. They offer a number of advantages, such as low cost, minimal care, and the availability of multiple well-established analytical tools. In recent years, as for TBI models, several of the commonly used rat SCI models have been adapted to the mouse, allowing the use of transgenic mice to address specific biological questions. Rats and mice display similar histopathologic, biochemical, and behavioral outcomes, with the sole exception of tissue cavitation and cyst formation, which are virtually absent in the mouse as opposed to the rat.

The animal models developed to study SCI can be broadly divided into contusion models and transection models, previously reviewed in great detail [3–5]. Because contusion and compression are the most common types of SCI in humans, contusion models are generally preferred for the evaluation of biochemical and behavioral outcomes, which are more representative of the human pathology. The most widely used contusion models in rodents are the New York University (NYU) impactor [6] and the Ohio State University (OSU) impactor [7], which deliver a single, rapid hit to the spinal cord by controlled weight drop, inducing injuries of different severity. Although both models reliably mimic the pathophysiological consequences of the human pathology, the OSU impactor, being fully computer operated, seems capable to induce more accurate, calibrated, and reproducible injuries [3]. As an alternative to these models, a number of methods of external compression (e.g., clip compression) of the spinal cord have been developed. They are relatively simple (do not require laminectomy) and produce consistent injuries whose severity is directly correlated with the duration of compression. Finally, transection models (complete or partial transection) have proven to be the choice to evaluate axonal regeneration strategies. Particularly, unlike contusion or partial transection models, complete transection models allow for a clear discrimination of truly regenerating axons from spared ones.

With respect to the role of cytokines in TBI and SCI, the vast majority of the studies has been carried out in direct impact models (TBI) and contusion models (SCI) in rodents, which offer the highest flexibility and reproducibility when evaluating biochemical parameters, such as receptor alterations, gene expression, and ionic changes.

II. TUMOR NECROSIS FACTOR (TNF) FAMILY

Among the many members of the TNF family of cytokines, TNF and Fas Ligand (FasL, also called APO-1L, CD95L) are the molecules that have been more intensively investigated in relation to CNS injury. TNF and FasL induce their biological effects through binding to specific membrane receptors, the TNF receptors (TNFR1 and TNFR2) and Fas (APO-1, CD95), respectively (described in greater detail in Chapters 4 and 5). Because of the broad range of biological functions regulated by TNF, often opposite to one another, the role of this cytokine in the pathophysiology of CNS trauma is a complex matter. If on one hand, some strategies designed to reduce TNF production or activity have resulted in anti-inflammatory effects leading to neuroprotection and improved functional recovery following injury, more recent studies in knock-out animals have indicated that the total abolition of TNF signaling is highly detrimental to CNS recovery after trauma, resulting in increased apoptotic cell death and significantly impaired functional outcome (Table 13.1 and Table 13.2). Furthermore, the temporal profile and cellular localization of TNF expression appear to vary significantly depending upon the type and severity of trauma. This may indicate that different cytokine-dependent mechanisms, either pathogenic or neuroreparative, are operative at different times after injury. Therefore, this must be taken into account while designing therapeutic strategies aimed at promoting neuroprotection or regeneration following trauma by targeting the TNF system.

A. TRAUMATIC BRAIN INJURY

TNF is upregulated following traumatic brain injury (TBI) and may serve as a mediator of inflammation and microvascular injury associated with trauma. TNF expression in the brain parenchyma after TBI is generally reported as an immediate and transient phenomenon. It is detected as early as 30 min after injury, peaks between 3 and 8 h, and returns to baseline levels within 24 h [8–11]. TNF levels in the cerebro-spinal fluid (CSF) increase following injury with a temporal profile overlapping the one described for the brain parenchyma [12]. As suggested by a study in a mouse model of wound trauma, TNF expression appears to correlate with the severity of injury [13].

With respect to the cellular source of TNF, Kita and colleagues, in a rat model of mild fluid percussion injury, observed TNF expression exclusively in macrophages at the lesion site and in glial cells surrounding the area (mainly microglia, but also astrocytes and oligodendrocytes) [14]. In contrast to this observation, however, Knoblach and coworkers [9] reported, in a similar model of severe fluid percussion trauma, an intense TNF immunoreactivity localized in neurons of the injured cortex. Only a small population of astrocytes, ventricular cells, and endothelial cells were TNF-positive, and macrophages did not exhibit any TNF immunoreactivity [9].

Differently from the widely documented early TNF expression following brain trauma, Holmin and coworkers [15], in a weight drop model of injury, observed a delayed production of TNF. TNF mRNA and protein were detected only 4 to 6 days after injury and were confined to astroglial cells both at the lesion site and in the controlateral uninjured hemisphere [15].

Similarly to SCI models, pharmacological treatments resulting in a reduction of tissue levels of TNF produced significant functional improvements and limited neurological dysfunctions associated to TBI. By inhibiting TNF production or activity, pentoxyfilline, HU-211 (a synthetic cannabinoid that inhibits TNF production post-transcriptionally), and TNF binding protein (TNF-BP, a physiological inhibitor of

TNF) all improved the clinical outcome after closed head injury, as measured by facilitated motor function recovery, reduction of brain edema, increased BBB preservation, and reduced hippocampal cell death [16]. The intraventricular administration of soluble TNF receptor fusion protein (sTNFR:Fc), which acts as a TNF antagonist, reduced the early neuronal expression of TNF following TBI, thereby improving the functional performance in a series of motor tasks [9].

In contrast with those reports, the therapeutic effects observed with pharmacological treatments were not replicated in studies on TNF or TNF receptor knockout animals, once again in line with data obtained in models of SCI. Even though in the acute phase of injury (up to 7 days) $TNF^{-/-}$ mice subjected to TBI exhibited less severe memory deficits and motor impairment than wild-type mice, in the long term (2–4 weeks) they showed a progressive worsening of the clinical outcome with respect to wild-type animals [17]. Ultimately, 4 weeks after TBI, $TNF^{-/-}$ mice showed no improvement in motor function recovery and greater cortical tissue loss than wild type. This indicates that although TNF may play a deleterious role in the acute phase of injury, it might instead have beneficial effects in the delayed, chronic response of the traumatized brain [17]. Mice lacking both TNFR1 and TNFR2 showed increased tissue damage and BBB disruption following moderate cortical impact when compared to wild-type mice [18]. This was paralleled by reduced expression of manganese superoxide dismutase (MnSOD) and reduced NF-κB activation, indicating that TNF may exert its neuroprotective effects by counteracting oxidative stress or preventing apoptosis through NF-κB-dependent mechanisms [18].

Fas and FasL expression was detected as early as 15 min after TBI and persisted up to 72 h [19,20]. Fas and FasL immunoreactivity was localized in apoptotic neurons and astrocytes at the lesion site, indicating a major role for these molecules in the induction of apoptotic neurodegeneration after TBI [19,20].

B. SPINAL CORD INJURY

Expression of TNF after SCI has been widely described. In a contusion model of SCI in the rat, Yakovlev and Faden [21] first reported a transient upregulation of TNF mRNA at the lesion site, which peaked 30 min after SCI and returned to control levels within 24 h. TNF expression was proportional to the severity of trauma. Interestingly, the authors observed a later accumulation of message at more distal segments rostral and caudal to the trauma site, likely reflecting a propagation of delayed damage along the spinal cord [21]. Similar findings were reported by Wang and colleagues [22], who detected an increase in tissue levels of TNF immediately after SCI at the site of injury but did not find any evidence of TNF expression in the CSF. In a mouse model of traumatic SCI, Bartholdi and Schwab [23] demonstrated by *in situ* hybridization the presence of transcripts for TNF, as well as IL-1β and the chemokines MIP-1α and MIP-1β, within 1 h after SCI. Differently from IL-1β, MIP-1α, and MIP-1β, TNF was expressed in a very narrow window of time following trauma.

The principal cellular sources of TNF were found to be neurons, astrocytes, oligodendrocytes, and microglia in addition to endothelial cells [24,25]. It appears clear, therefore, that the accumulation of TNF is a prompt and transient event in response to SCI, resulting from the contribution of resident CNS cells in the surroundings of the lesion (neurons, astrocytes, and oligodendrocytes [23,24]), rather than from peripheral leukocytes infiltrating into the spinal cord following injury. This supported the hypothesis that the local production of TNF immediately after injury could serve as an early signal for recruitment of blood cells into the lesion and propagation of the inflammatory response.

In parallel with the upregulation of TNF production following SCI, an increased expression and activation of molecules in the TNF signaling pathway has been documented. Immunoreactivity for TNFR1 and TNFR2 was detected in neurons, astrocytes, and oligodendrocytes shortly after traumatic SCI [26]. Expression levels peaked 8 h postinjury and returned to control levels within 3 days. Interestingly, no expression was observed in microglial cells, suggesting that although this cell type appears to be one of the major sources of TNF following injury, it is unlikely the target of TNF biological activity [26]. The transcription factor NF-κB, which represents the major effector of TNF inflammatory actions, is also highly activated following injury. NF-κB activation is detected as early as 30 min after injury and persists elevated after 24 to 72 h [27,28]. Interestingly, both the upregulation of TNF receptor expression (TNFR1, TNFR2) and the activation of NF-κB could be prevented by the administration of methylprednisolone (a synthetic glucocorticoid used clinically in the treatment of SCI) or a glucocorticoid receptor antagonist [26,28]. This indicated that the anti-inflammatory action of methylprednisolone might be in part dependent on the interference with TNF signaling.

In support of a negative role of TNF in SCI pathophysiology, treatments resulting in reduction of TNF production following injury proved to be effective in improving the functional outcome of animals subjected to SCI. The topical application of TNF antiserum at the site of injury, given either before or immediately after trauma, significantly attenuated swelling, edema formation, microvascular permeability disturbances, and cell injury induced by SCI [29]. TNF antiserum also reduced the SCI-induced upregulation of neuronal nitric oxide synthase (nNOS), indicating a possible beneficial role in preventing NO-mediated oxidative damage following injury. Administration of the anti-inflammatory cytokine IL-10 in a contusion model of injury in the rat reduced TNF expression and improved functional recovery [30]. Activated protein C (APC), a physiological anticoagulant implicated in the regulation of inflammation by inhibiting cytokine production in monocytes, was effective in reducing motor disturbances primarily by reducing the amount of TNF at the site of injury, thereby inhibiting neutrophil accumulation and the resulting damage to endothelial cells [31,32]. Similarly, thrombomodulin, an activator of APC, when given before or after the induction of SCI, markedly reduced tissue levels of TNF and accumulation of leukocytes at the injury site, leading to a significant improvement of motor function recovery [33]. More recently PGE1, a potent vasodilator reported to inhibit both neutrophil activation and monocytic production of TNF *in vitro*, was demonstrated to reduce motor deficits after SCI by inhibiting neutrophil activation directly or indirectly through the inhibition of TNF production at the injury site [34]. Antithrombin, whose anti-inflammatory properties are related to its ability to promote endothelial release of prostacyclin, was also effective in reducing spinal cord levels of TNF after injury and promoting functional recovery [35]. TNF expression was significantly reduced by the pretreatment of rats with the platelet activating factor (PAF) antagonist WEB2170 prior to SCI, implicating PAF in the induction of pro-inflammatory cytokines after injury. However, no functional data were produced to indicate effectiveness of this treatment in improving motor recovery [36]. A recent report indicated that nicotine, by specifically activating neuronal nicotinic receptors, markedly reduces oxidative and inflammatory reactions initiated by SCI, particularly TNF expression and activation of NF-κB [37].

The detrimental function of TNF in SCI has also been related to its ability to induce apoptotic cell death. In a crush injury model in the rat, the extent of apoptotic cell death (mainly in neurons and astrocytes surrounding the lesioned area), the upregulation of inducible NOS (iNOS) and the increase in NO production following injury were significantly reduced by local injection of a neutralizing antibody to TNF [38,39]. This suggested that TNF may function as an external signal initiating apoptosis in neurons and oligodendrocytes after SCI, partly by upregulating NOS [38,39]. Apoptosis occurring in supraspinal regions distant from the core of spinal cord damage has also been correlated with TNF expression [40]. Indeed, in the pontine reticular formation, the origin of pontine reticulospinal fibers, the injection of TNF antiserum markedly reduced TNF expression, upregulation of TNF-dependent caspase-3, and neuronal apoptosis after complete spinal cord transection [40].

In contrast with these reports, studies in TNF and TNF receptor (both TNFR1 and TNFR2) knock-out mice have indicated that the total abolition of TNF responses not only does not improve recovery after SCI, but can result in worsening of the clinical outcome. In a model of extradural compression, Farooque and coworkers [41] demonstrated that TNF$^{-/-}$ mice do not exhibit a recovery of motor function greater than wild-type mice following SCI, rather they present a trend to a reduction of white matter preservation [41]. Along the same line, mice lacking TNFR1 or TNFR2 displayed greater apoptotic cell death, larger lesion size and worse functional outcome than wild-type mice [42]. These findings support the argument that TNF signaling, most likely through NF-κB activation, is involved in limiting apoptotic cell death after SCI and hence plays a beneficial role in preventing neuronal/oligodendroglial degeneration and promoting regeneration following injury.

The transmembrane protein FasL and its receptor Fas, members of the TNF and TNF receptor families respectively, were also upregulated following SCI. Fas and FasL immunoreactivity was detected after injury rostral and caudal to the lesion in models of moderate and severe compression trauma in the rat [43]. Fas upregulation was demonstrated in apoptotic cells, predominantly oligodendrocytes, both in the gray and white matter, supporting the hypothesis that the Fas-FasL system is involved in apoptosis of oligodendrocytes following SCI [44,45]. In a recent study in a model of partial transection in the mouse, Demjen and coworkers [46] demonstrated expression of FasL on microglia, astrocytes and neurons, and of Fas on microglia, neurons and some oligodendrocytes 3 days after SCI [46]. These expression patterns paralleled the kinetics of apoptotic cell death. Neutralization of FasL by administration of a FasL-specific antibody lead to significant reduction of apoptosis, increased oligodendrocyte and neuronal survival, increased axonal regeneration, and improved functional outcome following SCI [46]. Surprisingly, the authors did not observe such a beneficial effect following TNF neutralization. In fact, inactivation of TNF did not result in reduction of apoptosis or improvement of locomotor performance in injured mice [46], suggesting that the Fas–FasL pathway, rather than the TNF pathway, is the major player in injury-induced apoptotic cell death. This result, however, appears to be in contrast with previous reports indicating instead a positive effect of TNF neutralization following injury [38–40]. The discrepancy could possibly be due to the different protocol of administration of the TNF antibody, which in this latest study is delivered systemically [46], rather than locally injected into the lesion [29,38,39].

III. INTERLEUKIN-1 (IL-1) FAMILY

The IL-1 family of cytokines includes IL-1β, IL-1α, and IL-18, which, as pro-inflammatory cytokines, have been largely implicated in acute and chronic CNS injury and disease [47]. As an additional indication of IL-1 (referred to as both IL-1β and IL-1α) role in CNS pathophysiology, IL-1 receptor antagonist (IL-1ra), a natural competitive antagonist of IL-1, exhibits a potent anti-inflammatory activity. Structurally related to IL-1, IL-1ra is also included in the IL-1 family of cytokines.

A. TRAUMATIC BRAIN INJURY

Following TBI, the time course of IL-1 mRNA and protein expression generally parallels the profile observed in SCI. Protein and mRNA levels increase within 15 min, peak at 3 to 4 h, and return to control levels by 24 to 48 h postinjury [48–52]. A slightly shifted profile was observed by Fassbender and coworkers, who detected a peak of protein expression only 2 days after TBI [53]. Similarly to the spinal cord, the major source of IL-1 in the injured brain appears to be microglia [52].

Hypothermia, which has been proposed as a treatment for CNS trauma due to its ability to attenuate inflammation, significantly reduces the increase in IL-1β mRNA and protein following TBI, suggesting that the efficacy of hypothermia may be attributed to the inhibition of pro-inflammatory cytokine production [49].

IL-1 expression following brain trauma appears to regulate molecules involved in the development of injury. Expression of matrix metalloproteinase-9 (MMP-9) and intracellular adhesion molecule-1 (ICAM-1), which both play a role in regulating leukocyte trafficking across the endothelium, are significantly attenuated by treatment with IL-1ra and a neutralizing antibody against IL-1β, respectively [54,55]. Furthermore, IL-1β$^{-/-}$ corticectomized mice exhibited a significant impairment in the restoration of the BBB integrity following injury. These studies suggest that IL-1 may be important in sustaining the injury process by compromising the BBB and recruiting peripheral blood leukocytes into the lesion, and indicate that strategies aimed at reducing IL-1 levels may be neuroprotective.

In support of this, administration of IL-1ra significantly reduced tissue damage following fluid percussion injury in the rat [56]. Furthermore, mice overexpressing the secreted form of IL-1ra in astrocytes displayed higher neurological recovery and lower cytokine production than wild-type mice [50]. On the other hand, Knoblach and Faden [48] reported that intracerebroventricular delivery of either IL-1ra or soluble IL-1R did not result in any improvement of functional recovery following TBI in rats [48]. This apparent discrepancy could be due in part to species differences (mouse vs. rat), or to the fact that IL-1ra levels were already elevated prior to injury in the transgenic model compared to the pharmacological approach where IL-1ra was administered postinjury.

In support of a beneficial role of IL-1, DeKosky and colleagues [57] reported that treatment of animals with IL-1ra-expressing fibroblasts prevented the upregulation of nerve growth factor (NGF) production following TBI [57]. Along the same line is a study by Herx and coworkers [52], who demonstrated in corticectomized IL-1β$^{-/-}$ mice that IL-1β is required for the production of ciliary neurotrophic factor (CNTF), a neuroprotective cytokine recognized as a survival factor for neurons and oligodendrocytes. These studies suggest that blocking IL-1β may be detrimental, due to the suppression of NGF- and CNTF-mediated reparative responses following trauma.

Inhibition of the pro-inflammatory cytokine IL-18 has been tested as a possible treatment for TBI. IL-18 is elevated in the mouse brain after injury and intracerebral administration of IL-18 binding protein (IL-18BP), a naturally occurring specific inhibitor of IL-18, significantly improved the neurological recovery 7 days posttrauma [58].

B. Spinal Cord Injury

Following traumatic injury to the spinal cord both IL-1α and IL-1β are rapidly induced. IL-1 mRNA is detected as early as 15 min after injury, with levels peaking at 6 to 12 h and returning to baseline within 24 to 72 h [51,59,60]. IL-1 protein expression in the spinal cord parenchyma is elevated at 6 h and returns to control levels by 72 h following injury [51]. According to a study by Wang and colleagues [61], IL-1β protein levels increase at the lesion site as early as 1 h after injury, peak at 8 h and remain elevated throughout at least 7 days posttrauma. Following SCI, IL-1 concentration is elevated also in serum and CSF, although to a lower extent than in the spinal cord [61]. The early production of IL-1 following injury appears to occur in resident cells, such as microglia and perivascular macrophages [23,62]. However, IL-1 expression can also be detected at a later stage in a population of polymorphonuclear granulocytes migrating into the spinal cord around 6 h following trauma [23]. Taken together, these reports suggest that expression of IL-1, similarly to that of other pro-inflammatory cytokines (e.g., TNF, IL-6), is an early and local reaction to trauma that could trigger secondary injury responses such as the recruitment of inflammatory cells into the lesion area.

Due to its potent anti-inflammatory properties, IL-1ra has recently been evaluated as a possible treatment in animal models of CNS injury. A continuous infusion of recombinant IL-1ra following SCI in rats significantly reduced caspase-3 activation and apoptotic cell death [51]. Even though promising, these studies were not extended to evaluate the effectiveness of this compound in improving motor function recovery or reducing tissue damage following SCI.

The recently acquired member of the IL-1 family IL-18 is a pro-inflammatory cytokine primarily involved in the response to viral infections by regulating innate and acquired immune responses [63]. Even though IL-18 is expressed by CNS cells, particularly microglia [63], suggesting a possible role for this molecule in CNS pathophysiology, very little is known about IL-18 implication in brain or spinal cord trauma. No published reports are available on IL-18 expression and function following SCI, however, preliminary studies in our laboratory have determined that IL-18 mRNA is constitutively present in the spinal cord of uninjured mice. Gene expression remains unaffected 6 h after contusive SCI, and it is significantly reduced by approximately 50% 24 h following injury (Brambilla and Bracchi-Ricard, unpublished observations). Studies to elucidate the biological significance of this reduction are currently ongoing in the laboratory.

IV. INTERLEUKIN-6 (IL-6) FAMILY

Together with IL-6, other members of the IL-6 family of cytokines, such as Leukemia Inhibitory Factor (LIF) and CNTF, have been implicated in traumatic injury to the CNS. Expression of these molecules, which is contained at low levels under normal conditions, is rapidly upregulated following trauma.

Similarly to other cytokines (e.g., TNF, IL-1), the role of IL-6 in the pathophysiology of CNS injury is rather complex. As a neurotrophic survival factor, IL-6 has been attributed a beneficial neuroprotective function, whereas as a mediator of inflammation, demyelination, and astrogliosis it exhibits neurodegenerative properties following injury. Recent studies in knock-out and transgenic mice lacking or overexpressing IL-6, respectively, have attempted to provide clearer insights into the function of IL-6 following spinal cord injury and brain trauma (Table 13.1 and Table 13.2).

A. TRAUMATIC BRAIN INJURY

Most of the studies in rodent models of TBI concur that IL-6 expression peaks around 8 h following injury [11,12,64,65]. In addition, IL-6 levels are higher in aged vs. young rats in response to brain injury, suggesting that age is an important factor to take into consideration when studying the inflammatory process following traumatic injury to the CNS [66].

As predicted by the well documented pro-inflammatory functions of IL-6, transgenic mice overexpressing IL-6 in astrocytes (GFAP-IL-6) display extensive brain damage (neurodegeneration, BBB disruption, learning deficit) caused by a persistent state of inflammation (astrocytosis, microgliosis, leukocyte recruitment, upregulation of inflammatory molecules like IL-1α, TNF, etc. [67]). However, although overexpression of IL-6 appears to be detrimental under physiologic conditions, it produces beneficial effects under pathologic conditions [67]. Following cryolesion to the frontoparietal cortex, GFAP-IL-6 mice exhibited accelerated tissue repair, decreased oxidative stress and reduced apoptosis compared to wild-type littermates [68–70]. The authors suggested that the neuroprotection afforded by the transgenic expression of IL-6 is related to the upregulation of metallothionein-I+II (MT-I+II), which in turn could modulate the inflammatory response, decrease oxidative stress and apoptosis, and increase brain tissue repair [71]. It is also possible that in GFAP-IL-6 mice neurons respond better to injury because they are preconditioned by the continuous level of inflammation already present under nonpathologic conditions.

In support of the neuroprotective role of IL-6, studies using IL-6$^{-/-}$ mice demonstrated that IL-6 is a crucial regulator of the glial and macrophage inflammatory response [70,72]. The inflammatory response to TBI in IL-6$^{-/-}$ mice was strongly compromised, and the rate of recovery and healing was comparatively slower [70,73,74]. IL-6 beneficial role in posttraumatic healing was also attributed to its ability to promote revascularization [70]. However, even though delayed, healing occurred also in IL-6$^{-/-}$ mice, suggesting that IL-6 it is not the only factor required for tissue repair [70]. Collectively, these studies demonstrated that IL-6 plays an important role in activation of microglia, recruitment of bone marrow-derived monocytes and activation of astrocytes around the lesion site.

In reactive astrocytes, IL-6 appears to be an essential regulator of GM-CSF, a potent microglial/macrophage mitogen [72]. IL-6 modulates brain expression of antioxidant and antiapoptotic metallothioneins under normal conditions as well as following trauma, in an isoform-specific fashion [72].

Like IL-6, LIF plays a key role in inflammation and can act in both a pro- and anti-inflammatory fashion (reviewed by Gadient and colleagues [75]). Although LIF has been extensively studied in a number of inflammatory conditions (such as rheumatoid arthritis), few studies have investigated the role of LIF in traumatic injury to the CNS. Following cortical stab wound injury, LIF mRNA was strongly upregulated within 1 day, predominantly in astrocytes present in the lesion penumbra [76]. In a sagittal cortical laceration model using LIF$^{-/-}$ mice, LIF was demonstrated to play a pro-inflammatory role by recruiting peripheral monocytes and inducing reactive gliosis [74]. This response was similar to that obtained with IL-6$^{-/-}$ mice. Surprisingly, there was no additive effect in IL-6$^{-/-}$-LIF$^{-/-}$ double knock-out mice, suggesting that LIF and IL-6 may act in series in this specific model. Whether the pro-inflammatory role of LIF was beneficial or detrimental was not addressed in this study.

In the adult brain CNTF is expressed at low levels by astrocytes, and CNTF receptor is widely expressed in motor-related areas. Following acute TBI in the adult rat brain, CNTF expression was upregulated 36 h postinjury both at the lesion site and in remote nontraumatized areas [77]. In a murine corticectomy model, CNTF expression increased 1 h postinjury and peaked after 12 h [52]. Herx and collaborators demonstrated that CNTF expression is regulated by IL-1β, whose peak of expression occurred 9 h before CNTF. The requirement of IL-1β for CNTF production was confirmed by blocking IL-1β activity with IL-1ra and by gene-targeted deletion of IL-1β [52].

Finally, in a rat model of entorhinal cortex lesion (a model of injury accompanied by loss of synapses), CNTF was upregulated in reactive astrocytes in response to deafferentation [78].

B. Spinal Cord Injury

IL-6 mRNA expression is upregulated early following contusive SCI and returns to control levels by 24 h postinjury [11,59,79]. Both in contusion [79] and hemisection [23] models of SCI, IL-6 production originates largely from resident astrocytes and microglial cells in the spinal cord parenchyma.

In support of the pro-inflammatory properties of IL-6, several studies have demonstrated that blocking IL-6 signaling leads to beneficial effects following SCI. In a murine contusion model, blockade of the IL-6 receptor with an anti-IL-6 receptor antibody injected intraperitoneally in a single dose suppressed reactive astrogliosis, decreased scar tissue formation at the lesion site and infiltration of immune cells, and ameliorated functional recovery [80]. Similarly, neutralization of IL-6 bioactivity with an anti-IL-6 antibody injected immediately after injury significantly attenuated iNOS activity and reduced secondary damage to the rat spinal cord, as observed 48 h following trauma [81]. In a hemisection model of injury, Lacroix and coworkers reported reduced axonal growth, increased infiltration of leukocytes and increased tissue damage in the presence of a bioactive IL-6/soluble IL-6 receptor fusion protein (called hyper-interleukin-6, H-IL-6), which directly stimulates the receptor gp130 [82]. H-IL-6, secreted by genetically engineered fibroblasts grafted at the site of lesion immediately after injury, induced amplification of the inflammatory response resulting in increased neutrophil infiltration accompanied by a significant elevation in reactive oxygen species. This appeared to overcome the potential axonal growth-promoting effects of IL-6 [82].

In contrast to peripheral nerve regeneration experiments where IL-6 was demonstrated to promote neuronal survival and outgrowth, many of the studies in the injured spinal cord suggested that the absence of IL-6 from the site of injury is beneficial. However, a recent study in a model of conditioning injury showed a failure of spinal axon regeneration in mice lacking IL-6 (IL-6$^{-/-}$) [83]. In this model, mice underwent unilateral axotomy of the left sciatic nerve 7 days before having their dorsal column projection crushed. Although in wild-type mice this procedure resulted in regeneration of dorsal column axons up and beyond the injury site, it failed to produce such an effect in IL-6$^{-/-}$ mice, due to a lack of growth-associated protein 43 (GAP43) upregulation in preinjured dorsal root ganglia. These results indicated that IL-6 might be an important factor in the regeneration process.

In contrast to IL-6, LIF treatment following SCI appears to be mostly neuroprotective. In a mouse model of SCI, intraperitoneal administration of LIF up to 24 h following hemisection produced beneficial effects, consisting in a statistically significant improvement of locomotor function along with a higher number of myelinated axons [84]. Similarly, in a rat model, LIF secreting fibroblasts grafted into the spinal cord immediately after dorsal hemisection injury promoted growth of corticospinal axons and increased the expression of the neurotrophin NT-3 [85].

Following SCI, expression of CNTF and CNTF receptor is increased in reactive astrocytes and motorneurons, respectively [86]. CNTF, thought to play a role in neuronal survival and axonal growth, when administered by continuous intrathecal infusion (10 days) in the rat contused spinal cord, exerted neuroprotective effects [87]. It reduced tissue damage, promoted preservation of the rubrospinal descending tracks and enhanced functional recovery [87]. Interestingly, CNTF also induced an increased gliosis, proving to have neurotrophic properties [87]. The mechanisms by which CNTF produces its neurotrophic effects *in vivo* are unclear. The authors suggested that CNTF could promote neuronal survival by affording protection against excitotoxicity and apoptosis through stabilization of neuronal calcium homeostasis. In addition, CNTF could enhance the expression of neurotrophic factors and survival of oligodendrocytes.

V. INTERLEUKIN-10 (IL-10)

IL-10 is a potent anti-inflammatory cytokine with proven efficacy in animal models of inflammatory diseases [88]. Due to this property, several studies have examined the potential therapeutic effect of IL-10 in reducing posttraumatic inflammation and consequently ameliorating functional recovery following spinal cord and brain injury.

A. TRAUMATIC BRAIN INJURY

The local application of IL-10 at the site of corticectomy in mice significantly attenuated astroglial reactivity, measured both as reduced proliferation and hypertrophy [89]. This effect appeared to be indirect through the inhibition of cytokine production, such as TNF, and could represent a potential explanation for the anti-inflammatory and neuroprotective properties of IL-10 [89]. In agreement with this study, Knoblach and Faden [90] demonstrated that systemic administration of IL-10 following fluid percussion injury in the rat significantly reduces TNF expression in the cortex and improves neurological recovery.

Controversial results emerged from a study evaluating the combination of IL-10 treatment with mild hypothermia following TBI in the rat [91]. Hypothermia alone was neuroprotective by enhancing both motor and cognitive functions and increasing survival of CA3 neurons. The systemic administration of IL-10 combined with hypothermia, however, did not provide synergistic neuroprotection, even though it was effective in reducing neutrophil accumulation. On the contrary, IL-10 treatment suppressed the beneficial effects of hypothermia alone on TBI.

B. SPINAL CORD INJURY

Bethea and coworkers reported that systemic administration of IL-10 in a contusion model of SCI in Sprague–Dawley rats resulted in a significant attenuation of TNF synthesis, reduction of lesion volume, and improvement of locomotor function [30]. In a similar contusion model in Fischer rats, IL-10 alone or in combination with methylprednisolone significantly reduced the volume of damaged tissue, without however producing any improvement in axonal preservation or motor function [92]. This apparent discrepancy with the report from Bethea and colleagues could possibly be explained by the different rat strains used in the two studies.

Systemic delivery of IL-10 was also neuroprotective against severe excitotoxic SCI, leading to a much greater gray matter preservation and reduction of lesion volume [93]. Furthermore, in the same model of excitotoxic SCI, IL-10 was effective in reducing pain and pain-related behavior. The onset of grooming behavior, an indication of pain, was significantly delayed, and the area of skin damage and grooming severity were reduced [94].

In a recent study, Abraham and colleagues [95] addressed the role of endogenous IL-10 in SCI using IL-10$^{-/-}$ mice. In the excitotoxic model of SCI, they observed a greater extent of tissue damage in animals lacking IL-10 only at early time points following injury (1, 7 days). By 14 days postinjury the extent of damage was comparable to the one measured in wild-type mice. The pronounced early damage was paralleled by a significant increase in peripheral neutrophils, an index of innate immune response to injury. This effect was abolished by the reconstitution of IL-10 with the intraperitoneal administration of 50 ng of recombinant IL-10. The study suggests that endogenous IL-10 is neuroprotective in the early phase of injury, indicating that a low dose of IL-10 given acutely following SCI may have a potential therapeutic application [95].

VI. TRANSFORMING GROWTH FACTOR BETA (TGFβ) FAMILY

The TGFβ family of growth factors is a large group of related molecules controlling the development and homeostasis of most tissues and organs [96]. Among the different members of this group, we will focus our attention on the TGFβ subfamily (composed of TGFβ1, TGFβ2, and TGFβ3) due to its implication in the pathophysiology of CNS injury. TGFβ has been attributed both anti-inflammatory and fibrotic properties that can lead to either beneficial (e.g., reduction of cytokine production) or detrimental (scar formation) effects following injury [97].

A. TRAUMATIC BRAIN INJURY

TGFβ1 is expressed in the normal brain at very low levels [98] and is upregulated following trauma. In a stab wound model of trauma in the rat, TGFβ1 mRNA was significantly increased 1 day after injury and persisted elevated at 7 days [99]. Expression was concentrated at the margin of the wound, primarily in microglial cells and macrophages. Logan and colleagues [100] observed that whereas TGFβ1 expression was localized at the edge of the lesion in the early phase of injury (up to 3 days posttrauma), in a later phase (7–14 days posttrauma) it was predominantly found in the region of the glial scar, expressed mainly by astrocytes and a few macrophages/microglial cells [100]. In agreement with the demonstrated fibrotic properties of TGFβ [101], these data support the hypothesis that TGFβ could be an important factor regulating the formation of the glial scar following CNS injury. In confirmation of this, injection of TGFβ1 into the injured brain dramatically increased the scarring response, and administration of TGFβ1 or TGFβ2 neutralizing antibodies significantly attenuated the deposition of scar tissue [102,103].

Expression of TGFβ2 was biphasically detected in a model of fluid percussion injury, with a first peak at 30 min and a second at 48 h after trauma, flanking the transient production of TNF and IL-6, suggesting a possible role of TGFβ2 in the regulation of the cytokine response following injury [104].

A more recent report evaluated the cellular response to TGFβ after injury by investigating the expression of TGFβ receptors (TGFβRI and TGFβRII). By immunohistochemistry, TGFβRI and TGFβRII were detected at very low levels on endothelial cells of the uninjured brain [105]. Four days after traumatic injury (a time marking the transition from acute to chronic injury in this experimental model), TGFβRII was significantly upregulated, and, in parallel, the expression of key molecules in the signaling pathway of TGFβ (SMAD3, SMAD4, SMAD7) [105]. The differential expression of the two receptors could support the biphasic role of TGFβ in brain injury.

B. SPINAL CORD INJURY

Following contusion injury in the rat, TGFβ1 expression is detected within 1 h at the lesion site. The protein is both secreted and present in astrocytes, motor neurons, and endothelial cells [106]. The upregulation of TGFβ1 mRNA peaks at 7 days after SCI and is still elevated after 10 days [107].

Subarachnoidal administration of TGFβ1 after SCI significantly reduced the expression of iNOS, which is associated with the production of neurotoxic NO [108]. Treatment with TGFβ1 resulted in better recovery of motor function only in the short term (first 5 days after injury), whereas no difference in functional recovery was observed 5 weeks after SCI with respect to untreated animals [108]. These results are in line with the reported double function (positive and negative) of TGFβ in the pathophysiology of CNS injury.

Tyor and coworkers [109] described that treatment with TGFβ1, administered by tail injection 30 min after contusion injury, resulted in a 50% reduction of lesion volume 48 h after SCI and decreased accumulation of mononuclear phagocytes around the injury site, potentially limiting the contribution of these cells to secondary injury [109].

VII.   OTHER CYTOKINES

A.   Erythropoietin (EPO)

EPO is well known for its function as a primary mediator of the normal physiologic response to hypoxia. EPO exerts its actions through activation of the EPO receptor (EPO-R), which is abundantly expressed in the CNS of humans and rodents under normal conditions, particularly on brain capillaries [110]. In humans, EPO production within the CNS is triggered not only by hypoxia, but by a number of other pathological stimuli, such as mitochondrial reactive oxygen species generated in response to metabolic disturbances, including hypoglycemia and strong neuronal depolarization [111]. Following injury, EPO is found both in the brain parenchyma and in the CSF [112]. It is produced by astrocytes and neurons [113].

In SCI as well as in TBI, EPO exhibits neuroprotective functions [110,114]. In a model of hypoxic–ischemic brain injury in newborn rats, a 20 units intracerebroventricular injection of recombinant human EPO (r-Hu-EPO) administered immediately following injury significantly reduced the extent of brain damage [115]. Similarly, in a mouse model of moderate blunt trauma to the frontal cortex, Brines and collaborators demonstrated the neuroprotective role of r-Hu-EPO administered systemically 24 h before or up to 6 h after injury [110]. In this study, the authors delivered a high concentration of r-Hu-EPO (5,000 units/kg) and demonstrated that EPO could cross the BBB through binding to its receptor, which is abundantly expressed on brain capillaries. In addition, EPO-treated animals exhibited a decreased infiltration of immune cells into the lesion area, suggesting that EPO may have anti-inflammatory functions [110]. A similar role for EPO was observed in both a contusion and a compression model of SCI in the rat [114]. Administration of r-Hu-EPO immediately after injury resulted in increased recovery of motor function as early as 12 h after SCI (compression model) and in a reduction of secondary inflammation and cavitation. These data suggest that EPO provides neuroprotection both in the early and secondary phases of damage [114]. In a model of ischemic SCI in the rabbit, EPO treatment significantly prevented motorneuron apoptosis and improved the neurological outcome [113]. Finally, in a model of contusive SCI in the rat, EPO was reported to be effective in reducing trauma-induced lipid peroxidation, which contributes to tissue damage following injury [116].

B. ENDOTHELIAL MONOCYTE-ACTIVATING POLYPEPTIDE II (EMAP II)

Increased expression of EMAP II, a pro-inflammatory antiangiogenic cytokine, was described following both SCI and TBI [117,118]. In a stab wound injury model in the rat, EMAP II positive microglia/macrophages had already accumulated 1 day after injury, reached the maximum peak at 5 days, and were still detectable 21 days postinjury. Immunoreactivity was present at the necrotic lesion site in areas of developing secondary damage adjacent to the lesion core and in perivascular spaces [117]. A similar profile of expression was observed following SCI in the rat, where EMAP II positive microglia/macrophages accumulated at the lesion site and in perivascular areas. Accumulation reached a maximum peak 3 days postinjury and gradually declined thereafter until day 28, still remaining elevated above uninjured controls [118]. In both studies, the authors postulate that EMAP II expression during the acute phase postinjury indicates the participation of this molecule in the inflammatory response following CNS trauma. However, besides the profile of expression, no data are provided as to the mechanisms by with EMAP II may exert its proinflammatory function.

VIII. CLINICAL STUDIES

The animal studies conducted in *in vivo* models of traumatic brain and spinal cord injury have clearly demonstrated the key role played by cytokines in the pathophysiology of CNS injury. In the attempt to validate the knowledge acquired in those studies and to translate it into feasible therapeutic strategies for CNS trauma, considerable effort has been made to evaluate the inflammatory response to injury in humans. Due to obvious practical limitations, measurement of cytokine levels in patients with brain trauma and, even more, spinal cord trauma has been challenging. In most cases, clinical studies have been able to determine the concentrations of these molecules in CSF and plasma but not in the parenchyma. Just recently, using a novel technique of intracranial microdialysis, it has been possible to measure cytokine levels in the brain tissue of traumatized individuals; however, no tools are available for such measurement in spinal cord injury patients. The objective of these studies is to identify which cytokines can be correlated with the severity of injury and be predictive of the final outcome.

A. TRAUMATIC BRAIN INJURY

TBI in humans is characterized by high mortality, which is largely determined by the extent of the initial mechanical trauma to the brain and by the development of secondary injury. One of the characteristics of the secondary injury process is the intrathecal production of cytokines. With a certain degree of variability, due to the total lack of uniformity of the experimental subjects in the studies (different severity, localization, and modality of trauma), cytokines are generally elevated following traumatic injury to the brain. Increased levels of IL-6 [119–123], IL-1 [124], IL-8

[125–127], IL-12 [128], TNF [129], and TGF-β [130] were found in the CSF and serum of patients suffering TBI. With respect to the anti-inflammatory cytokine IL-10, a study by Maier and colleagues revealed increased levels in the plasma [119] but not in the CSF, whereas other authors found it elevated both in plasma and CSF [120,131]. In patients with severe TBI, soluble Fas (sFas) and FasL (sFasL) were detected in the CSF only [132,133]. In most of the studies, cytokine concentrations were higher in the CSF or jugular venous plasma than in the systemic plasma or serum, suggesting that the main source of cytokines following trauma is represented by resident cells of the brain. In the cases of IL-6 [119], Fas [133], IL-10 [131], and IL-8 [119], most of the studies agree that the cytokine distribution is not influenced by the integrity of the BBB. However for IL-8, Kossmann and coworkers reported that highly elevated IL-8 concentrations in the CSF correlated with a severe dysfunction of the BBB [127].

Several studies have attempted to correlate the levels of cytokines, in particular IL-6, with the severity of head trauma in order to identify a possible prognostic indicator of the clinical outcome of patients. However, the results have been inconsistent. Singhal and collaborators observed that peak levels of IL-6 in the CSF were higher in individuals with improved Glasgow Outcome Scale (GOS), a rating scale for the assessment of clinical outcome in TBI patients [121], hypothesizing that IL-6 could be used as a predictive marker of TBI outcome. However, in contrast with this report, other studies suggested that elevated IL-6 plasma levels 1 day after TBI seemed to be correlated with a short-term prognosis and infectious complications [122] as IL-6 plasma levels were significantly higher in case of nonsurviving patients [125,134]. In a recent study using an innovative method of intracranial microdialysis, Winter and collaborators were able to measure IL-6 levels in the brain parenchyma of TBI patients [123]. They reported higher levels of IL-6 in patients who survived severe head injury, suggesting that IL-6 acts as an endogenous neuroprotective cytokine [123]. According to a study by Tasci and coworkers [124], initial serum concentrations of IL-1β correlated with the extent of tissue damage, which in turn related to the prognosis of the patient's outcome.

Ertel and coworkers proposed that soluble FasL could be used as an index of injury severity because they demonstrated that CSF levels of this molecule were significantly correlated to the severity of head trauma in their experimental cohort [132]. On the other hand, Lenzlinger and colleagues did not find any correlation between serum or plasma levels of sFas and extent of trauma-related disturbances, such as BBB disruption [133]. However, maximal CSF levels of sFas were correlated with the early peak of neuron-specific enolase (a marker of neuronal cell destruction), indicating that activation of the Fas-mediated pathway of apoptosis may be in part the direct result of the initial trauma [133].

The discrepancy among the results of different clinical studies and the difficulty to correlate cytokine levels to clinical outcome is likely dependent on the wide variety of brain injuries (contusion, stab injury, diffuse axonal injury, haematoma, etc.) and on the many variables in the patient cohort (age, gender, immunological state, presence of additional pathologies). In relation to this point, a recent study provided evidence that gender has an effect on the severity of injury [135], with men displaying greater severity than women as reflected by their lower Glasgow Coma Score (GCS) and longer posttraumatic amnesia. Interestingly, no significant difference between genders was observed with respect to outcome measures (GOS, mortality rate) [135].

B. Spinal Cord Injury

Similarly to rodent models of SCI, elevated serum titers of pro-inflammatory cytokines (TNF, IL-2) were detected in patients with chronic SCI [136]. Whether this is beneficial or detrimental to the clinical outcome remains to be determined. No significant increase over uninjured individuals was detected in the expression of anti-inflammatory cytokines such as IL-10 and IL-4 [136]. A recent study on postmortem samples from acute SCI cases revealed that IL-1β, IL-6, and TNF are produced early following SCI by endogenous cells in the spinal cord, mainly neurons and microglia, rather than by blood-borne leukocytes [137].

IX. CONCLUSIONS

As evident from the vast literature accumulated over the past 10 years on *in vivo* models of CNS trauma, neuroinflammation is a major and complex component of the injury process. It encompasses distinct molecular and cellular events occurring at specific times and locations following injury, among which the increased cytokine production appears to be a prominent phenomenon. It emerges clear from the studies reviewed in this chapter that most cytokines, even though classified as pro- or anti-inflammatory, contribute to the pathophysiology of CNS trauma with both beneficial and detrimental effects, on one hand, promoting repair and regeneration of the injured tissue and, on the other, causing additional damage to the nervous system by activating cell death programs or inducing the release of neurotoxic compounds. Consequently, interfering with cytokine functions or signaling cascades has produced conflicting results. As an example, recent studies in knockout models have indicated that the complete abolition of cytokines traditionally considered to be pro-inflammatory (e.g., TNF, IL-6) is detrimental to functional recovery following trauma. This reinforces the concept that modulating the inflammatory response for therapeutic purposes can represent a double-edged sword and indicates that further investigations need to be undertaken. In particular, the time frame of cytokine production and activity at the different stages of injury appear to be an important factor to take into consideration, as the same molecule may have detrimental functions in the acute phase and reparative functions in the chronic phase of injury.

With this knowledge at hand, it will be possible to design effective cytokine-based strategies for the treatment of CNS trauma, allowing a safe transition from the experimental work on animal models to the clinical arena.

ACKNOWLEDGMENTS

This work was supported by the NIH grant NS37130 to JRB and by The Miami Project To Cure Paralysis.

REFERENCES

1. Gennarelli, T.A., Animate models of human head injury. *J Neurotrauma* **11**, 357-68 (1994).
2. Finnie, J.W. & Blumbergs, P.C., Traumatic brain injury. *Vet Pathol* **39**, 679–89 (2002).
3. Stokes, B.T. & Jakeman, L.B., Experimental modelling of human spinal cord injury: a model that crosses the species barrier and mimics the spectrum of human cytopathology. *Spinal Cord* **40**, 101–9 (2002).
4. Kwon, B.K., Oxland, T.R. & Tetzlaff, W., Animal models used in spinal cord regeneration research. *Spine* **27**, 1504–10 (2002).
5. Talac, R. et al., Animal models of spinal cord injury for evaluation of tissue engineering treatment strategies. *Biomaterials* **25**, 1505–10 (2004).
6. Gruner, J.A., A monitored contusion model of spinal cord injury in the rat. *J Neurotrauma* **9**, 123–6; discussion 126–8 (1992).
7. Stokes, B.T., Experimental spinal cord injury: a dynamic and verifiable injury device. *J Neurotrauma* **9**, 129–31; discussion 131–4 (1992).
8. Fan, L. et al., Experimental brain injury induces differential expression of tumor necrosis factor-alpha mRNA in the CNS. *Brain Res Mol Brain Res* **36**, 287–91 (1996).
9. Knoblach, S.M., Fan, L. & Faden, A.I., Early neuronal expression of tumor necrosis factor-alpha after experimental brain injury contributes to neurological impairment. *J Neuroimmunol* **95**, 115–25 (1999).
10. Rostworowski, M., Balasingam, V., Chabot, S., Owens, T. & Yong, V.W., Astrogliosis in the neonatal and adult murine brain post-trauma: elevation of inflammatory cytokines and the lack of requirement for endogenous interferon-gamma. *J Neurosci* **17**, 3664–74 (1997).
11. Taupin, V., Toulmond, S., Serrano, A., Benavides, J. & Zavala, F., Increase in IL-6, IL-1 and TNF levels in rat brain following traumatic lesion. Influence of pre- and post-traumatic treatment with Ro5 4864, a peripheral-type (p site) benzodiazepine ligand. *J Neuroimmunol* **42**, 177–85 (1993).
12. Stover, J.F., Schoning, B., Beyer, T.F., Woiciechowsky, C. & Unterberg, A.W., Temporal profile of cerebrospinal fluid glutamate, interleukin-6, and tumor necrosis factor-alpha in relation to brain edema and contusion following controlled cortical impact injury in rats. *Neurosci Lett* **288**, 25–8 (2000).
13. Kamei, H. et al., Severity of trauma changes expression of TNF-alpha mRNA in the brain of mice. *J Surg Res* **89**, 20–5 (2000).
14. Kita, T., Tanaka, T., Tanaka, N. & Kinoshita, Y., The role of tumor necrosis factor-alpha in diffuse axonal injury following fluid-percussive brain injury in rats. *Int J Legal Med* **113**, 221–8 (2000).
15. Holmin, S. et al., Delayed cytokine expression in rat brain following experimental contusion. *J Neurosurg* **86**, 493–504 (1997).
16. Shohami, E., Gallily, R., Mechoulam, R., Bass, R. & Ben-Hur, T., Cytokine production in the brain following closed head injury: dexanabinol (HU-211) is a novel TNF-alpha inhibitor and an effective neuroprotectant. *J Neuroimmunol* **72**, 169–77 (1997).
17. Scherbel, U. et al., Differential acute and chronic responses of tumor necrosis factor-deficient mice to experimental brain injury. *Proc Natl Acad Sci U.S.A.* **96**, 8721–6 (1999).
18. Sullivan, P.G. et al., Exacerbation of damage and altered NF-kappaB activation in mice lacking tumor necrosis factor receptors after traumatic brain injury. *J Neurosci* **19**, 6248–56 (1999).

19. Beer, R. et al., Expression of Fas and Fas ligand after experimental traumatic brain injury in the rat. *J Cereb Blood Flow Metab* **20**, 669–77 (2000).

20. Qiu, J. et al., Upregulation of the Fas receptor death-inducing signaling complex after traumatic brain injury in mice and humans. *J Neurosci* **22**, 3504–11 (2002).

21. Yakovlev, A.G. & Faden, A.I., Sequential expression of c-fos protooncogene, TNF-alpha, and dynorphin genes in spinal cord following experimental traumatic injury. *Mol Chem Neuropathol* **23**, 179–90 (1994).

22. Wang, C.X., Nuttin, B., Heremans, H., Dom, R. & Gybels, J., Production of tumor necrosis factor in spinal cord following traumatic injury in rats. *J Neuroimmunol* **69**, 151–6 (1996).

23. Bartholdi, D. & Schwab, M.E., Expression of pro-inflammatory cytokine and chemokine mRNA upon experimental spinal cord injury in mouse: an *in situ* hybridization study. *Eur J Neurosci* **9**, 1422–38 (1997).

24. Yan, P. et al., Glucocorticoid receptor expression in the spinal cord after traumatic injury in adult rats. *J Neurosci* **19**, 9355–63 (1999).

25. Yan, P. et al., Cellular localization of tumor necrosis factor-alpha following acute spinal cord injury in adult rats. *J Neurotrauma* **18**, 563–8 (2001).

26. Yan, P. et al., Expression of the type 1 and type 2 receptors for tumor necrosis factor after traumatic spinal cord injury in adult rats., *Exp Neurol* **183**, 286–97 (2003).

27. Bethea, J.R. et al., Traumatic spinal cord injury induces nuclear factor-kappaB activation. *J Neurosci* **18**, 3251–60 (1998).

28. Xu, J. et al., Methylprednisolone inhibition of TNF-alpha expression and NF-kB activation after spinal cord injury in rats. *Brain Res Mol Brain Res* **59**, 135–42 (1998).

29. Sharma, H.S. et al., Topical application of TNF-alpha antiserum attenuates spinal cord trauma induced edema formation, microvascular permeability disturbances and cell injury in the rat. *Acta Neurochir Suppl* **86**, 407–13 (2003).

30. Bethea, J.R. et al., Systemically administered interleukin-10 reduces tumor necrosis factor-alpha production and significantly improves functional recovery following traumatic spinal cord injury in rats. *J Neurotrauma* **16**, 851–63 (1999).

31. Taoka, Y. et al., Activated protein C reduces the severity of compression-induced spinal cord injury in rats by inhibiting activation of leukocytes. *J Neurosci* **18**, 1393–8 (1998).

32. Hirose, K. et al., Activated protein C reduces the ischemia/reperfusion-induced spinal cord injury in rats by inhibiting neutrophil activation. *Ann Surg* **232**, 272–80 (2000).

33. Taoka, Y., Schlag, M.G., Hopf, R. & Redl, H., The long-term effects of pre-treatment with activated protein C in a rat model of compression-induced spinal cord injury. *Spinal Cord* **38**, 754–61 (2000).

34. Naruo, S. et al., Prostaglandin E1 reduces compression trauma-induced spinal cord injury in rats mainly by inhibiting neutrophil activation. *J Neurotrauma* **20**, 221–8 (2003).

35. Hirose, K. et al., Antithrombin reduces the ischemia/reperfusion-induced spinal cord injury in rats by attenuating inflammatory responses. *Thromb Haemost* **91**, 162–70 (2004).

36. Hostettler, M.E. & Carlson, S.L., PAF antagonist treatment reduces pro-inflammatory cytokine mRNA after spinal cord injury. *Neuroreport* **13**, 21–4 (2002).

37. Ravikumar, R., Flora, G., Geddes, J.W., Hennig, B. & Toborek, M., Nicotine attenuates oxidative stress, activation of redox-regulated transcription factors and induction of proinflammatory genes in compressive spinal cord trauma. *Brain Res Mol Brain Res* **124**, 188–98 (2004).

38. Lee, Y.B. et al., Role of tumor necrosis factor-alpha in neuronal and glial apoptosis after spinal cord injury. *Exp Neurol* **166**, 190–5 (2000).

39. Yune, T.Y. et al., Increased production of tumor necrosis factor-alpha induces apoptosis after traumatic spinal cord injury in rats. *J Neurotrauma* **20**, 207–19 (2003).

40. Wu, K.L., Chan, S.H., Chao, Y.M. & Chan, J.Y., Expression of pro-inflammatory cytokine and caspase genes promotes neuronal apoptosis in pontine reticular formation after spinal cord transection. *Neurobiol Dis* **14**, 19–31 (2003).

41. Farooque, M., Isaksson, J. & Olsson, Y., Improved recovery after spinal cord injury in neuronal nitric oxide synthase-deficient mice but not in TNF-alpha-deficient mice. *J Neurotrauma* **18**, 105–14 (2001).

42. Kim, G.M. et al., Tumor necrosis factor receptor deletion reduces nuclear factor-kappaB activation, cellular inhibitor of apoptosis protein 2 expression, and functional recovery after traumatic spinal cord injury. *J Neurosci* **21**, 6617–25 (2001).

43. Li, G.L., Farooque, M. & Olsson, Y., Changes of Fas and Fas ligand immunoreactivity after compression trauma to rat spinal cord. *Acta Neuropathol (Berl)* **100**, 75–81 (2000).

44. Casha, S., Yu, W.R. & Fehlings, M.G., Oligodendroglial apoptosis occurs along degenerating axons and is associated with FAS and p75 expression following spinal cord injury in the rat. *Neuroscience* **103**, 203–18 (2001).

45. Zurita, M., Vaquero, J. & Zurita, I., Presence and significance of CD-95 (Fas/APO1) expression after spinal cord injury. *J Neurosurg* **94**, 257–64 (2001).

46. Demjen, D. et al., Neutralization of CD95 ligand promotes regeneration and functional recovery after spinal cord injury. *Nat Med* **10**, 389–95 (2004).

47. Rothwell, N., Interleukin-1 and neuronal injury: mechanisms, modification, and therapeutic potential. *Brain Behav Immun* **17**, 152–7 (2003).

48. Knoblach, S.M. & Faden, A.I., Cortical interleukin-1 beta elevation after traumatic brain injury in the rat: no effect of two selective antagonists on motor recovery. *Neurosci Lett* **289**, 5–8 (2000).

49. Kinoshita, K. et al., Interleukin-1beta messenger ribonucleic acid and protein levels after fluid-percussion brain injury in rats: importance of injury severity and brain temperature. *Neurosurgery* **51**, 195–203; discussion 203 (2002).

50. Tehranian, R. et al., Improved recovery and delayed cytokine induction after closed head injury in mice with central overexpression of the secreted isoform of the interleukin-1 receptor antagonist. *J Neurotrauma* **19**, 939–51 (2002).

51. Nesic, O. et al., IL-1 receptor antagonist prevents apoptosis and caspase-3 activation after spinal cord injury. *J Neurotrauma* **18**, 947–56 (2001).

52. Herx, L.M., Rivest, S. & Yong, V.W., Central nervous system-initiated inflammation and neurotrophism in trauma: IL-1 beta is required for the production of ciliary neurotrophic factor. *J Immunol* **165**, 2232–9 (2000).

53. Fassbender, K. et al., Temporal profile of release of interleukin-1beta in neurotrauma. *Neurosci Lett* **284**, 135–8 (2000).

54. Vecil, G.G. et al., Interleukin-1 is a key regulator of matrix metalloproteinase-9 expression in human neurons in culture and following mouse brain trauma *in vivo*. *J Neurosci Res* **61**, 212–24 (2000).

55. Shibayama, M., Kuchiwaki, H., Inao, S., Yoshida, K. & Ito, M., Intercellular adhesion molecule-1 expression on glia following brain injury: participation of interleukin-1 beta. *J Neurotrauma* **13**, 801–8 (1996).

56. Toulmond, S. & Rothwell, N.J., Interleukin-1 receptor antagonist inhibits neuronal damage caused by fluid percussion injury in the rat. *Brain Res* **671**, 261–6 (1995).

57. DeKosky, S.T. et al., Interleukin-1 receptor antagonist suppresses neurotrophin response in injured rat brain. *Ann Neurol* **39**, 123–7 (1996).

58. Yatsiv, I. et al., Elevated intracranial IL-18 in humans and mice after traumatic brain injury and evidence of neuroprotective effects of IL-18-binding protein after experimental closed head injury. *J Cereb Blood Flow Metab* **22**, 971–8 (2002).

59. Pan, J.Z. et al., Cytokine activity contributes to induction of inflammatory cytokine mRNAs in spinal cord following contusion. *J Neurosci Res* **68**, 315–22 (2002).

60. Nakamura, M., Houghtling, R.A., MacArthur, L., Bayer, B.M. & Bregman, B.S., Differences in cytokine gene expression profile between acute and secondary injury in adult rat spinal cord. *Exp Neurol* **184**, 313–25 (2003).

61. Wang, C.X., Olschowka, J.A. & Wrathall, J.R., Increase of interleukin-1beta mRNA and protein in the spinal cord following experimental traumatic injury in the rat. *Brain Res* **759**, 190–6 (1997).

62. Hayashi, M., Ueyama, T., Nemoto, K., Tamaki, T. & Senba, E., Sequential mRNA expression for immediate early genes, cytokines, and neurotrophins in spinal cord injury. *J Neurotrauma* **17**, 203–18 (2000).

63. Gracie, J.A., Robertson, S.E. & McInnes, I.B., Interleukin-18. *J Leukoc Biol* **73**, 213–24 (2003).

64. Yan, H.Q., Banos, M.A., Herregodts, P., Hooghe, R. & Hooghe-Peters, E.L., Expression of interleukin (IL)-1 beta, IL-6 and their respective receptors in the normal rat brain and after injury. *Eur J Immunol* **22**, 2963–71 (1992).

65. Woodroofe, M.N. et al., Detection of interleukin-1 and interleukin-6 in adult rat brain, following mechanical injury, by *in vivo* microdialysis: evidence of a role for microglia in cytokine production. *J Neuroimmunol* **33**, 227–36 (1991).

66. Kyrkanides, S., O'Banion, M.K., Whiteley, P.E., Daeschner, J.C. & Olschowka, J.A., Enhanced glial activation and expression of specific CNS inflammation-related molecules in aged versus young rats following cortical stab injury. *J Neuroimmunol* **119**, 269–77 (2001).

67. Campbell, I.L. et al., Neurologic disease induced in transgenic mice by cerebral overexpression of interleukin 6. *Proc Natl Acad Sci U.S.A.* **90**, 10061–5 (1993).

68. Penkowa, M. et al., Metallothionein-I overexpression alters brain inflammation and stimulates brain repair in transgenic mice with astrocyte-targeted interleukin-6 expression. *Glia* **42**, 287–306 (2003).

69. Penkowa, M. et al., Astrocyte-targeted expression of interleukin-6 protects the central nervous system during neuroglial degeneration induced by 6-aminonicotinamide. *J Neurosci Res* **73**, 481–96 (2003).

70. Swartz, K.R. et al., Interleukin-6 promotes post-traumatic healing in the central nervous system. *Brain Res* **896**, 86–95 (2001).

71. Penkowa, M. et al., Astrocyte-targeted expression of IL-6 protects the CNS against a focal brain injury. *Exp Neurol* **181**, 130–48 (2003).

72. Penkowa, M. et al., Strongly compromised inflammatory response to brain injury in interleukin-6-deficient mice. *Glia* **25**, 343–57 (1999).

73. Klein, M.A. et al., Impaired neuroglial activation in interleukin-6 deficient mice. *Glia* **19**, 227–33 (1997).

74. Sugiura, S. et al., Leukaemia inhibitory factor is required for normal inflammatory responses to injury in the peripheral and central nervous systems *in vivo* and is chemotactic for macrophages *in vitro*. *Eur J Neurosci* **12**, 457–66 (2000).

75. Gadient, R.A. & Patterson, P.H., Leukemia inhibitory factor, Interleukin 6, and other cytokines using the GP130 transducing receptor: roles in inflammation and injury. *Stem Cells* **17**, 127–37 (1999).

76. Banner, L.R., Moayeri, N.N. & Patterson, P.H., Leukemia inhibitory factor is expressed in astrocytes following cortical brain injury. *Exp Neurol* **147**, 1–9 (1997).

77. Oyesiku, N.M. et al., Regional changes in the expression of neurotrophic factors and their receptors following acute traumatic brain injury in the adult rat brain. *Brain Res* **833**, 161–72 (1999).

78. Xia, X.G., Hofmann, H.D., Deller, T. & Kirsch, M., Induction of STAT3 signaling in activated astrocytes and sprouting septal neurons following entorhinal cortex lesion in adult rats. *Mol Cell Neurosci* **21**, 379–92 (2002).

79. Streit, W.J. et al., Cytokine mRNA profiles in contused spinal cord and axotomized facial nucleus suggest a beneficial role for inflammation and gliosis. *Exp Neurol* **152**, 74–87 (1998).

80. Okada, S. et al., Blockade of interleukin-6 receptor suppresses reactive astrogliosis and ameliorates functional recovery in experimental spinal cord injury. *J Neurosci Res* **76**, 265–76 (2004).

81. Tuna, M. et al., Effect of anti-rat interleukin-6 antibody after spinal cord injury in the rat: inducible nitric oxide synthase expression, sodium- and potassium-activated, magnesium-dependent adenosine-5'-triphosphatase and superoxide dismutase activation, and ultrastructural changes. *J Neurosurg* **95**, 64–73 (2001).

82. Lacroix, S., Chang, L., Rose-John, S. & Tuszynski, M.H., Delivery of hyper-interleukin-6 to the injured spinal cord increases neutrophil and macrophage infiltration and inhibits axonal growth. *J Comp Neurol* **454**, 213–28 (2002).

83. Cafferty, W.B. et al., Conditioning injury-induced spinal axon regeneration fails in interleukin-6 knock-out mice. *J Neurosci* **24**, 4432–43 (2004).

84. Zang da, W. & Cheema, S.S., Leukemia inhibitory factor promotes recovery of locomotor function following spinal cord injury in the mouse. *J Neurotrauma* **20**, 1215–22 (2003).

85. Blesch, A. et al., Leukemia inhibitory factor augments neurotrophin expression and corticospinal axon growth after adult CNS injury. *J Neurosci* **19**, 3556–66 (1999).

86. Oyesiku, N.M., Wilcox, J.N. & Wigston, D.J., Changes in expression of ciliary neurotrophic factor (CNTF) and CNTF-receptor alpha after spinal cord injury. *J Neurobiol* **32**, 251–61 (1997).

87. Ye, J. et al., The effects of ciliary neurotrophic factor on neurological function and glial activity following contusive spinal cord injury in the rats. *Brain Res* **997**, 30–9 (2004).

88. Bethea, J.R. & Dietrich, W.D., Targeting the host inflammatory response in traumatic spinal cord injury. *Curr Opin Neurol* **15**, 355–60 (2002).

89. Balasingam, V. & Yong, V.W., Attenuation of astroglial reactivity by interleukin-10. *J Neurosci* **16**, 2945–55 (1996).

90. Knoblach, S.M. & Faden, A.I., Interleukin-10 improves outcome and alters proinflammatory cytokine expression after experimental traumatic brain injury. *Exp Neurol* **153**, 143–51 (1998).

91. Kline, A.E. et al., Acute systemic administration of interleukin-10 suppresses the beneficial effects of moderate hypothermia following traumatic brain injury in rats. *Brain Res* **937**, 22–31 (2002).

92. Takami, T. et al., Methylprednisolone and interleukin-10 reduce gray matter damage in the contused Fischer rat thoracic spinal cord but do not improve functional outcome. *J Neurotrauma* **19**, 653–66 (2002).

93. Brewer, K.L., Bethea, J.R. & Yezierski, R.P., Neuroprotective effects of interleukin-10 following excitotoxic spinal cord injury. *Exp Neurol* **159**, 484–93 (1999).

94. Plunkett, J.A., Yu, C.G., Easton, J.M., Bethea, J.R. & Yezierski, R.P., Effects of interleukin-10 (IL-10) on pain behavior and gene expression following excitotoxic spinal cord injury in the rat. *Exp Neurol* **168**, 144–54 (2001).

95. Abraham, K.E., McMillen, D. & Brewer, K.L., The effects of endogenous interleukin-10 on gray matter damage and the development of pain behaviors following excitotoxic spinal cord injury in the mouse. *Neuroscience* **124**, 945–52 (2004).

96. Massague, J., TGF-beta signal transduction. *Annu Rev Biochem* **67**, 753–91 (1998).

97. Chin, D., Boyle, G.M., Parsons, P.G. & Coman, W.B., What is transforming growth factor-beta (TGF-beta)? *Br J Plast Surg* **57**, 215–21 (2004).

98. Wilcox, J.N. & Derynck, R., Developmental expression of transforming growth factors alpha and beta in mouse fetus. *Mol Cell Biol* **8**, 3415–22 (1988).

99. Lindholm, D., Castren, E., Kiefer, R., Zafra, F. & Thoenen, H., Transforming growth factor-beta 1 in the rat brain: increase after injury and inhibition of astrocyte proliferation. *J Cell Biol* **117**, 395–400 (1992).

100. Logan, A., Frautschy, S.A., Gonzalez, A.M., Sporn, M.B. & Baird, A., Enhanced expression of transforming growth factor beta 1 in the rat brain after a localized cerebral injury. *Brain Res* **587**, 216–25 (1992).

101. Ihn, H., Pathogenesis of fibrosis: role of TGF-beta and CTGF. *Curr Opin Rheumatol* **14**, 681–5 (2002).

102. Logan, A. et al., Effects of transforming growth factor beta 1 on scar production in the injured central nervous system of the rat. *Eur J Neurosci* **6**, 355–63 (1994).

103. Moon, L.D. & Fawcett, J.W., Reduction in CNS scar formation without concomitant increase in axon regeneration following treatment of adult rat brain with a combination of antibodies to TGFbeta1 and beta2. *Eur J Neurosci* **14**, 1667–77 (2001).

104. Rimaniol, A.C. et al., Biphasic transforming growth factor-beta production flanking the pro-inflammatory cytokine response in cerebral trauma. *Neuroreport* **7**, 133–6 (1995).

105. Fee, D.B. et al., Traumatic brain injury increases TGFbetaRII expression on endothelial cells. *Brain Res* **1012**, 52–9 (2004).

106. O'Brien, M.F., Lenke, L.G., Lou, J., Bridwell, K.H. & Joyce, M.E., Astrocyte response and transforming growth factor-beta localization in acute spinal cord injury. *Spine* **19**, 2321–9; discussion 2330 (1994).

107. Semple-Rowland, S.L. et al., Analysis of TGF-beta 1 gene expression in contused rat spinal cord using quantitative RT-PCR. *J Neurotrauma* **12**, 1003–14 (1995).

108. Hamada, Y. et al., Effects of exogenous transforming growth factor-beta 1 on spinal cord injury in rats. *Neurosci Lett* **203**, 97–100 (1996).

109. Tyor, W.R., Avgeropoulos, N., Ohlandt, G. & Hogan, E.L., Treatment of spinal cord impact injury in the rat with transforming growth factor-beta. *J Neurol Sci* **200**, 33–41 (2002).

110. Brines, M.L. et al., Erythropoietin crosses the blood–brain barrier to protect against experimental brain injury. *Proc Natl Acad Sci U.S.A.* **97**, 10526–31 (2000).

111. Chandel, N.S. et al., Mitochondrial reactive oxygen species trigger hypoxia-induced transcription. *Proc Natl Acad Sci U.S.A.* **95**, 11715–20 (1998).

112. Juul, S.E., Stallings, S.A. & Christensen, R.D., Erythropoietin in the cerebrospinal fluid of neonates who sustained CNS injury. *Pediatr Res* **46**, 543–7 (1999).

113. Celik, M. et al., Erythropoietin prevents motor neuron apoptosis and neurologic disability in experimental spinal cord ischemic injury. *Proc Natl Acad Sci U.S.A.* **99**, 2258–63 (2002).

114. Gorio, A. et al., Recombinant human erythropoietin counteracts secondary injury and markedly enhances neurological recovery from experimental spinal cord trauma. *Proc Natl Acad Sci U.S.A.* **99**, 9450–5 (2002).

115. Aydin, A. et al., Erythropoietin exerts neuroprotective effect in neonatal rat model of hypoxic-ischemic brain injury. *Brain Dev* **25**, 494–8 (2003).

116. Kaptanoglu, E. et al., Erythropoietin exerts neuroprotection after acute spinal cord injury in rats: effect on lipid peroxidation and early ultrastructural findings. *Neurosurg Rev* **27**, 113–20 (2004).

117. Mueller, C.A., Schluesener, H.J., Conrad, S., Meyermann, R. & Schwab, J.M., Lesional expression of a proinflammatory and antiangiogenic cytokine EMAP II confined to endothelium and microglia/macrophages during secondary damage following experimental traumatic brain injury. *J Neuroimmunol* **135**, 1–9 (2003).

118. Mueller, C.A., Schluesener, H.J., Conrad, S., Meyermann, R. & Schwab, J.M., Spinal cord injury induces lesional expression of the proinflammatory and antiangiogenic cytokine EMAP II. *J Neurotrauma* **20**, 1007–15 (2003).

119. Maier, B. et al., Differential release of interleukines 6, 8, and 10 in cerebrospinal fluid and plasma after traumatic brain injury. *Shock* **15**, 421–6 (2001).

120. Bell, M.J. et al., Interleukin-6 and interleukin-10 in cerebrospinal fluid after severe traumatic brain injury in children. *J Neurotrauma* **14**, 451–7 (1997).

121. Singhal, A. et al., Association between cerebrospinal fluid interleukin-6 concentrations and outcome after severe human traumatic brain injury. *J Neurotrauma* **19**, 929–37 (2002).

122. Woiciechowsky, C. et al., Early IL-6 plasma concentrations correlate with severity of brain injury and pneumonia in brain-injured patients. *J Trauma* **52**, 339–45 (2002).

123. Winter, C.D., Pringle, A.K., Clough, G.F. & Church, M.K., Raised parenchymal interleukin-6 levels correlate with improved outcome after traumatic brain injury. *Brain* **127**, 315–20 (2004).

124. Tasci, A., Okay, O., Gezici, A.R., Ergun, R. & Ergungor, F., Prognostic value of interleukin-1 beta levels after acute brain injury. *Neurol Res* **25**, 871–4 (2003).

125. Kushi, H., Saito, T., Makino, K. & Hayashi, N., IL-8 is a key mediator of neuroinflammation in severe traumatic brain injuries. *Acta Neurochir Suppl* **86**, 347–50 (2003).

126. Mussack, T. et al., Serum S-100B and interleukin-8 as predictive markers for comparative neurologic outcome analysis of patients after cardiac arrest and severe traumatic brain injury. *Crit Care Med* **30**, 2669–74 (2002).

127. Kossmann, T. et al., Interleukin-8 released into the cerebrospinal fluid after brain injury is associated with blood–brain barrier dysfunction and nerve growth factor production. *J Cereb Blood Flow Metab* **17**, 280–9 (1997).

128. Stahel, P.F., Kossmann, T., Joller, H., Trentz, O. & Morganti-Kossmann, M.C., Increased interleukin-12 levels in human cerebrospinal fluid following severe head trauma. *Neurosci Lett* **249**, 123–6 (1998).

129. Goodman, J.C., Robertson, C.S., Grossman, R.G. & Narayan, R.K., Elevation of tumor necrosis factor in head injury. *J Neuroimmunol* **30**, 213–7 (1990).

130. Morganti-Kossmann, M.C. et al., TGF-beta is elevated in the CSF of patients with severe traumatic brain injuries and parallels blood–brain barrier function. *J Neurotrauma* **16**, 617–28 (1999).

131. Csuka, E. et al., IL-10 levels in cerebrospinal fluid and serum of patients with severe traumatic brain injury: relationship to IL-6, TNF-alpha, TGF-beta1 and blood–brain barrier function. *J Neuroimmunol* **101**, 211–21 (1999).

132. Ertel, W. et al., Detectable concentrations of Fas ligand in cerebrospinal fluid after severe head injury. *J Neuroimmunol* **80**, 93–6 (1997).

133. Lenzlinger, P.M., Marx, A., Trentz, O., Kossmann, T. & Morganti-Kossmann, M.C., Prolonged intrathecal release of soluble Fas following severe traumatic brain injury in humans. *J Neuroimmunol* **122**, 167–74 (2002).

134. Arand, M., Melzner, H., Kinzl, L., Bruckner, U.B. & Gebhard, F., Early inflammatory mediator response following isolated traumatic brain injury and other major trauma in humans. *Langenbecks Arch Surg* **386**, 241–8 (2001).

135. Slewa-Younan, S., Green, A.M., Baguley, I.J., Gurka, J.A. & Marosszeky, J.E., Sex differences in injury severity and outcome measures after traumatic brain injury. *Arch Phys Med Rehabil* **85**, 376–9 (2004).

136. Hayes, K.C. et al., Elevated serum titers of proinflammatory cytokines and CNS autoantibodies in patients with chronic spinal cord injury. *J Neurotrauma* **19**, 753–61 (2002).

137. Yang, L. et al., Early expression and cellular localization of proinflammatory cytokines interleukin-1beta, interleukin-6, and tumor necrosis factor-alpha in human traumatic spinal cord injury. *Spine* **29**, 966–71 (2004).

138. Marino, M.W. et al., Characterization of tumor necrosis factor-deficient mice. *Proc Natl Acad Sci USA* **94**, 8093–8 (1997).

139. Airaksinen, M.S. et al., Ataxia and altered dendritic calcium signaling in mice carrying a targeted null mutation of the calbindin D28k gene. *Proc Natl Acad Sci U.S.A.* **94**, 1488–93 (1997).

140. Zheng, H. et al., Resistance to fever induction and impaired acute-phase response in interleukin-1 beta-deficient mice. *Immunity* **3**, 9–19 (1995).

141. Kopf, M. et al., Impaired immune and acute-phase responses in interleukin-6-deficient mice. *Nature* **368**, 339–42 (1994).

142. Penkowa, M. & Hidalgo, J., IL-6 deficiency leads to reduced metallothionein-I+II expression and increased oxidative stress in the brain stem after 6-aminonicotinamide treatment. *Exp Neurol* **163**, 72–84 (2000).

143. Stewart, C.L. et al., Blastocyst implantation depends on maternal expression of leukaemia inhibitory factor. *Nature* **359**, 76–9 (1992).

144. Pasparakis, M., Alexopoulou, L., Episkopou, V. & Kollias, G., Immune and inflammatory responses in TNF alpha-deficient mice: a critical requirement for TNF alpha in the formation of primary B cell follicles, follicular dendritic cell networks and germinal centers, and in the maturation of the humoral immune response. *J Exp Med* **184**, 1397–411 (1996).

145. Rothe, J. et al., Mice lacking the tumour necrosis factor receptor 1 are resistant to TNF-mediated toxicity but highly susceptible to infection by Listeria monocytogenes. *Nature* **364**, 798–802 (1993).

146. Erickson, S.L. et al., Decreased sensitivity to tumour-necrosis factor but normal T-cell development in TNF receptor-2-deficient mice. *Nature* **372**, 560–3 (1994).

14 Cytokine-Based Therapeutics for Intrinsic CNS Disease Processes

Heinz Wiendl and Reinhard Hohlfeld

CONTENTS

I. INTRODUCTORY REMARKS

Multiple sclerosis (MS) is the prototype intrinsic inflammatory CNS disease process. Autoreactive T lymphocytes directed against components of central myelin are believed to orchestrate the pathophysiological process, eventually leading to demyelination and axonal damage (Hohlfeld, 1997; Hemmer et al., 2002; Noseworthy et al., 2000). MS is a disease of young adults and the most common of the inflammatory neurological disorders leading to neurological disability. MS and its animal model, experimental autoimmune encephalomy-elitis (EAE), are associated with inflammatory, delayed-type hypersensitivity (Th1) responses, with cytokines playing a prominent role as mediators of inter-cellular signaling and effector function. Cell adhesion molecules, costimulatory ligands, and matrix metalloproteinsases are also upregulated during the autoim-mune response and are implicated in leukocyte infiltration to the CNS. The inflammatory responses that arise as a result of tissue injury can contribute to further injury as well as repair. As additional immune mediators and regulators are identified, it is necessary to establish their precise role in contributing to or protecting the CNS from disease-induced damage. The expression or upregula-tion of mediators has been associated with autoimmune disease and correlated with relapse and remission, leading to the "guilt by association" hypothesis. Strategies designed to block these mediators using antibodies and pharmacolog-ical inhibitors have been informative, but they are always subject to concerns about specificity, side effects, and access of the inhibitory reagent into the CNS. Ultimate proof of causation must come from disease induction through CNS-specific expression of the mediator in question. The use of transgenic technology to overexpress or prevent expression of genes encoding molecules related to inflammation has allowed direct examination of their role in experimental disease (Owens et al., 2001). Thus, numerous experiments involving transgenic animals have identified causative roles for some cytokines but have also raised questions about the strains of animals used in the study and expression kinetics (Steinman, 1997).

Novel insights into the immunopathological processes, advances in biotechnol-ogy, and the development of powerful MRI technologies together with improvements in clinical trial design have led to a variety of evaluable therapeutical approaches in this respect. Cytokine-based therapeutics have traditionally been regarded as appeal-ing and promising strategies for modifying the disease course of MS, because the key features in the disease pathogenesis are considered to be causally related to cytokine imbalances. The advent of the Betainterferons in the early 1990s repre-sented a tremendous advancement in the therapeutical options for modifying the natural course of the disease. However, their partial effectiveness in reducing the relapse rate and the disease progression in treated patients demands superior or complementary therapies. However, up until now, they have represented the only cytokine-based therapeutic strategy that has successfully entered clinical standard therapy.

II. THERAPY WITH CYTOKINES AND ANTI-CYTOKINES: GENERAL CONSIDERATIONS

The cytokine-based immunetherapies of intrinsic disease processes that can be envisaged for MS follow one of three basic strategies. The first strategy is to administer "downregulatory" cytokines, such as interferon-ß, IL-10, IL-4, or TGF-β. It should be kept in mind, however, that cytokines with exclusively downregulatory (or exclusively proinflammatory) properties do not exist. The second strategy is to administer inhibitors of proinflammatory cytokines, the most prominent example being TNF-α. In principle, this inhibition could be achieved by using genetically engineered ("humanized") monoclonal antibodies or, alternatively, by designing recombinant receptor constructs for neutralizing the circulating cytokines. Third, one could make therapeutic use of naturally occurring cytokine inhibitors or antagonists, such as the IL-1 receptor antagonist, a naturally occurring competitive inhibitor of the proinflammatory cytokine IL-1 (Dinarello and Thompson, 1991).

One consequence of the complexity of the cytokine network is that it is virtually impossible to predict the overall effect of a given cytokine or cytokine inhibitor *in vivo*. Although it is obviously essential to test the safety and efficacy of cytokine-based therapies in animal models, totally unexpected, severe adverse effects may still occur in human clinical trials. This is exemplified by a tragic episode that occurred in 1995, when genetically engineered IL-12 severely harmed several patients with renal cell carcinoma. It was particularly disturbing that the cytokine had been previously tested not only in animal experiments but also in humans in another pilot trial without adverse effects. It appears that a slight difference in mode of application was critical. This instance serves as a reminder that any therapy is ultimately empirical, regardless of how solid its theoretical foundations appear to be. Previous experience has shown that, despite rational therapeutical concepts, convincing preliminary animal experiments, and positive experiences with other autoimmune diseases, several initial cytokine-based therapy studies failed because of lack of efficacy or unforeseen side effects. Whereas the positive trials usually make it into prestigious journals, many negative trials are published merely as abstracts or not at all (Hohlfeld and Wiendl, 2001). This is unfortunate, because a lot can be learned from a negative result, and critical reflection is highly important for understanding human MS-immunepathogenesis and appropriate trial design.

III. INTERFERONS

A. BACKGROUND

Interferon-γ (IFN-γ) is considered a prototype "proinflammatory" cytokine. It is produced by activated T cells and natural killer cells and acts as a potent activator of macrophages and monocytes inducing an array of inflammatory mediators in these cells. Among many, an important function is to increase the expression of class I and class II MHC molecules on a variety of cell types, including both specialized and more adaptable antigen-presenting cells. This increase in MHC expression facilitates antigen presentation, thereby augmenting and accelerating immune responses. Apart from augmenting T- and natural killer cell responses, IFN-γ has potent antiviral properties. IFN-γ expression is observed in MS and EAE lesions (Owens et al., 1994; Popko et al., 1997) and direct injection of IFN-γ into the CNS induces inflammatory pathology in rats and mice (Simmons and Willenborg, 1990; Sethna and Lampson, 1991). Transgenic- IFN-γ-overexpression in the CNS has been achieved using the MBP-promoter. Phenotypes ranged from lethal "jimpy"-like mice with hypomyelinated neurons (Corbin et al., 1996) to mice with progressive demyelinating disease (Horwitz et al., 1997). The mechanism underlying the hypomyelinated, tremoring phenotype may be due to the induction of MHC expression in oligodendrocytes. However, several studies have also suggested an immune suppressive role for IFN-γ in EAE: IFN-γ blockade consistently enhances EAE (Billiau et al., 1988), and IFN-γ treatment has been reported to confer resistance to EAE (Krakowski and Owens, 1996). Further, mice deficient in either IFN-γ or its receptor showed a similar phenotype with regard to EAE development, pronounced neutrophilia and lethal outcome, regardless of strain background (Willenborg et al., 1996). Thus it has become increasingly clear that in addition to its well-known proinflammatory effects, IFN-γ exhibits potent immunosuppressive properties, which have been attributed to a shift in the chemokine profile (Tran et al., 2000).

The type I Interferons, interferon-α (IFN-α) and interferon-β (IFN-β), both consist of 166 amino acids and are produced by almost all mammalian cells on stimulation. One (but not the only) inducer of interferon synthesis is double-stranded RNA signaling via certain toll-like receptors (especially TLR3). Like IFN-γ, type I interferons increase expression of MHC class I molecules and thereby enhance the ability of virus-infected cells to present viral peptides to CD8 T cells. In contrast to INF-γ, however, IFN-β and IFN-α do not induce the synthesis of MHC class II proteins; moreover, they suppress it. Type I interferons are effective in EAE (Abreu, 1985; Brod et al., 1995; Yu et al., 1996). Administration of IFN-β results in the downregulation of delayed-type hypersensitivity responses and of the secretion of IFN-γ and TNF-α/β. The expression of MHC antigens decreases as does the proliferative response of T cells. In addition, a rise in the expression of the immunomodulatory cytokine IL-10 was noted not only *in vitro*, but also during the treatment of MS patients *in vivo*. As a consequence, many of the mechanisms that are involved in immune activation during an exacerbation as well as those that are part of the effector phase of lesion development (secretion of IFN-γ and TNF-α/β) are counteracted by IFN-β. The pleiotropic effects of these immunomodulatory interferons include proinflammatory as well as antiinflammatory properties. It is important to note that the efficacy in animal models as well as in clinical settings results from an "immunological net effect" that seems clearly antiinflammatory. Antiviral and antiproliferative actions might contribute to the overall clinical effect in MS, although immunomodulatory effects such reduction of MHC class II molecules are considered more important. The mechanism(s) of the complex *in vivo* effects of type I interferons are unknown and are obviously difficult to investigate *in vitro*. Thus it has remained it is not known exactly how IFN-β changes the clinical course of MS. Putative mechanisms of action include (i) inhibition of T-cell activation and proliferation; (ii) regulation of cytokine expression; and (iii) protection of the bloodbrain barrier via the interference with cell adhesion, migration, and matrix metalloproteinase activity (Arnason et al., 1996; Yong, 2002).

B. Studies

Interferons were first used in MS because of their anti-viral activities. At first, no distinction was placed on the type of interferon, and each was assessed after administration by a systemic or intrathecal route. A series of pilot studies, mostly uncontrolled and involving small numbers of patients, was performed in the 1980s. These involved IFN-α (Knobler et al., 1984; Camenga et al., 1986), IFN-ß given systemically or by the intrathecal route (Jacobs et al., 1981) and IFN-γ (Panitch et al., 1987). The details of these studies are now more of historical interest (Jacobs and Johnson, 1994).

It is noteworthy, however, that in contrast to the beneficial type I interferons, IFN–γ dramatically exacerbated the disease activity in MS: This well-known trial performed as a phase II study had to be terminated after few months of treatment because of a sharp increase of exacerbations in the treated patients (Panitch et al., 1987). This was attributed to the proinflammatory properties executed by IFN-γ; clearly, IFN-γ has pro-inflammatory effects when expressed in the CNS. However, this cytokine also has immunomodulatory effects, regulating both T-cell induction and leukocyte homeostasis. Because these effects seem predominant in the periphery, this underlines the value of CNS-targeted approaches in further studies of IFN-γ.

A number of positive trials have been and are being performed with IFN-α. The first positive trial was reported by Knobler et al. (1984), who administered IFN-α intramuscularly in 24 patients with relapsing disease. Camenga et al. (1986) tested the effect of recombinant IFN-α, given by self-administered subcutaneous injection three times weekly for 1 year (98 patients). The response with respect to relapse rate was positive. Durelli et al. (1994) repeated the study of IFN-α given by intramuscular injection on alternate days in relapsing-remitting patients. Again, patients receiving the verum had a lower exacerbation rate. However, disability was unaffected. In a follow-up study, Durelli et al. (1996) examined the resumption of clinical, MRI, and immunological activity in patients who had to discontinue IFN-α (after 6 months of treatment) for administrative and financial reasons. MRI results suggest a transient reduction in activity that reverses after discontinuation of treatment with IFN-α. In summary, despite partly encouraging results from smaller trials, there is not much activity in developing IFN-α further for use in patients with MS.

Following the initial observation of a clinical benefit from intrathecal IFN-β (Jacobs and Johnson, 1994) a number of large multicenter trials have been conducted during the last decade in various stages of MS, and these have firmly established IFN-β as the treatment of choice in relapsing-remitting MS, secondary progressive-MS with inflammatory activity, and recently also in monosymptomatic stages of MS (Comi et al., 2001; Jacobs et al., 2000), relapsing remitting MS (IFNB MS Study Group, 1993, 1995; Jacobs et al., 1996; OWIMS Study Group, 1999; PRISMS Study Group, 1998, 2001), and secondary progressive MS (European MS Study Group, 1998; Cohen et al., 2001; Goodkin and Group, 2000; Kappos et al., 2001; Li et al., 2001; SPECTRIMS Study Group, 2001). Three preparations of IFN-β are currently available: IFN-β 1b (recombinant unglycosylated IFN-β with a substitution of cysteine in position 17 by serine and removal of the aminoterminal methionine; IFN-β-ser-17 = Betaseron®, or Betaferon®); and two preparations of IFN-β 1a (recombinant, glycosylated IFN-β produced in chinese hamster ovary cells), i.e., Avonex® and Rebif®. Two doses of IFN-β 1b (1.6 and 8 Mill. international units, i.U.) s.c. every other day were compared with placebo over a period of three years (IFNB MS Study Group, 1993). The high dose reduced the attack frequency by one-third and cut serious attacks by 50%. The decline in attacks started early (after 2 months of onset) and lasted throughout the trial. The number of MRI lesions dropped dramatically (80% for the medians of activity; 83% for the appearance of active and 75% for the number of new lesions, whereas an increase of MRI-documented disease burden was observed in the placebo arm) (IFNB MS Study Group, 1995). The effects with the lower dose were smaller but still notable. When the EDSS score was used as an outcome measure, the differences between the treatment and placebo arm were less dramatic, in part due to the mild disease of the patients and the short follow-up (for early RR-MS). Based on the reduction of attack and MRI lesion frequencies, it is expected that the clinical differences will become more pronounced over time. Similar results were observed with the two glycosylated IFN-β 1a preparations (Avonex® and Rebif®; the former injected as 6×10^6 IU once weekly i.m.; the latter injected three times a week at 6×10^6 or 12×10^6 IU s.c.). Sustained worsening of disability was chosen as primary outcome measure in the IFN-β 1a; Avonex® trial, and a reduction of progression could be demonstrated in RR MS (Jacobs et al., 1996). Both IFN-β 1a (Rebif®) and IFN-β 1b (Betaseron®) were also tested in RR- and SP MS, and following the demonstration in a large scale, controlled, randomized, double-blind and multicenter phase III trial of IFN-β 1b (Betaseron®) in Europe (European MS Study Group, 1998) that this compound slows progression in SP MS, IFN-β 1b (Betaseron®) was also approved by the European regulatory agencies (EMEA) for SP MS. This effect on SP MS was not confirmed by a similarly large U.S. phase III trial in SP MS, and after careful comparison of the study results and patient populations, it was concluded that the patients in the U.S. study were slightly further advanced in their disease course, and consequently that IFN-β 1b (Betaseron®) is probably effective during earlier stages of SP MS but less so at later stages, when inflammation declines and the disease course is mainly driven by degenerative aspects and oligodendrocyte death or scar formation. Finally, both IFN-β 1a preparations have been tested in monosymptomatic demyelinating disease or early MS (CHAMPS and ETOMS) (Comi et al., 2001; Jacobs et al., 2000) and documented that the onset of clinically definite MS can be delayed for 10 and 9 months, respectively.

C. Commentary

Betainterferons represent the paradigm for a successful cytokine-based therapy now firmly established as a standard disease-modifying treatment for MS (Goodin et al., 2002) (see Table 14.1). The major short- and medium-term side effects observed thus far in MS trials of IFN-ß include flu-like symptoms, (usually mild) laboratory abnormalities, and (rarely) skin necrosis at the injection sites (Walther and Hohlfeld, 1999).

Although numerous large and well-controlled trials have been performed with interferons, a number of critical questions are still unresolved. Beta-interferons reduce clinical and MRI indicators of inflammatory disease activity (clinical relapses, new and active MRI lesions) although for most of the published studies the period of follow-up is short (less than 3 years). There is a growing hope, but as yet only limited evidence, that early use of the currently available, partially effective IFNs may modify the course of the disease in the long term. Inherently linked to this is the key question of whether early treatment will prevent or reduce irreversible injury to axons and oligodendrocytes and thereby eliminate or delay the inevitable and irreversible clinical worsening that hallmarks the secondary phase of the disease.

Interferons have their clearest effect in reducing the relapse rate and signs of inflammation on MRIs. It is interesting to note that only one out of four trials of Beta-Interferons in SP-MS was positive in terms of primary clinical endpoints (European and MS, 1998). However, each of the four large studies demonstrated that the beta-IFNs continue to reduce clinical and MRI markers of inflammatory activity even during the SP phase of the illness. Regrettably, the clinical (disability progression) and late MRI counterparts of axonal degeneration (e.g., atrophy) are less clearly altered by IFN treatment started during the SP phase of the illness.

There is a lively debate about which of the available interferons is most effective and how its effects compare to the effectiveness of other available immune modulatory therapies, namely glatiramer acetate. These issues are and have been addressed in a variety of comparison trials (e.g., INCOMIN, EVIDENCE trial) (Durelli et al., 2002;) Panitch et al., 2002). There is still a considerable number of trials testing the (comparative) effectiveness, the optimum dosing regimen, the effectiveness, practicability, safety of alternative modes of application or formulations, or the relevance of neutralizing antibodies of the various interferons (IFNβ-1a and IFNβ-1b, IFNα-1a, IFNα-1b [PEG-intron], recombinant tau). Because of the partial effectiveness of all currently approved disease modifying agents in MS, the theoretical and practical profits of combination therapies are the subjects of intense study and debate (Maurer and Rieckmann, 2000).

Another major unsolved problem of interferon therapy concerns the development of antibodies against interferons, which can appear in a relevant number of patients. The subset of antibodies that can neutralize IFN-β activity is called neutralizing antibodies. When IFN-betas were first used, neutralizing antibodies were not considered important. However, recent clinical, biological, and immunological data have demonstrated that they reduce or abolish the therapeutic efficacy of IFN-β in 10–20% of patients (Bertolotto, 2004; Sorensen et al., 2003). Quantification of antibodies using various biologic methods makes it difficult to compare among different laboratories, and, hence, standardization of assay procedures is necessary (Hartung et al., 2004). Despite these technical difficulties, data consistently show differences in immunogenicity among the different IFN-β products and suggest negative effects of neutralizing antibodies on the clinical efficacy of IFN-β. Because the therapeutic action of IFN-β depends on activation of IFN-inducible genes, new methods for the quantification of the biologic activity of IFN-βεταo have been developed, and a good correlation has been found between the presence of neutralizing antibodies and the abrogation of IFN-β bioactivity. Thus, the quantification of neutralizing

TABLE 14.1
Summary of Approved Cytokine-Based Therapies for the Treatment of Multiple Sclerosis

Agent	Characteristics: Preparation, Dose Regimen	MS Course	Reference	Outcome/ MRI	Effect Clinical	Problems
Beta-IFNs	IFNβ-1b (Betaseron®): 8 MIU sc every other day IFNβ-1a (Avonex®): 30 µg im 1x/wk IFNβ-1a (Rebif®): 22 or 44 µg sc 3x/we	CIS	Jacobs et al., 2000 Comi et al., 2000	Reduction in lesion volume, activity	Delayed conversion to CDMS	Limited follow-up, patient unblinding likely,
		RRMS	IFNB MS study group 1993, IFNB MS study group 1995 Jacobs et al., 1996 The PRISMS study group, 1998 The PRISMS study group, 2001 The OWIMS study group, 1999	Reduced MRI activity and progression of T2 volume	Appr. 30% reduction of relapse rate, possibly modest slowing in disability	Yet unanswered effects on long-term disability, correlation of long term disability with MRI effect

| SPMS | The European MS study group, 1998; Kappos et al., 2001; Goodkin et al., 2000 The SPECTRIMS study group, 2001; Li et al., 2001; Cohen et al., 2001 | Reduced MRI activity and progression of T2 volume. Limited or no effect on brain atrophy measures. | Moderate reduction in relapse rate; overall not consistent benefit on slowing of disability (depending on outcome measures) | Yet unanswered efffects on long-term disability; limited effect on delaying brain atrophy, |

Legend: IFN = interferon; CIS = clinically isolated syndrome; RRMS = relapsing remitting MS; SPMS = secondary progressive MS; MRI = magnetic resonance imaging; CDMS = clinically definitive MS; im = intramuscular; sc = subcutaneous; we = week;

TABLE 14.2
Examples of Recent Cytokine-Based Therapies in Multiples Sclerosis Trials That Did Not Show a Convincing Clinical Benefit

Agent Cytokines or Cytokine Modulators	Characteristics	MS-type Patients (n) Trial Duration []	Reference	Outcome/ MRI	Effect Clinical	Problems
Lenercept (RO-452081)	Soluble TNF-receptor p55: inhibition of TNF-α-functions	RRMS (168) [11 months]	Arnason et al., 1999	No effect	Worsening	Paradoxical effect of TNF-α; discrepancy between MRI and clinical effects
Infliximab (cA2)	TNF-α neutralizing antibody; human/murine chimeric IgG1: inhibition of TNF-α-functions	SPMS (2) [2 months]	Van Oosten et al., 1996	Worsening	No effect on EDSS	Paradoxical effect of TNF-α
TGF-β2 (transforming growth factor-β2)	Immune suppression, pleiotropic growth factor	SPMS (11) [6 months]	Calabresi et al., 1998	No effect	No effect	Bioavailability in the CNS?; nephrotoxicity
Il-10 (Interleukin-10)	Recombinant cytokine: inhibition of macrophage APC-function, upregulation of Th2-cells	RRMS, SPMS [terminated]	unpublished	—	—	Insufficient efficacy; possible induction of exacerbations

Il-4 (Interleukin-4, BAY 36-1677)	Recombinant cytokine: mutein with 2 AA exchanges and selectivity for T-, B-cells and monocytes, upregulation of Th2-cells	[terminated]	unpublished	—	—	Insufficient efficacy
IFN-γ	Cytokine with antiviral and preferably proinflammatory properties	RRMS (18) [4 weeks]	Panitch et al., 1987	Not done	Worsening	Exacerbations associated with activation of the immune system

Legend RRMS = relapsing remitting MS; SPMS = secondary chronic progressive MS; EDSS = expanded disability status scale (Kurtzke); APC = Antigen presenting cell; MBP = Myelin basic protein; Th = T helper cell; Th1/2 = T helpe cell type]; IFN-γ = Interferon-γ; TNF-α = Tumor necrosis factor-α; AA = amino acid antibodies and the *in vivo* bioactivity of IFN-β through IFN-β-inducible gene products, such as Myxovirus protein A, may offer valuable information on therapeutic efficacy. Important topics, such as the optimal therapeutic strategy for managing neutralizing antibodies positive patients, require further studies.

Finally, treatment response to interferons differs considerably among patients. This is not surprising, given the complexity of MS as a disease, where several pathophysiological processes (including inflammation, demyelination, axonal damage, and repair mechanism) take place, often in dissemination of time and space. Thus, the distinction between responders and (primary or secondary) nonresponders would be an ultimate goal. The characterization of individual patient profiles before and during Interferon therapy (for example, with DNA-array technologies) may help to define criteria that allow one to predict the individual treatment response (Wandinger et al., 2001). This is embedded in the concept of biomarkers and their potential use as surrogate endpoints in making treatment predictions and decisions (Bielekova and Martin, 2004).

IV. TNF-α-ANTAGONISTS

A. BACKGROUND

Numerous investigations have identified TNF-α as an essential pathogenetic factor in different models of experimental allergic encephalomyelitis (EAE) and MS. The cytokine has been detected in inflammatory CNS lesions; in active lesions it is involved in pathological tissue damage (inflammation as well as demyelination); and *in vitro* TNF-α is cytotoxic for oligodendrocytes (Cannella and Raine, 1995; Selmaj et al., 1991). The elimination of TNF-producing macrophages, as well as the antagonization with TNF antibodies, the administration of different therapeut drugs affecting the TNF-α production (e.g., Thalidomide, Pentoxifylline, Rolipram), or doses of soluble TNF receptor (Lenercept) clearly showed a positive effect on pathogenesis and demyelination in various animal models (Klinkert et al., 1997; Körner et al., 1997).

A series of studies with MS patients showed a correlation of TNF levels in blood, serum, or CSF with the clinical course or disease activity (Beck et al., 1988; Chofflon et al., 1992; Imamura et al., 1993; Rieckmann et al., 1995; Rudick and Ransohoff, 1992; Sharief and Hentges, 1991; Van Oosten et al., 1998).

B. STUDIES

1. Infliximab (cA2)

In an open phase I study two patients with a severe secondary chronic progressive (SPMS) form of MS were treated with a monoclonal antibody against TNF-α (Infliximab, cA2) (van Oosten et al., 1996). Inflammatory activity as measured by MRI, CSF lymphocytic pleocytosis, and IgG index was clearly increased after receiving the infusions. After 2 to 3 weeks values dropped back to their initial level, and the EDSS ("expanded disability status scale") was not altered.

2. Lenercept (RO-452081)

In a phase II study (168 patients with mainly relapsing-remitting MS), the effect of the soluble TNF receptor-immunoglobulin-fusion protein Lenercept on the development of new lesions in MRI was examined (Arnason et al., 1999). In this four-armed study, patients received 10, 50, or 100 mg of the drug every 4 weeks (up to 12 months). A baseline MRI was taken as a reference and followed up every 4 weeks (up to week 24 of the study).

The MRIs showed no significant differences between the verum and placebo (primary endpoint: cumulative number of new active lesions). However, the number of clinical exacerbations was significantly higher in the Lenercept group (annual relapse rate placebo 0.98 versus Lenercept (50 mg) 1.64; $p = 0.007$). Exacerbations occurred earlier ($p = 0.006$), the duration of the relapses was longer, the time until clinical exacerbations appeared was shortened, and neurological deficits appeared to be more serious (not significant; EDSS not altered). Side effects such as headaches, nausea, abdominal pain, or hot flashes occurred more frequently with Lenercept. Antibodies against Lenercept were detected in 88–100% of the patients. These antibodies did not interfere with the neutralization of TNF but accelerated the elimination of the drug. A first evaluation after all the patients were treated for 24 weeks led to an early termination of the study.

C. COMMENTARY

Blockage of TNF-α, a putative key cytokine in MS, failed in two trials. In contrast, anti-TNF strategies showed impressive beneficial effects in other T-cell–mediated autoimmune diseases, above all rheumatoid arthritis (RA) (Elliott et al., 1993; Feldmann and Maini, 2001; Lovell et al., 2000; Maini et al., 1999; Weinblatt et al., 1999). The unexpected and surprising negative results for Infliximab and Lenercept require careful analysis, particularly as they seem to question some of the current concepts of MS pathogenesis. The discrepancy between clinical exacerbations and MRI findings in the Lenercept study is remarkable. Whereas an increase in clinical exacerbation rate was overt, the MRI findings showed only a trend toward increased activity during therapy. The detection of new active lesions in MRI is assumed to be a highly sensitive indicator for disease activity (approximately 10 new lesions correspond to about one clinical exacerbation). MRI assessment as a primary endpoint in MS studies therefore can help in rapidly gaining information but using lower patient numbers (Miller et al., 1996). However, an exact correlation of "positive" MRI findings with clinical parameters is not guaranteed. In the Lenercept study the number of new lesions did not differ significantly from the lesion number in the placebo group. It is not clear if the MRI as the primary endpoint parameter for the efficacy-assessment was wrong or if "technical reasons" in the study protocol could account for the difference. The MRI scans were taken before each intravenous infusion, in other words, 4 weeks after the last infusion. In contrast, in the Infliximab study, where the MRI activity was increased, examinations were conducted shortly after the infusions of the antibodies. After 2 to 3 weeks, MRI activity dropped back to its initial level. Obviously Lenercept promoted the formation of clinically relevant lesions without detectable correlates in MRI.

In almost all patients antibodies against the TNF receptor construct were detected. Although they did not inhibit the binding to TNF, they accelerated the elimination and thereby may have shortened the duration of the drug effect. On the other hand, it has been suggested that anti-TNF antibodies that bind TNF may per se act like a "TNF-sink," which may then release TNF later and induce exacerbation of disease. Lenercept itself also possesses the Fc part of IgG immunoglobulins. The Ig backbone may increase the longevity of the antibody in circulation and increase Fc binding events. It is likely that Fc parts stimulated the formation of immune complexes as well as the activation of Fc receptors on lymphocytes. This effect could potentially be a trigger for inflammatory events. However, the use of Fab fragments or the extracellular portions of receptors linked by a polyglycine/serine linker may confer a short half-life and reduce unwanted peripheral side effects. These points may help give a future direction to research in this field.

Experimental studies in EAE models often investigate the effects of therapeutic agents applied prior to disease onset. Far less animal studies addressed the therapeutic efficacy on established disease or prophylactic effects during remission. In the case of TNF, a study delivering a dimeric TNF receptor to mice with EAE demonstrated that the anti-TNF treatment was more efficacious when delivered prior to onset rather than during remission although therapy at both time-points could significantly inhibit disease severity (Croxford et al., 2000). Although mechanisms for this were not demonstrated, it was suggested that the secondary relapse phase may be less dependent on TNF than the initial priming event. This is also suggested by *Tnf* gene–deleted mice where the onset of disease was delayed compared to wild-type mice but EAE developed in the absence of TNF with a comparable severity (Sean Riminton et al., 1998).

It is becoming increasingly clear that, in addition to its well-established proinflammatory effects, TNF exhibits potent immunosuppressive properties, providing one possible explanation for the immune and disease-activating effect of anti-TNF treatment of MS. In TNF-deficient mice, myelin-specific T-cell reactivity fails to regress and the expansion of activated/memory T cells is abnormally prolonged, leading to exacerbated EAE (Kassiotis and Kollias, 2001). Strikingly, immunosuppression by TNF and protection against EAE does not require the p55 TNFR, whereas the same receptor is necessary for the detrimental effects of TNF during the acute phase of the disease (Kassiotis and Kollias, 2001). In another study, TNF-deficient animals developed severe EAE after immunization with myelin-oligodendrocytes-glycoprotein (MOG), also suggesting a protective role of this cytokine (Liu et al., 1998). TNF contributes to the elimination of inflammatory infiltrates by signaling effects via the TNF receptor p75 (TNFRII). In the pathogenetic cascade of MS it therefore has the potential to trigger both "On" and "Off" signals (Eugster et al., 1999; Probert et al., 2000). Such an immune-suppressive role for TNF has also been proposed in several other models of systemic and organ-specific autoimmune diseases (Campbell et al., 2001; Cope, 1998; Grewal et al., 1996; Weishaupt et al., 2000). A lesson to learn from the TNF studies is the fact that MRI effects and clinical results can diverge considerably. Furthermore, MRI assessment of some more subtle facets of the disease process (such as remyelination, gliosis, neuronal and axonal damage) is possibly insufficient.

V. TRANSFORMING GROWTH FACTOR-β2 (TGF-β2)

A. BACKGROUND

The transforming growth factor β is a multifunctional polypeptidic growth factorthat is involved as a mediator in many biological processes, ranging from inflammation, development, and wound healing to tumorgenesis. Three highly homologous iso-forms exist in mammalians (TGF-β1, TGF-β2, and TGF-β3) (Stavnezer, 1995), which — depending on cell type and growth conditions — differ in their efficacy potential and biological activity. Almost all cells express the three types of TGF-β receptors. In general, TGF-β has a stimulatory effect on cells of mesenchymal origin and an inhibitory effect on cells of epithelial or neuroectodermal origin.

An inhibitory effect of TGF-β1 or TGF-β2 on the development of EAE has been shown in several investigations. On the other hand, neutralizing antibodies against TGF-β1 worsened the disease progress (Fabry et al., 1995; Johns et al., 1991; Kuruvilla et al., 1991; Racke et al., 1994; Schluesener and Lider, 1989; Stevens et al., 1994).

B. STUDIES

No significant effect on either the EDSS score or MRI lesion load was shown for TGF-β2 in an open dose escalation and toxicity study (Calabresi et al., 1998). Eleven patients with SPMS (on an average EDSS 7.5; age on the average of 44.5 years) received the cytokine intravenously three times a week for 4 weeks. The most important adverse effect was a reversible reduction of the glomerular filtration rate observed in five patients. Consistent with the observation that TGF-β is able to block cell adhesion and migration into the CNS (Fabry et al., 1995), a trend toward a reduction of CSF-pleocytosis was observable during the treatment period.

C. COMMENTARY

The cytokine immune deviation with TGF-β2 in SPMS did not lead to a slowing of disease progression or reduction of lesion load in the CNS. It is not clear if local TGF-β2 bioavailability in the CNS is sufficient for exertion of antiinflammatory effects. However, direct renal toxicity at doses applied here limits the use of TGF-β2 in humans. Furthermore, application for longer periods increases the danger of extended tissue fibrosis and deposition of the extracellular matrix (Wahl, 1994).

VI. INTERLEUKIN-10 (IL-10)

A. BACKGROUND

Interleukin-10 (Il-10) is secreted by T helper-cells (Th2), macrophages, and other cells of the immune system and has various effects on different cell types *in vitro*. It suppresses the production of proinflammatory cytokines, prevents the expression of costimulation and adhesion molecules in antigen presenting cells (APC), inhibits T-cell proliferation, and leads to a long-lasting antigen-specific anergy in CD4-T cells (Moore et al., 2001). In contrast, Il-10 stimulates B-cell growth and the production of antibodies. The Th2 response is upregulated, which suggests a central role in the regulation of both Th1-cells and Th2-cells.

The clinical effect of Il-10 in rheumatoid arthritis (RA), inflammatory bowel disease (IBD), adult respiratory distress syndrome (ARDS), HIV infectionm, and psoriasis is being examined. Application of Il-10 showed positive effects on the development and severity of EAE (Crisi et al., 1995; Rott et al., 1994), although controversial results exist (Cannella et al., 1996). For a comprehensive overview see Owens et al. (2001). The therapeutical effect of β-interferons at least partially implies the promotion of Il-10 secretion by monocytes (Porrini et al., 1995; Rudick et al., 1996). Serum levels of Il-10 are lower in MS patients than in healthy individuals (Salmaggi et al., 1996).

B. Studies

A phase I study on the tolerance of intravenous Il-10 in healthy individuals showed the typical cytokine side effects such as headaches, fever, and myalgia as well as transient neutrophilia, monocytosis, and lymphopenia. The secretion of TNF-α and Il-1β was inhibited (Chernoff et al., 1995). A dose escalation study with subcutaneous Il-10 in MS was initiated but prematurely terminated by the company. Based on single-case observations, Il-10 might have induced exacerbations and lacked efficacy. However, no published information on these data is available to date.

VII. INTERLEUKIN-4

A. Background

Interleukin-4 (IL-4) exhibits many biological and immunoregulatory functions on B lymphocytes, monocytes, dendritic cells, and fibroblasts (Paul and Seder, 1994). The IL-4 gene is located on chromosome 5 and displays several cell-specific regulatory sequences in its promoter, which explains its restricted secretion pattern to activated T cells and mast cells. Like IL-10, this cytokine is mainly produced by Th2 cells. IL-4 promotes T helper cell type 2 (Th2) differentiation and stability and inhibits Th1-cell differentiation. IL-4 also affects antibody class switch and expression of Fc receptors, which strongly influences the effector mechanisms following antibody production. The IL-4 receptor is multimeric and is constituted by at least IL-4Ralpha, a chain common to other cytokine receptors. Two types of IL-4 receptors have been defined: one constituted by the IL-4Ralpha and the gamma(c) chain, and a second constituted by the IL-4Ralpha and the IL-13Ralpha, which is able to transduce both IL-4 and IL-13 signals (de Vries et al., 1999). Numerous studies have demonstrated the key regulatory role of IL-4 in allergic responses as well as its anti-inflammatory and anti-tumor effects, i.e., growth suppression of acute lymphoblastic leukemia (ALL) cells (Manabe et al., 1994; Srivannaboon et al., 2001). Of note is the fact that therapeutic administration of IL-4 showed impressive clinical improvement in psoriasis, another T-cell–mediated autoimmune disease (Ghoreschi et al., 2003).

The clinical signs of T-cell transfer EAE could be suppressed in SJL mice by injection of exogenous Il-4 together with the encephalitogenic MBP-specific T cells, although the number and distribution of infiltrating cells were similar to those in nontreated controls (Racke et al., 1994). Transgenic overexpression of Il-4 did not have an effect on EAE (Bettelli et al., 1998). In one study, Il-4 gene ablation did not have an effect on EAE induction or progression (Bettelli et al., 1998). In contrast, Falcone et al. reported enhanced severity of EAE and increased proinflammatory cytokine parameters in Il-4 deficient mice (Falcone et al., 1998). In MS patients a positive and significant correlation was found between the carriage rate of the IL-4 B1 allele and the age of disease onset (Vandenbroeck et al., 1997), whereas Il-4 alleles were not associated with disease progression, sex, or ethnic background (Vandenbroeck et al., 1997). The IL-4 B1 allele therefore might be associated with late onset of MS and therefore potentially represents a modifier of age of onset. Il-4 production in T cells is a well-accepted marker for description of a Th2 phenotype. Il-4 together with IFN-γ therefore is often used to monitor T helper subset polarization (Th1/Th2) in MS and to correlate disease course or response to immunomodulatory therapy in antigen-specific T cells (Duda et al., 2000; Farina et al., 2001; Gran et al., 2000; Neuhaus et al., 2000).

B. Studies

The effects of interleukin-4 (IL-4) on immune cells, such as the control of T helper cell balance and suppression of inflammatory cytokines, have been hypothesized to offer therapeutic benefits as treatment for MS. BAY 36-1677 (Il-4SA, IL-4 Selective Agonist) is a recombinant protein differing from natural human interleukin-4 at two amino acid positions (Shanafelt et al., 1998). The compound was engineered to selectively exhibit IL-4 activities on T cells, B cells, and monocytes but not on endothelial cells and fibroblasts. A clinical study with BAY 36-1677 (IL-4 Selective Agonist) in patients with MS failed to show the desired pharmacological effects. Due to potentially superimposed immunological effects, the *in vitro* selectivity of BAY 36-1677 did not translate into a clinical advantage. The project has been discontinued (Bayer Company, personal communication). Clinical usefulness seems to be of greater relevance in the therapy of acute lymphoblastic leukemia cells, because Interleukin-4 variant (BAY 36-1677) has been shown to selectively induce apoptosis in lymphoblastic cells (Srivannaboon et al., 2001).

As mentioned above, IL-4 is successfully applied in other T-cell–mediated autoimmune diseases, most importantly psoriasis (Ghoreschi et al., 2003).

VIII. OTHER CYTOKINES

In addition to the cytokines discussed in the previous sections, several other cytokine and anti-cytokine treatments may be considered as attractive candidates for MS therapy. For example, therapeutic strategies based on the inhibition of IL-12 have received long-standing interest. More recent findings, however, indicate that some of the proinflammatory effects previously attributed to IL-12 are in fact mediated by IL-23, which shares a p40 subunit with IL-12. The IL-23/IL-12 system with its ligands and receptors might be an interesting target for therapeutic inhibition (Trinch-ieri, 2003; Becher et al., 2003).

Another interesting development in the field of cytokine therapeutics relates to the surprisingly close interrelationship between the traditional cytokines (produced by cells of the immune system) and the neuropoetic cytokines and neurotrophic factors (produced by cells of the nervous system). The intense "cross-talk" between these cytokine networks opens new possibilities for immunotherapy, because it allows elegant combinations of immunomodulatory and neuroprotective therapeutic actions (Kerschensteiner et al., 2003).

IX. GENERAL COMMENTS AND CONCLUDING REMARKS

Our initial hope that the success of interferon-β would usher in a multitude of cytokine therapies has faded. None of the other cytokine-based therapies tested so far in pilot trials were truly promising, let alone superior to interferon-β.

What conclusions can be drawn from these experiences for future strategies concerning cytokine-base therapies for intrinsic CNS disease processes (see Table 14.2)? Clinical trials testing the effects of cytokines or cytokine inhibitors on MS patients have often been derived from studies involving the EAE model. However, it is also important to consider findings from transgenic animal studies in designing therapeutic approaches for MS (Owens et al., 2001). Assuming the inherent lim-itations with transgenic approaches, transgene expression that is induced tran-siently or in response to specific stimuli with control by a cell-specific promoter will address relevant questions about the role of certain candidate molecules in the context of CNS inflammation or degeneration. Furthermore, it should be considered whether EAE experiments better represent RR-MS versus SP-MS, in that most EAE-directed therapies affect the clonal expansion of T cells, the first immune cells to enter the CNS (Owens et al., 2001). Several examples now show that cytokines represent parts of a complex network exerting dichotomic functions, depending on the local mileu and kinetics (Townsend and McKenzie, 2000). Therapeutic application of any one of the cytokines or the respective inhibitory components could be expected to disturb this balance in a complicated way. Similar considerations also translate to the increasing family of chemokines and their possible utilization as therapeutic targets (Arimilli et al., 2000). Additionally it has to be assumed that, in contrast to animal models, there is great heterogeneity in the immunopathology, the clinical phenotype, and in the therapeutic response of MS patients (Lassmann et al., 2001). Along this line, the tremendous advance-ments in our knowledge of "key factors" (or immunopathogenetic cascades) con-tributing to lesion pathogenesis and chronicity in intrinsic CNS disease processes has clearly led us to revisit the "classical" Th1/Th2 paradigm (Lassmann and Ransohoff, 2004). Moreover one has to consider that immune reactions in the CNS differ considerably from other parenchymal organs. The suitability of certain cytokine-based therapeutic regimens in other T-cell–mediated autoimmune dis-eases (e.g., TNF-α antagonization in RA, IL-4 in psoriasis) contrasted by its failure

in MS only strengthens this remark. In the instance of autoimmune CNS demyelination the dissemination of lesions in time and space further complicate the
reflection. Thus, destructive and reparative mechanisms are likely to take place at
the same time, in different places. Theoretically highly specific (anti-)cytokine
and also (anti-)chemokine therapies would even seem more suitable for the treatment of acute exacerbations than for long-term therapy.

REFERENCES

Abreu SL. Interferon in experimental autoimmune encephalomyelitis (EAE): effects of exo-
geneous interferon on the antigen-enhanced adoptive transfer of EAE. *Int Arch Allergy
Appl Immunol* 1985; 76: 302–7.

Arimilli S, Ferlin W, Solvason N, Deshpande S, Howard M, Mocci S. Chemokines in autoimmune diseases. *Immunol Rev* 2000; 177: 43–51.

Arnason BG, Dayal A, Qu ZX, Jensen MA, Genc K, Reder AT. Mechanisms of action of
interferon-beta in multiple sclerosis. Springer Semin *Immunopathol* 1996; 18:
125–48.

Arnason BGW, Jacobs G, Hanlon M, Clay BH, Noronha ABC, Auty A, et al. TNF neutralization in MS—Results of a randomized, placebo controlled multicenter study. *Neurology* 1999; 53: 457–465.

Becher B, Durell BG, Noelle RJ. IL-23 produced by CNS-resident cells controls T cell
encephalitogenicity during the effector phase of experimental autoimmune encephalomyelitis. *J Clin Invest* 2003; 112: 1186–91.

Beck J, Rondot P, Catinot L, Falcoff E, Kirchner H, Wietzerbien J. Increased production of
interferon gamma and tumor necrosis factor preceds clinical manifestation in multiple
sclerosis: do cytokines trigger off exacerbations? *Acta Neurol Scand* 1988; 78:
318–323.

Bertolotto A. Neutralizing antibodies to interferon beta: implications for the management of
multiple sclerosis. *Curr Opin Neurol* 2004; 17: 241–6.

Bettelli E, Das MP, Howard ED, Weiner HL, Sobel RA, Kuchroo VK. IL-10 is critical in the
regulation of autoimmune encephalomyelitis as demonstrated by studies of IL-10-
and IL-4-deficient and transgenic mice. *J Immunol* 1998; 161: 3299–306.

Bielekova B, Martin R. Development of biomarkers in multiple sclerosis. *Brain* 2004; 127:
1463–78.

Billiau A, Heremans H, Vandekerckhove F, Dijkmans R, Sobis H, Meulepas E, et al. Enhancement of experimental allergic encephalomyelitis in mice by antibodies against IFN-
gamma. *J Immunol* 1988; 140: 1506–10.

Brod SA, Khan M, Kerman RH, Pappolla M. Oral administration of human or murine
interferon alpha suppresses relapses and modifies adoptive transfer in experimental
autoimmune encephalomyelitis. J Neuroimmunol 1995; 58: 61–9.

Calabresi PA, Fields NS, Maloni HW, Hanham A, Carlino J, Moore J, et al. Phase-1 trial of
transforming-growth-factor-beta-2 in chronic progressive MS. *Neurology* 1998; 51:
289–292.

Camenga DL, Johnson KP, Alter M, Engelhardt CD, Fishman PS, Greenstein JI, et al. Systemic
recombinant alpha-2 interferon therapy in relapsing multiple sclerosis. *Arch Neurol*
1986; 43: 1239–46.

Campbell IK, O'Donnell K, Lawlor KE, Wicks IP. Severe inflammatory arthritis and lym-
phadenopathy in the absence of TNF. *J Clin Invest* 2001; 107: 1519–1527.

Cannella B, Gao YL, Brosnan C, Raine CS. Il-10 fails to abrogate experimental autoimmune encephalomyelitis. *J Neurosci* Res 1996; 45: 735–746.

Cannella B, Raine CS. The adhesion molecule and cytokine profile of multiple sclerosis lesions. *Ann Neurol* 1995; 37: 424–435.

Chernoff AE, Granowitz EV, Shapiro L, Vannier E, Lonnemann G, Angel JB, et al. A randomized controlled trial of Il-10 in humans: inhibition of inflammatory cytokine production and immune responses. *J Immunol* 1995; 154: 5292–5499.

Chofflon M, Juillard C, Juillard P, Gauthier G, Grau GE. Tumor necrosis factor alpha production as a possible predictor of relapse in patients with multiple sclerosis. *Eur Cytokine Netw* 1992; 3: 523–531.

Cohen JA, Goodman AD, Heidenreich FR. Results of IMPACT, a phase 3 trial of interferon beta-1a in secondary progressive MS. *Neurology* 2001; 56: A148–A149.

Comi G, Filippi M, Barkhof F, Durelli L, Edan G, Fernandez O, et al. Effect of early interferon treatment on conversion to definite multiple sclerosis: a randomised study. *Lancet* 2001; 357: 1576–82.

Cope AP. Regulation of autoimmunity by proinflammatory cytokines. *Curr Opin Immunol* 1998; 10: 669–76.

Corbin JG, Kelly D, Rath EM, Baerwald KD, Suzuki K, Popko B. Targeted CNS expression of interferon-gamma in transgenic mice leads to hypomyelination, reactive gliosis, and abnormal cerebellar development. *Mol Cell Neurosci* 1996; 7: 354–70.

Crisi GM, Santambrogio L, Hochwald GM, Smith SR, Carlino JA, Thorbecke GJ. Staphylococcus enterotoxin B and tumor-necrosis factor-alpha induced relapses of experimental allergic encephalomyelitis: protection by transforming growth factor-ß and interleukin-10. *Eur J Immunol* 1995; 25: 3035–3040.

Croxford JL, Triantaphyllopoulos KA, Neve RM, Feldmann M, Chernajovsky Y, Baker D. Gene therapy for chronic relapsing experimental allergic encephalomyelitis using cells expressing a novel soluble p75 dimeric TNF receptor. *J Immunol* 2000; 164: 2776–81.

de Vries JE, Carballido JM, Aversa G. Receptors and cytokines involved in allergic TH2 cell responses. *J Allergy Clin Immunol* 1999; 103: S492–6.

Dinarello CA, Thompson RC. Blocking IL-1: interleukin 1 receptor antagonist in vivo and in vitro. *Immunol Today* 1991; 12: 404–10.

Duda PW, Schmied MC, Cook SL, Krieger JI, Hafler DA. Glatiramer acetate (Copaxone) induces degenerate, Th2-polarized immune responses in patients with multiple sclerosis. *J Clin Invest* 2000; 105: 967–76.

Durelli L, Bongioanni MR, Cavallo R, Ferrero B, Ferri R, Ferrio MF, et al. Chronic systemic high-dose recombinant interferon alfa-2a reduces exacerbation rate, MRI signs of disease activity, and lymphocyte interferon gamma production in relapsing-remitting multiple sclerosis. *Neurology* 1994; 44: 406–13.

Durelli L, Bongioanni MR, Ferrero B, Ferri R, Imperiale D, Bradac GB, et al. Interferon alpha-2a treatment of relapsing-remitting multiple sclerosis: disease activity resumes after stopping treatment. *Neurology* 1996; 47: 123–9.

Durelli L, Verdun E, Barbero P, Bergui M, Versino E, Ghezzi A, et al. Every-other-day interferon beta-1b versus once-weekly interferon beta-1a for multiple sclerosis: results of a 2-year prospective randomised multicentre study (INCOMIN). *Lancet* 2002; 359: 1453–60.

Elliott MJ, Maini RN, Feldmann M, Long-Fox A, Charles P, Katsikis P, et al. Treatment of rheumatoid arthritis with chimeric monoclonal antibody to tumor necrosis factor alpha. *Arthritis Rheum* 1993; 36: 1681–1690.

Eugster HP, Frei K, Bachmann R, Bluethmann H, Lassmann H, Fontana A. Severity of symptoms and demyelination in MOG-induced EAE depends on TNFR1. *Eur J Immunol* 1999; 29: 626–632.

The European Study Group of Interferon beta-1b in secondary progressive MS. Placebo-controlled multicentre randomised trial of interferon beta-1b in treatment of secondary progressive multiple sclerosis. European Study Group on interferon beta-1b in secondary progressive MS. *Lancet* 1998; 352: 1491–7.

Fabry Z, Topham DJ, Fee D, Herlein J, Carlino JA, Hart MN, et al. TGF-β2 decreases migration of lymphocytes in vitro and homing of cells into the central nervous system in vivo. *J Immunol* 1995; 155: 325–332.

Falcone M, Rajan AJ, Bloom BR, Brosnan CF. A critical role for IL-4 in regulating disease severity in experimental allergic encephalomyelitis as demonstrated in IL-4-deficient C57BL/6 mice and BALB/c mice. *J Immunol* 1998; 160: 4822–30.

Farina C, Then Bergh F, Albrecht H, Meinl E, Yassouridis A, Neuhaus O, et al. Treatment of multiple sclerosis with Copaxone (COP): Elispot assay detects COP-induced interleukin-4 and interferon-gamma response in blood cells. *Brain* 2001; 124: 705–19.

Feldmann M, Maini RN. Anti-TNF alpha therapy of rheumatoid arthritis: what have we learned? *Annu Rev Immunol* 2001; 19: 163–96.

Ghoreschi K, Thomas P, Breit S, Dugas M, Mailhammer R, van Eden W, et al. Interleukin-4 therapy of psoriasis induces Th2 responses and improves human autoimmune disease. *Nat Med* 2003; 9: 40–6.

Goodin DS, Frohman EM, Garmany GP, Jr., Halper J, Likosky WH, Lublin FD, et al. Disease modifying therapies in multiple sclerosis: report of the Therapeutics and Technology Assessment Subcommittee of the American Academy of Neurology and the MS Council for Clinical Practice Guidelines. *Neurology* 2002; 58: 169–78.

Goodkin DE, Group NASS. The North American Study of interferon beta-1b in secondary progressive multiple sclerosis. 52nd Annual Meeting of the American Academy of Neurology, San Diego, CA, 2000 2000; Abstract #LBN.002.

Gran B, Tranquill LR, Chen M, Bielekova B, Zhou W, Dhib-Jalbut S, et al. Mechanisms of immunomodulation by glatiramer acetate. *Neurology* 2000; 55: 1704–14.

Grewal IS, Grewal KD, Wong FS, Picarella DE, Janeway CA, Jr., Flavell RA. Local expression of transgene encoded TNF alpha in islets prevents autoimmune diabetes in nonobese diabetic (NOD) mice by preventing the development of auto-reactive islet-specific T cells. *J Exp Med* 1996; 184: 1963–74.

Hartung HP, Schellekens H, Munschauer FE, 3rd. Neutralizing antibodies to interferon beta in patients with multiple sclerosis: scientific background and clinical implications. *J Neurol* 2004; 251 Suppl 2: II1–II3.

Hemmer B, Archelos JJ, Hartung HP. New concepts in the immunopathogenesis of multiple sclerosis. *Nat Rev Neurosci* 2002; 3: 291–301.

Hohlfeld R. Biotechnological agents for the immunotherapy of multiple sclerosis: Principles, problems and perspectives. *Brain* 1997; 120: 865–916.

Hohlfeld R, Wiendl H. The ups and downs of multiple sclerosis therapeutics. *Ann Neurol* 2001; 49: 281–4.

Horwitz MS, Evans CF, McGavern DB, Rodriguez M, Oldstone MB. Primary demyelination in transgenic mice expressing interferon-gamma. *Nat Med* 1997; 3: 1037–41.

The IFNB MS study group. Interferon beta-1b is effective in relapsing-remitting multiple sclerosis. I. Clinical results of a multicenter, randomized, double-blind, placebo-controlled trial. The IFNB Multiple Sclerosis Study Group. *Neurology* 1993; 43: 655–61.

The IFNB MS study group. Interferon beta-1b in the treatment of multiple sclerosis: final outcome of the randomized controlled trial. The IFNB Multiple Sclerosis Study Group and The University of British Columbia MS/MRI Analysis Group. *Neurology* 1995; 45: 1277–85.

Imamura K, Suzumura A, Hayashi F, Marunouchi R. Cytokine production by peripheral blood monocytes/macrophages in multiple sclerosis patients. *Acta Neurol Scand* 1993; 87: 281–285.

Jacobs L, Johnson KP. A brief history of the use of interferons as treatment of multiple sclerosis. *Arch Neurol* 1994; 51: 1245–52.

Jacobs L, O'Malley J, Freeman A, Ekes R. Intrathecal interferon reduces exacerbations of multiple sclerosis. *Science* 1981; 214: 1026–8.

Jacobs LD, Beck RW, Simon JH, Kinkel RP, Brownscheidle CM, Murray TJ, et al. Intramuscular interferon beta-1a therapy initiated during a first demyelinating event in multiple sclerosis. CHAMPS Study Group. *N Engl J Med* 2000; 343: 898–904.

Jacobs LD, Cookfair DL, Rudick RA, Herndon RM, Richert JR, Salazar AM, et al. Intramuscular interferon beta-1a for disease progression in relapsing multiple sclerosis. The Multiple Sclerosis Collaborative Research Group (MSCRG). *Ann Neurol* 1996; 39: 285–94.

Johns LD, Flanders KC, Ranges GE, Sriram S. Successful treatment of experimental allergic encephalomyelitis with transforming growth factor-ß1. *J Immunol* 1991; 147: 1792–1796.

Kappos L, Polman C, Pozzilli C, Thompson A, Beckmann K, Dahlke F. Final analysis of the European multicenter trial on IFNbeta-1b in secondary-progressive MS. *Neurology* 2001; 57: 1969–75.

Kassiotis G, Kollias G. Uncoupling the proinflammatory from the immunosuppressive properties of tumor necrosis factor (TNF) at the p55 TNF receptor level: implications for pathogenesis and therapy of autoimmune demyelination. *J Exp Med* 2001; 193: 427–34.

Kerschensteiner M, Stadelmann C, Dechant G, Wekerle H, Hohlfeld R. Neurotrophic crosstalk between the nervous and immune systems: implications for neurological diseases. *Ann Neurol* 2003; 53: 292–304.

Klinkert WEF, Kojima K, Lesslauer W, Rinner W, Lassmann H, Wekerle H. TNF-alpha receptor fusion protein prevents experimental auto-immune encephalomyelitis and demyelination in Lewis rats: an overview. *J Neuroimmunol* 1997; 72: 163–168.

Knobler RL, Panitch HS, Braheny SL, Sipe JC, Rice GP, Huddlestone JR, et al. Systemic alpha-interferon therapy of multiple sclerosis. *Neurology* 1984; 34: 1273–9.

Körner H, Lemckert FA, Chaudhri G, Etteldorf S, Sedgwick JD. Tumor necrosis factor blockade in actively induced experimental autoimmune encephalomyelitis prevents clinical disease despite activated T cell infiltration to the central nervous system. *Eur J Immunol* 1997; 27: 1973–1981.

Krakowski M, Owens T. Interferon-gamma confers resistance to experimental allergic encephalomyelitis. *Eur J Immunol* 1996; 26: 1641–6.

Kuruvilla AP, Shah R, Hochwald GM, Liggitt HD, Palladino MA, Thorbecke GJ. Protective effect of transforming growth factor-β1 on experimental autoimmune diseases in mice. *Proc Natl Acad Sci USA* 1991; 88: 2918–2921.

Lassmann H, Bruck W, Lucchinetti C. Heterogeneity of multiple sclerosis pathogenesis: implications for diagnosis and therapy. *Trends Mol Med* 2001; 7: 115–21.

Lassmann H, Ransohoff RM. The CD4-Th1 model for multiple sclerosis: a crucial reappraisal. *Trends Immunol* 2004; 25: 132–7.

Li DK, Zhao GJ, Paty DW. Randomized controlled trial of interferon-beta-1a in secondary progressive MS: MRI results. *Neurology* 2001; 56: 1505–13.

Liu J, Marino MW, Wong G, Grail D, Dunn A, Bettadapura J, et al. TNF is a potent anti-inflammatory cytokine in autoimmune-mediated demyelination. *Nat Med* 1998; 4: 78–83.

Lovell DJ, Giannini EH, Reiff A, Cawkwell GD, Silverman ED, Nocton JJ, et al. Etanercept in children with polyarticular juvenile rheumatoid arthritis. Pediatric Rheumatology Collaborative Study Group. *N Engl J Med* 2000; 342: 763–9.

Maini R, St Clair EW, Breedveld F, Furst D, Kalden J, Weisman M, et al. Infliximab (chimeric anti-tumour necrosis factor alpha monoclonal antibody) versus placebo in rheumatoid arthritis patients receiving concomitant methotrexate: a randomised phase III trial. ATTRACT Study Group. *Lancet* 1999; 354: 1932–9.

Manabe A, Coustan-Smith E, Kumagai M, Behm FG, Raimondi SC, Pui CH, et al. Interleukin-4 induces programmed cell death (apoptosis) in cases of high-risk acute lymphoblastic leukemia. *Blood* 1994; 83: 1731–7.

Maurer M, Rieckmann P. What is the potential of combination therapy in MS. *BioDrugs* 2000.

Miller DH, Albert PS, Barkhof F, Francis G, Frank JA, Hodgkinson S, et al. Guidelines for the use of magnetic resonance techniques in monitoring the treatment of multiple sclerosis. US National MS Society Task Force. *Ann Neurol* 1996; 39: 6–16.

Moore KW, de Waal Malefyt R, Coffman RL, O'Garra A. Interleukin-10 and the interleukin-10 receptor. *Annu Rev Immunol* 2001; 19: 683–765.

Neuhaus O, Farina C, Yassouridis A, Wiendl H, Then Bergh F, Dose T, et al. Multiple sclerosis: comparison of copolymer-1- reactive T cell lines from treated and untreated subjects reveals cytokine shift from T helper 1 to T helper 2 cells. *Proc Natl Acad* Sci USA 2000; 97: 7452–7.

Noseworthy JH, Lucchinetti C, Rodriguez M, Weinshenker BG. Multiple sclerosis. *N Engl J Med* 2000; 343: 938–52.

Owens T, Renno T, Taupin V, Krakowski M. Inflammatory cytokines in the brain: does the CNS shape immune responses? *Immunol Today* 1994; 15: 566–71.

Owens T, Wekerle H, Antel J. Genetic models for CNS inflammation. *Nat Med* 2001; 7: 161–6.

The OWIMS study group. Evidence of interferon beta-1a dose response in relapsing-remitting MS: the OWIMS Study. The Once Weekly Interferon for MS Study Group. *Neurology* 1999; 53: 679–86.

Panitch H, Goodin DS, Francis G, Chang P, Coyle PK, O'Connor P, et al. Randomized, comparative study of interferon beta-1a treatment regimens in MS: The EVIDENCE Trial. *Neurology* 2002; 59: 1496–506.

Panitch HS, Hirsch RL, Schindler J, Johnson KP. Treatment of multiple sclerosis with gamma interferon: exacerbations associated with activation of the immune system. *Neurology* 1987; 37: 1097–102.

Paul WE, Seder RA. Lymphocyte responses and cytokines. *Cell* 1994; 76: 241–51.

Popko B, Corbin JG, Baerwald KD, Dupree J, Garcia AM. The effects of interferon-gamma on the central nervous system. *Mol Neurobiol* 1997; 14: 19–35.

Porrini AM, Gambi D, Reder AT. Interferon effects on interleukin-10 secretion: mononuclear cell response to interleukin-10 is normal in multiple sclerosis patients. *J Neuroimmunol* 1995; 61: 27–34.

The PRISMS study group. A randomized, placebo-controlled, double-blind study of interferon-beta 1a in relapsing-remitting multiple sclerosis. *Lancet* 1998; 352: 1498–1504.

The PRISMS study group. PRISMS-4: Long-term efficacy of interferon-beta-1a in relapsing MS. *Neurology* 2001; 56: 1628–36.

Probert L, Eugster HP, Akassoglou K, Bauer J, Frei K, Lassmann H, et al. TNFR1 signalling is critical for the development of demyelination and the limitation of T-cell responses during immune-mediated CNS disease. *Brain* 2000; 123: 2005–19.

Racke MK, Bonomo A, Scott DE, Cannella B, Levine A, Raine CS, et al. Cytokine-induced immune deviation as a therapy for inflammatory autoimmune disease. *J Exp Med* 1994; 180: 1961–1966.

Rieckmann P, Albrecht M, Kitze B, Weber T, Tumani H, Broocks A, et al. Tumor-Necrosis-Factor-Alpha messenger-RNA Expression in patients with Relapsing-Remitting Multiple-Sclerosis is associated with disease-activity. *Ann Neurol* 1995; 37: 82–88.

Rott O, Fleischer B, Cash E. Interleukin-10 prevents experimental allergic encephalomyelitis in rats. *Eur J Immunol* 1994; 24: 1434–1440.

Rudick RA, Ransohoff RM. Cytokine scretion by multiple sclerosis monocytes. Relationship to disease activity. *Arch Neurol* 1992; 49: 265–270.

Rudick RA, Ransohoff RM, Peppler R, VanderBrug Medendorp S, Lehmann P, Alam J. Interferon beta induces interleukin-10 expression: relevance to multiple sclerosis. *Ann Neurol* 1996; 40: 618–27.

Salmaggi A, Dufour A, Eoli M, Corsini E, LaMantia L, Massa G, et al. Low serum interleukin-10 levels in multiple sclerosis: further evidence for decreased systemic immunosuppression? *J Neurol* 1996; 243: 13–17.

Schluesener HJ, Lider O. Transforming growth factors ß1 and ß2: cytokines with identical immunosuppressive effects and a potential role in the regulation of autoimmune T cell function. *J Neuroimmunol* 1989; 24: 249–258.

Sean Riminton D, Korner H, Strickland DH, Lemckert FA, Pollard JD, Sedgwick JD. Challenging cytokine redundancy: inflammatory cell movement and clinical course of experimental autoimmune encephalomyelitis are normal in lymphotoxin-deficient, but not tumor necrosis factor-deficient, mice. *J Exp Med* 1998; 187: 1517–28.

Selmaj K, Raine CS, Cannella B, Brosnan CF. Identification of lymphotoxin and tumor necrosis factor in multiple sclerosis lesions. *J Clin Invest* 1991; 87: 949–954.

Sethna MP, Lampson LA. Immune modulation within the brain: recruitment of inflammatory cells and increased major histocompatibility antigen expression following intracerebral injection of interferon-gamma. *J Neuroimmunol* 1991; 34: 121–32.

Shanafelt AB, Forte CP, Kasper JJ, Sanchez-Pescador L, Wetzel M, Gundel R, et al. An immune cell-selective interleukin 4 agonist. *Proc Natl Acad Sci USA* 1998; 95: 9454–8.

Sharief MK, Hentges R. Association between tumor necrosis factor alpha and disease progression in patients with multiple sclerosis. *N Engl J Med* 1991; 325: 467–472.

Simmons RD, Willenborg DO. Direct injection of cytokines into the spinal cord causes autoimmune encephalomyelitis-like inflammation. *J Neurol Sci* 1990; 100: 37–42.

Sorensen PS, Ross C, Clemmesen KM, Bendtzen K, Frederiksen JL, Jensen K, et al. Antibodies to IFN-beta: the Danish National IFN-beta Project. *Neurology* 2003; 61: S27–S28.

The SPECTRIMS study group. Randomized controlled trial of interferon- beta-1a in secondary progressive MS: Clinical results. *Neurology* 2001; 56: 1496–504.

Srivannaboon K, Shanafelt AB, Todisco E, Forte CP, Behm FG, Raimondi SC, et al. Interleukin-4 variant (BAY 36-1677) selectively induces apoptosis in acute lymphoblastic leukemia cells. *Blood* 2001; 97: 752–8.

Stavnezer J. Regulation of antibody production and class switching by TFG-ß. *J Immunol* 1995; 155: 1647–1651.

Steinman L. Some misconceptions about understanding autoimmunity through experiments with knockouts. *J Exp Med* 1997; 185: 2039–41.

Townsend MJ, McKenzie AN. Unravelling the net ? cytokines and diseases. *J Cell Sci* 2000; 113: 3549–50.

Stevens DB, Gould KE, Swanborg RH. Transforming growth factor-ß1 inhibits tumor necrosis factor-alpha/lymphotoxin production and adoptive transfer of disease by effector cells of autoimmune encephalomyelitis. *J Neuroimmunol* 1994; 51: 77–83.

Tran EH, Prince EN, Owens T. IFN-gamma shapes immune invasion of the central nervous system via regulation of chemokines. *J Immunol* 2000; 164: 2759–68.

Trinchieri G. Interleukin-12 and the regulation of innate resistance and adaptive immunity. *Nat Rev Immunol* 2003; 3: 133–46.

Van Oosten BW, Barkhof F, Scholten PET, Mary B, Von Blomberg E, Ader HJ, et al. Increased production of tumor necrosis factor alpha, and not of interferon gamma, preceding disease activity in patients with multiple sclerosis. *Arch Neurol* 1998; 55: 793–798.

van Oosten BW, Barkhof F, Truyen L, Boringa JB, Bertelsmann FW, von Blomberg BM, et al. Increased MRI activity and immune activation in two multiple sclerosis patients treated with the monoclonal anti-tumor necrosis factor antibody cA2. *Neurology* 1996; 47: 1531–4.

Vandenbroeck K, Martino G, Marrosu M, Consiglio A, Zaffaroni M, Vaccargiu S, et al. Occurrence and clinical relevance of an interleukin-4 gene polymorphism in patients with multiple sclerosis. *J Neuroimmunol* 1997; 76: 189–92.

Wahl SM. Transforming growth factor β: the good, the bad, and the ugly [review]. *J Exp Med* 1994; 180: 1587–1590.

Walther EU, Hohlfeld R. Multiple sclerosis: side effects of interferon beta therapy and their management. *Neurology* 1999; 53: 1622–7.

Wandinger KP, Sturzebecher CS, Bielekova B, Detore G, Rosenwald A, Staudt LM, et al. Complex immunomodulatory effects of interferon-beta in multiple sclerosis include the upregulation of T helper 1-associated marker genes. *Ann Neurol* 2001; 50: 349–57.

Weinblatt ME, Kremer JM, Bankhurst AD, Bulpitt KJ, Fleischmann RM, Fox RI, et al. A trial of etanercept, a recombinant tumor necrosis factor receptor: Fc fusion protein, in patients with rheumatoid arthritis receiving methotrexate. *N Engl J Med* 1999; 340: 253–259.

Weishaupt A, Gold R, Hartung T, Gaupp S, Wendel A, Bruck W, et al. Role of TNF-alpha in high-dose antigen therapy in experimental autoimmune neuritis: inhibition of TNF-alpha by neutralizing antibodies reduces T-cell apoptosis and prevents liver necrosis. *J Neuropathol Exp Neurol* 2000; 59: 368–76.

Willenborg DO, Fordham S, Bernard CC, Cowden WB, Ramshaw IA. IFN-gamma plays a critical down-regulatory role in the induction and effector phase of myelin oligodendrocyte glycoprotein-induced autoimmune encephalomyelitis. *J Immunol* 1996; 157: 3223–7.

Yong VW. Differential mechanisms of action of interferon-beta and glatiramer aetate in MS. Neurology 2002; 59: 802–8.

Yu M, Nishiyama A, Trapp BD, Tuohy VK. Interferon-beta inhibits progression of relapsing-remitting experimental autoimmune encephalomyelitis. *J Neuroimmunol* 1996; 64: 91–100.

Index